Methods in Enzymology

Volume 330
HYPERTHERMOPHILIC ENZYMES
Part A

METHODS IN ENZYMOLOGY

EDITORS-IN-CHIEF

John N. Abelson   Melvin I. Simon

DIVISION OF BIOLOGY
CALIFORNIA INSTITUTE OF TECHNOLOGY
PASADENA, CALIFORNIA

FOUNDING EDITORS

Sidney P. Colowick and Nathan O. Kaplan

Methods in Enzymology

Volume 330

Hyperthermophilic Enzymes Part A

EDITED BY

Michael W. W. Adams

THE UNIVERSITY OF GEORGIA
ATHENS, GEORGIA

Robert M. Kelly

NORTH CAROLINA STATE UNIVERSITY
RALEIGH, NORTH CAROLINA

ACADEMIC PRESS
San Diego   London   Boston   New York   Sydney   Tokyo   Toronto

Academic Press
A Harcourt Science and Technology Company
525 B Street, Suite 1900, San Diego, California 92101-4495, USA
http://www.academicpress.com

Academic Press
Harcourt Place, 32 Jamestown Road, London NW1 7BY, UK
http://www.academicpress.com

International Standard Book Number: 0-12-182231-1

Table of Contents

Section I. Enzyme Discovery

Section II. Saccharolytic Enzymes

Section III. Proteolytic Enzymes

Contributors to Volume 330

Article numbers are in parentheses following the names of contributors.
Affiliations listed are current.

MICHAEL W. W. ADAMS (3, 30), *Department of Biochemistry and Molecular Biology, Center for Metalloenzyme Studies, University of Georgia, Athens, Georgia 30602*

IAIN J. ANDERSON (5), *Integrated Genomics, Inc., Chicago, Illinois 60612*

GARABED ANTRANIKIAN (17), *Institute of Technical Microbiology, Technische Universität Hamburg-Harburg, Hamburg 21073, Germany*

ROBERT D. BARBER (28), *Department of Biochemistry and Molecular Biology, Pennsylvania State University, University Park, Pennsylvania 16802*

MICHAEL W. BAUER (16, 22), *Novartis Agricultural Biotechnology Research Institute, Research Triangle Park, North Carolina 27709*

MARKE M. BEERTHUYZEN (21), *NIZO Food Research, Ede NL-6701 BA, The Netherlands*

PETER L. BERGQUIST (19), *Department of Biological Sciences, Macquarie University, Sydney, New South Wales 2109, Australia, and Department of Molecular Medicine, University of Auckland Medical School, Auckland 92019, Australia*

COSTANZO BERTOLDO (17), *Institute of Technical Microbiology, Technische Universität Hamburg-Harburg, Hamburg 21073, Germany*

JAMES R. BROWN (7), *Department of Bioinformatics, SmithKline Beecham Pharmaceuticals, Collegeville, Pennsylvania 19426-0989*

SUSAN G. CADY (22), *Department of Chemical Engineering, North Carolina State University, Raleigh, North Carolina 27695-7905*

WALTER CALLEN (13, 15, 22, 33), *Diversa Corporation, San Diego, California 92121-7032*

LARA S. CHANG (27, 32), *Department of Chemical Engineering, North Carolina State University, Raleigh, North Carolina 27695-7905*

STEPHEN T. CHANG (16), *MedImmune, Inc., Gaithersburg, Maryland 20878*

SWAPNIL R. CHHABRA (13, 14), *Department of Chemical Engineering, North Carolina State University, Raleigh, North Carolina 27695-7905*

MARIA CIARAMELLA (11), *Institute of Protein Biochemistry and Enzymology, CNR, 80125 Naples, Italy*

PEDRO M. COUTINHO (10), *Centre for Biological and Chemical Engineering, Instituto Superior Técnico, 1049-001 Lisboa, Portugal*

ROY M. DANIEL (19), *Department of Biological Sciences, Waikato University School of Science and Technology, Hamilton, New Zealand*

EDWARD F. DELONG (1), *Monterey Bay Aquarium Research Institute, Moss Landing, California 95039*

WILLEM M. DE VOS (21, 24, 25), *Laboratory of Microbiology, Wageningen Agricultural University, Wageningen, NL-6703 CT, The Netherlands, and Wageningen Centre for Food Sciences, Wageningen NL-670 AM, The Netherlands*

MARCEL DIJKGRAAF (25), *Laboratory of Microbiology, Wageningen Agricultural University, Wageningen, NL-6703 CT, The Netherlands*

JOCELYNE DIRUGGIERO (7), *Center of Marine Biotechnology, University of Maryland Biotechnology Institute, Baltimore, Maryland 21202*

DIANNE M. DUNN (7), *Department of Human Genetics, University of Utah, Salt Lake City, Utah 84112*

JONATHAN A. EISEN (9), *The Institute for Genomic Research, Rockville, Maryland 20850*

JAMES G. FERRY (28), *Department of Biochemistry and Molecular Biology, Pennsylvania State University, University Park, Pennsylvania 16802*

CLAIRE M. FRASER (9), *The Institute for Genomic Research, Rockville, Maryland 20850*

SHINSUKE FUJIWARA (20), *Department of Biotechnology, Graduate School of Engineering, Osaka University, Osaka 565-0871, Japan*

TOSHIAKI FUKUI (20), *Department of Synthetic Chemistry and Biological Chemistry, Graduate School of Engineering, Kyoto University, Kyoto 606-8501, Japan*

MOUSUMI GHOSH (30), *Department of Biochemistry, West Virginia University, Morgantown, West Virginia 26505*

MORELAND D. GIBBS (19), *Department of Biological Sciences, Macquarie University, Sydney, New South Wales 2109, Australia*

DAVID E. GRAHAM (5), *Department of Biochemistry, Virginia Polytechnical Institute & State University, Blacksburg, Virginia 24061-0308*

AMY M. GRUNDEN (30), *Department of Microbiology, North Carolina State University, Raleigh, North Carolina 27695*

VALERIE J. HARWOOD (31), *Department of Biology, University of South Florida, Tampa, Florida 33620*

BERNARD HENRISSAT (10), *Architecture et Fonction des Macromolécules Biologiques, CNRS-IFR1, Marseille, Cedex 20, France*

PAULA M. HICKS (27, 32), *Cargill Biotechnology Research, Excelsior, Minnesota 55331*

ROBERT HUBER (2), *Lehrstuhl für Mikrobiologie und Archaeenzentrum, Universität Regensburg, Regensburg D-93053, Germany*

TADAYUKI IMANAKA (20, 29), *Department of Synthetic Chemistry and Biological Chemistry, Graduate School of Engineering, Kyoto University, Kyoto 606-8501, Japan*

THIJS KAPER (21, 24), *Laboratory of Microbiology, Wageningen Agricultural University, Wageningen, NL-6703 CT, The Netherlands*

YUTAKA KAWARABAYASI (6), *Biotechnology Center, National Institute of Technology and Evaluation, Tokyo 151-0066, Japan*

ROBERT M. KELLY (4, 12, 13, 14, 15, 16, 22, 27, 32, 33), *Department of Chemical Engineering, North Carolina State University, Raleigh, North Carolina 27606*

SERVÉ W. M. KENGEN (21, 24), *Laboratory of Microbiology, Wageningen Agricultural University, Wageningen, NL-6703 CT, The Netherlands*

LEON D. KLUSKENS (21, 25), *Laboratory of Microbiology, Wageningen Agricultural University, Wageningen, NL-6703 CT, The Netherlands*

NIKOS KYRPIDES (5), *Integrated Genomics, Inc., Chicago, Illinois 60612*

DAVID LAM (13, 14, 15), *Diversa Corporation, San Diego, California 92121-7032*

JOYCE H. G. LEBBINK (21, 24), *Laboratory of Microbiology, Wageningen Agricultural University, Wageningen, NL-6703 CT, The Netherlands*

WOLFGANG LIEBL (15, 18), *Institut für Mikrobiologie und Genetik, Georg-August-Universität, Göttingen D-37077, Germany*

J. A. LITTLECHILD (26), *Schools of Chemistry and Biological Sciences, University of Exeter, Exeter EX4 4QD, United Kingdom*

DENNIS L. MAEDER (7), *Center of Marine Biotechnology, University of Maryland Biotechnology Institute, Baltimore, Maryland 21202*

ERIC J. MATHUR (13, 14, 15, 22, 33), *Diversa Corporation, San Diego, California 92121-7032*

ANGELI LAL MENON (3), *Department of Biochemistry and Molecular Biology, Center for Metalloenzyme Studies, University of Georgia, Athens, Georgia 30602*

Edward S. MILLER, JR. (15), *DuPont Central Research and Development Experimental Station, Wilmington, Delaware 19880-0328*

MARCO MORACCI (11), *Institute of Protein Biochemistry and Enzymology, CNR, 80125 Naples, Italy*

MASAAKI MORIKAWA (29), *Department of Material and Life Science, Graduate School of Engineering, Osaka University, Osaka 565-0871, Japan*

DANIEL D. MORRIS (19), *Department of Biological Sciences, Macquarie University, Sydney, New South Wales 2109, Australia*

KAREN E. NELSON (9), *The Institute for Genomic Research, Rockville, Maryland 20850*

ROSS OVERBEEK (5), *Integrated Genomics, Inc., Chicago, Illinois 60612*

KIMBERLEY N. PARKER (13, 14, 15, 16), *DuPont Chemical Research and Development Experimental Station, Wilmington, Delaware 19880-0328*

MARYBETH A. PYSZ (4), *Department of Chemical Engineering, North Carolina State University, Raleigh, North Carolina 27606*

KRISTINA D. RINKER (4), *Department of Chemical Engineering, Colorado State University, Fort Collins, Colorado 80523-1370*

FRANK T. ROBB (7), *Center of Marine Biotechnology, University of Maryland Biotechnology Institute, Baltimore, Maryland 21202*

MOSE ROSSI (11), *Institute of Protein Biochemistry and Enzymology, CNR, Naples 80125, Italy, and Dipartimento di Chimica Organica e Biologica, Universita di Napoli, Naples 80134, Italy*

ALEXEI SAVCHENKO (23), *Banting and Best Department of Medical Research, C. H. Best Institute, Toronto, Ontario M5GIL6, Canada*

HAROLD J. SCHREIER (31), *Center of Marine Biotechnology, University of Maryland Biotechnology Institute, Baltimore, Maryland 21202*

GERRIT J. SCHUT (3), *Department of Biochemistry and Molecular Biology, Center for Metalloenzyme Studies, University of Georgia, Athens, Georgia 30602*

A. C. SEHGAL (33), *Department of Chemical Engineering, North Carolina State University, Raleigh, North Carolina 27606*

KEITH R. SHOCKLEY (4), *Department of Chemical Engineering, North Carolina State University, Raleigh, North Carolina 27606*

JAY M. SHORT (13, 15, 22, 33), *Diversa Corporation, San Diego, California 92121-7032*

ROLAND J. SIEZEN (25), *NIZO Food Research, Ede NL-6701 BA, The Netherlands*

M. R. SINGLETON (26), *Sir William Dunn School of Pathology, University of Oxford, Oxford OX1 3RE, United Kingdom*

MARJORY A. SNEAD (13, 14, 15, 22), *Diversa Corporation, San Diego, California 92121-7032*

DINLAKA SRIPRAPUNDH (12), *Department of Biochemistry and Molecular Biology, Michigan State University, East Lansing, Michigan 48824-1319*

KARL O. STETTER (2), *Lehrstuhl für Mikrobiologie und Archaeenzentrum, Universität Regensburg, Regensburg D-93053, Germany*

MARK D. STUMP (7), *Department of Human Genetics, University of Utah, Salt Lake City, Utah 84112*

RONALD V. SWANSON (8), *Department of Molecular Biology, Syrrx, Inc., San Diego, California 92121*

DION R. THOMPSON (19), *Department of Biological Sciences, Waikato University School of Science and Technology, Hamilton, New Zealand*

ANDREAS M. UHL (19), *Department of Biological Sciences, Waikato University School of Science and Technology, Hamilton, New Zealand*

JOHN VAN DER OOST (21, 24, 25), *Laboratory of Microbiology, Wageningen Agricultural University, Wageningen, NL-6703 CT, The Netherlands*

JOHAN F. T. VAN LIESHOUT (21), *Laboratory of Microbiology, Wageningen Agricultural University, Wageningen, NL-6703 CT, The Netherlands*

MARC F. J. M. VERHAGEN (3), *Allergan, Inc., Irvine, California 92612*

CORNÉ H. VERHEES (21), *Laboratory of Microbiology, Wageningen Agricultural University, Wageningen, NL-6703 CT, The Netherlands*

CLAIRE VIEILLE (12, 23), *Department of Biochemistry and Molecular Biology, Michigan State University, East Lansing, Michigan 48824-1319*

WILFRIED G. B. VOORHORST (25), *Laboratory of Microbiology, Wageningen Agricultural University, Wageningen, NL-6703 CT, The Netherlands*

DON E. WARD (21), *Laboratory of Microbiology, Wageningen Agricultural University, Wageningen, NL-6703 CT, The Netherlands*

ROBERT B. WEISS (7), *Department of Human Genetics, University of Utah, Salt Lake City, Utah 84112*

WILLIAM B. WHITMAN (5), *Department of Microbiology, University of Georgia, Athens, Georgia 30602-2605*

RAYMOND K. YEH (7), *Department of Human Genetics, University of Utah, Salt Lake City, Utah 84112*

J. GREGORY ZEIKUS (12, 23), *Michigan Biotechnology Institute International, Lansing, Michigan 48909*

Preface

More than thirty years ago, the pioneering work of Thomas Brock of the University of Wisconsin on the microbiology of hot springs in Yellowstone National Park alerted the scientific community to the existence of microorganisms with optimal growth temperatures of 70°C and even higher. In the early 1980s, the known thermal limits of life were expanded by the seminal work of Karl Stetter and colleagues at the University of Regensburg, who isolated from a marine volcanic vent the first microorganisms that could grow at, and even above, the normal boiling point of water. Subsequent work by Stetter and several other groups have led to the discovery in a variety of geothermal biotopes of more than twenty different genera that can grow optimally at or above 80°C. Such organisms are now termed *hyperthermophiles*.

Initial efforts to explore the enzymology of hyperthermophiles were impeded by the difficulty of culturing the organisms on a scale large enough to allow the purification of specific proteins in sufficient quantities for characterization. This often meant processing hundreds of liters of nearly boiling fermentation media under anaerobic conditions. In addition, relatively low biomass yields were typically obtained. Nevertheless, the first "hyperthermophilic enzymes" were purified in the late 1980s. It was demonstrated that they are, indeed, extremely stable at high temperatures, that this is an intrinsic property, and that they exhibit no or very low activity at temperatures below the growth conditions of the organism from which they were obtained. At that time it was difficult to imagine how quickly the tools of molecular biology would make such a dramatic impact on the world of hyperthermophiles. In fact, it was unexpected that the recombinant forms of hyperthermophilic enzymes would, to a large extent, correctly achieve their active conformation in mesophilic hosts grown some 70°C below the enzyme's source organism's normal growth temperature. This approach provided a much-needed alternative to large-scale hyperthermophile cultivation. With the ever-expanding list of genomes from hyperthermophiles that have been or are being sequenced, molecular biology provides universal access to a treasure chest of known and putative proteins endowed with unprecedented levels of thermostability.

In Volumes 330, 331, and 334 of *Methods in Enzymology,* a set of protocols has been assembled that for the first time describe the methods involved in studying the biochemistry and biophysics of enzymes and proteins from hyperthermophilic microorganisms. As is evident from the various chapters, hyperthermophilic counterparts to a range of previously studied but less thermostable enzymes exist. In addition, the volumes include descriptions of many novel enzymes that were first identified and, in most cases, are still limited to, hyperthermophilic organisms. Also included in these volumes are genomic analyses from selected hyperthermophiles that provide some perspective on what remains to be investigated in terms of hyperthermophilic enzymology. Specific chapters address the basis for extreme levels of thermostability and special considerations that must be taken into account in defining experimentally the biochemical and biophysical features of hyperthermophilic enzymes.

There are many individuals whose pioneering efforts laid the basis for the work discussed in these volumes. None was more important than the late Holger Jannasch of Woods Hole Oceanographic Institute. His innovation and inspiration opened a new field of microbiology in deep-sea hydrothermal vents and provided the research world access to a biotope of great scientific and technological promise. Holger will be remembered in many ways, and it is a fitting tribute that the first genome of a hyperthermophile to be sequenced should bear his name: *Methanocaldococcus jannaschii.* We wish to recognize Holger's pioneering efforts by dedicating these volumes to him.

MICHAEL W. W. ADAMS
ROBERT M. KELLY

METHODS IN ENZYMOLOGY

VOLUME 91. Enzyme Structure (Part I)
Edited by C. H. W. HIRS AND SERGE N. TIMASHEFF

VOLUME 92. Immunochemical Techniques (Part E: Monoclonal Antibodies and General Immunoassay Methods)
Edited by JOHN J. LANGONE AND HELEN VAN VUNAKIS

VOLUME 93. Immunochemical Techniques (Part F: Conventional Antibodies, Fc Receptors, and Cytotoxicity)
Edited by JOHN J. LANGONE AND HELEN VAN VUNAKIS

VOLUME 94. Polyamines
Edited by HERBERT TABOR AND CELIA WHITE TABOR

VOLUME 95. Cumulative Subject Index Volumes 61–74, 76–80
Edited by EDWARD A. DENNIS AND MARTHA G. DENNIS

VOLUME 96. Biomembranes [Part J: Membrane Biogenesis: Assembly and Targeting (General Methods; Eukaryotes)]
Edited by SIDNEY FLEISCHER AND BECCA FLEISCHER

VOLUME 97. Biomembranes [Part K: Membrane Biogenesis: Assembly and Targeting (Prokaryotes, Mitochondria, and Chloroplasts)]
Edited by SIDNEY FLEISCHER AND BECCA FLEISCHER

VOLUME 98. Biomembranes (Part L: Membrane Biogenesis: Processing and Recycling)
Edited by SIDNEY FLEISCHER AND BECCA FLEISCHER

VOLUME 99. Hormone Action (Part F: Protein Kinases)
Edited by JACKIE D. CORBIN AND JOEL G. HARDMAN

VOLUME 100. Recombinant DNA (Part B)
Edited by RAY WU, LAWRENCE GROSSMAN, AND KIVIE MOLDAVE

VOLUME 101. Recombinant DNA (Part C)
Edited by RAY WU, LAWRENCE GROSSMAN, AND KIVIE MOLDAVE

VOLUME 102. Hormone Action (Part G: Calmodulin and Calcium-Binding Proteins)
Edited by ANTHONY R. MEANS AND BERT W. O'MALLEY

VOLUME 103. Hormone Action (Part H: Neuroendocrine Peptides)
Edited by P. MICHAEL CONN

VOLUME 104. Enzyme Purification and Related Techniques (Part C)
Edited by WILLIAM B. JAKOBY

VOLUME 105. Oxygen Radicals in Biological Systems
Edited by LESTER PACKER

VOLUME 106. Posttranslational Modifications (Part A)
Edited by FINN WOLD AND KIVIE MOLDAVE

VOLUME 107. Posttranslational Modifications (Part B)
Edited by FINN WOLD AND KIVIE MOLDAVE

Section I

Enzyme Discovery

[1] A Phylogenetic Perspective on Hyperthermophilic Microorganisms

By EDWARD F. DeLONG

Introduction

Hyperthermophilic microorganisms grow optimally, and sometimes exclusively, at temperatures greater than 80°.[1] Their existence has been known for only a few decades, and much remains to be learned about their physiology, biochemistry, and natural history. In fact, the upper temperature limit for microbial life is still uncertain, but it reaches to at least 113° for cultivated isolates.[2] Extreme thermophiles are found in a wide variety of high temperature habitats, both terrestrial and marine. Hyperthermophilic microbes have a remarkable variety of lifestyles and physiologies, ranging from aerobic autotrophy to strictly anaerobic heterotrophy. The biochemical adaptations and physiological diversity of hyperthermophiles are fascinating and extremely productive topics of current research. Equally as fascinating, if less well developed, are the details concerning the natural history, ecology, and evolution of hyperthermophilic microbial life.

The microbial world extends well within habitats with temperatures greater than 100° and is providing considerable insight into the possible scenarios for life's origin and evolution. What are the evolutionary relationships of extant microorganisms, and what does this tell us about the past? What were the likely properties of extant life's most recent common ancestor? What evidence supports the notion that hyperthermophiles may have arisen early in life's history? Alternatively, is there evidence for a later arrival of hyperthermophiles over the course of life's evolution? These questions are central to deciphering how early cellular life on Earth may have evolved. Developments in molecular phylogenetic analyses and, more recently, genomic science are providing some of the most provocative but still controversial answers to these questions. Interestingly, hyperthermophilic microorganisms are providing some central clues on what early life on Earth may have looked like.

---

[1] K. O. Stetter, *FEMS Microbiol. Rev.* **18,** 149 (1996).
[2] E. Blochl, R. Rachel, S. Burggraf, D. Hafenbradl, H. W. Jannasch, and K. O. Stetter, *Extremophiles* **1,** 14, (1997).

Foundations of Microbial Evolution

Until recently, evolutionary relationships among microorganisms were virtually unknown and undescribed. There are few conspicuous morphological features that can be used to systematically differentiate microorganisms or infer their evolutionary relationships. Additionally, the microbial fossil record is neither extensive nor informative enough to provide much insight into ancestral microbial life. Until the mid-1960s, microbiologists had to be content with simply distinguishing prokaryotes (that do not possess membrane-bound nuclei) from eukaryotes (with true nuclear organelles). This situation changed dramatically when physical chemists first recognized that molecules can serve as documents for evolutionary history.[3] Evolutionary relationships could now be deduced from sequence differences observed between homologous macromolecules. This development has transformed biological disciplines in profound ways. For the first time a universal evolutionary tree, encompassing all cellular life forms, is a reality.

Carl Woese was the first to fully exploit the power of molecular phylogenetics for microbiology, as he and his colleagues sought to create a unified picture of evolutionary relationship among prokaryotes in the 1970s. With remarkable foresight, he selected perhaps the single most optimal macromolecule for this work, 16S ribosomal RNA (rRNA). Woese realized that the optimal macromolecule for constructing global phylogenies should have a universal distribution, high conservation, some moderate variability, and minimal lateral genetic transfer. Initially, ribosomal RNA oligonucleotide catalogs were used to infer relationships among disparate phylogenetic groups. Later, advanced nucleic acid sequencing techniques allowed direct acquisition and comparison of rRNA sequences. One of the first and most dramatic results was the discovery of a new prokaryotic kingdom, the Archaea (then called archaebacteria).[4] These prokaryotes are as evolutionarily distant from common bacteria as they are from eukaryotes. The archaea known then were a fairly bizarre collection of microbes: sulfur-respiring thermophiles, extreme halophiles, and obligately anaerobic methanogens.[5] The distinctiveness of the Archaea eventually led to a still controversial recategorization of life into three major domains, consisting of Eucarya (all eukaryotes) and the two exclusively microbial domains, Archaea and Bacteria.[6]

The fruit of Woese's efforts, a universal tree depicting the evolutionary

---

[3] E. Zuckerkandl and L. Pauling, *J. Theor. Biol.* **8,** 357 (1965).

[4] C. R. Woese and G. E. Fox, *Proc. Natl. Acad. Sci. U.S.A* **74,** 5088 (1977).

[5] C. R. Woese, *Microbiol. Rev.* **51,** 221 (1987).

[6] C. R. Woese, O. Kandler, and M. L. Wheelis, *Proc. Natl. Acad. Sci. U.S.A.* **87,** 4576 (1990).

relationships of all life (Fig. 1), is simple in form but profound in its implications.[5] For instance, a casual glance at the tree shows that the lion's share
of phylogenetic diversity resides in the microbial world. Macroscopic organisms occupy a small terminal node on the tree of extant life. Given the 3.8
billion years for microbial evolution, this is perhaps not so surprising.
Another brief look at the tree in Fig. 1 reveals the endosymbiotic origins
of several organelles: chloroplasts within the cyanobacterial cluster (near
Synechococcus) and mitochondria within the alpha Proteobacteria lineage
(near *Agrobacterium*). The major bifurcations of the tree clearly show
three discrete, major domains, two of which are prokaryotic (Archaea and
Bacteria). The dotted line ("origin" in Fig. 1) indicates the approximate
position of the common ancestor that gave rise to all extant cellular life.
The true position of this common ancestor is still uncertain and has been
inferred mainly from analyses of duplicated protein-encoding genes (elongation factors, ATPases) that allow rooting of the universal tree.[6,7] The
major point to recognize when considering these molecular phylogenies is
that the genealogy of the organism is extrapolated from the evolutionary
trajectory of single genes. Depending on the gene used, its particular history
(including lateral transfer), and the magnitude of evolutionary distances
considered, these extrapolations will have varying degrees of accuracy.
Despite all of the uncertainties and artifacts associated with single gene
phylogenies, the major features of Woese's ribosomal RNA tree, and its
lessons for biology, are weathering the test of time well.

Phylogenetic Diversity in Cultured Hyperthermophiles

Hyperthermophilic microorganisms are found in both prokaryotic domains: Bacteria and Archaea (Fig. 1). Although the search has been ongoing
for a number of years, no hyperthermophilic eukaryote has yet been discovered, but the geothermal world may still harbor secrets about Eucarya in
high temperature environments. In the universal tree of life, hyperthermophilic Bacteria and Archaea occupy a pivotal position. Hyperthermophilic microbes are found in the most basal positions in the universal tree
(e.g., are the most deeply rooted) in both Archaea and Bacteria (bold lines
in Fig. 1). In Bacteria, two of the most deeply branching groups contain
well-known hyperthermophilic genera: *Thermotoga* and *Aquifex* (Fig. 1).
Archaea, made up of the two major lineages, Euryarchaeota and Crenarchaeota, show a similar pattern. The deepest lineages of the Euryarchaeota
contain known hyperthermophilic groups (Fig. 1, *Methanopyrus, Ther-*

---

[7] N. Iwabe, K. Kuma, M. Hasegawa, S. Osawa, and T. Miyata, *Proc. Natl. Acad. Sci. U.S.A*
86, 9355 (1989).

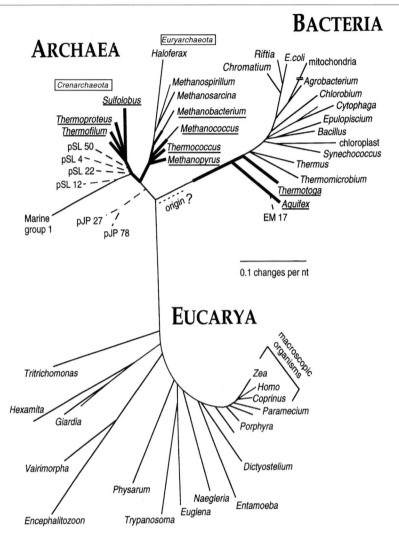

FIG. 1. The ribosomal RNA universal tree of life. The wide dark branches and underlined text indicate hyperthermophilic groups. Two groups depicted, *Methanococcus* and *Methanobacterium,* contain both high and low temperature-adapted species. Dashed lines indicate rRNA genes recovered from environmental samples from uncultivated microorganisms. Adapted, with permission, from N. Pace, *ASM News* **62,** 463 (1996).

ARCHAEA

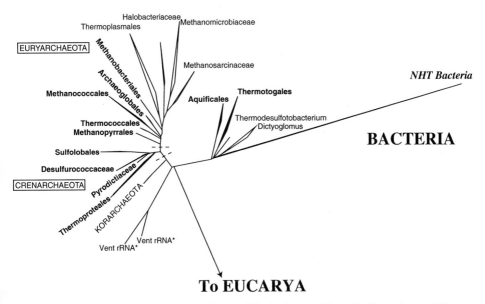

FIG. 2. Families and orders of hyperthermophilic Archaea and Bacteria. Hyperthermophilic groups are indicated in bold text. Korarchaeota represents rRNA sequences cloned from hot springs. Vent rRNA represents cloned rRNA sequences from submarine hydrothermal vents. NHT, nonhyperthermophilic bacteria.

mococcus, and *Methanococcus*). Likewise, the cultivated Crenarchaeota are well represented by genera containing known hyperthermophiles (*Sulfolobus, Thermoproteus,* and *Thermofilum*). The rRNAs of these hyperthermophiles appear to have accumulated on average fewer mutations than those of other mesophilic organisms, relative to the most recent common ancestor. The conspicuous deep rooting and short branch lengths of hyperthermophiles in the tree of life (Fig. 1) have therefore led to much speculation about their potential similarities to ancient life forms.[8]

Considerable phylogenetic diversity is found among the hyperthermophilic Archaea (Fig. 2). This evolutionary diversity manifests itself in the the varied morphologies, physiologies, and biochemistries found among the hyperthermophilic Archaea. The domain Archaea has been subdivided into two major kingdoms: the Crenarchaeota and the Euryarchaeota[6] (Fig. 2). Hyperthermophiles are found in both these archaeal kingdoms.

Cultivated members of the Crenarchaeota consist entirely of extreme

---

[8] N. R. Pace, *Cell* **65,** 531 (1991).

thermophiles and hyperthermophiles. However, culture-independent surveys indicate that psychrophilic and mesophilic crenarchaeotes are abundant in nature as well.[9] Representative Crenarchaeota include both autotrophs and heterotrophs, as well as strictly anaerobic or microaerophilic species. Hyperthermophilic genera contained within the Crenarchaeota include *Sulfolobus, Desulfurococcus, Pyrodictium, Pyrolobus, Thermofilum, Thermoproteus,* and *Pyrolobus* (Fig. 2). The current record holder for high temperature growth in a pure culture (113° !) is *Pyrolobus fumarii,* in the family Pyrodictiaceae of the Crenarchaeota[2] (Fig. 2). The kingdom Crenarchaeota encompasses hyperthermophiles with the deepest and shortest branch lengths in the rRNA-based trees.

Euryarchaeota are more cosmopolitan in their ecological distribution than cultivated Crenarchaeota. Hyperthermophiles in the Euryarchaeota include sulfur-reducing heterotrophs, CO_2-reducing methanogens, and sulfate-reducing hydrogen oxidizers and heterotrophs.[1] The Methanopyrales, a hyperthermophilic methanogenic group, represents the most deeply rooted euryarchaeotal lineage in the rRNA tree (Fig. 2). Another methanogenic order, Methanococcales, includes both hyperthermophiles and mesophiles (Figs. 1 and 2). The first full genome sequence from an archaeon came from a member of this group, the hyperthermophilic methanogen *Methanococcus jannaschii.*[10] The "lab rats" of archaeal world also reside in the euryarchaeotal subdivision, in the order Thermococcales. The rapid generation times and general hardiness of these hyperthermophiles have made them excellent model systems for studying the physiology and enzymology of hyperthermophilic archaea. The Methanobacteriales includes one hyperthermophilic genus, *Methanothermus,* as well as many other species adapted to lower temperatures. Another euryarchaeaotal order, the Archaeoglobales (Fig. 2), includes hyperthermophilic sulfate reducers (*Archaeoglobus* spp.) and iron oxidizers (*Ferroglobus* spp.).

Less phylogenetic diversity is found in cultivated hyperthermophilic Bacteria than in Archaea. Only two hyperthermophilic groups are currently known in Bacteria: Aquificales and Thermotogales. The most deeply branching hyperthermophilic bacterial group is the Aquificales (Fig. 2). These hydrogen-oxidizing, hyperthermophilic bacteria have growth temperature maxima of about 95°.[1] The other hyperthermophilic bacterial group is the Thermotogales. These strictly anaerobic heterotrophs are named for their characteristic sheath, or toga. Thermotogales are found in a wide variety of marine and terrestrial habitats, including hot springs, hydrothermal vents, and geothermally heated subsurface oil reservoirs. Recent analy-

[9] E. F. DeLong, *Science* **280,** 542 (1998).
[10] C. J. Bult *et al., Science* **273,** 1058 (1996).

ses of the *Thermotoga maritima* genome has provided some interesting surprises. In preliminary analyses, nearly 24% of all the genes identified in *T. maritima* have highest similarity to homologs found within the Archaea.[11]

Origins of Life: Hot or Not?

The exact physicochemical environment of the Earth near the time of life's origin (ca. 3.8 billion years ago) is uncertain. It is clear, however, that the early Earth was a harsh place to be compared to today's global environment. Mild to strongly reducing atmospheric conditions, high vapor pressure, extreme heat (or, alternatively, extreme cold), and frequent bolide bombardment were all potential conditions encountered on an average archaean day.[8] Whatever the scenario, global uniformity at the time of life's origin seems unlikely. The exact physical setting for life's origin may in fact have been quite different from an extrapolated global average deduced from geological and astronomical evidence. Major deviations from average conditions ranging from cold to hot, and mildly oxidizing to strongly reducing, were all likely represented in various microenvironments. Exactly which archaean niche(s) early life evolved in is largely a matter of conjecture. Nevertheless, many speculate that life evolved at high temperatures that reflect a putative archaean global average. Alternatively, high temperatures for early evolution could have been found at volcanic sites, even if a cold early Earth model is correct. Regardless of global temperature average, most agree that free oxygen in the early archaean was negligible. If speculation about hot origins for life are correct, Earth's earliest ecosystems may have physicochemically resembled contemporary geothermal habitats, where numerous hyperthermophilic microorganisms thrive today. Extant hyperthermophilic microbes might therefore provide important clues about the nature of Earth's earliest life.

In the universal rRNA tree (Fig. 1), phyla that contain hyperthermophilic bacteria or archaea are usually the most deeply rooted and have the shortest branch lengths. Similarly, within specific orders containing both hyperthermophilic and cold-adapted microorganisms (e.g., Methanococcales, Methanobacteriales), hyperthermophiles are usually the basal species and also have the shortest branch lengths in the group.[5,8] If the short branch lengths in rRNA trees accurately reflect organismal evolutionary relationships, then hyperthermophilic microorganisms might most closely resemble the last common ancestor of all contemporary life. This speculation, however, is by no means certain. Several authors have argued alterna-

[11] K. E. Nelson *et al., Nature* **399,** 323 (1999).

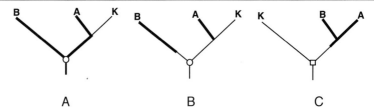

Fɪɢ. 3. Potential scenarios for the evolution of extant life. B, Bacteria; A, Archaea; K, Eucarya. The circle and square represent a prokaryotic or eukaryotic common ancestor, respectively. (a) Thermophilic prokaryotic origin, bacterially rooted, with Archaea and Eucarya as sister lineages. (b) Nonthermophilic prokaryotic origin, bacterially rooted, with Archaea and Eucarya as sister lineages. (c) Nonthermophilic eukaryotic origin, with Archaea and Bacteria as sister lineages.

tive hypotheses,[12,13] some of which are depicted in Fig. 3. Phylogenetic trees derived from highly conserved gene families (e.g., protein elongation factors) suggest that the true root of the tree of life may be located in Bacteria.[6,7] The tree shown in Fig. 3A depicts the hypothetical situation where hyperthermophilic members of Bacteria represent the ancestral root of the tree. In this scenario, Archaea forms a sister group with Eucarya, consistent with gene phylogenies derived from many genes involved in DNA replication, transcription, or translation.[7] Alternative scenarios (Fig. 3B) depict a "cold" bacterial origin for life, implying that the hyperthermophilic phenotype is a derived trait that arose independently within Bacteria and Archaea. A more controversial hypothesis,[12] illustrated in Fig. 3C, depicts Eucarya as the nonthermophilic common ancestor of all extant life. Although data are scant, one study[13] developed a model attempting to back-extrapolate the guanine plus cytosine (G + C) content of the rRNA of the last common ancestor of extant life. In contemporary hyperthermophiles, G + C content tends to be high (>58%) and so is a potential indicator of thermophily. These authors concluded that the rRNA G + C content of the ancestral group rooting the universal tree was about 54.0%, suggesting that the common ancestor was not a thermophile.[13] The veracity of their model, however, as well as the reliability of rRNA G + C content for inferring thermophily, are both matters of some uncertainty.

Conclusion

The phylogenetic diversity of microorganisms that grow optimally at temperatures greater than 80° is impressive, considering that such organisms

[12] P. Forterre, *Cell* **85,** 789 (1996).
[13] N. Galtier, N. Tourasse, and M. Gouy, *Science* **283,** 220 (1999).

have only been known since the 1980s. Cultivation-independent surveys have revealed that our view of microbial diversity has been somewhat myopic.[9,14] In many environments, less than 1% of naturally occurring microorganisms can be cultivated readily. For example, abundant environmental evidence shows that in the deep sea and Antarctic oceans, close relatives of hyperthermophilic crearchaea thrive at temperatures as low as $-1.8°$, yet no cold-adapted crenarchaeote has yet been isolated in culture. Preliminary evidence also shows that novel yet-uncultivated hyperthermophilic archaea exist that root more deeply in the tree of life than any currently cultivated group (Figs. 1 and 2). One group, represented by rRNA gene clones pJP27 and pJP28 (Fig. 1), has been dubbed the Korarchaeota (Figs. 1 and 2) and appears be ancestral to the Crenarchaeota and Euryarchaeota (Fig. 1). Another new group roots even more deeply in the tree than the Korarchaeota, and is represented by the two Vent rRNA clones[15] shown in Fig. 2. So far these new microbes have been detected solely by phylogenetic analysis of rRNA gene fragments, so their identity as hyperthermophiles remains speculative.

Hyperthermophilic microbial habitats are difficult to sample exhaustively, and recreating hyperthermophilic niches in the laboratory is nontrivial. It is likely that much hyperthermophilic microbial diversity remains to be discovered and described. In addition, new lineages of cold-adapted microbial species, specifically related to hyperthermophilic microorganisms, continue to be discovered.[9] Further collaboration among microbial ecologists, evolutionists, microbial physiologists, and enzymologists promises to significantly refine our understanding of hyperthermophilic life. Much exciting comparative biochemistry and enzymology is evident in this volume, and yet more looms on the horizon, as the potential of some of these newer discoveries begins to be realized.

[14] N. R. Pace, *Science* **276,** 734, (1996).
[15] K. Takai and K. Horikoshi, *Genetics* **152,** 1285 (1999).

[2] Discovery of Hyperthermophilic Microorganisms

By ROBERT HUBER and KARL O. STETTER

Introduction

A variety of hyperthermophilic Archaea and Bacteria have been isolated from water-containing terrestrial, subterranean, and submarine

0076-6879/00 $35.00

high-temperature ecosystems.[1,2] Hyperthermophiles grow fastest at temperatures between 80° and 106°. In contrast to moderate thermophiles, they are unable to propagate below about 60°.[2,3] *Pyrolobus fumarii,* the most extreme hyperthermophile, is even unable to grow below 90° and exhibits the highest growth temperature observed of all at 113°.[4] Hyperthermophiles are very divergent, both in terms of their phylogeny and physiological and biochemical properties.[5,6] Due to their metabolic flexibility and their outstanding heat resistance, hyperthermophiles are as interesting for basic research as they are for biotechnological applications.[5,7,8]

Biotopes

So far, hyperthermophiles have been isolated mainly from water-containing terrestrial and marine high-temperature areas, where they form microbial communities.[8] On land, natural biotopes of hyperthermophiles are water-containing areas, such as hot springs and solfataric fields, with a wide range of pH values (pH 0.5–9.0) and usually low salinity (0.1–0.5%). Marine biotopes of hyperthermophiles are shallow hydrothermal systems, abyssal hot vents ("black smokers"), and active seamounts, such as Teahicya and Macdonald in the Tahiti area, Polynesia. These biotopes are characterized by high concentrations of salt (about 3%) and pH values that are slightly acidic to slightly alkaline (pH 5.0–8.5). Samples taken from the active crater and the cooled-down open ocean plume of the erupting Macdonald seamount contained communities of hyperthermophiles with up to 10^6 viable cells per liter.[9] Artificial biotopes include smouldering coal refuse piles (e.g., Ronneburg, Thüringen, Germany)[10] and hot outflows from geothermal and nuclear power plants. Other suitable biotopes for hyperthermophiles are deep subterranean, geothermally heated oil reser-

[1] K. O. Stetter, *J. Chem. Technol. Biotechnol.* **42**, 315 (1988).
[2] K. O. Stetter, *in* "Life at the Upper Temperature Border" (J. & K. Trân Thanh Vân, J. C. Mounolou, J. Schneider, and C. McKay, eds.), p. 195. Editions Frontières, Gif-sur-Yvette, France, 1992.
[3] K. O. Stetter, *FEBS Lett.* **452**, 22 (1999).
[4] E. Blöchl, R. Rachel, S. Burggraf, D. Hafenbradl, H. W. Jannasch, and K. O. Stetter, *Extremophiles* **1**, 14 (1997).
[5] K. O. Stetter, *ASM News* **61**, 285 (1995).
[6] K. O. Stetter, *in* "Extremophiles" (K. Horikoshi and W. D. Grant, eds.), p. 1. Wiley, New York, 1999.
[7] H. Huber and K. O. Stetter, *J. Biotechnol.* **64**, 39 (1988).
[8] K. O. Stetter, *in* "Volcanoes and the Environment" (J. Marti and G. J. Ernst, eds.). Cambridge Univ. Press, Cambridge, 2000.
[9] R. Huber, P. Stoffers, J. L. Cheminée, H. H. Richnow, and K. O. Stetter, *Nature* **345**, 179 (1990).
[10] T. Fuchs, H. Huber, K. Teiner, S. Burggraf, and K. O. Stetter, *System. Appl. Microbiol.* **18**, 560 (1996).

voirs about 3500 m below the bottom of the North Sea and the permafrost soil at the North Slope, northern Alaska.[11] The production fluids contained up to 10^4 to 10^7 viable cells of different species of hyperthermophiles per liter at Thistle platform, North Sea, and Prudehoe Bay, Alaska, respectively.[11,12] Many of these hyperthermophiles are sulfidogens and may, therefore, participate in "reservoir souring" at temperatures previously considered too high for biochemical reactions.[13] In active volcanic environments, large amounts of steam are formed, containing carbon dioxide, hydrogen sulfide, and variable quantities of hydrogen, methane, nitrogen, carbon monoxide, and traces of ammonia or nitrate. Furthermore, sulfur, thiosulfate, and sulfate are the major sulfur compounds and are present in solfatara fields as well as in marine hot areas.

Taxonomy and Classification

To date, about 70 species of hyperthermophiles are known (Table I).[3,8] Based on a recent reclassification, they are grouped into 29 genera in 10 orders (Table I).[10,14] With the exception of the two hyperthermophilic bacterial genera *Thermotoga* (Fig. 1A) and *Aquifex,* which exhibit maximal growth temperatures at 90° and 95°C, respectively, hyperthermophiles belong exclusively to the domain of the Archaea (Table I). Organisms with the highest growth temperatures within the Archaea (between 103 and 113°) are members of the genera *Methanopyrus, Pyrococcus, Pyrodictium, Pyrobaculum,* and *Pyrolobus* (Table I; Figs. 1B–1D). Cultures with (vegetative) cells of *Pyrolobus* and *Pyrodictium* are even able to survive autoclaving.[3]

The 16S rRNA-based phylogenetic tree shows a tripartite division of the living world consisting of the three domains Bacteria, Archaea, and Eucarya.[15,16] Around the root, hyperthermophiles form a cluster and occupy all the short deep phylogenetic branches.[2,17] This is true for both bacterial and archaeal domains. Short phylogenetic branches indicate a rather slow rate of evolution, and deep branching points are evidence for an early

[11] K. O. Stetter, R. Huber, E. Blöchl, M. Kurr, R. D. Eden, M. Fielder, H. Cash, and I. Vance, *Nature* **365,** 743 (1993).
[12] K. O. Stetter and R. Huber, *in* "Microbial Biosystems: New Frontiers" (C. R. Bell, M. Brylinsky, and P. Johnson-Green, eds.), *Proceedings of the 8th International Symposium on Microbial Ecology.* Halifax, Canada, 2000.
[13] S. D. Marshland, R. A. Dawe, and G. H. Marsland, *Trans. Inst. Chem. Engineers* **68,** 357 (1990).
[14] S. Burggraf, H. Huber, and K. O. Stetter, *Int. J. Syst. Bacteriol.* **47,** 657 (1997).
[15] C. R. Woese and G. E. Fox, *Proc. Natl. Acad. Sci. U.S.A.* **74,** 5088 (1977).
[16] C. R. Woese, O. Kandler, and M. L. Wheelis, *Proc. Natl. Acad. Sci. U.S.A* **87,** 4576 (1990).
[17] K. O. Stetter, *in* "The Molecular Origins of Life: Assembling Pieces of the Puzzle" (A. Brack, ed.), p. 315. Cambridge Univ. Press, Cambridge, 1999.

TABLE I

TAXONOMY AND UPPER GROWTH TEMPERATURES OF HYPERTHERMOPHILIC ARCHAEA AND BACTERIA[a]

Order	Place of first discovery of genus[b]	Genus	Species[c]	T_{max} (°)
Domain: Bacteria				
Thermotogales	Vulcano$_P$	Thermotoga	T. maritima[T] G!	90
			T. neapolitana	90
			T. thermarum	84
			T. subterranea	75[d]
			T. elfii	72[d]
			T. hypogea	90
	Djibouti	Thermosipho	T. africanus[T]	77[d]
	Rotorua	Fervidobacterium	F. nodosum[T]	80[d]
			F. islandicum	80[d]
Aquificales	Kolbeinsey	Aquifex	A. pyrophilus[T]	95
			A. aeolicus G!	93
Domain: Archaea				
I. Kingdom: Crenarchaeota				
Sulfolobales	Yellowstone	Sulfolobus	S. acidocaldarius[T]	85
			S. solfataricus G!	87
			S. shibatae	86
			S. metallicus	75[d]
	Naples$_P$	Metallosphaera	M. sedula[T]	80[d]
			M. prunae	80[d]
	Naples$_S$	Acidianus	A. infernus[T]	95
			A. brierleyi	75[d]
			A. ambivalens	95
	Furnas	Stygiolobus	S. azoricus[T]	89
Thermoproteales	Krafla$_S$	Thermoproteus	T. tenax[T]	97
			T. neutrophilus	97
			T. uzoniensis	97
	Krafla$_G$	Pyrobaculum	P. islandicum[T]	103
			P. organotrophum	103
			P. aerophilum G!	104
	Iceland	Thermofilum	T. pendens[T]	95
			T. librum	95
	Noji-onsen	Thermocladium	T. modestius[T]	82[d]
	Mt Maquiling	Caldivirga	C. maquilingensis[T]	92
Desulfurococcales	Iceland	Desulfurococcus	D. mucosus[T]	97
			D. mobilis	95
			D. saccharovorans	97
			D. amylolyticus	97
	Vulcano$_F$	Staphylothermus	S. marinus[T]	98
	Hveragerthi	Sulfophobococcus	S. zilligii[T]	95
	Milos	Stetteria	S. hydrogenophila[T]	102
	Kodakara	Aeropyrum	A. pernix[T] G!	100
	Kolbeinsey	Igneococcus	I. islandicus[T]	103
	Yellowstone	Thermosphaera	T. aggregans[T]	90
	Vulcano$_P$	Thermodiscus	T. maritimus[T]	98
	Vulcano$_P$	Pyrodictium	P. occultum[T]	110
			P. brockii	110
			P. abyssi	110

TABLE I (*continued*)

Order	Place of first discovery of genus[b]	Genus	Species[c]	T_{max} (°)
	Sao Miguel	*Hyperth-ermus*	*H. butylicus*[T]	108
	Mid-Atlantic	*Pyrolobus*	*P. fumarii*[T]	113
II. Kingdom: Euryarchaeota				
Thermococcales	Vulcano[P]	*Ther-mococcus*	*T. celer*[T]	93
			T. litoralis	98
			T. stetteri	98
			T. profundus	90
			T. alcaliphilus	90
			T. chitonophagus	93
			T. fumicolans	103
			T. peptonophilus	100
			T. barophilus	100
			T. waiotapuensis	90
	Vulcano[P]	*Pyrococcus*	*P. furiosus*[T] G!	103
			P. woesei	103
			P. abyssi G!	102
			P. horikoshii G!	102
Archaeoglobales	Vulcano[P]	*Archaeog-lobus*	*A. fulgidus*[T] G!	92
			A. profundus	92
	Vulcano[P]		*A. veneficus*	88
		Ferroglobus	*F. placidus*[T]	95
Methanobacteriales	Kerlingarfjöll	*Methanoth-ermus*	*M. fervidus*[T]	97
			M. sociabilis	97
Methanococcales	Naples[N]	*Methano-coccus*	*M. thermolithotrophicus*[e]	70
			M. jannaschii G!	86
			M. igneus	91
Methanopyrales	Guaymas	*Methano-pyrus*	*M. kandleri*[T]	110

[a] From Ref. 6.

[b] Sites of isolation: Djibouti, marine hot vents, Obock, Djibouti, Africa; Furnas, Furnas Caldeiras, Sao Miguel, Azores; Guaymas, hot deep sea sediments, Guaymas Basin, Mexico; Hveragerthi, hot alkaline spring, Hveragerthi, Iceland; Iceland, solfataric hot spring, Iceland; Kerlingarfjöll, Kerlingarfjöll Solfataras, Iceland; Kodakara, coastal solfataric vent, Kodakara-Jima Island, Japan; Kolbeinsey, submarine hot vents, Kolbeinsey, Iceland; Krafla[G], Krafla Geothermal Power Plant, Iceland; Krafla[S], Krafla Solfataras, Iceland; Mid-Atlantic, TAG site hot deep sea vents, Mid-Atlantic Ridge; Milos, shallow submarine hot brine seeps, Paleohori Bay, Milos, Greece; Mt Maquiling, hot spring, Mt Maquiling, Laguna, Philippines; Naples[N], Stufe di Nerone marine hot vents, Naples, Italy; Naples[P], Pisciarelli Solfatara, Naples, Italy; Naples[S], Solfatara Crater, Pozzuoli, Naples, Italy; Noji-onsen, solfataric mud from Noji-onsen, Fukushima, Japan; Rotorua, hot spring Rotorua, New Zealand; Sao Miguel, submarine hot vents, Sao Miguel, Azores; Vulcano[F], Faraglione marine hot vents, Vulcano, Italy; Vulcano[P], Porto di Levante marine hot vents, Vulcano, Italy; Yellowstone, Yellowstone National Park hot springs, Wyoming, USA.

[c] T, type species; G!, genome sequenced or under sequencing.

[d] Extreme thermophiles related to hyperthermophiles.

[e] First thermophilic member of the genus.

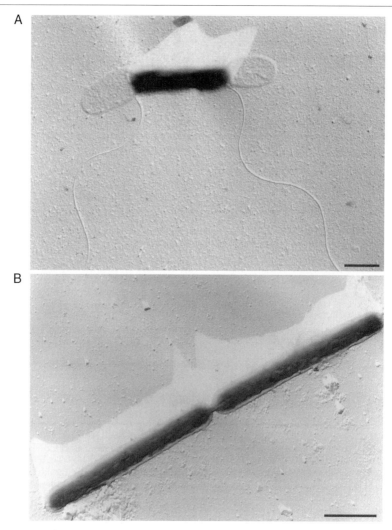

Fɪɢ. 1. Electron micrographs of hyperthermophilic Bacteria and Archaea, platinum shadowed. (A) Single flagellated cell of *Thermotoga maritima* surrounded by a sheath-like outer structure ("toga"). Bar: 1 μm. (B) Dividing cell of *Methanopyrus kandleri*. Bar: 1 μm. (C) *Pyrodictium abyssi*, network of cannulae with cells. Bar: 2 μm. (D) Flagellated cell of *Pyrococcus furiosus*. Bar: 1 μm.

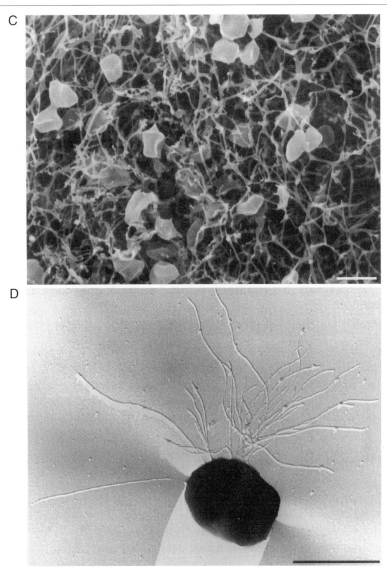

FIG. 1. (*continued*)

separation of two groups. The deepest and earliest branching point so far is the separation of the Bacteria from the Archaea–Eucarya lineage. Based on this observation, hyperthermophiles appear to be the most primitive organisms that still exist and the last common ancestor may have been a hyperthermophile.[2,17]

TABLE II
ENERGY CONSERVATION IN CHEMOLITHOAUTOTROPHIC HYPERTHERMOPHILES

Energy-yielding reaction	Genera
$H_2 + \frac{1}{2}O_2 \rightarrow H_2O$	*Aquifex, Sulfolobus,[a] Acidianus,[a] Metallosphaera,[a] Pyrobaculum[a]*
$H_2 + HNO_3 \rightarrow HNO_2 + H_2O$	*Aquifex, Pyrobaculum[a]*
$4H_2 + HNO_3 \rightarrow NH_4OH + 2H_2O$	*Pyrolobus*
$2FeCO_3 + HNO_3 + 5H_2O \rightarrow 2Fe(OH)_3 + HNO_2 + 2H_2CO_3$	*Ferroglobus*
$H_2 + 6Fe(OH)_3 \rightarrow 2Fe_3O_4 + 10H_2O$	*Pyrobaculum*
$4H_2 + H_2SO_4 \rightarrow H_2S + 4H_2O$	*Archaeoglobus[a]*
$H_2 + S^0 \rightarrow H_2S$	*Igneococcus, Acidianus, Stygiolobus, Pyrodictium,[a] Pyrobaculum,[a] Thermoproteus[a]*
$4H_2 + CO_2 \rightarrow CH_4 + 2H_2O$	*Methanopyrus, Methanothermus, Methanococcus*
$2S^0 + 3O_2 + 2H_2O \rightarrow 2H_2SO_4$	*Aquifex, Sulfolobus,[a] Acidianus,[a] Metallosphaera[a]*
$(2FeS_2 + 7O_2 + 2H_2O \rightarrow 2FeSO_4 + 2H_2SO_4) = $ "metal leaching"	

[a] Facultatively heterotrophic.

Meanwhile, several genomes of hyperthermophiles are already sequenced or are being sequenced (Table I). Comparison of the genome sequences of *Aquifex pyrophilus, Thermotoga maritima,* and *Archaeoglobus fulgidus* suggests that lateral gene transfer occurred even between hyperthermophilic members of the Archaea and Bacteria domain (Table I)[18–20]

General Metabolic Potential

Hyperthermophiles are well adapted to their biotopes, growing at high temperatures and extremes of pH, salinity, and redox potential. In their high-temperature ecosystems, they form complex food webs, which consist of primary producers and consumers of organic material. Primary producers are chemolithoautotrophs, which gain energy through chemical oxidation, using a variety of anorganic compounds as electron donors and acceptors (Table II). Hydrogen is a very important electron donor in their energy-yielding reactions. Cellular carbon is obtained through the reduction of carbon dioxide via the reductive tricarboxylic acid cycle, the reductive

[18] H.-P. Klenk *et al., Nature* **390,** 364 (1997).
[19] G. Deckert *et al., Nature* **392,** 353 (1998).
[20] K. E. Nelson *et al., Nature* **399,** 323 (1999).

TABLE III
ENERGY CONSERVATION IN HETEROTROPHIC HYPERTHERMOPHILES

Type of metabolism	External electron acceptor	Energy-yielding reaction	Genera
Respiration	S^0	$2[H] + S^0 \rightarrow H_2S$	*Pyrodictium, Thermoproteus, Pyrobaculum, Thermofilum, Desulfurococcus, Thermodiscus*
	$SO_4^{2-} (S_2O_3^{2-}; SO_3^{2-})$	$8[H] + SO_4^{2-} + 2H^+ \rightarrow H_2S + 4H_2O$	*Archaeoglobus*
	O_2	$2[H] + \frac{1}{2}O_2 \rightarrow H_2O$	*Sulfolobus, Metallosphaera, Acidianus*
Fermentation	—	Peptides $\rightarrow$ isovalerate, isobutyrate, butanol, CO_2, H_2, etc.	*Pyrodictium, Hyperthermus, Thermoproteus, Desulfurococcus, Staphylothermus, Thermococcus*
	—	Pyruvate $\rightarrow$ L(+)-lactate + acetate + H_2 + CO_2	*Thermotoga, Thermosipho, Fervidobacterium*
		Pyruvate $\rightarrow$ acetate + H_2 + CO_2	*Pyrococcus*

acetyl-CoA pathway, or the 3-hydroxypropionate cycle.[21-23] Several chemolithoautotrophs are facultatively heterotrophic, using organic matter for aerobic or anaerobic respiration (Table III). Furthermore, a variety of hyperthermophiles grow by fermentation, forming short-chain fatty acids, carbon dioxide, and hydrogen as major products (Table III).[8]

Hydrogen is a potent inhibitor of growth for different hyperthermophilic Archaea and Bacteria.[24,25] This inhibition can be overcome by the addition of elemental sulfur with the formation of hydrogen sulfide.[24,25] The fermentatively growing members of the bacterial genus *Thermotoga* are hydrogen forming. Several species are able to use sulfur, cystine, or thiosulfate to eliminate inhibitory hydrogen by H_2S formation. The respiratory growth of *T. maritima* has been reported, with hydrogen as an electron donor and

[21] O. Kandler, *in* "Early Life on Earth: Nobel Symposium 84" (S. Bengtson, ed.), p. 152. Columbia Univ. Press, New York, 1993.

[22] M. Ishii, T. Miyake, T. Satoh, H. Sugiyama, Y. Oshima, T. Kodama, and Y. Igarashi, *Arch. Microbiol.* **111,** 368 (1997).

[23] C. Menendez, Z. Bauer, H. Huber, N. Gad'on, K. O. Stetter, and G. Fuchs, *J. Bacteriol.* **181,** 1088 (1999).

[24] G. Fiala and K. O. Stetter, *Arch. Microbiol.* **145,** 56 (1986).

[25] R. Huber, T. A. Langworthy, H. König, M. Thomm, C. R. Woese, U. B. Sleytr, and K. O. Stetter, *Arch. Microbiol.* **144,** 324 (1986).

Fe(III) as the electron acceptor, forming Fe(II) as the final product.[26] In contrast, we have observed growth of *T. maritima* with similar doubling times and final cell densities in the absence of both hydrogen and Fe(III) in the same culture medium (Huber, Moissl, and Stetter, unpublished results, 1998). When we cultivated *T. maritima* in the presence of hydrogen (100 kPa) and Fe(III), Fe(II) was formed without obvious growth stimulation. These results indicate that *T. maritima* may not gain energy by iron respiration, but may use Fe(III) in place of sulfur as an additional electron sink to get rid of inhibitory hydrogen during fermentation.[25,27]

Sampling, Enrichment, and Isolation

Hot waters, soils, muds, sediments, or pieces of rocks can be collected and used as primary material for the enrichment of hyperthermophiles. Strongly acidic samples may be neutralized by the addition of calcium carbonate. During sampling, special care has to be taken to avoid contamination of the samples with oxygen, which is toxic to many anaerobic hyperthermophiles at high temperatures. In laboratory experiments, it was shown that all cells of an exponentially growing culture of *Pyrobaculum organotrophum* were killed after exposure to oxygen at 100° for 40 min.[28] In contrast, at low temperatures (e.g., 4°), anaerobic hyperthermophiles can survive for several years, even in the presence of oxygen. They can even be isolated from cold, oxygenated sea water after concentration by ultrafiltration. In view of these results, shock cooling and/or immediate reduction of the introduced oxygen by the injection of appropriate concentrations of dithionate or H_2S is recommended after filling the samples in anaerobic, stoppered glass bottles.[29] Transportation and shipping of the original samples can be done at ambient temperature.

In the laboratory, samples can be stored either at ambient temperature within an anaerobic chamber or at 4°. In general, samples stored at 4° can be used for at least 10 years for successful enrichments of hyperthermophiles. However, for oil–water samples from hot oil reservoirs, enrichment attempts failed already after a storage time of about half a year, indicating that the organisms had died out during the short storage period at 4° in the presence of crude oil.

In consideration of the varying geochemical and physical compositions of the habitats, enrichment attempts for different species of hyperthermo-

[26] M. Vargas, K. Kashefi, E. L. Blunt-Harris, and D. R. Lovley, *Nature* **395,** 65 (1998).
[27] C. Schröder, M. Selig, and P. Schönheit, *Arch. Microbiol.* **161,** 460 (1994).
[28] R. Huber, J. K. Kristjansson, and K. O. Stetter, *Arch. Microbiol.* **149,** 95 (1987).
[29] K. O. Stetter, *Nature* **300,** 258 (1982).

philes can be carried out on various potential substrates under anaerobic, microaerophilic, or aerobic culture conditions and at the approximate *in situ* temperature. A primary enrichment of hyperthermophiles can be already achieved by injecting specific substrates *in situ* into the hot biotope. For initial enrichments of hyperthermophiles in the laboratory, media should be inoculated with original sample material in an anaerobic hood. Depending on the initial cell concentration and the doubling time of the organisms, positive enrichment cultures of hyperthermophiles can be identified within 1–7 days by phase-contrast microscopy. After staining the enrichments with a fluorescent dye such as 4′,-6′-Diamidino-2-phenylindole (DAPI), fluorescence microscopy is used to identify organisms, sticking to solid surfaces.[30] For the enrichment of hyperthermophiles from deep sea and deep subterranean samples, incubation at high temperatures may be performed in a high-pressure chamber at elevated pressure.[31]

Isolation Procedures

Plating

For a better understanding of organisms, the study of pure cultures is an imperative prerequisite. Enrichments of hyperthermophiles may be purified by plating on solid surfaces. However, most solidifying agents are too unstable at temperatures above about 75° for long incubation periods. Agar is not suitable because its melting point is only about 95°. For plating, more thermostable self-gelling agents such as polysilicate or gellan gum, Gelrite (Roth, Karlsruhe, Germany) are used. *Methanothermus fervidus* was cloned on a medium solidified by polysilicate at 85°.[32] Gelrite, the most commonly used agent to prepare solid media, is a linear, anionic heteropolysaccharide (PS-60) produced by a *Pseudomonas* species.[33] The gel stability of Gelrite plates is highly dependent on the Gelrite concentration, the soluble salt concentration, and the type of salt added. Gelrite solidifies rapidly in the presence of divalent ions, such as magnesium or calcium, and, to a lesser extent, in the presence of monovalent ions.[33] In view of the different chemical composition of the plating media used, the Gelrite concentration should be varied (e.g., from 0.8 to 2%) to explore optimal gel strength. After autoclaving, the hot plating medium (about 90°)

[30] H. Huber, G. Huber, and K. O. Stetter, *System. Appl. Microbiol.* **6,** 105 (1985).

[31] H. Huber, R. Huber, H.-D. Lüdemann, and K. O. Stetter, *Sci. Drill.* **4,** 127 (1994).

[32] K. O. Stetter, *in* "Bergey's Manual of Systematic Bacteriology" (J. T. Staley, M. P. Bryant, N. Pfennig, and J. G. Holt, eds.), p. 2183. Williams & Wilkins, Baltimore, 1989.

[33] K. S. Kang, G. T. Veeder, P. J. Mirrasoul, T. Kaneko, and I. W. Cottrell, *Appl. Environ. Microbiol.* **43,** 1086 (1982).

is transferred immediately into the anaerobic chamber and is poured into glass petri dishes. In order to prepare plates for hyperthermophilic acidophiles, double-strength medium and double-strength Gelrite solution should be prepared and autoclaved separately. After autoclaving, both solutions should be mixed immediately. After pouring, the surface of the Gelrite plates should be dried overnight in a stainless steel cylinder[34] at the temperature of cultivation. After cooling, the enrichment culture is streaked in an anaerobic chamber in a serial dilution on the surface. Incubation of the plates is again carried out in a stainless steel cylinder, after pressurization with the desired gas phase.[34] For the plating of strict anaerobes, the addition of 1% H_2S (v/v) may be helpful in maintaining a low redox potential in the cylinder. Depending on the organism, colonies will be visible within 2–14 days of incubation. Using Gelrite, different hyperthermophiles, such as *Methanopyrus kandleri* or *Pyrolobus fumarii,* have been propagated successfully at incubation temperatures up to 100°.[4,35]

In several cases, however, plating may not be successful, most likely due to the inability of the organism to grow on solid surfaces.[6] If plating is not possible, repeated serial dilutions (at least three) may serve as an alternative but less reliable cultivation method.

Selected Single Cell Isolation Using "Optical Tweezers Trap"

A novel isolation procedure has been developed that allows the cloning of single cells from enrichments without plating through the use of a laser microscope.[36] The laser microscope consists of a computer-controlled inverted microscope, equipped with a continuously working neodymium-doped–yttrium aluminum garnet laser (Nd:YAG laser). The laser has a wavelength of 1064 nm and a maximum output power of 1 W. Using a high numerical aperture oil-immersion objective, the laser can be focused to a spot size of less than a micron in diameter. Due to the strong intensity of the laser light forces, single cells of micronsize can be trapped optically within the focus point ("optical tweezers trap," "laser trap") and manipulated in three dimensions.[37,38] The micromanipulation of the cells is performed by keeping the laser beam at fixed position and moving the motor-driven *xy* table.

[34] W. E. Balch, G. E. Fox, L. J. Magrum, C. R. Woese, and R. S. Wolfe, *Microbiol. Rev.* **43,** 260 (1979).

[35] M. Kurr, R. Huber, H. König, H. W. Jannasch, H. Fricke, A. Trincone, J. K. Kristjansson, and K. O. Stetter, *Arch. Microbiol.* **156,** 239 (1991).

[36] R. Huber, S. Burggraf, T. Mayer, S. M. Barns, P. Rossnagel, and K. O. Stetter, *Nature* **376,** 57 (1995).

[37] A. Ashkin, J. M. Dziedzic, and T. Yamane, *Nature* **330,** 769 (1987).

[38] A. Ashkin and J. M. Dziedzic, *Science* **235,** 1517 (1987).

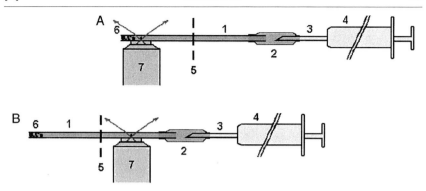

Fig. 2. Isolation of a single cell from a mixed culture using a laser microscope; schematic drawing, reprinted with permission from Spektrum Akademischer Verlag, R. Huber, *BIOspektrum* **4,** 289 (1999).[39] (A) A single cell is trapped optically within the focus of the laser beam. (B) The single cell is separated from the mixed culture into the sterile compartment. (1) Microslide, (2) tube, (3) needle, (4) syringe, (5) cutting line, (6) mixed culture, and (7) objective.

The cell separation unit for the isolation of single cells consists of a rectangular microslide as the observation and separation chamber (inside dimension: 0.1×1 mm^2; length: 10 cm). The microslide is connected with a tube to the needle of a 1-ml syringe (Fig. 2A).[36,39] A cutting line separates the microslide in two compartments (Fig. 2A). From the open end, about 90% of the microslide is filled with sterile medium. Afterward, the mixed culture is soaked into the microslide, which is fixed to the microscope stage. From the culture, inspected under 1000-fold magnification, the selected single cell is trapped optically in the laser beam (Fig. 2A). By moving the microscope stage, the cell is separated within 3–10 min (depending on the organism) by at least 6 cm from the mixed culture into the sterile compartment (Fig. 2B).[39] The microslide is broken gently at the cutting line, and the single cell is flushed into sterile medium (this is referred to as the selected cell cultivation technique).

Comparative studies have shown that the selected cell cultivation technique can be applied successfully for hyperthermophilic Archaea and Bacteria, belonging to the different phylogenetic branches of the 16S rRNA tree.[39] Depending on the organism used in the isolation experiments, 20–100% of the single isolated cells are growing afterward in the culture medium, indicating high efficiency of the new method. Furthermore, the technique is safe, reliable, and fast. Depending on the generation time of the isolated

[39] R. Huber, *BIOspektrum* **4,** 289 (1999).

cell, pure cultures (cell titer of at least 10^7/ml) are available within 1–5 days. Cultivation efficiency can be enhanced further when viable cells are selected by employing the membrane potential-sensitive fluorescent dye bis(1, 3-dibutylbarbituric acid)trimethine oxonol [DiBAC$_4$(3)].[40]

The selected cell cultivation technique can be applied to very divergent groups of microorganisms. In addition to hyperthermophiles, isolated single cells from mesophilic, psychrophilic, or extreme halophilic prokaryotes or unicellular eukaryotes (yeasts, amebas) have already been cultivated.[39]

A New Isolation Strategy

An unexpectedly diverse variety of hyperthermophiles have been isolated from high temperature environments (Table I). This diversity is evident in terms of unusual physiological and metabolic properties as well as by 16S rRNA phylogeny. However, microscopic inspection of different biotopes indicates that there are many more morphotypes yet to be cultivated. Moreover, a direct *in situ* 16S rRNA sequence analysis from samples from a hot pond in Yellowstone National Park demonstrated the presence of a rich phylogenetic diversity of still uncultivated microorganisms with unknown physiology.[41] A new isolation strategy has been developed to obtain pure cultures of such unknown organisms harboring novel 16S rRNA sequences.[36] This procedure combines *in situ* 16S rRNA sequence analysis[41] and specific whole cell hybridization ("phylogenetic staining")[42] with enrichments and isolation of the identified, single cell by the "optical tweezers trap."[36]

Maintenance of Stock Cultures

Depending on the organism, stock cultures of hyperthermophiles stored at 4° can serve as an inoculum for a certain time period ranging from about 1 week up to some years. For long-time preservation, the storage of cultures in the presence of 5% dimethyl sulfoxide over liquid nitrogen (−130°) is recommended. No loss of cell viability of various hyperthermophiles was observed after a storage for at least 5 years.

Acknowledgment

We thank Reinhard Rachel for electron micrographs.

[40] P. Beck and R. Huber, *FEMS Microbiol. Lett.* **147,** 11 (1997).
[41] S. M. Barns, R. E. Fundyga, M. W. Jeffries, and N. R. Pace, *Proc. Natl. Acad. Sci. U.S.A.* **91,** 1609 (1994).
[42] E. F. DeLong, G. S. Wickham, and N. R. Pace, *Science* **243,** 1360 (1989).

[3] *Pyrococcus furiosus:* Large-Scale Cultivation and Enzyme Purification

By MARC F. J. M. VERHAGEN, ANGELI LAL MENON, GERRIT J. SCHUT, and MICHAEL W. W. ADAMS

Introduction

Pyrococcus furiosus is currently one of the most studied of the hyperthermophilic microorganisms mainly because of the relative ease with which it can be cultivated and good growth yields. The organism was originally isolated from a geothermally heated shallow sea vent off the coast of Italy by Stetter and co-workers.[1] It grows at temperatures ranging from 70° to 105° with an optimum at 100° using complex peptide mixtures or sugars, e.g., starch, glycogen, maltose, and cellobiose as carbon and energy sources.[1] The sugars are degraded through a modified version of the Embden–Meyerhof (EM) pathway in which H_2, acetate, CO_2, and alanine are the main fermentation products.[2] Unlike many other heterotrophic hyperthermophiles, significant growth of *P. furiosus* is not obligately dependent on elemental sulfur (S°), although if it is added to the growth medium it is reduced by the organism to H_2S. This article presents an overview of the methods used to grow *P. furiosus* up to the 600-liter scale and to prepare cell-free extracts for the large-scale purification of proteins. In addition, a summary is provided of the elution patterns of different enzymes when the cell-free extract is applied to an anion-exchange chromatography column. The further purification of these specific enzymes is discussed in more detail in other articles in this volume, although the method can also be used to purify other enzymes.

Growth of *Pyrococcus furiosus*

Pyrococcus furiosus (DSM 3638) is grown in a modified version of the medium described previously.[1,3,4] The different components of the medium are prepared as separate stock solutions and are stable for several weeks when stored as sterile solutions. A low concentration of yeast extract (0.05%,

[1] G. Fiala and K. O. Stetter, *Arch. Mircobiol.* **145,** 56 (1996).
[2] S. W. M. Kengen and A. J. M. Stams, *Arch. Microbiol.* **161,** 168 (1994)
[3] S. W. Kengen, E. J. Luesink, A. J. Stams, and A. J. Zehnder, *Eur. J. Biochem,* **213,** 305 (1993).
[4] F. O. Bryant and M. W. W. Adams, *J. Biol. Chem.* **264,** 5070 (1989).

w/v) is included in the medium to avoid adding vitamins. This amount is not sufficient to sustain significant growth without the addition of either peptides (casein) or carbohydrate (maltose).

Stock Solutions

Salt stock: this contains (per liter) NaCl, 140.00 g; $MgSO_4 \cdot 6H_2O$, 17.50 g; $MgCl_2 \cdot 6H_2O$, 13.50 g; KCl, 1.65 g; NH_4Cl, 1.25 g; $CaCl_2 \cdot 2H_2O$, 0.70 g.

Sodium tungstate stock (100 mM): this contains (per 100 ml) $Na_2WO_4 \cdot 2H_2O$, 3.30 g.

Trace minerals stock: this contains (per liter) Na_4EDTA, 0.50 g, HCl (concentrated), 1.00 ml; $FeCl_3$, 2.00 g; H_3BO_3, 0.05 g; $ZnCl_2$, 0.05 g; $CuCl_2 \cdot 2H_2O$, 0.03 g; $MnCl_2 \cdot 4H_2O$, 0.05 g; $(NH_4)_2MoO_4$, 0.05 g; $AlK(SO_4) \cdot 2H_2O$, 0.05 g; $CoCl_2 \cdot 6H_2O$, 0.05 g; $NiCl_2 \cdot 6H_2O$, 0.05 g.

These three stock solutions are used to prepare the basic salts medium (see later) and are not autoclaved.

Potassium phosphate stock (1.0 M): this contains (per 100 ml) KH_2PO_4, 6.8 g and K_2HPO_4, 8.71 g (pH 6.8).

Yeast extract (10%): this contains (per 750 ml) yeast extract (Difco, Detroit, MI), 75 g.

Casein (10%): this contains (per 750 ml) casein hydrolyzate, enzymatic digest (US Biochemicals, Cleveland, OH), 75 g.

These three stock solutions are added to the sterile medium (see later) and must be autoclaved in bottles that can be fitted with crimp-sealed, rubber stoppers. On removal from the autoclave the bottles are immediately sealed and connected to a vacuum manifold using a sterile needle fitted with a sterile 0.22-μm filter (Millipore, Bedford, MA).[4] The bottles are subjected to three cycles of vacuum and flushing with argon. The headspaces are then flushed with a steady stream of argon until the solutions have cooled to room temperature. The stock solutions are stored at 4°.

Maltose (50%): this contains (per 150 ml) maltose (Sigma, St. Louis, MO), 75 g.

The maltose stock solution should not be autoclaved and is sterilized using a 0.22-μm bottle top filter. After filtration the solution is degassed, flushed with argon, and stored at 4°.

Preparation of Media

Basic salts medium (per liter): combine 200 ml salt stock, 0.1 ml sodium tungstate stock (100 mM), 1.0 ml trace minerals stock, and 50 μl resazurin (5 mg/ml). To this add, with stirring, cysteine hydrochloride, 0.5 g; Na_2S, 0.5 g; and $NaHCO_3$, 1.0 g, but wait between additions until the salts are dissolved. Let the pH stabilize and adjust to pH 6.8 with 1 N HCl. Immediately filter sterilize the solution using a 500-ml (0.22-μm) filter and a sterile 1-liter screw-cap bottle. Aseptically transfer the medium into sterile serum bottles (50 ml of solution per 120-ml bottle). Cap and seal the bottles using sterile stoppers and aluminum seals. Connect to a vacuum manifold using a sterile needle and 0.22-μm filter (Millipore) and quickly cycle three times between vacuum and argon. The basic salts medium is stored at 4° under slight overpressure of argon. Prior to inoculation, any excess pressure is released from the bottles by inserting a sterile needle into the rubber stopper.

"Defined" medium: To 50 ml of the basic salts medium add 0.25 ml yeast extract stock (10%, w/v), 0.5 ml maltose stock (50%, w/v), and 0.05 ml potassium phosphate stock (1.0 M, pH 6.8).

"Rich" medium: To 50 ml of the basic salts medium add 2.5 ml yeast extract stock (10%, w/v), 0.5 ml maltose stock (50%, w/v), 0.05 ml potassium phosphate stock (1.0 M, pH 6.8), and 2.5 ml casein stock (10%, w/v).

Elemental sulfur (S°) can be added to either the *defined* or the *rich* medium and this is done before inoculation. The powder (1 mg/ml of medium) is added to the barrel of a 1-ml syringe fitted with a 23-gauge needle from which the plunger is removed. The plunger is then replaced and slowly pressed down on the S° to remove all the air from the syringe. The syringe is inserted into the stopper of the bottle containing the medium and a small amount of medium is drawn into the syringe. This is pushed back into the bottle together with the S° as a slurry. The procedure is repeated several times until all the S° has been introduced into the bottle.

Growth of *P. furiosus* can be scaled up from 100 ml to 1-liter bottles, to 15-liter carboys, and then to a 600-liter fermentor (500-liter working volume) using the media described earlier, but with some modifications. For example, for growth in 15-liter carboys, separate solutions of cysteine hydrochloride (7.5 g in 50 ml distilled H_2O) and Na_2S (7.5 g in 50 ml distilled H_2O) are freshly made and added to the medium, rather than using the dry compounds. Adjust to ~pH 6.8 (at room temperature) with freshly made 1 M NaOH or HCl, and add 15 ml 1 M potassium phosphate stock (pH 6.8). The carboy is then immediately transferred to an incubator, sparged with N_2/CO_2 (80/20, v/v), and heated to 95°. When at the desired temperature (after ~2 hr), the carboy is inoculated with ~400 ml fresh inoculum (containing ~3 × 10^8 cm cells/ml).

For the 600-liter fermentor, the medium contains (per liter): NaCl, 140 g; yeast extract, 5 g; and casein hydrolyzate, 5 g. This is autoclaved and cooled to 28° while sparging with argon and stirred at 123 rpm. The following are then added (per liter): maltose, 5 g; $MgSO_4 \cdot 6H_2O$, 3.5 g; $MgCl_2 \cdot 6H_2O$, 2.7 g; KCl, 0.33 g; NH_4Cl, 0.25 g; $CaCl_2 \cdot 2H_2O$, 0.14 g; 100 mM $Na_2WO_4 \cdot 2H_2O$, 0.1 ml; and trace minerals stock, 1 ml. Separate, freshly prepared solutions of cysteine hydrochloride (250 g in 2 liters) and Na_2S (250 g in 2 liters) are then added, and the medium is adjusted to pH ~6.8 (at room temperature) with freshly made 1 M NaOH. To this is slowly added 500 ml 1 M potassium phosphate stock (pH 6.8), and the argon flow is discontinued and the headspace is flushed with N_2/CO_2 (80/20, v/v). The temperature is then increased to 90°. A 5% inoculum grown to ~3 × 10^8 cells/ml is used to inoculate the fermenter, and the pH is maintained at 6.8 by the automatic addition of $NaHCO_3$ (10%, w/v). Cell growth is monitored by cell counts. Cells are harvested in the late log phase (after 12–14 hr) after cooling to 28° while sparging with argon and flash frozen in liquid N_2. The frozen cell paste is stored at −80° until required.

Anaerobic Purification of Oxidoreductase-Type Enzymes and Related Redox Proteins

Preparation of Cell-Free Extract

Frozen cells (~500 g) are added to 1500 ml of anaerobic buffer under a continuous flow of argon. The buffer is 50 mM Tris–HCl, pH 8.0, which is made anaerobic by several vacuum/argon cycles with the subsequent addition of 2 mM sodium dithionite, 2 mM dithiothreitol, and 0.5 μg/ml DNase I. The suspension is incubated in a water bath at 37° for 1 hr. During this time most of the cells are lysed due to the combination of a freeze–thaw cycle and the osmotic shock. The extract is transferred anaerobically to centrifuge tubes and spun for 1 hr at 30,000g and 15° to separate soluble proteins from the cell debris. The clear supernatant is decanted and used as the cell-free extract.

DEAE-Sepharose Chromatography

The cell-free extract is diluted threefold with 50 mM Tris–HCl containing 2 mM sodium dithionite and 2 mM dithiothreitol and loaded onto a column (10 × 20 cm) of DEAE-Sepharose FF (Pharmacia Biotech, Piscataway, NJ) equilibrated with the same buffer. To elute the bound proteins, a linear gradient (15 liters) from 0 to 0.5 M NaCl in the equilibration buffer is applied and 125-ml fractions are collected. The regions of the

TABLE I

ELUTION BEHAVIOR OF VARIOUS OXIDOREDUCTASE-TYPE ENZYMES OF *P. furiosus* ON
ANION-EXCHANGE CHROMATOGRAPHY

Elution position (mM NaCl)[a]	Enzyme[b]	Refs.
100–150	Glyceraldehyde-3-phosphate Fd oxidoreductase (GAPOR)	6, 7
130–190	Formaldehyde Fd oxidoreductase (FOR)	6, 8
160–200	Acetyl-CoA synthetase I (ACS I)	9, 10
150–200	Indolepyruvate Fd oxidoreductase (IOR)	11, 12
180–240	Ferredoxin:NADP oxidoreductase (FNOR)	5, 13
	Pyruvate Fd oxidoreductase (POR)	11, 14
	NADPH:rubredoxin oxidoreductase (NROR)	15, 16
190–260	Aldehyde Fd oxidoreductase (AOR)	6, 17
200–260	Prolidase	18, 19
	Isovalerate Fd oxidoreductase (VOR)	6, 20
240–280	Hydrogenase I	4, 21
	Hydrogenase II	21, 22
280–300	Rubredoxin (Rd)	23, 24
290–320	Ferredoxin (Fd)	25, 26

[a] The elution position indicates the approximate range of NaCl concentration that results in elution of the indicated protein from a DEAE-Sepharose FF column (at pH 8.0).

[b] Fd, Ferredoxin; Rd, rubredoxin.

gradient in which different enzymes elute are indicated in Table I. All of these enzymes can be purified from the same batch of cells if desired, although some require a second chromatography step to separate them from one or more of the other proteins listed in Table I.

General Principle of Oxidoreductase-Type Assays

Assays for the different oxidoreductases listed in Table I are performed in 3-ml glass cuvettes using a spectrophotometer equipped with a thermostatted cuvette holder. For most enzymes the preparation for the assay follows the same procedure, and the specificity for an enzyme of interest is determined by the choice of substrate. Cuvettes are sealed with serum stoppers and made anaerobic by several cycles of vacuum and argon. At the same time, 50 mM EPPS, pH 8.4 (at room temperature), is degassed in a vacuum sidearm flask in a similar way. When the buffer and the cuvettes are anaerobic, 2 ml of buffer is transferred to each cuvette using a gas-tight syringe. The cuvette is incubated in the holder of the spectrophotometer until it reaches the required temperature, substrates are added, and the reaction is started by the addition of enzyme. In some cases the substrate is thermolabile so the enzyme must be added first and the reaction is then started by addition of the substrate. The most commonly used dyes to monitor the activity of these enzymes are benzyl viologen ($E'_m = -350$ mV; $\varepsilon_{580} = 7{,}800\ M^{-1}\ cm^{-1}$) and methyl viologen ($E'_m = -442$ mV; $\varepsilon_{600} = 12{,}000\ mM^{-1}\ cm^{-1}$).[5] The activities of the enzymes are expressed as micromole of substrate oxidized per minute. The specific assays used for each of the enzymes listed in Table I are given in the indicated references.

Acknowledgements

This research was supported by grants from the Department of Energy, the National Institutes of Health, and the National Science Foundation.

[5] K. Ma and M. W. W. Adams, *J. Bacteriol,* **176,** 6509 (1994).

[6] R. Roy, A. L. Menon, and M. W. W. Adams, *Methods Enzymol.* **331** [11] (2001).

[7] S. Mukund and M. W. W. Adams, *J. Biol. Chem.* **270,** 8389 (1995).

[8] R. Roy, S. Mukund, G. J. Schut, D. M. Dunn, R. Weiss, and M. W. W. Adams, *J. Bacteriol.* **181,** 1171 (1999).

[9] A Hutchins, X. Mai, and M. W. W. Adams, *Methods Enzymol.* **331** [13] (2001).

[10] X. Mai and M. W. W. Adams, *J. Bacteriol.* **178,** 5897 (1996).

[11] G. J. Schut, A. L. Menon, and M. W. W. Adams, *Methods Enzymol.* **331** [12] (2001).

[12] X. Mai and M. W. W. Adams, *J. Biol. Chem.* **269,** 16726 (1994).

[13] K. Ma and M. W. W. Adams, *Methods Enzymol.* **334** [4] (2001).

[14] J. M. Blamey and M. W. W. Adams, *Biochim. Biophys. Acta* **1161,** 19 (1993).

[15] K. Ma and M. W. W. Adams, *Methods Enzymol.* **334** [6] (2001).

[16] K. Ma and M. W. W. Adams, *J. Bacteriol.* **181,** 5530 (1999).

[17] S. Mukund and M. W. W. Adams, *J. Biol. Chem.* **266,** 14208 (1991).

[18] A. M. Grunden, M. Gosh, and M. W. W. Adams, *Methods Enzymol.* **330** [30] (2001) (this volume).

[19] M. Ghosh, A. Grunden, R. Weiss, and M. W. W. Adams, *J. Bacteriol.* **180,** 4781 (1998).

[20] J. Heider, X. Mai, and M. W. W. Adams, *J. Bacteriol.* **178,** 780 (1996).

[21] K. Ma and M. W. W. Adams, *Methods Enzymol.* **331** [18] (2001).

[22] K. Ma, R. Weiss, and M. W. W. Adams, *J. Bacteriol.* **182,** 1864 (2000).

[23] F. E. Jenney, Jr., and M. W. W. Adams, *Methods Enzymol.* **334** [5] (2001).

[24] P. R. Blake, J. B. Park, F. O. Bryant, S. Aono, J. K. Magnuson, E. Eccleston, J. B. Howard, M. F. Summers, and M. W. W. Adams, *Biochemistry* **30,** 10885 (1991).

[25] C. H. Kim P. S. Brereton, M. F. J. M. Verhagen, and M. W. W. Adams, *Methods Enzymol.* **334** [3] (2001).

[26] S. Aono, F. O. Bryant, and M. W. W. Adams, *J. Bacteriol.* **171,** 3433 (1989).

[4] Continuous Cultivation of Hyperthermophiles

By MARYBETH A. PYSZ, KRISTINA D. RINKER, KEITH R. SHOCKLEY, and ROBERT M. KELLY

Introduction

In general, a significant amount of biomass must be generated from pure cultures of microorganisms in order to purify sufficient quantities of enzymes for detailed analyses. A number of cultivation techniques have been used for hyperthermophilic organisms ranging from glass serum bottles to large-scale fermentors constructed of ceramic or stainless steel to dialysis membrane reactors.[1] Even if the gene encoding a given hyperthermophilic enzyme can be successfully expressed in a mesophilic host,[2] often there is the need to examine the native version. There is always the question of whether the recombinant and native versions have precisely the same properties, and native biomass is the only sourse of enzymes whose genes are not readily expressed in mesophilic hosts. An additional motivation is that a large fraction of open reading frames identified in the genomes of hyperthermophiles have not yet been assigned a biochemical function.[3,4] Thus, the ability to cultivate hyperthermophiles remains an important element of research efforts focusing on the biology and biotechnology of these organisms.

A continuous culture system is described here that can generate biomass from hyperthermophiles on a scale suitable for enzyme purification. High-temperature chemostats have several advantages over large-scale batch systems. Long-term, stable, steady-state operation (arising from minimal problems with contamination) can provide biomass generated from exponential growth phase, i.e., balanced growth.[5] Because of the smaller operating volumes, continuous systems are inexpensive to construct and minimize problems with handling toxic and explosive gas substrates and products (e.g., H_2S, H_2, CH_4). Small (i.e., 1–10 liter) operating volumes also minimize problems associated with the growth of sulfide-producing anaerobes and

[1] P. M. Hicks and R. M. Kelly, *in* "Encyclopedia of Bioprocess Technology: Fermentation, Biocatalysis, and Bioseparation" (M. C. Flickinger and S. W. Drew, eds.), p. 2536. Wiley, New York, 1999.

[2] M. W. W. Adams and R. M. Kelly, *Trends Biotechnol.* **16**, 329 (1998).

[3] R. Huber and K. O. Stetter, *Methods Enzymol.* **330** [2] (2001) (this volume).

[4] M. W. W. Adams, *Methods Enzymol.* 1999.

[5] H. W. Blanch and D. S. Clark, "Biochemical Engineering." Dekker, New York, 1996.

thermoacidophiles in terms of choosing a proper material for reactor construction (e.g., glass, gold). Continuous cultivation has also been useful for studying the bioenergetics and physiology of hyperthermophiles and for developing media formulations that induce enzyme expression.[6–11]

High-temperature continuous cultivation systems were originally used to grow *Pyrococcus furiosus* (a fermentative anaerobe and facultative sulfur reducer, T_{opt} 98°),[12] but other hyperthermophiles, including *Thermococcus litoralis* (a fermentative anerobe and facultative sulfur reducer, T_{opt} 88°), *Methanococcus jannaschii* (methanogen, T_{opt} 85°), and *Thermotoga maritima* (a fermentative anerobe and facultative sulfur reducer, T_{opt} 80°), have also been grown successfully.[7–10,12–17]

Materials

Reactor System

A schematic of a typical configuration for the high-temperature continuous culture system is shown in Fig. 1. Although operating volumes up to 10 liters have been used, usually a 2-liter, five-neck round-bottom flask (Ace Glass, Vineland, NJ) is chosen to process a 1-liter (operating volume) culture. A Graham condenser (Ace Glass) is used to reduce water content in the headspace effluent gas of the reactor; the gas may then be analyzed by gas chromatography (GC), mass spectrometry (MS), or vented (after scrubbing with 1 N NaOH for the reduction of odors from sulfidic gases). The culture vessel is heated by an insulated mantle (Glas-Col, Terre Haute,

[6] R. M. Kelly, I. I. Blumentals, L. J. Snowden, and M. W. W. Adams, *Ann. N.Y. Acad. Sci.* **665**, 309 (1992).

[7] K. D. Rinker and R. M. Kelly, *Appl. Environ. Microbiol.* **62**, 4478 (1996).

[8] K. D. Rinker, "Growth Physiology and Bioenergetics of the Hyperthermophilic Archaeon *Thermococcus litoralis* and Bacterium Thermotoga maritima." North Carolina State University, Raleigh, 1998.

[9] K. D. Rinker, C. J. Han, and R. M. Kelly, *J. Appl. Microbiol.* **85**, 118S (1999).

[10] R. N. Schicho, K. Ma, M. W. W. Adams, and R. M. Kelly, *J. Bacteriol.* **175**, 1823 (1993).

[11] R. N. Schicho, L. J. Snowden, S. Mukund, J. B. Park, M. W. W. Adams, and R. M. Kelly, *Arch. Microbiol.* **159**, 380 (1993).

[12] S. H. Brown and R. M. Kelly, *Appl. Environ. Microbiol.* **55**, 2086 (1989).

[13] I. I. Blumentals, S. H. Brown, R. N. Schicho, A. K. Skaja, H. R. Constantino, and R. M. Kelly, *Ann. N.Y. Acad. Sci.* **589**, 301 (1990).

[14] M. A. Pysz and R. M. Kelly, unpublished data (1999).

[15] N. D. H. Raven and R. J. Sharp, *FEMS Microbiol. Lett.* **146**, 135 (1997).

[16] R. N. Schicho, S. H. Brown, I. I. Blumentals, T. L. Peeples, G. D. Duffaud, and R. M. Kelly, *in* "Archaea: A Laboratory Manual" (F. T. Robb and A. R. Place, eds.), p. 31, 167. Cold Spring Harbor Laboratory Press, Plainview, 1995.

[17] C. J. Han, S. H. Park, and R. M. Kelly, *Appl. Environ. Microbiol.* **63**, 2391 (1997).

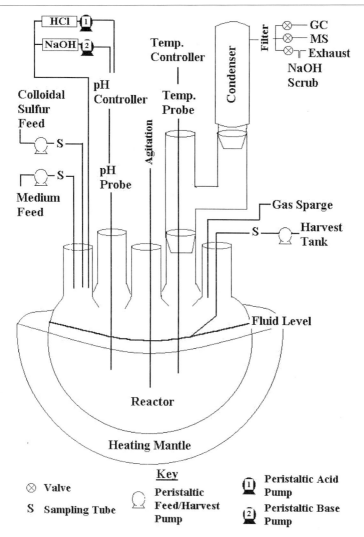

FIG. 1. Schematic of continuous cultural vessel for growth of hyperthermophiles.

IN) that is connected to a temperature controller (Cole-Parmer, Vernon Hills, IL) and J thermocouple (Cole-Parmer). The pH of the culture can be monitored and adjusted with a pH controller (Cole-Parmer) and an autoclavable, double-junction pH electrode (Cole-Parmer). Fixed rate peristaltic pumps (Cole-Parmer) with Norprene Masterflex tubing (Cole-Parmer) are used with the pH controller to add 1 N HCl or NaOH as needed. Norprene Masterflex tubing is used as it is autoclavable, durable, and has low O_2 permeability (an important feature of an anaerobic system). Teflon adapters with Viton O rings (Cole-Parmer) are used for insertion of the pH electrode, thermocouple, and other tubing into the culture vessel.

There are several ways to agitate the culture, including a mechanical impeller in the center of the flask, a magnetic egg-shaped stir bar placed in the center of the culture, or by sparging with gas directly into the culture volume. A mechanical impeller (Model SL 300, Fisher Scientific, Pittsburgh, PA) works well, but care must be taken to avoid excessive shearing. A magnetic stir bar may be difficult to control through the heating mantle, although it is less expensive than an impeller. Sparging with an inert gas (e.g., argon, nitrogen) also sufficiently mixes reactor contents and maintains anaerobic conditions. Gas may be sparged into the culture (as well as into the culture medium, cell collection, and colloidal sulfur feed tanks) with glass tubing connections (inserted into rubber stoppers) and controlled with a flowmeter.

The continuous feed (i.e., culture media and sulfur) and harvest collection (i.e., reactor effluent) are pumped to and from the reactor with Masterflex tubing and a Masterflex peristaltic pump (Cole-Parmer). Polypropylene carboys (Nalge, available through Cole-Parmer), are used to store the feed and harvest liquids. Pump heads (Masterflex, Cole-Parmer) varying in size from 14, 15, and 16 can be used for the colloidal-sulfur feed, product stream, and media feed, respectively. A single peristaltic pump (Masterflex, Cole-Parmer) should be used for both inlet streams (i.e., the medium and colloidal sulfur, if used) and the outlet stream, such that any constant dilution rate will prevent the total feed from exceeding the amount of harvest removed. The media feed, sulfur feed, and harvest tanks should be sparged with an inert gas (monitored by a flowmeter) to maintain the anaerobic conditions (where applicable) and also have 0.2-μm filters (Gelman Acro 50, Fisher Scientific) attached to rubber stoppers to relieve excess pressure from the carboys.

Gas Sparging

The choice of gas for the headspace depends on the hyperthermophile that is being grown. Methanogens (e.g., *M. jannaschii*) and other autotrophs (e.g., *Archaeoglobus fulgidus*) use an 80:20 (v/v) mixture of $H_2:CO_2$. Typically, N_2 is used for heterotrophs (e.g., *P. furiosus*).

Media for Growth of Hyperthermophiles

Table I lists several types of medium that have been used to grow various hyperthermophiles (for a more complete listing of hyperthermo-

TABLE I

HYPERTHERMOPHILES GROWN IN CONTINUOUS CULTURE

Hyperthermophile	T_{opt} (°)	Physiological and growth characteristics	Suggested growth medium and conditions
Archaeoglobus fulgidus	83	Archaeon Obligate chemolithoautotroph (H_2/CO_2 or N_2 in presence of lactic acid) Sulfate/thiosulfate reducer[21,22]	STL medium[21,23] Modified marine medium[24,25] Medium[21,26]
Methanococcus jannaschii	85	Archaeon Methanogen (H_2/CO_2) Higher pressures enhanced growth[27,28]	Medium[23,28–32]
Pyrococcus furiosus	100	Archaeon Anaerobic chemoorganotroph Facultative S° reducer Enhanced growth on S°[10,33–35]	ASW medium[34,36] Modified ASW medium[34,35] Modified SME medium[36,37] Defined and minimal media[15] Modified *M. jannaschii* medium[8] Modified *A. fulgidus* medium[14,24]
Thermococcus barossii	83	Archaeon Anaerobic chemolithoorganotroph Obligate S° reducer	Medium[18]
T. litoralis	88	Archaeon Anaerobic chemolithoorganotroph Facultative S° reducer Enhanced growth energetics on S°[38]	Marine medium[39,40] 2216 marine broth[38,41] ASW medium[38,42]
Thermotoga maritima	80	Bacterium Anaerobic chemoorganotroph Facultative S° reducer S° has neutral effect on growth energetics[44]	MMS medium[43] Modified SME medium[36,43] Medium[43,44]
T. neapolitana	80	Bacterium Anaerobic chemoorganotroph Facultative S° reducer S° enhances growth[46]	TB medium[45,46] Modified TB medium[45,47] Common marine medium[23,45] Medium[43,47]

philes isolated to date, see [2], this volume).[3] If adding a carbohydrate source (e.g., maltose or cellobiose), the medium should be concentrated 1.2-fold. Prepare 1 liter of the desired concentration of carbohydrate and filter-sterilize through 0.2-μm filters (Whatman; Fisher Scientific) in vacuum-filtration units. The medium (10 liters) should be autoclaved just prior to starting the continuous feed. Sterilized carbohydrate solutions should then be added to the hot medium, if required. For anaerobic conditions, reduce the hot medium using 8 to 10 ml of reducing agent (e.g., 100 g/liter each of $Na_2S \cdot 9H_2O$ and L-cysteine prepared in a chemical hood and kept under N_2 gas). The medium should be sparged until oxygen levels are reduced (e.g., if resazarin is used as an oxygen indicator, the medium should turn from a pink to a golden color). The anaerobic feed is then ready to be pumped into the reactor system. The aerobic medium should be aerated while heating, and any inorganic substrates (e.g., 10 g/liter $FeSO_4 \cdot 7H_2O$, elemental sulfur, or FeS_2) should be passed through a 0.2-μm filter (Falcon bottle-top filters, Fisher Scientific) to avoid contamination.

Sulfur Feed

Many hyperthermophiles, either obligately and facultatively, reduce sulfur.[3,4] Sulfur may be fed to a continuous cultivation system as periodic additions of elemental sulfur[18] or by a continuous feed of either a colloidal sulfur suspension[10,16] or of a soluble polysulfide.[13] The colloidal sulfur must be added separately from the medium to avoid precipitation in the presence of 10 g/liter or more of NaCl.[16] Polysulfides may accumulate in the reactor under sulfur-limiting conditions from the presence of H_2S, a product of sulfur metabolism.[13,19]

The colloidal sulfur feed is prepared as follows.[10,16] All work should be performed in a chemical fume hood, and gloves and safety glasses should be worn for protection from toxic sulfides. For a sterilized feed, autoclave the feed bottle or tank and enough water to dilute the colloid (see later). Prepare two solutions: A, containing 64 g of $Na_2S \cdot 9H_2O$ in 500 ml of H_2O, and B, containing 36 g of Na_2SO_3 in 500 ml of distilled H_2O. Add 15 ml of solution B to solution A. Slowly add 80–100 ml of diluted H_2SO_4 (15 ml of concentrated acid with 100 ml of distilled H_2O) to solution A until turbidity persists and add 30 ml of concentrated H_2SO_4 to solution B. Add solution A to solution B (over a time period of 30 to 160 sec) with vigorous stirring and let stand for 1 hr. Add H_2O to give a total volume of 2 liters and then add 20 g of NaCl (which promotes settling) with stirring until it dissolves. Let the solution stand for at least 4 hr, during which time the colloid should settle to less than a 50-ml volume. Remove the cloudy solution by aspiration and repeat the settling procedure three times using 2 liters of H_2O and 20 g of NaCl each time. The colloid is resuspended in the solution (using a magnetic stirrer) each time. Finally, resuspend the colloid in 1 liter of H_2O and store at room temperature. The suspension should be diluted with H_2O to a final turbidity of $A_{850} = 1.25$ for use as a

[18] G. D. Duffaud, O. B. d'Hennezel, A. S. Peek, A. L. Reysenbach, and R. M. Kelly, *Syst. Appl. Microbiol.* **21**, 40 (1998).

[19] L. K. Poulsen, G. Ballard, and D. A. Stahl, *Appl. Environ. Microbiol.* **59**, 1354 (1993).

continuous feed. If the H_2O was autoclaved prior to dilution, it must be cooled to room temperature before the concentrated colloid is added. The colloid can be stored for up to a month without particle agglomeration. It is sparged with gas and stirred prior to addition.

Cultivation Methods

Prestart-up

Prior to operation, it is important to make sure that the pumps, pH meter, and temperature controller are working properly. The pump speed needs to be correlated with the dilution rate, *D*, which is equal to the flow rate divided by the working culture volume. The pH probe is best calibrated just before start-up, recognizing that the pH measurement is sensitive and temperature dependent.

A growth curve should be prepared for the inoculum to be used. Approximately 50 ml of a dense, late log phase culture (10^8 cell/ml) is used to inoculate the reactor. The batch inoculum is prepared in 125-ml serum bottles (Fisher Scientific) containing 50 ml of medium (and carbohydrate source and/or sulfur, if applicable). Heat the medium at 98° and the starter culture at its optimum growth temperature (e.g., *P. furiosus* at 98° or *A. fulgidus* at 83°) in oil baths for an hour prior to inoculation. For anaerobic cultures, reduce the medium by adding 2 to 3 drops of $Na_2S \cdot 9H_2O$ (or another reducing agent such as L-cysteine hydrochloride) with a syringe while sparging the system with gas. Inoculate the medium (while continuously sparging with gas) with enough stock to yield a final culture density of about 10^8 cells/ml. Quickly seal the serum bottle with butyl rubber stoppers and aluminum crimp seals (Fisher Scientific). Place the culture bottle in an oil bath at the appropriate growth temperature for the organism and grow to late log phase. The batch culture bottles for many hyperthermophiles can be stored at room temperature for several months and at 4° for several years.

Start-up

Prior to continuous operation, the reactor should be operated as a batch culture. Fill the culture vessel with medium 50% of its volume and heat to 98°. Add the carbohydrate source (e.g., maltose) and/or sulfur (e.g., elemental form), if necessary. Agitate the medium using the mechanical stirrer, sparging gas, or magnetic stir bar. Reduce the medium by adding the reducing agent, e.g., $Na_2S \cdot 9H_2O$, dropwise while sparging until the medium is deoxygenated. Adjust the heat of the culture vessel to the appropriate growth temperature of the organism (e.g., *A. fulgidus* 83°) and inoculate the reactor using the 50-ml batch culture that was grown as described earlier. Be sure all reactor ports are effectively sealed with rubber stoppers. Grow the batch culture until it reaches late log phase (e.g., for *P. furiosus* about 8 or 9 hr) and then turn on the pump for the feed/harvest (e.g., for *P. furiosus,* an optimal dilution rate is 0.24).

Culture stability should be checked periodically. This can be done in a variety of ways, including measurements of cell density, cell dry/wet weights, total protein assays, sugar assays of the medium and harvest, gas production profile, and so on. These measurements can be used to determine when the reactor reaches a biological steady state, i.e., the cells are in the state of balanced growth.

Enzyme Collection

Processing Harvest

Collection volumes for a culture volume of 1 liter in a 2-liter vessel are typically on the order of 20 liters in a couple of days, depending on the dilution rate. Higher dilution rates (i.e., greater than the growth rate; however, be careful not to wash out the cells) lead to higher biomass productivities but require more frequent media preparation and cell processing. Large volumes of biomass can be difficult and time-consuming to centrifuge directly unless appropriate equipment (e.g., a continuous centrifuge) is available. Alternatively, harvested cells can be concentrated into smaller volumes (e.g., about 2 liters for a 20-liter harvest) with a crossflow membrane filter concentrator. For example, a Millipore concentrator (Millipore, Bedford, MA) with Durapore Microporous, GVPP/GVLP Pellicon cassette membranes (filter pore size of 0.22 μm; Millipore) has been used successfully for a variety of hyperthermophiles such as *P. furiosus, T. maritima,* and *T. litoralis.* The concentrated cells from the cross-flow step are of a volume such that typical laboratory scale units at 8000 rpm for 20 min can be used to collect the cell pellet.

Extracellular Enzyme Collection

The extracellular proteins in the culture supernatant can also be concentrated by filtration using polyethersulfone, PTGC 10-kDa membranes (Millipore). The Millipore concentrator (Millipore) will concentrate 20 liters down to 1 liter in a few hours. The extracellular proteins and material can then be processed in a smaller concentrator to about 200 ml for purification efforts.

Intracellular Enzyme Collection

Centrifuged cell pellets can be lysed using a French-pressure cell (18,000 psi, SLM Aminco, Urbana, IL). If working with bacteria (e.g., *T. maritima* or *Aquifex aeolicus*), lysosyme can also be used in conjunction with French press or sonication (on ice for 5 min of 30-sec pulse, 30-sec rest) to break open the cells.[20] The lysed cell material can then be spun down at 10,000g for 30 min at 4°. The resulting crude extract can be processed using small concentrators, in addition to separation (e.g., DEAE chromatography, hydrophobic interaction chromatography, hydroxylapatite chromatography), for enzyme purification. Typical enzyme purification protocols for *P. furiosus* are described elsewhere in this volume.[20a]

Concluding Comments

Continuous culture systems have proved to be useful for hyperthermophiles and producing biomass in part because they can be operated for

[20] E. S. Miller, Jr., K. N. Parker, W. Liebl, and R. M. Kelly, *Methods Enzymol.* (1999).

[20a] M. F. J. M. Verhagen, A. L. Menon, G. J. Schut, and M. W. W. Adams, *Methods Enzymol.* (1999).

[21] W. E. Balch, G. E. Fox, L. J. Magrum, C. R. Woese, and R. S. Wolfe, *Microbiol. Rev.* **43**, 260 (1979).

[22] C. LaPaglia and P. L. Hartzell, *Appl. Environ. Microbiol.* **63**, 3158 (1997).

[23] G. Zellner, E. Stackebrandt, H. Kneifel, P. Messner, U. B. Sleytr, E. C. Demacario, H. P. Zabel, K. O. Stetter, and J. Winter, *Syst. Appl. Microbiol.* **11**, 151 (1989).

[24] J. Beeder, R. K. Nilsen, J. T. Rosnes, T. Torsvik, and T. Lien, *Appl. Environ. Microbiol.* **60**, 1227 (1994).

[25] K. O. Stetter, G. Lauerer, M. Thomm, and A. Neuner, *Science* **236**, 822 (1987).

[26] C. Dahl, N. M. Kredich, R. Deutzmann, and H. G. Truper, *J. Gen. Microbiol.* **139**, 1817 (1993).

[27] J. H. Tsao, S. M. Kaneshiro, S. S. Yu, and D. S. Clark, *Biotechnol. Bioeng.* **43**, 258 (1994).

[28] P. C. Michels and D. S. Clark, *Appl. Environ. Microbiol.* **63**, 3985 (1997).

[29] G. Ferrante, J. C. Richards, and G. D. Sprott, *Biochem. Cell Biol.* **68**, 274 (1990).

[30] G. D. Sprott, I. Ekiel, and G. B. Patel, *Appl. Environ. Microbiol.* **59**, 1092 (1993).

[31] J. F. Miller, N. N. Shah, C. M. Nelson, J. M. Ludlow, and D. S. Clark, *Appl. Environ. Microbiol.* **54**, 3039 (1988).

[32] M. Kurr, R. Huber, H. Konig, H. W. Jannasch, H. Fricke, A. Trincone, J. K. Kristjansson, and K. O. Stetter, *Arch. Microbiol.* **156**, 239 (1991).

[33] J. van der Oost, G. Schut, S. W. M. Kengen, W. R. Hagen, M. Thomm, and W. M. de Vos, *J. Biol. Chem.* **273**, 28149 (1998).

[34] S. W. M. Kengen, F. A. M. Debok, N. D. Vanloo, C. Dijkema, A. J. M. Stams, and W. M. Devos, *J. Biol. Chem.* **269**, 17537 (1994).

[35] K. O. Stetter, H. Konig, and E. Stackerbrandt, *Syst. Appl. Microbiol.* **4**, 535 (1983).

[36] W. J. Jones, M. J. B. Paynter, and R. Gupta, *Arch. Microbiol.* **135**, 91 (1983).

[37] S. W. M. Kengen and A. J. M. Stams, *Arch. Microbiol.* **161**, 168 (1994).

[38] S. Belkin, C. O. Wirsen, and H. W. Jannasch, *Appl. Environ. Microbiol.* **51**, 1180 (1986).

[39] A. Neuner, H. W. Jannasch, A. Belkin, and K. O. Stetter, *Arch. Microbiol.* **153**, 205 (1990).

[40] M. Windholz (ed.), "The Merck Index." Merck and Co., Rahway, 1976.

long periods of time with a low potential for contamination and can be constructed with inexpensive equipment. In addition, continuous cultures have distinct advantages over batch cultures in exploring transcriptional and translational phenomena. For example, the effects of specific medium components can be studied to observe changes in protein and mRNA expression levels (e.g., NH_4Cl on *T. litoralis* biofilms).[9]

Acknowledgments

This work was supported in part by grants from the National Science Foundation and the Department of Energy.

[41] E. G. Ruby, C. O. Wirsen, and H. W. Jannasch, *Appl. Environ. Microbiol.* **42,** 317 (1981).
[42] J. M. Gonzalez, C. Kato, and K. Horikoshi, *Arch. Microbiol.* **164,** 159 (1995).
[43] T. Kobayashi, Y. S. Kwak, T. Akiba, T. Kudo, and K. Horikoshi, *Syst. Appl. Microbiol.* **17,** 232 (1994).
[44] S. E. Childers, M. Vargas, and K. Noll, *Appl. Environ. Microbiol.* **58,** 3949 (1992).
[45] M. Vargas and K. M. Noll, *Arch. Microbiol.* **162,** 357 (1994).
[46] M. Y. Galperin, K. M. Noll, and A. H. Romano, *Appl. Environ. Microbiol.* **63,** 969 (1997).
[47] E. Windberger, R. Huber, A. Trincone, H. Fricke, and K. O. Stetter, *Arch. Microbiol.* **151,** 506 (1989).

[5] Genome of *Methanocaldococcus* (*Methanococcus*) *jannaschii*

By DAVID E. GRAHAM, NIKOS KYRPIDES, IAIN J. ANDERSON, ROSS OVERBEEK, and WILLIAM B. WHITMAN

Introduction

Methanocaldococcus (*Methanococcus*) *jannaschii* strain JAL-1 is a hyperthermophilic methanogenic archaeon that was isolated from surface material collected at a "white smoker" chimney at a depth of 2600 m in the East Pacific Rise near the western coast of Mexico.[1] Cells are irregular cocci possessing polar bundles of flagella.[1] The cell envelope is composed of a cytoplasmic membrane and a protein surface layer.[2] Similar isolates have also been obtained from hydrothermally active

[1] W. J. Jones, J. A. Leigh, F. Mayer, C. R. Woese, and R. S. Wolfe, *Arch. Microbiol.* **136,** 254 (1983).
[2] E. Nusser and H. Konig, *Can. J. Microbiol.* **33,** 256 (1987).

sediments in the Guaymas Basin and the Mid-Atlantic Ridge,[3,4] and related species have been found at other marine hydrothermal vents.[5–8] Because these hyperthermophilic species are very different from the mesophilic methanococci, they have been reclassified into a new family, Methanocaldococcaceae, and two new genera, *Methanocaldococcus* and *Methanotorris*.[9] The characteristics of the source material for these isolates suggest that they possess adaptations for growth at high temperature and pressure as well as moderate salinity. The water chemistry of the sites suggests an environment rich in sulfide, H_2, CO_2, Fe^{2+}, and Mn^{2+}.[10] This anaerobic environment would be well suited for a H_2-utilizing methanogen that reduces CO_2 to methane. Fixed nitrogen, either as NH_3 or NO_2^-, is not abundant. In addition, small amounts of CO are present. Thus, it is possible that CO could be utilized as an electron donor.

From its growth characteristics and what is known about its biochemistry, *M. jannaschii* appears to be typical of a H_2-utilizing, autotrophic methanogen. These archaea carry out an anaerobic respiration with CO_2 as the terminal electron acceptor according to the general equation:

$$4H_2 + CO_2 \rightarrow CH_4 + 2H_2O$$

So far, all isolated methanogens appear to be obligate methanogens. As expected, *M. jannaschii* does not grow in a rich heterotrophic medium in the absence of H_2, and related methanococci do not metabolize glucose or most amino acids. However, the current evidence does not exclude alternative but minor pathways of energy metabolism or cometabolism. For instance, *M. jannaschii* produces glycogen as an intracellular storage material[11] and presumably possesses the pathways to utilize this carbohydrate.[12]

The pathway of methanogenesis from CO_2 is complex and requires five

[3] S. Burggraf, H. Fricke, A. Neuner, J. Kristjansson, P. Rouviere, L. Mandelco, C. R. Woese, and K. O. Stetter, *Syst. Appl. Microbiol.* **13**, 263 (1990).

[4] C. Jeanthon, S. L'Haridon, N. Pradel, and D. Prieur, *Int. J. System. Bacteriol.* **49**, 591 (1999).

[5] H. Zhao, A. G. Wood, F. Widdel, and M. P. Bryant, *Arch. Microbiol.* **150**, 178 (1988).

[6] W. J. Jones, C. E. Stugard, and H. W. Jannasch, *Arch. Microbiol.* **151**, 314 (1989).

[7] C. Jeanthon, S. L'Haridon, A. L. Reysenbach, M. Vernet, P. Messner, U. B. Sleytr, and D. Prieur, *Int. J. System. Bacteriol.* **48**, 913 (1998).

[8] C. Jeanthon, S. L'Haridon, A. L. Reysenbach, E. Corre, M. Vernet, P. Messner, U. B. Sleytr, and D. Prieur, *Int. J. System. Bacteriol.* **49**, 583 (1999).

[9] W. B. Whitman, D. R. Boone, and Y. Koga, *in* "Bergey's Manual of Systematic Bacteriology" (G. M. Garrity, ed.), in press.

[10] H. W. Jannasch and M. J. Mottl, *Science* **229**, 717 (1986).

[11] H. Konig, E. Nusser, and K. O. Stetter, *FEMS Microbiol. Lett.* **28**, 265 (1985).

[12] J.-P. Yu, J. Ladapo, and W. B. Whitman, *J. Bacteriol.* **176**, 325 (1994).

unusual coenzymes: methanofuran, tetrahydromethanopterin (H_4MPT), coenzyme M (HS-CoM), 7-mercaptoheptanoylthreonine phosphate (HS-HTP), and coenzyme F_{430}.[13,14] Putting it simply, the pathway involves the stepwise reduction of CO_2 with H_2 as the ultimate electron donor. It contains three coupling sites to the protonmotive force (PMF). In the first, the PMF is utilized to drive the endergonic reduction of CO_2 to the formyl level. The second and third coupling sites generate the PMF by coupling exergonic steps in CO_2 reduction to proton or sodium pumps. Each of the three coupled reactions is catalyzed by a membrane protein complex. In addition, the methylreductosome is a large complex attached to the interior of the cytoplasmic membrane that contains at least one enzyme of the pathway.

Electron carriers for many of the reactions in methanogenesis are not known with certainty. It is likely that Fe/S proteins are utilized for many steps. For some reactions, the deazaflavin coenzyme F_{420} is utilized. Methanococci also contain NAD(P)H and flavins, although cytochromes and ubiquinone or menaquinone are believed to be absent. Many H_2-utilizing methanogens grow with formate in place of H_2. Although *M. jannaschii* JAL-1 will not grow with formate, cell extracts oxidize formate, and closely related isolates grow with formate. Therefore, *M. jannaschii* probably has a limited capacity to metabolize formate.

The proton motive force generated during methanogenesis is utilized for ATP synthesis, transport, motility, and other cellular functions. In the related mesophile *Methanococcus voltae,* the sodium motive force is probably the major component of the membrane potential.[15] It is coupled to ATP synthesis by a Na^+-translocating ATPase and the proton gradient by a Na^+/H^+ antiporter.[16–18] Similarly, transport is dependent on sodium.[19,20] Presumably, other bioenergetic processes in methanococci such as motility will prove to be coupled to the sodium motive force.

M. jannaschii grows autotrophically, and there is little evidence that it assimilates organic compounds. Thus, it must biosynthesize all its cellular components from CO_2. The likely path of carbon flow is from CO_2 to acetyl-CoA via a modified Ljungdahl–Wood pathway. From acetyl-CoA, sugars are formed from pyruvate via gluconeogenesis, and amino acids are formed from a reductive, incomplete tricarboxylic acid (TCA) cycle. Aside

[13] R. K. Thauer, *Microbiol.* **144,** 2377 (1998).
[14] G. Schäfer, M. Engelhard, and V. Müller, *Microbiol. Mol. Biol. Rev.* **63,** 570 (1999).
[15] K. F. Jarrell and G. D. Sprott, *Can. J. Microbiol.* **31,** 851 (1985).
[16] S. W. Carper and J. R. Lancaster, Jr., *FEBS Lett.* **200,** 177 (1986).
[17] M. Dybas and J. Konisky, *J. Bacteriol.* **174,** 5575 (1992).
[18] W. Chen and J. Konisky, *J. Bacteriol.* **175,** 5677 (1993).
[19] I. Ekiel, K. F. Jarrell, and G. D. Sprott, *Eur. J. Biochem.* **149,** 437 (1985).
[20] M. Dybas and J. Konisky, *J. Bacteriol.* **171,** 5866 (1989).

from the specialized intermediary metabolism, the biosynthesis of most amino acids, nucleosides, and hexoses in methanogens probably occur by pathways common in bacteria and eukarya.[21] Some noteworthy features are described later. Lysine is made by the diaminopimelic pathway, isoleucine is made by the citramalate pathway, and the methyl group of methionine is probably derived from an intermediate of methanogenesis. Similarly, although pyrimidines and purines appear to be derived from common pathways, C-1 units required for purine biosynthesis may be derived from intermediates of methanogenesis instead of serine.[22]

Overview of Annotation

M. jannaschii was the first archaeon whose genome was completely sequenced,[23] and its annotation has been the subject of significant discussion and revision (see Appendix).[24–27] Why is the annotation of *M. jannaschii* of special interest? Because the concept of the Archaea is relatively modern,[28] there is a relatively small legacy of empirical studies to draw on to deduce gene function. Similarly, *M. jannaschii* is also a hyperthermophile, strict anaerobe, and obligate lithotroph. The physiology of these types of organisms is also poorly described. Thus, it is a tremendous challenge to integrate and reconcile incomplete physiological and biochemical data with genomic data during annotation. Nevertheless, studies of the *M. jannaschii* genome promise new insights into prokaryotic evolution and diversity. This section outlines the annotation process and describes the current list of gene function assignments. Even though significant progress has been made since 1996, a large number of ambiguities remain, and the function of many of the genes remains unknown.

Genomics is inherently comparative—one looks for similarities and differences between genes and genomes from different organisms to explain

[21] E. Selkov, N. Maltsev, G. J. Olsen, R. Overbeek, and W. B. Whitman, *Gene* **197,** GC11 (1997).
[22] C. G. Choquet, J. C. Richards, G. B. Patel, and G. D. Sprott, *Arch. Microbiol.* **161,** 471 (1994).
[23] C. J. Bult, O. White, G. J. Olsen, L. Zhou, R. D. Fleischmann, G. G. Sutton, J. A. Blake, L. M. FitzGerald *et al., Science* **273,** 1058 (1996).
[24] N. C. Kyrpides, G. J. Olsen, H.-P. Klenk, O. White, and C. R. Woese, *Microb. Comp. Genomics* **1,** 329 (1996).
[25] M. Andrade, G. Casari, A. de Daruvar, C. Sander, R. Schneider *et al., Comput. Appl. Biosci.* **13,** 481 (1997).
[26] E. V. Koonin, A. R. Mushegian, M. Y. Galperin, and D. R. Walker, *Mol. Microbiol.* **25,** 619 (1997).
[27] N. C. Kyrpides and C. A. Ouzounis, *Mol. Microbiol.* **32,** 886 (1999).
[28] C. R. Woese and G. E. Fox, *Proc. Natl. Acad. Sci. U.S.A.* **74,** 5088 (1977).

how cells work. As the current rate of DNA sequencing far exceeds the capacity to test predicted gene functions in the laboratory, our goal is to leverage experimental data to make the most useful predictions. While functional assignments of protein based on comparative analysis are indispensable, they are all hypotheses. The most important and speculative predictions must still be tested *in vitro* using biochemical techniques and *in vivo* by genetic analysis.[29,30]

Although genomic annotation can begin at any stage of a sequencing project, it becomes most productive when the average assembled contigs are large enough to encode more than one gene and the DNA sequence is sufficiently accurate to allow predictions of open reading frame (ORF) start and stop coordinates. A large contig size provides important contextual information about genes, whereas high levels of accuracy facilitate comparative searches and identication of nonfunctional pseudogenes. Functional assignments reach a pinnacle once the complete genomic sequence is deduced and a full inventory of genes is obtained.

The premise that the function of archaeal genes can be determined by comparison with previously characterized genes relies on the proposition that structurally similar genes will have similar functions and mechanisms. Thus gene function annotators have three intertwined jobs: (1) Recognize similarities between newly sequenced genes and previously characterized genes. (2) Interpet these similarities. (3) Integrate hypotheses about gene function into realistic models of metabolism and ecology.

Recognizing Sequence Similarities

This first task is now relegated to computers. The most informative searches usually make pairwise comparisons of a deduced protein sequence against all sequences in a database. The algorithms and statistics for these searches are well reviewed elsewhere.[31,32] Similarity search programs generate a list of sequences in the database that align most closely to the query sequence. These are ranked according to the expectation value of each alignment, the estimated frequency of randomly encountering an equally good alignment in the database. Sequences from the database that align with the query sequence with low expectations ($E < 10^{-5}$) are probably homologs. However, there are several caveats. Because these searches rely on sequence length-dependent scoring, a good alignment of two short,

[29] W. W. Metcalf, *Methods Microbiol.* **29**, 277 (1999).
[30] D. L. Tumbula and W. B. Whitman, *Mol. Microbiol.* **33**, 1 (1999).
[31] W. R. Pearson, *Methods Mol. Biol.* **132**, 185 (2000).
[32] S. Karlin, P. Bucher, V. Brendel, and S. F. Altschul, *Annu. Rev. Biophys. Biophys. Chem.* **20**, 175 (1991).

homologous protein sequences will have a higher expectation value than an alignment of two long protein sequences.[33] A good alignment of a query sequence versus a protein from a small database will have a lower expectation value than an equivalent alignment from a large database. Finally, sequences containing regions of low complexity (e.g., runs of highly acidic or basic residues) can produce spurious hits in the database. Using filters to mask degenerate sequence regions alleviates this problem. These artifacts can compromise automated search strategies that attempt to identify homologs by setting thresholds on acceptable expectation values.

Although it is certainly possible to compare ORF nucleotide sequences rather than their translated amino acid sequences, DNA sequences change much more rapidly through evolution than protein sequences due to degeneracy in the genetic code. The resulting "noise" thwarts similarity searches and seldom improves search results for archaeal genes, which are frequently highly diverged from the bacterial and eukaryal sequences that comprise most of the databases.

Search algorithms are only as good as their target databases. Large, nonredundant protein sequence databases compiled by the European Molecular Biology Laboratory (EMBL), National Center of Biotechnology Information (NCBI), or DNA Data Bank of Japan (DDBJ) are most popular, and the overall quality of sequences deposited is very good. However, there are significant advantages to searching smaller, actively curated databases such as SWISS-PROT,[34] WIT,[35] or local equivalents. In the preliminary annotation of new archaea, the genomes were queried to a local database of archaeal genes.[36] For instance, to assign functions to genes of *Pyrococcus abyssi,* the third publicly available *Pyrococcus* spp. genome, each gene was queried against a local database of archaeal genes, including *Pyrococcus* sp. OT3 and *Pyrococcus furiosus.* Because most genes in *P. abyssi* have close homologs in the other *Pyrococcus* spp., genes were rapidly annotated while maintaining a consistent nomenclature. Genes with no homologs were then queried against a large nonredundant database. It is usually appropriate to maintain updated copies of these databases on local computers to speed up searches, to protect private data, and to limit the load on publicly available servers.

[33] J. F. Collins, A. F. Coulson, and A. Lyall, *Comput. Appl. Biosci.* **4,** 67 (1988).
[34] A. Bairoch and R. Apweiler, *Nucleic Acids Res.* **28,** 45 (2000).
[35] R. Overbeek, N. Larsen, G. D. Pusch, M. D'Souza, E. Selkov, Jr., N. Kyrpides, M. Fonstein, N. Maltsev, and E. Selkov, *Nucleic Acids Res.* **28,** 123 (2000).
[36] D. Graham, unpublished results.

Interpreting Sequence Similarities

Although computers are particularly adept at comparing, ranking, and sorting database search reports, a human annotator is invaluable for evaluating these reports. The annotator must decide whether the best database hit is homologous to the query sequence, whether the function assigned by that gene's depositors is credible, and whether the query sequence likely has the same function.

Recognizing Homologs. Two genes are considered homologs if they evolved from a common ancestral gene. If two genes diverged recently, then one would expect that the alignment shows a high degree of sequence similarity that spans the full length of the proteins and includes all conserved catalytic residues. If these criteria are met, then one would predict that the two proteins have similar functions. Conversely, if two genes are anciently diverged or unrelated, one might expect low overall sequence similarity, alignment over only part of the protein sequences, and significant changes in secondary structural elements. Homologous genes that are greatly diverged in structure may also perform different functions.

Evaluating Gene Functions Described in Databases. Evaluating published claims of gene function requires a significant knowledge of the literature. Fortunately, hyperlinks to online abstracts and the full text of journals expedite such research. Most functional annotations are now based on comparative sequence analysis, so the annotator must usually look beyond the closest homolog of the query sequence in order to find original laboratory research characterizing the protein family. Homologs in microbial model systems, such as *Escherichia coli, Saccharomyces cerevisiae,* or *Bacillus subtilis,* are usually the best-characterized members of each protein family. Nevertheless, experimental evidence is not flawless. Epistasis, polarity, suppressors, pleiotropic mutations, and uncertainties in gene mapping sometimes confuse genetic analysis. Impure reagents and the investigators' choice of substrates and reaction direction may affect biochemical assays: many enzymes act on numerous substrates *in vitro,* and most enzymatic reactions are reversible. Thus, a combination of genetic and biochemical characterization is most desirable to confirm a gene's function. Due to the time involved in researching these claims, references to characterized homologs should be stored alongside each gene function assignment. Curated databases such as SWISS-PROT are most helpful because they include evaluations of the evidence for gene function and are updated when new research becomes available.[34]

Assigning Gene Functions. Once we have established that the primary structure, i.e., sequence, of our query protein is similar to a previously characterized protein, we must ask whether it also functions similarly. As postulated earlier, structural similarity usually implies mechanistic and functional similarity, but families and superfamilies of proteins can frustrate this extrapolation. ATP-binding subunits of ABC transporters, aldehyde dehydrogenases, and pyridoxal phosphate-dependent aminotransferases are all families of homologous proteins whose members catalyze similar reactions on disparate substrates. Changing a single residue of lactate dehydrogenase (Gln102Arg) converts that enzyme into a catalytically efficient malate dehydrogenase.[37] Even larger groups of proteins, such as the enolase superfamily, can catalyze a bewildering array of reactions that share only a common mechanism (abstracting the α-proton from a carboxylic acid to produce an enolic intermediate).[38]

The most efficient way to confirm the function of a protein in one of these families is to examine the list of sequences returned by a similarity search program (e.g., BLAST[39]). Is the next functionally distinct homolog on the list significantly less similar to the query sequence? If so, then one can usually assign the function of the first homolog to the unknown protein. In other cases, the annotator cannot discriminate among possible functions, but should record that the protein is a member of a certain family. Specificity can sometimes be discerned through phylogenetic analysis—clustering families and subfamilies of proteins with similar function. Alternatively, if the family is well studied, then signature sequences or active site residues may be found that are characteristic of a specific enzyme. These can be rapidly searched for by pattern matching or hidden Markov models. Finally, one can sometimes use contextual information in the genome or even in other genomes containing the homolog. Because genes in bacterial and archaeal genomes are frequently organized into operons alongside genes of related function, a gene's neighbors often provide clues to its function.[40]

Despite the vast amount of knowledge about gene functions that is recorded in the databases, a large number of archaeal proteins are not similar to any characterized protein. A confusing collection of names has emerged to describe relationships between uncharacterized proteins: "hypothetical," "unknown," and "putative." These are simply placeholders to record that a protein has no known function, but has a homolog elsewhere. In the updated *M. jannachii* gene list, homologs are indicated in brackets.

---

[37] H. M. Wilks, K. W. Hart, R. Feeney, C. R. Dunn, H. Muirhead, W. N. Chia, D. A. Barstow, T. Atkinson, A. R. Clarke, and J. J. Holbrook, *Science* **242,** 1541 (1988).

[38] P. C. Babbitt, M. S. Hasson, J. E. Wedekind, D. R. Palmer, W. C. Barrett, G. H. Reed, I. Rayment, D. Ringe, G. L. Kenyon, and J. A. Gerlt, *Biochemistry* **35,** 16489 (1996).

[39] S. F. Altschul, W. Gish, W. Miller, E. W. Myers, and D. J. Lipman, *J. Mol. Biol.* **215,** 403 (1990).

[40] R. Overbeek, M. Fonstein, M. D'Souza, G. D. Pusch, and N. Maltsev, *Proc. Natl. Acad. Sci. U.S.A.* **96,** 2896 (1999).

Another set of words qualifies the annotator's functional assignments: "similar to," "related," "putative," "potential," "probable." Like motifs or patterns observed in a sequence, these qualified statements provide clues to a protein's function but are not true functional assignments. Although these terms may sometimes be confusing there is a real need to record levels of confidence in assignments. Ideally, these qualitative terms could be standardized and used to convey confidence in the similarity of two proteins, the likelihood of their sharing a common function, and the certainty that their function is fully understood.

Assigning Aminotransferases. The assignment of function to aminotransferases is a concrete example of how these factors were utilized in the current annotation. This gene family is complex and composed of multiple subgroups with overlapping functions.[41] The availability of the *Archaeoglobus fulgidus*[42] and *Methanobacterium thermoautotrophicum*[43] genomic sequences were particularly useful because they facilitated the discrimination of orthologs and paralogs. First, on comparison of 47 groups of likely orthologs from outside the aminotransferase family, the mean Protdist distances ($\pm$ standard deviation) were found to be 0.68 ± 0.21, 0.75 ± 0.26, and 0.79 ± 0.26 for comparisons between *M. jannaschii* and *M. thermoautotrophicum*, *M. jannaschii* and *A. fulgidus*, and *M. thermoautotrophicum* and *A. fulgidus*, respectively. These mean distances correspond to amino acid identities of 52–56%. For comparisons of unidentified ORFs, ORFs were considered likely orthologs if their Protdist distances were equal to or less than the mean plus two standard deviations. Even though this simple analysis neglected different rates of change in different genes, it was very useful. For instance, the *M. jannaschii* genome encodes 11 potential aminotransferases. Four of these were originally annotated as aspartate aminotransferase.[23] By the criteria defined earlier, only one of these, MJ0959, is likely to be orthologous to the aspartate aminotransferase identified in *M. thermoautotrophicum*[44] and correctly annotated. Two of the remaining ORFs (MJ0001 and MJ684) are closely related to each other and possess likely orthologs in *M. thermoautotrophicum* and *A. fulgidus*. However, there is no additional evidence for their function. The last ORF, MJ1391, possesses a homolog in *A. fulgidus*, but the Protdist distance is 1.40, suggesting that they are not orthologs. There is no evidence for the function of either of these aminotransferases. Thus, of the 11 potential aminotransferases in the

[41] P. K. Mehta, T. I. Hale, and P. Christen, *Eur. J. Biochem.* **214,** 549 (1993).

[42] H. P. Klenk, R. A. Clayton, J. Tomb, O. White, K. E. Nelson *et al., Nature* **390,** 364 (1997).

[43] D. R. Smith, L. A. Doucette-Stamm, C. Deloughery, H. Lee, J. Dubois, T. Aldredge, R. Bashirzadeh, D. Blakely, R. Cook *et al., J. Bacteriol.* **179,** 7135 (1997).

[44] T. Tanaka, S. Yamamoto, T. Moriya, M. Taniguchi, H. Hayashi, H. Kagamiyama, and S. Oi, *J. Biochem.* **115,** 309 (1994).

M. jannaschii genome, there is evidence for a specific biochemical function in only five cases. MJ0959 is identified as the aspartate aminotransferase by its similarity to the *M. thermoautotrophicum* protein. MJ0955 is identified as the histinol-phosphate aminotransferase based on its similarity to the *Haloferax volcanii* gene.[45] MJ0603 is proposed to be glutamate-1-semialdehyde 2,1-aminomutase based on its similarity to AF0080. The identification of AF0080 is supported by its location in the genome with other tetrapyrrole biosynthetic genes. MJ1300 is proposed to be adenosylmethionine-8-amino-7-oxononanoate aminotransferase based on its similarity to the bacterial genes and its location with other biotin biosynthetic genes. MJ1420 is proposed to be glucosamine-fructose-6-phosphate aminotransferase based on its high similarity to the bacterial genes. Because *N*-acetylglucoamine is a component of lipids of *M. jannaschii*,[46] this assignment seems reasonable. In contrast, MJ1008 was proposed to be the branched-chain amino acid aminotransferase,[23] and likely orthologs are present in both *M. thermoautotrophicum* and *A. fulgidus*. While this ORF possesses homology to the bacterial subgroup III aminotransferases,[41] this subgroup contains both branched-chain amino acid and alanine aminotransferases, and there is insufficient basis to decide between these alternative assignments. For this reason, it is annotated only as a subgroup III aminotransferase.

Amino Acid Composition

In some cases, amino acid composition can provide evidence for the function of the ORFs. This is particularly true for ORFs predicted to contain large amounts of certain functional classes of amino acids. For instance, the average cysteinyl composition of *M. jannaschii* ORFs is 1.56%. Three polyferredoxins, MJ0934, MJ1193, and MJ1303, have the highest cysteinyl content, 11.3–11.5%. Many of the other cysteine-rich ORFs encode subunits of iron–sulfur proteins and are likely to function as electron carriers or redox proteins. Therefore, unidentified ORFs with a cysteinyl content of ≥5.0% were annotated. Likewise the average composition of basic amino acids, acidic amino acids, proline, and glycine are 15, 13, 4, and 6%. Many of the ORFs with a high content of basic amino acids are ribosomal proteins or encode enzymes that interact with nucleic acids, with the highest being MJ0655 (30%), which encodes ribosomal protein L34. Unidentified ORFs with an basic amino acid content of ≥20.8% were noted in the gene list. The function of the ORF, MJ0280, with the highest content of acidic amino acids, 26.3%, is unknown. Unidentified ORFs with an acidic amino acid

[45] R. K. Conover and W. F. Doolittle, *J. Bacteriol.* **172,** 3244 (1990).
[46] Y. Koga, H. Morii, M. Akagawa-Matsushita, and M. Ohga, *Biosci. Biotechnol. Biochem.* **62,** 230 (1998).

content $\geq 20.0\%$ were annotated in the gene list. Similarly, unidentified ORFs with a proline and glycine content of ≥ 7.8 and $\geq 12.0\%$ were annotated.

Integrating Gene Functions into a Metabolic Map

Attempting to assign gene functions to several thousand unknown archaeal proteins using comparative analysis is a laborious effort, but it is only the first step in reconstructing the organism's metabolic scheme. Complete genomic sequences contain the full inventory of every gene in the cell, encoding every RNA, protein, and DNA regulatory sequence necessary for the organism's growth and reproduction. This integrative stage in gene function annotation proofreads the comparative analysis and highlights essential enzymes missing from the gene list.[21]

Proofreading. Despite the best efforts of the annotator, some gene functions cannot be identified without a complete view of the organism's physiology and metabolism. For example, the ferredoxin-dependent 2-oxoacid oxidoreductases from the heterotroph *P. furiosus* function catabolically,[47] whereas homologous enzymes in the autotroph *M. thermoautotrophicum* probably function anabolically.[48] These differences are not discernible by sequence alignment alone. At this stage, many gene functions assigned by genetics or functional genomics, such as "virulence factor" or "resistance protein," are revealed as inappropriate.

Identifying Missing Genes. When a key enzyme in an essential metabolic pathway appears to be missing from a genome there are several possible explanations: the homologous enzyme is so highly diverged as to be unrecognizable by similarity search, an analogous enzyme catalyzes the same reaction, or the organism uses an entirely different pathway. The archaeal type *S*-adenosylmethionine synthetase shares only active site residues with its bacterial/eucarayal homolog.[49] These distant relatives can sometimes be detected by profiling searches such as PSI-BLAST[50] or by modeling protein secondary structures.[51] The ATP-dependent and ADP-dependent phosphofructokinases are examples of analogous enzymes. While most bacteria and eukarya use the ATP-dependent version, many archaea use

[47] K. Ma, A. Hutchins, S.-J. S. Sung, and M. W. W. Adams, *Proc. Natl. Acad. Sci. U.S.A.* **94,** 9608 (1997).

[48] A. Tersteegen, D. Linder, R. K. Thauer, and R. Hedderich, *Eur. J. Biochem.* **244,** 862 (1997).

[49] D. E. Graham, C. L. Bock, C. Schalk-Hihi, Z. Lu, and G. D. Markham, *J. Biol. Chem.*, in press.

[50] S. F. Altschul, T. L. Madden, A. A. Schaffer, J. Zhang, Z. Zhang, W. Miller, and D. J. Lipman, *Nucleic Acids Res.* **25,** 3389 (1997).

[51] H. Xu, R. Aurora, G. D. Rose, and R. H. White, *Nature Struct. Biol.* **6,** 750 (1999).

the unrelated ADP-dependent analog.[52] Finally, some archaea lack the canonical pathway for 2-oxobutyrate biosynthesis (a precursor to isoleucine); instead they have a citramalate pathway.[53]

Essential Pathways. The availability of the complete genomic sequence of *M. jannaschii* (and subsequently of other archaeal organisms) provided new insights into the nature and organization of the information processing in the third domain of life. Of the three major information-processing subsystems (transcription, translation, and replication), our view on transcription regulation was the most affected from new data. Before the publication of the *M. jannaschii* genome, it was generally accepted that Archaea and Eukarya shared a transcriptional machinery that was very different from that of Bacteria.[54,55] However, as more archaeal genomes became available, it became clear that transcription in Archaea has a mixed character, with a multitude of components distinctly found either in the bacterial or in the eukaryotic domain.[56–58] In essence, the basic transcription machinery (initiation of transcription) is similar to that of the eukarya (i.e., the RNA polymerase subunits and the transcription factors TBP and TFIIB), whereas most of the transcriptional regulators are similar to those of the bacteria (i.e., AsnC, LysR, MerR DtxR, HypF, PhoU, DegT),[57] thus making archaea a unique system to study this process. In contrast to transcription, our view on archaeal protein translation did not change considerably with the availability of the genome sequences.[59,60] Here, most of the archaeal proteins are either similar to eukarya to the exclusion of the bacteria (i.e., many ribosomal proteins and translation initiation factors) or similar to proteins found in both eukarya and bacteria (i.e., aminoacyl-tRNA synthetases). Moreover, analyses of archaeal translation support a now familiar theme seen in other cellular systems; i.e., the archaeal version fills the gap between bacterial and eukaryotic processes and reveals relationships previously unobserved.[61,62] In contrast, archaeal replication appears distinctively eukaryotic. The complete genome of *M. jannaschii* revealed only a

[52] J. E. Tuininga, C. H. Verhees, J. van der Oost, S. W. Kengen, A. J. M. Stams, and W. M. de Vos, *J. Biol. Chem.* **274,** 21023 (1999).

[53] B. Eikmanns, D. Linder, and R. K. Thauer, *Arch. Microbiol.* **136,** 111 (1983).

[54] D. Langer, J. Hain, P. Thuriaux, and W. Zillig, *Proc. Natl. Acad. Sci. U.S.A* **92,** 5768 (1995).

[55] M. Thomm, *FEMS Microbiol. Rev.* **18,** 159 (1996).

[56] N. C. Kyrpides and C. A. Ouzounis, *J. Mol. Evol.* **45,** 706 (1997).

[57] N. C. Kyrpides and C. A. Ouzounis, *Proc. Natl. Acad. Sci. U.S.A.* **96,** 8545 (1999).

[58] L. Aravind and E. V. Koonin, *Nucleic Acids Res.* **27,** 4658 (1999).

[59] R. Amils, P. Cammarano, and P. Londei, *in* "The Biochemistry of Archaea (Archaebacteria)" (M. Kates *et al.,* eds.), p. 393. Elsevier, Amsterdam, 1993.

[60] P. P. Dennis, *Cell* **89,** 1007 (1997).

[61] P. Baumann and S. P. Jackson, *Proc. Natl. Acad. Sci. U.S.A.* **93,** 6726 (1996).

[62] N. C. Kyrpides and C. R. Woese, *Proc. Natl. Acad. Sci. U.S.A.* **95,** 224 (1998).

single typical alpha-like (family B) DNA polymerase (MJ0885) and a number of replication factors, which were all clearly similar to their eukaryotic counterparts.[23] Subsequently, a number of additional proteins involved in replication have been identified experimentally, including a new two-subunit DNA polymerase (Pol II; MJ0702 and MJ1630)[63] and a DNA primase (MJ0839).[64] A number of minichromosome maintenance (MCM) proteins are also present in the genome of *M. jannaschii* (MJ0363, MJ0961, and MJ1489), but the Cdc6 protein, which is responsible for loading them onto replication origins, has not been identified (although it is present in the genomes of *A. fulgidus* and *M. thermoautotrophicum*).[65] In contrast, many features of metabolism appear to be shared with bacteria. Ubiquitous anabolic and catabolic pathways in the Archaea include glycolysis/gluconeogenesis, derivatives of the TCA cycle, amino acid and nucleotide biosynthesis, which are described in standard references for microbial metabolism.[66,67] From this perspective, unique archaeal processes, such as methanogenesis, might be considered specializations similar to those found in certain physiological groups of bacteria and are not necessarily indicative of ancient metabolic pathways. Online databases of metabolic pathways, such as WIT, EMP/MPW, KEGG, MetaCyc, and UM-BBD, are described in the annual summary of databases published in *Nucleic Acids Research*.[68]

Aromatic Amino Acid Biosynthesis. Metabolic reconstruction was useful in annotation of the pathways for chorismate, phenylalanine and tyrosine, and tryptophan biosynthesis. While all the genes required for tryptophan synthesis from chorismate have been found in the sequenced euryarchaeotes, gaps still remain within the other pathways.

Chorismate is biosynthesized by a series of seven enzymes in most organisms, proceeding through the intermediates 3-dehydroquinate and shikimate.[69] These two intermediates are likely to be present in the *M. jannaschii* pathway because the sequenced euryarchaeotes, *M. jannaschii, M. thermoautotrophicum, A. fulgidus,* and *P. furiosus,* all possess homologs of the enzymes needed to convert one to the other: 3-dehydroquinate dehydratase and shikimate dehydrogenase. In fact, recognizable homologs of all of the enzymes needed for the conversion of shikimate to chroismate

[63] I. K. Cann and Y. Ishino, *Genetics* **152,** 1249 (1999).

[64] G. Desogus, S. Onesti, P. Brick, M. Rossi, and F. M. Pisani, *Nucleic Acids Res.* **27,** 4444 (1999).

[65] R. Bernander, *Mol. Microbiol.* **29,** 955 (1998).

[66] G. Gottschalk, "Bacterial Metabolism." Springer- Verlag, New York, 1986.

[67] G. Michal, "Biochemical Pathways: An Atlas of Biochemistry and Molecular Biology." Wiley, New York, 1999.

[68] A. D. Baxevanis, *Nucleic Acids Res.* **28,** 1 (2000).

[69] R. Bentley, *Crit. Rev. Biochem. Mol. Biol.* **25,** 307 (1990).

are present except for shikimate kinase, which is lacking in all four organisms. A candidate for the *M. jannaschii* shikimate kinase, MJ1440, was identified based on its sequence similarity to other kinases and the association of homologs in other archaea with chorismate biosynthetic genes. The *M. thermoautotrophicum* ORF (MTH805) is adjacent to chorismate mutase, and the *P. furiosus* ORF (PF1410) is found within a large operon of chorismate biosynthetic genes. A homolog of this kinase is also present in the crenarchaeote *Aeropyrum pernix* as part of a chorismate biosynthetic operon.

The first steps of the pathway are less well established. *P. furiosus* is the only sequenced euryarchaeote possessing homologs of the first two enzymes of the pathway: 2-dehydro-3-deoxyphosphoheptonate aldolase (or DAHP synthase) and 3-dehydroquinate synthase. While there are a number of possible explanations for the absence of homologs in the other euryarchaeotes, it is most likely that these organisms use a different pathway for the formation of 3-dehydroquinate. Labeling studies in the mesophile *Methanococcus maripaludis* indicate that erythrose-4-phosphate is not a precursor for chorismate.[70] Also the enzyme activity for DAHP synthase was not found in another methanogen, *Methanohalophilus mahii*.[71] It is interesting to note that the genes for the first two enzymes of chorismate synthesis have also not been found in *Aquifex aeolicus*.

The syntheses of phenylalanine and tyrosine from chorismate each require three reactions, two of which are shared: chorismate mutase and an aromatic aminotransferase. The unique enzymes for phenylalanine and tyrosine synthesis are prephenate dehydratase and prephenate dehydrogenase, respectively. Chorismate mutase is often found as part of bifunctional enzymes; however, the chorismate mutase of *Methanococcus jannaschii* is monofunctional,[72] and homologs can be found in *M. thermoautotrophicum* and *P. furiosus*. *A. fulgidus*, however, has a (so far) unique trifunctional enzyme containing chorismate mutase, prephenate dehydrogenase, and prephenate dehydratase.[72] Except for *A. fulgidus*, monofunctional prephenate dehydrogenases and prephenate dehydratases have been identified in the euryarchaeotes based on sequence comparisons. Biosynthesis of phenylalanine and tyrosine also requires an aminotransferase. While aromatic amino acid aminotransferase activity has been identified in the mesophile *Methanococcus aeolicus*,[73] the gene or genes corresponding to this

[70] D. L. Tumbula, Q. Teng, M. G. Bartlett, and W. B. Whitman, *J. Bacteriol.* **179,** 6010 (1997).
[71] R. S. Fischer, C. A. Bonner, D. R. Boone, and R. A. Jensen, *Arch. Microbiol.* **160,** 440 (1993).
[72] G. MacBeath, P. Kast, and D. Hilvert, *Biochemistry* **37,** 10062 (1998).
[73] R. Xing and W. B. Whitman, *J. Bacteriol.* **174,** 541 (1992).

enzyme cannot be identified by sequence comparison (see earlier discussion). Even though a promising candidate (PF1399) has been identified in the *P. furiosus* genome based on its association with other aromatic amino acid biosynthetic genes, *M. jannaschii* does not possess a homolog with a high degree of sequence similarity to this gene.

The tryptophan biosynthetic operon has previously been identified in *M. thermoautotrophicum*, and the functions of several genes were confirmed by complementation of *Escherichia coli* mutants.[74,75] All of the *trp* genes can be identified for the euryarchaeotes and, with the exception of *M. jannaschii*, all are found grouped in potential operons. The only multifunctional enzyme is found in *A. fulgidus*, where indoleglycerol-phosphate synthetase and anthranilate phosphoribosyltransferase are fused into one ORF, AF1604. This is the only known example of a bifunctional enzyme containing these two activities.

There is an interesting variety in the way these four organisms have arranged their aromatic amino acid biosynthetic genes. At one extreme, *M. jannaschii* has only two genes adjacent to each other: the two subunits of tryptophan synthase. At the other extreme, *P. furiosus* has all of the genes for aromatic amino acid biosynthesis, except prephenate dehydratase, clustered into two apparent operons. In *A. fulgidus* and *M. thermoautotrophicum*, all the genes for tryptophan biosynthesis are grouped together, while most of the remaining genes are scattered on the genome. The reasons for these disparate forms of organization may relate to the organisms' evolutionary histories and to their different needs to coordinately regulate these genes.

Like the functional assignments themselves, metabolic reconstructions undergo continuous refinement. Gene annotations from *M. jannaschii* have changed significantly since the original publication of that genome sequence. We anticipate that these functional assignments will continue to improve as more experimental work is carried out in the archaea. A similar gene list by one of us (N.K.) is also available at: http://geta.life.uiuc.edu/~nikos/MJannotations.html.

Conclusion

A useful exercise to gain perspective on the *M. jannaschii* genomic sequence is to compare the observed number of ORFs to the number

[74] L. Meile, R. Stettler, R. Banholzer, M. Kotik, and T. Leisinger, *J. Bacteriol.* **173,** 5017 (1991).
[75] L. Sibold and M. Henriquet, *Mol. Gen. Genet.* **214,** 439 (1988).

that might be expected for an autotrophic methanogen. Two approaches were utilized to estimate the number of genes essential for an autotrophic methanogen. The first approach began with the composition of the cell and the assumption that it was growing under steady-state conditions. The number of required biosynthetic genes was estimated from the pathways required to make the cellular components. In cases where the pathways were not known, such as for biosynthesis of some of the coenzymes, each reaction was assumed to require one enzyme corresponding to one gene. One transport system was also assumed to be required for each inorganic nutrient. Likewise, known essential genes for translation, transcription, replication, protein secretion, and cell division were included. Based on these analyses, about 650 or 37% of the ORFs would be expected to be essential. In the second approach, essential genes were assumed to be highly conserved. These genes could then be identified as homologs in the related prokaryotes *M. thermoautotrophicum* and *A. fulgidus*. In good agreement with the first approach, about 500 ORFs are conserved in all three euryarchaeotes. In conclusion, only a small fraction of the genome is required to encode known essential functions. This answer is not intuitive. The nutritional and growth requirements for *M. jannaschii* are extremely simple, yet the number of genes known to be required for this chemolithotrophic life style is not inordinately large.

What is the function of the remaining genes? About 200 or 12% of the ORFs were estimated to encode functions nonessential for growth but known to be present in *M. jannaschii*, such as flagellar biosynthesis and chemotaxis. However, the genome is not particularly redundant, and a large proportion of the genome is not composed of duplicates of essential genes. From this we conclude that most of the ORFs encode functions whose importance is not known, either from the known properties of the organism or by implication from the general properties of other prokaryotes.

Finally, it is interesting to compare the distribution of homologs to the genes of *M. jannaschii*. Of the 1792 genes proposed, 332 do not possess homologs in the databases and are probably unique to this organism and its close relatives. An additional 381 genes are found only in the archaea. There are 110 genes that possess homologs in eukaryotes but not bacteria and 610 genes that possess homologs in bacteria but not eukaryotes. There are also 359 genes that appear to be universal and possess homologs in all three domains. Thus, nearly 40% of the genes are found exclusively in the archaea and emphasizes the distinctiveness of *M. jannaschii*.

Appendix: Features of the *Methanocaldococcus* (*Methanococcus*) *jannaschii* JAL-1 Genome

The *M. jannaschii* DNA is composed of three contigs: the 1.66-mbp chromosome, a small 16-kbp plasmid (MJECS), and a large 58-kbp plasmid (MJECL). Open reading frames are named according to the conventions of Bult *et al.*[23] When new ORFs are inserted, they are given a decimal number. "Dirct" refers to the coding strand, positive (+) or negative (−).

Name	Length	Dirct	Gap	Start	Annotated function
Contig: chromosome					
LR1_16	394	+	12	12	repeat
SR1_108	27	+	-27	379	repeat
SR1_107	27	+	43	449	repeat
SR1_106	27	+	40	516	repeat
SR1_105	27	+	40	583	repeat
SR1_104	27	+	43	653	repeat
SR1_103	27	+	40	720	repeat
SR1_102	27	+	40	787	repeat
SR1_101	27	+	43	857	repeat
SR1_100	27	+	40	924	repeat
SR1_99	27	+	45	996	repeat
SR1_98	27	+	43	1066	repeat
SR1_97	27	+	41	1134	repeat
SR1_96	27	+	41	1202	repeat
SR1_95	27	+	39	1268	repeat
SR1_94	27	+	44	1339	repeat
SR1_93	27	+	44	1410	repeat
SR1_92	27	+	41	1478	repeat
SR1_91	27	+	43	1548	repeat
SR1_90	27	+	44	1619	repeat
SR1_89	27	+	44	1690	repeat
SR1_88	27	+	41	1758	repeat
SR1_87	27	+	43	1828	repeat
SR1_86	27	+	41	1896	repeat
SR1_85	27	+	41	1964	repeat
SR1_84	27	+	41	2032	repeat
MJ0001	1124	-	160	3343	aminotransferase (subgroup I), [MJ0684, MTH1894, AF2366, AF2129, AF1623]
MJ0002	728	-	0	4071	hypothetical protein [Streptomyces coelicolor A3(2)]

MJ0002.5	107	+	385	4456	
MJ0003	401	+	414	4977	
MJ0004	728	+	0	5378	activator of (R)-2-hydroxyglutaryl-CoA dehydratase
MJ0005	1136	-	8	7250	formate dehydrogenase (EC 1.2.1.2), subunit beta
MJ0006	1133	-	91	8474	formate dehydrogenase (EC 1.2.1.2), subunit alpha
MJ0007	1115	+	413	8887	2-hydroxyglutaryl-CoA dehydratase (EC 4.2.1.-)
MJ0008	590	+	142	10144	endonuclease IV related protein (end4 [Mycoplasma pneumoniae])
MJ0009	827	+	9	10743	
MJ0010	1283	+	11	11581	similar to phosphopentose mutase [MTH1591]; M.Y. Galperin et al., Protein Sci. 7, 1829 (1998)
MJ0011	590	+	5	12869	
MJPSEO1	1409	-	-32	14836	pseudogene, hypothetical (partial) protein [MJ1635]
MJ0014	569	-	-16	15389	recombinase related protein
MJ0015	197	+	218	15607	similar to the N-terminus of histidinol-phosphatase
MJ0015.5	374	-	62	16240	transposase [Insertion sequence IS982], C-terminal 60%
IS_A1	662	+	551	16791	repeat
MJ0017	641	+	-647	16806	transposase
MJ0018	1511	+	271	17718	12.1% gly
MJ0019	1889	-	3	21121	glutamyl-tRNA amidotransferase, subunit B related protein
MJ0020	1250	-	30	22401	L-asparagine amidohydrolase (EC 3.5.1.1) (L-asparaginase)
MJ0021	1124	+	361	22762	hypothetical protein [AF1217]
MJ0022	1085	+	117	24003	cobalamin biosynthesis protein cbiD [Salmonella typhimurium]
MJ0023	353	+	196	25284	hypothetical protein [MJ1072]
MJ0024	416	-	52	26105	hypothetical protein, 23% acidic aa [*Mycobacterium tuberculosis*]
MJ0025	1013	-	4	27122	RNA 3'-terminal phosphate cyclase (EC 6.5.1.4)
MJ0026	794	-	15	27931	proliferating-cell nucleolar antigen
MJ0027	518	-	90	28539	
MJ0028	947	-	-22	29464	thiamin-monophosphate kinase (EC 2.7.4.16) (ThiL [S. typhimurium])
MJ0029	1229	+	389	29853	coenzyme F420-reducing hydrogenase (EC 1.12.99.1), alpha subunit; contains SeC residue

MJ0030	548	+	173	31255	coenzyme F420-reducing hydrogenase (EC 1.12.99.1), delta subunit (membrane anchor protein)
MJ0031	680	+	5	31808	coenzyme F420-reducing hydrogenase (EC 1.12.99.1), gamma subunit
MJ0032	845	+	38	32526	coenzyme F420-reducing hydrogenase (EC 1.12.99.1), beta subunit
MJ0033	1616	+	258	33629	fumarate reductase (flavoprotein subunit) (EC 1.3.99.1)
MJ0034	947	-	195	36387	ABC transporter, substrate binding protein [PH0883]
MJ0035	737	-	145	37269	ABC transporter ATP-binding protein
MJ0036	458	+	282	37551	SSU ribosomal protein S15P
MJ0037	713	+	64	38073	hypothetical protein [PH1310]
MJ0038	650	-	7	39443	hypothetical archaeal protein [MTH1325]
MJ0039	320	-	211	39974	DNA-directed RNA polymerase (EC 2.7.7.6), subunit F (rpoF), T.J. Darcy et al., J. Bacteriol. 181, 4424 (1999)
MJ0040	293	-	127	40394	LSU ribosomal protein L21E
MJ0041	1361	-	83	41838	zinc-finger hypothetical protein-[*Acidianus ambivalens*]
MJ0042	644	-	45	42527	hypothetical zinc-binding protein
McjBs_hyp	1175	+	612	43139	intein
MJ0043	2354	+ -	1559	42755	similar to UDP-glucose dehydrogenase [+Intein: aa 128-520]
MJ0044	779	-	11	45899	acetylglutamate kinase (*argB [E. coli]*) related protein
MJ0045	662	-	8	46569	hypothetical protein [*Mycobacterium leprae*]
MJ0046	779	-	0	47348	adenine-specific methyltransferase
MJ0047	1283	-	25	48656	mRNA cleavage and polyadenylation specificity factor related, N-terminal 70%
SR1_83	26	-	142	48824	repeat
SR2_52	28	-	123	48975	repeat
SR1_82	27	-	42	49044	repeat
SR1_81	27	-	43	49114	repeat
LR1_15	422	-	-27	49509	repeat
MJ0048	677	-	10	50196	translation initiation factor eIF-6 [*Homo sapiens*]
MJ0049	245	-	13	50454	LSU ribosomal protein L31E
MJ0050	1187	-	596	52237	pyridoxal-phosphate-dependent amino acid decarboxylase
MJ0051	344	+	110	52347	hypothetical protein [MTH1697]
MJ0052	665	-	18	53374	hypothetical protein, *E. coli ybbB*
MJ0053	653	-	14	54041	

MJ0054	842	-	118	55001	hypothetical protein [MTH1574]
MJ0055	680	-	82	55763	3,4-dihydroxy-2-butanone 4-phosphate synthase (*ribB*)
MJ0056	395	-	-4	56154	hypothetical protein [AF2106]
MJ0057	1277	-	31	57462	Na+/H+ exchanger (NAH1 [*Bos taurus*])
MJ0058	1154	-	129	58745	ammonium transporter
MJ0059	335	+	234	58979	nitrogen regulatory protein P-II
MJ0060	755	+	153	59467	methylthioadenosine phosphorylase (EC 2.4.2.28)
MJ0061	239	+	141	60363	ferredoxin
MJ0062	620	+	16	60618	hypothetical protein (yciO [E. coli])
MJ0063	407	+	210	61448	
MJ0064	566	+	33	61888	
MJ0065	1088	-	9	63551	hypothetical protein [MTH1863]
MJ0066	1421	-	106	65078	similar to 3'-phosphoadenosine 5'-phosphosulfate sulfotransferase
MJ0067	299	+	91	65169	hypothetical protein, 12.8% gly [PH1115]
MJ0068	344	-	49	65861	hypothetical protein [AF0282]
MJ0069	887	+	158	66019	acetylglutamate kinase (EC 2.7.2.8)
MJ0070	245	+	60	66966	
MJ0071	269	+	0	67211	hypothetical protein [PHs013]
MJ0072	131	+	82	67562	
MJ0073	278	+	36	67729	
MJ0074	1073	-	9	69089	ATPase
MJ0075	1088	-	147	70324	ATPase
MJ0077	1145	-	70	71539	hypothetical protein [MJ0811]
MJ0076	338	+	81	71620	hypothetical protein [MTH1141]
MJ0078	620	-	96	72674	
MJ0079	1127	-	-7	73794	similar to *moxR* transcriptional regulator
MJ0080	380	-	8	74182	hypothetical protein, family with MJ0549/MTH0739/AF0743
MJ0081	1340	+	488	74670	methyl coenzyme M reductase II (EC 1.8.-.-), subunit beta
MJ0082	797	+	16	76026	methyl coenzyme M reductase II (EC 1.8.-.-), subunit gamma
MJ0083	1655	+	3	76826	methyl coenzyme M reductase II (EC 1.8.-.-), subunit alpha
MJ0084	746	+	265	78746	cell division inhibitor minD
MJ0085	1118	+	138	79630	iron transport system binding protein
MJ0086	1115	+	40	80788	hypothetical O-methyltransferase [AF0429]
MJ0087	1046	+	27	81930	iron(III) dicitrate transport system, permease protein (*fecC* [E. coli])

MJ0088	539	+	22	82998	21% basic aa
MJ0089	752	+	13	83550	ferric enterobactin transport ATP-binding protein
MJ0090	1211	+	93	84395	microbial collagenase (EC 3.4.24.3)
MJ0091	905	+	99	85705	similar to fragment of Na+/Ca+ exchanger protein
MJ0092	1466	+	3	86613	bifunctional protein N-terminal 211 aa homolog of fumarate reductase (iron-sulfur subunit) (EC 1.3.1.6) (frdB [E. coli]); C-terminal portion homolog of Methanosarcina-type heterodisulfide reductase.
MJ0093	425	-	13	88517	21% acidic aa
MJ0094	917	-	47	89481	hypothetical protein [MTH1180]
MJ0095	377	-	87	89945	
MJ0096	473	-	22	90440	
MJ0097	428	-	608	91476	translation initiation factor eIF-2, subunit beta
MJ0098	182	-	199	91857	LSU ribosomal protein L37E
SR1_80	27	-	443	92327	repeat
SR2_51	28	-	38	92393	repeat
SR2_50	28	-	40	92461	repeat
SR2_49	28	-	41	92530	repeat
LR1_14	394	-	-28	92896	repeat
MJ0099	392	-	9	93297	ferredoxin
MJ0100	1526	-	0	94823	hypothetical protein with 2 CBS domains
MJ0101	1352	-	230	96405	signal recognition particle, 54 kDa subunit
MJ0102	551	-	97	97053	phenylacrylic acid decarboxylase
MJINTR002	34	-	412	97499	intron
tRNA_	108	-	-70	97537	Met (CAT)
tRNA_	84	+	92	97629	Leu (GAG)
MJ0103	1295	+	248	97961	molybdopterin cofactor synthesis protein
MJ0104	1958	+	17	99273	DNA helicase, putative
MJ0105	341	-	8	101580	hypothetical protein [PH0127]
MJ0106	701	-	260	102541	hypothetical protein [AF0525]
MJ0107	1562	+	192	102733	hypothetical protein, similar to dihydropteroate synthase [AF1414]
MJ0108	1328	+	22	104317	pyruvate kinase (EC 2.7.1.40)
MJ0109	755	-	19	106419	inositol monophosphatase (EC 3.1.3.25), L. Chen and M.F. Roberts, Appl. Environ. Microbiol. 64, 2609 (1998)
MJ0110	266	-	195	106880	
MJ0111	1187	+	78	106958	protein-export membrane protein secD [Escherichia coli]
SR1_79	27	-	179	108351	repeat

LR1_13	321	-	-27	108645	repeat
MJ0112	1463	-	49	110157	carbon monoxide dehydrogenase/acetyl-CoA synthase (EC 1.2.99.2), gamma subunit [Methanosarcina thermophila]
MJ0113	1184	-	174	111515	carbon monoxide dehydrogenase/acetyl-CoA synthase (EC 1.2.99.2), delta subunit [Methanosarcina thermophila]
tRNA	88	0	251	111766	Selenyl-Cys (TCA)
MJ0114	908	+	20	111874	
MJ0115	464	-	3	113249	hypothetical protein [MTH1014]
MJ0116	674	-	8	113931	hypothetical protein, 5.7% cys [MTH651]
MJ0116.5	179	-	45	114155	hypothetical protein, family with PAB7213/MTH1307/AF0526
MJ0117	788	-	10	114953	translation initiation factor, eIF-2, subunit alpha
MJ0118	479	+	262	115215	methyl coenzyme M reductase II operon, protein D
MJ0119	641	-	32	116367	hypothetical ferredoxin protein [MTH1671]
MJ0120	698	-	5	117070	ATP-binding putative nickel incorporation protein (UreG [E. coli])
MJ0121	782	-	0	117852	ABC transporter, ATP-binding protein
MJ0122	911	-	382	119145	translation initiation factor aIF-2B, subunit II
MJ0123	329	-	50	119524	hypothetical protein [MJ1213]
MJ0124	3476	-	-10	122990	type I site-specific deoxyribonuclease (EC 3.1.21.3), restriction subunit
MJ0125	347	-	41	123378	hypothetical protein, 20% acidic aa [MJ0127]
MJ0126	293	-	14	123685	hypothetical protein [MJ0128]
MJ0127	362	-	-13	124034	hypothetical protein, 21% acidic aa [MJ0125]
MJ0128	293	-	14	124341	hypothetical protein [MJ0126]
MJ0129	464	+	191	124532	hypothetical protein [MJ0554]
MJ0130	1274	-	186	126456	type I restriction enzyme (EcoR124/3 I) (EC 3.1.21.3), specificity subunit
MJ0131	308	-	19	126783	
MJ0132	659	-	30	127472	site-specific DNA-methyltransferase (adenine-specific) (EC 3.1.21.3) (hsdM)
MJ0132.5	1034	-	32	128538	site-specific DNA-methyltransferase (adenine-specific) (EC 3.1.21.3) (hsdM)
MJ0133	818	-	71	129427	some similarity with endonuclease IV
MJ0133.5	80	-	-4	129503	
MJ0134	845	-	66	130414	hypothetical methyltransferase protein [AF0216]
MJ0135	689	-	8	131111	ribonuclease HII (rnhB), M. Haruki et al., J. Bacteriol. 180, 6207 (1998)
MJ0136	1082	+	358	131469	hypothetical protein-[*Methanococcus vannielii*]
LR1_12	392	+	9	132560	repeat

SR1_78	27	+	-27	132925	repeat
SR1_77	27	+	38	132990	repeat
SR1_76	27	+	50	133067	repeat
SR1_75	27	+	40	133134	repeat
SR1_74	27	+	39	133200	repeat
SR1_73	27	+	43	133270	repeat
SR1_72	27	+	39	133336	repeat
SR1_71	27	+	41	133404	repeat
SR1_70	27	+	45	133476	repeat
SR1_69	27	+	41	133544	repeat
SR1_68	27	+	43	133614	repeat
SR1_67	27	+	46	133687	repeat
SR1_66	27	+	40	133754	repeat
SR1_65	27	+	41	133822	repeat
SR1_64	27	+	50	133899	repeat
MJ0137	842	-	193	134961	hypothetical membrane protein [MJ1495]
MJ0138	1445	-	160	136566	cobyric acid synthase related protein[MTH1497]
MJ0139	593	+	83	136649	putative potassium channel
MJ0138.5	998	+	4	137246	similar to potassium channel protein
tRNA	72	-	105	138421	Val (CAC)
IS_B1	363	-	53	138837	repeat
MJ0140	617	+	85	138922	similar to the N-terminus of sirohaem synthase [probable NADH-oxidoreductase]
MJ0141	296	+	-10	139529	hypothetical protein [MJ0126]
MJ0142	425	+	-13	139812	hypothetical protein [AF0298]
MJ0143	1175	+	-16	140221	glutamyl-tRNA reductase (EC 1.2.1.-) (hemA [E. coli])
MJ0144	716	-	3	142115	hypothetical protein [MTH1016]
MJ0145	803	-	73	142991	hypothetical protein [MTH1017]
MJ0146	206	-	212	143409	2-oxoglutarate ferredoxin oxidoreductase, delta subunit, 8.8 % cys
MJ0147	1100	-	292	144801	ATPase
MJ0148	998	-	22	145821	tRNA pseudouridine 55 synthase
MJ0149	173	-	9	146003	
MJ0150	503	+	81	146084	hypothetical protein [MJ1642]
MJ0151	443	+	8	146595	transcriptional regulatory protein [AsnC/Lrp Helix-Turn-Helix family], D. Charlier et al., Gene 201, 63 (1997)
MJ0152	2210	+	293	147331	C-terminus similar to carbon monoxide dehydrogenase/acetyl-CoA synthase (EC 1.2.99.2), beta subunit [*Methanosarcina thermoaceticum*]

MJ0153	2321	+	243	149784	carbon monoxide dehydrogenase/acetyl-CoA synthase (EC 1.2.99.2), alpha subunit [*Methanosarcina thermoaceticum*]
MJ0154	437	+	47	152152	carbon monoxide dehydrogenase/acetyl-CoA synthase (EC 1.2.99.2), epsilon subunit [*Methanosarcina thermophila*]
MJ0155	452	+	5	152594	iron-sulfur protein (4Fe-4S) (CooF [*Rhodospirillum rubrum*])
MJ0156	1406	+	26	153072	carbon monoxide dehydrogenase/acetyl-CoA synthase (EC 1.2.99.2), beta subunit [*Methanosarcina thermoaceticum*]
LSU_rRNA_1	2889	-	283	157650	
tRNA	7	2 -	199	157921	Ala (TGC)
SSU_rRNA_1	1475	-	66	159462	
MJ0157	278	+	345	159807	hypothetical protein, 22% acidic aa [MTH814]
MJ0158	1121	+	70	160155	hypothetical pyridoxal 5-phosphate binding protein [MTH1914]
MJ0159	1616	-	154	163046	hypothetical protein [MTH1569]
MJ0160	1403	-	50	164499	glutamyl-tRNA(Gln) amidotransferase, subunit B (GatB [B.subtilis])
MJ0161	503	-	160	165162	acetolactate synthase (EC 4.1.3.18), small subunit, T.L. Bowen et al., Gene 188, 77 (1997)
MJ0162	1262	+	375	165537	similar to Cleavage and Polyadenylation specificity factor (CPSF) /family with MJ0047/MJ1236
MJ0163	560	-	19	167378	
MJ0164	1184	-	52	168614	22% acidic aa
MJ0165	767	-	13	169394	similar to air-carboxylase
MJ0166	764	-	36	170194	hypothetical protein [AF0625]
MJ0167	488	-	5	170687	molybdenum cofactor biosynthesis moaB protein [Escherichia coli]
MJ0168	200	-	126	171013	archaeal histone
MJ0169	779	+	240	171253	septum site-determining protein (minD [Escherichia coli])
MJ0170	1043	-	144	173219	hypothetical protein [AF1546]
MJ0171	1718	-	89	175026	DNA ligase (ATP) (EC 6.5.1.1)
MJ0172	644	-	160	175830	protein-L-isoaspartate(D-aspartate) O-methyltransferase (EC 2.1.1.77) (L-isoaspartyl protein carboxyl methyltransferase)
MJ0173	410	+	101	175931	Tfx transcription regulator [MTH0916], family with MJ0529/MJ0272/MJ1641/MJ1243, A. Hochheimer et al., Mol. Microbiol.31,641 (1999)
MJ0174	1040	+	4	176345	cell division protein

MJ0175	614	-	90	178089	similar to Translin associated protein X (TRAX) [AF2260]
MJ0176	1004	+	334	178423	LSU ribosomal protein L3P
MJ0177	755	+	45	179472	LSU ribosomal protein L4P
MJ0178	257	+	37	180264	LSU ribosomal protein L23P
MJ0179	725	+	63	180584	LSU ribosomal protein L2P
MJ0180	455	+	89	181398	SSU ribosomal protein S19P
MJ0181	707	-	65	182625	
MJ0182	581	-	105	183311	
MJ0183	143	-	37	183491	
MJ0184	221	+	115	183606	
MJ0185	146	+	59	183886	
MJ0186	1205	-	189	185426	glutamate N-acetyltransferase (EC 2.3.1.35) (ornithine acetyltransferase) (ornithine transacetylase)
MJ0187	434	-	14	185874	hypothetical protein [MTH722]
MJ0188	794	-	6	186674	hypothetical protein with 2 CBS domains
tRNA	85	+	307	186981	Ser (GGA)
MJ0189	461	+	63	187129	SSU ribosomal protein S13P
MJ0190	560	+	61	187651	SSU ribosomal protein S4P
MJ0191	386	+	29	188240	SSU ribosomal protein S11P
MJ0192	572	+	15	188641	DNA-directed RNA polymerase (EC 2.7.7.6), subunit D (rpoD)
MJ0193	362	+	81	189294	LSU ribosomal protein L18E
MJ0194	410	+	35	189691	LSU ribosomal protein L13P
MJ0195	407	+	13	190114	SSU ribosomal protein S9P
MJ0196	218	+	61	190582	DNA-directed RNA polymerase (EC 2.7.7.6), subunit N
tRNA	73	+	35	190835	Pro (GGG)
MJ0197	170	+	36	190944	DNA-directed RNA polymerase (EC 2.7.7.6), subunit K (rpoK)
MJ0198	941	+	204	191318	enolase related protein (N-terminus)
MJ0199	185	+	126	192385	ferredoxin
MJ0200	245	-	52	192867	hydrogenase expression/formation protein C
MJ0201	479	-	140	193486	hypothetical protein (ybfM [Bacillus subtilis])
MJ0202	875	+	93	193579	hypothetical protein [MTH1744]
MJ0203	1049	+	124	194578	phosphoribosylformylglycinamidine cyclo-ligase (EC 6.3.3.1)
MJ0204	1412	+	103	195730	amidophosphoribosyltransferase (EC 2.4.2.14)
MJ0205	1043	-	268	198453	aspartate-semialdehyde dehydrogenase (EC 1.2.1.11)
MJ0206	404	-	14	198871	hypothetical protein [AF1068]

MJ0207	452	+	96	198967	
MJ0208	713	-	10	200142	hypothetical archaeal [4Fe-4S] cluster protein
MJ0209	785	-	17	200944	hypothetical protein [MTH435]
MJ0210	1091	+	99	201043	succinate--CoA ligase (ADP-forming) (EC 6.2.1.5), beta subunit
MJ0210.5	494	+	68	202202	hypothetical protein [MJ0803]
MJ0211	914	+	19	202715	UDP-glucose 4-epimerase (EC 5.1.3.2)
MJ0212	260	+	130	203759	DNA binding protein 10b (Sac10b [Sulfolobus acidocaldarius])
MJ0213	446	+	118	204137	hypothetical protein, 20% acidic aa [AF2318]
MJ0214	371	+	93	204676	hydrogenase accessory protein
MJ0215	446	-	143	205636	hypothetical protein tyrosine phosphatase [MJECL20]
MJ0216	1394	-	110	207140	ATP synthase (EC 3.6.1.34), subunit B (atpB [Sulfolobus acidocaldarius])
MJ0217	1760	-	50	208950	ATP synthase (EC 3.6.1.34), subunit A (atpA [Sulfolobus acidocaldarius])
MJ0218	293	-	64	209307	V-type sodium ATP synthase, subunit G (Na+-translocating ATPase, subunit G)
MJ0219	1196	-	14	210517	V-type sodium ATP synthase (EC 3.6.1.34), subunit C
MJ0220	608	-	107	211232	V-type ATP synthase, subunit E (epsilon chain)
MJ0221	659	-	69	211960	V-type sodium ATP synthase, subunit K, C. Ruppert et al., J. Bacteriol. 180, 3448 (1998)
MJ0222	2075	-	107	214142	ATP synthase (EC 3.6.1.34), subunit I
MJ0223	311	-	21	214474	similar to ATP synthase, subunit E
MJ0224	506	-	92	215072	hypothetical protein [AF0407]
tRNA	84	+	141	215213	Leu (TAA)
MJ0225	779	+	164	215461	hypothetical protein [PH0128]
MJ0226	554	+	178	216418	non-standard nucleotide triphosphatase [MJ0226]
MJ0226.4	578	-	-11	217539	
MJ0226.6	365	-	262	218166	hypothetical protein [MJ0803]
MJ0227	878	+	55	218221	hypothetical protein [MTH743]
MJ0228	1730	+	8	219107	glycine--tRNA ligase (EC 6.1.1.14) (glycyl-tRNA synthetase)
MJ0229	284	-	15	221136	transcriptional regulatory protein (ArsR Helix-Turn-Helix family)
MJ0230	242	-	8	221386	hypothetical protein [APE0597]
MJ0231	1247	-	161	222794	similar to TLDD protein (involved in the control of DNA gyrase)
MJ0232	1280	-	135	224209	enolase (EC 4.2.1.11) (2-phosphoglycerate dehydratase)

MJ0233	830	+	72	224281	hypothetical protein [MJ1189]
MJ0234	1007	+	0	225111	anthranilate phosphoribosyltransferase (EC 2.4.2.18)
MJ0235	245	+	6	226124	hypothetical protein with repeats similar to those of the ROPE protein
MJ0236	1250	+	20	226389	hydroxymethylpyrimidine kinase (EC 2.7.1.49)/phosphomethylpyrimidine kinase (EC 2.7.4.7) (thiD) (thiJ) bifunctional enzyme, T. Mizote et al., Microbiology 145, 495 (1999)
tRNA	73	-	71	227783	Ala (GGC)
MJ0237	1697	+	240	228023	arginine--tRNA ligase (EC 6.1.1.19) (arginyl-tRNA synthetase)
MJ0238	575	+	150	229870	anthranilate synthase (EC 4.1.3.27), component II (glutamine amidotransferase)
MJ0239	482	+	61	230506	
MJ0240	524	-	106	231618	adenylate cyclase (class IV) (AhaC2 [*Aeromonas hydrophila*]), O. Sismeiro et al., J. Bacteriol. 180, 3339 (1998)
MJ0241	434	-	10	232062	
MJ0242	104	-	71	232237	hypothetical protein [RPO00416]
MJ0243	245	-	81	232563	L-glutamyl-tRNA(Gln)-dependent amidotransferase, subunit C [MTH415], A.W. Curnow et al., Proc.Natl.Acad.Sci.USA 94, 11819 (1997)
MJ0244	866	-	22	233451	dihydrodipicolinate synthase (EC 4.2.1.52)
MJ0245	185	-	14	233650	SSU ribosomal protein S17BE
MJ0246	296	-	51	233997	chorismate mutase (EC 5.4.9.5), G. MacBeath et al., Biochemistry 37, 10062 (1998)
MJ0247	740	+	302	234299	proliferating-cell nuclear antigen
MJ0248	509	+	103	235142	
MJ0249	281	+	-19	235632	LSU ribosomal protein L44E
MJ0250	182	+	74	235987	SSU ribosomal protein S27E
LR1_11	394	+	32	236201	repeat
SR1_63	27	+	-27	236568	repeat
SR1_62	27	+	45	236640	repeat
SR1_61	27	+	39	236706	repeat
SR1_60	27	+	42	236775	repeat
SR1_59	27	+	41	236843	repeat
SR1_58	27	+	39	236909	repeat
SR1_57	27	+	54	236990	repeat
SR1_56	27	+	41	237058	repeat
SR1_55	27	+	41	237126	repeat
SR1_54	27	+	40	237193	repeat

SR1_53	27	+	40	237260	repeat
SR1_52	27	+	40	237327	repeat
SR1_51	27	+	40	237394	repeat
SR1_50	27	+	40	237461	repeat
SR1_49	27	+	40	237528	repeat
SR1_48	27	+	40	237595	repeat
SR1_47	27	+	111	237733	repeat
SR1_46	27	+	40	237800	repeat
SR1_45	27	+	42	237869	repeat
SR1_44	27	+	41	237937	repeat
SR1_43	27	+	50	238014	repeat
SR1_42	27	+	50	238091	repeat
MJ0251	440	-	170	238728	ferredoxin
MJ0252	638	+	121	238849	orotidine 5'-phosphate decarboxylase (EC 4.1.1.23)
MJ0253	461	+	8	239495	F420-reducing hydrogenase, delta subunit (membrane anchor protein) (hydD [Wolinella succinogenes])
MJ0254	635	+	13	239969	paralog of RAD51
MJ0255	752	-	3	241359	hypothetical protein [MTH1674]
MJ0255.5	506	+	156	241515	hypothetical protein [MTH1206]
MJ0256	563	+	17	242038	similar to C-terminus of fom2 phosphonopyruvate decarboxylase, family with MTH1207
MJ0257	932	+	163	242764	hypothetical protein [MTH1039]
MJ0258	1199	-	144	245039	hypothetical protein [MTH873]
MJ0259	521	-	73	245633	
MJ0260	608	-	84	246325	microsomal signal peptidase (SPC21 [Canis familiaris])
MJ0261	650	-	98	247073	hypothetical protein [MTH1897]
McjIF2	1637	+	261	247334	intein
MJ0262	3464	+ -	1727	247244	G-protein, member of the FUN12/bIF-2 family
MJ0263	959	-	19	251686	hypothetical protein with TPR domain-like repeats
MJ0264	431	-	242	252359	similar to pyruvate ferredoxin oxidoreductase, cys-rich subunit (EC 1.2.7.1)
MJ0265	491	-	72	252922	pyruvate ferredoxin oxidoreductase, cys-rich subunit (EC 1.2.7.1)
MJ0266	884	-	52	253858	pyruvate ferredoxin oxidoreductase, beta subunit (EC 1.2.7.1)
MJ0267	1157	-	40	255055	pyruvate ferredoxin oxidoreductase, alpha subunit (EC 1.2.7.1)

MJ0268	257	-	34	255346	pyruvate ferredoxin oxidoreductase, delta subunit (EC 1.2.7.1)
MJ0269	533	-	22	255901	pyruvate ferredoxin oxidoreductase, gamma subunit (EC 1.2.7.1)
MJ0270	233	-	287	256421	
MJ0271	539	+	481	256902	
MJ0272	197	+	11	257452	HTH DNA-binding protein /family with MJ0529/MJ1641/MJ0173/MJ1243
MJ0273	293	+	470	258119	transcriptional regulatory protein (AsnC family)
MJ0274	1559	-	407	260378	hypothetical protein [PH0094]
MJ0275	605	-	138	261121	
MJ0275.5	443	-	6	261570	
MJ0276	1097	-	64	262731	2-oxoglutarate ferredoxin oxidoreductase, alpha subunit
MJ0277	1772	-	123	264626	acetolactate synthase (EC 4.1.3.18), large subunit, T.L. Bowen et al., Gene 188, 77 (1997)
MJ0278	461	-	187	265274	peptidyl-prolyl cis-trans isomerase, FKBP-type (EC 5.2.1.8) (rotamase), M. Furutani et al., J. Bacteriol. 180, 388 (1998)
MJ0279	836	+	104	265378	4-hydroxybenzoate octaprenyltransferase related protein
MJ0280	383	+	161	266375	hypothetical protein, 26% acidic aa [MTH177]
MJ0281	530	-	3	267291	zinc metalloproteinase
MJ0282	410	+	86	267377	hypothetical protein [MTH413]
MJ0283	869	+	142	267929	nucleotide-binding protein
MJ0284	656	-	376	269830	LSU ribosomal protein L11P methyltransferase
MJ0285	440	-	72	270342	small heat-shock protein (HSPi6.5), R. Kim et al., Proc.Natl.Acad.Sci.USA 95, 9129 (1998)
MJ0286	350	-	157	270849	
MJ0287	290	-	21	271160	hypothetical protein [MJ0290]
MJ0288	533	-	62	271755	
MJ0289	956	-	46	272757	
MJ0290	632	-	364	273753	hypothetical protein [MJ0287]
MJ0291	1226	+	138	273891	cell division protein (FtsY [B. subtilis])
MJ0292	266	-	20	275403	energy converting hydrogenase, ehbC [MTH1249] A. Tersteegen and R. Hedderich, Eur. J. Biochem. 264, 930 (1999)
MJ0293	551	+	211	275614	thymidylate kinase (EC 2.7.4.9)
IS_B4	362	+	7	276172	repeat
MJ0294	2522	-	10	279066	ATP-dependent RNA helicase LHR (Large Helicase-Related protein)
MJ0295	680	-	-9	279737	formate dehydrogenase (EC 1.2.1.2), fdhD protein [Wolinella succinogenes]

MJ0296	593	+	30	279767	hypothetical protein [AF1748]
MJ0297	749	-	46	281155	
MJ0298	449	+	135	281290	protoporphyrinogen oxidase, (*hemG* [E. coli])
MJ0299	1157	-	22	282918	hypothetical protein [MTH1686]
MJ0300	887	+	362	283280	LysR family transcriptional regulator
MJ0301	875	-	53	285095	pteroate synthase, Xu et al., Nature Structural. Biol. 6, 750 (1999)
MJ0302	284	-	58	285437	phosphoribosyl-ATP pyrophosphohydrolase (EC 3.6.1.31)
MJ0303	413	-	121	285971	riboflavin synthase (EC 2.5.1.9), beta subunit (6,7-dimethyl-8-ribityllumazine synthase) (ribH) (risB)
MJ0304	476	-	57	286504	carbonic anhydrase (gamma family of Zn(II)-dependent enzymes)
MJ0305	1178	+	96	286600	similar to voltage-gated chloride channel protein
MJ0306	188	-	40	288006	
MJ0307	254	-	23	288283	protein disulfide oxidoreductase
MJ0308	698	-	103	289084	hypothetical protein [MTH1418]
MJ0309	851	-	200	290135	agmatinase (EC 3.5.3.11) (agmatine ureohydrolase), Perozich et al., Biochim. Biophys. Acta 1382, 23 (1998)
MJ0310	341	-	133	290609	hypothetical protein [MJ1340]
MJ0311	329	-	43	290981	hypothetical protein [MTH1836]
MJ0312	563	-	247	291791	
MJ0313	878	-	769	293438	spermidine synthase (EC 2.5.1.16)
MJ0314	788	+	143	293581	similar to rRNA intron-encoded homing endonuclease [PH0309, MJ0398], Nomura et al., J. Bacteriol. 180, 3635 (1998)
MJ0315	371	-	86	294826	S-adenosylmethionine decarboxylase (EC 4.1.1.50) (dcaM [Escherichia coli])
MJ0316	494	-	138	295458	hypothetical protein [AF2431]
MJ0317	671	-	275	296404	
MJ0318	902	-	134	297440	formylmethanofuran--tetrahydromethanopterin N-formyltransferase (EC 2.3.1.101)
MJ0319	290	+	172	297612	hypothetical protein [AF1549]
MJ0320	644	+	99	298001	GTP-binding protein
MJ0321	365	+	30	298675	
MJ0322	302	-	103	299445	SSU ribosomal protein S10P
MJ0324	1283	-	86	300814	translation elongation factor, EF-1 alpha
MJ0325	923	-	358	302095	elongation factor 1-alpha related protein
MJ0326	1301	+	156	302251	ATP-binding protein
MJ0327	302	+	73	303625	hypothetical protein [MJ0580]

tRNA	86	+	68	303995	Ser (TGA)
MJ0328	437	-	237	304755	
MJ0329	1847	-	5	306607	
MJ0330	1643	-	13	308263	
MJ0331	365	-	3	308631	
MJ0332	329	-	47	309007	
MJ0332.5	377	-	18	309402	
MJ0333	287	-	8	309697	
MJ0334	296	+	119	309816	21% acidic aa
MJ0335	740	+	67	310179	
MJ0336	356	+	13	310932	12.3% gly
MJ0337	782	+	14	311302	
MJ0338	302	+	16	312100	20% acidic aa
MJ0339	296	+	-4	312398	
MJ0340	698	+	6	312700	
MJ0341	359	+	13	313411	
MJ0342	368	+	148	313918	
MJ0343	2537	+	-16	314270	
MJ0344	539	+	13	316820	
MJ0345	686	+	219	317578	
MJ0346	302	+	13	318277	
MJ0347	452	+	14	318593	hypothetical protein, *E. coli yfdL*
MJ0347.5	548	+	21	319066	
MJ0348	2375	+	6	319620	
MJ0349	314	-	58	322367	hypothetical protein [MJ0023]
MJ0350	263	-	51	322681	23% basic aa
MJ0351	407	-	24	323112	
MJ0352	716	-	73	323901	
MJ0353	251	-	-10	324142	
MJ0354	173	-	-19	324296	
MJ0355	326	-	78	324700	
MJ0356	290	-	-3	324987	
MJ0356.5	224	+	314	325301	
MJ0357	464	-	418	326407	component of prot secretion

MJ0358	383	-	6	326796	
MJ0359	623	-	12	327431	
MJ0360	356	-	339	328126	12.1% gly
MJ0361	320	-	1056	329502	similar to arsenical resistance operon repressor
MJ0362	188	+	157	329659	hypothetical protein [AF2091]
MJ0363	2276	+	18	329865	DNA replication initiation protein (MCM family)
MJ0364	332	+	22	332163	
MJ0365	527	+	8	332503	similar to 3' to 5' exonuclease, Moser et al., Nucl. Acids Res. 25, 5110 (1997)
MJ0366	257	+	21	333051	hypothetical protein-[Methanococcus maripaludis]
MJ0367	989	+	-16	333292	lambdoid integrase/recombinase (xerC [E. coli])
MJ0368	305	+	300	334581	22% acidic aa
MJ0369	1106	-	48	336040	type VI topoisomerase (subunit A) (TOPVIA [Sulfolobus shibatae])
MJ0370	1091	+	274	336314	cell division ftsZ protein [Escherichia coli]
MJ0371	221	+	13	337418	protein transport protein (SecE/SEC61-gamma)
MJ0372	440	+	159	337798	transcription elongation factor (SPT5/NusG family)
MJ0373	482	+	101	338339	LSU ribosomal protein L11P
MJ0374	971	-	63	339855	hypothetical protein [fusion of AQ2066 and AQ1706]
MJ0375	749	+	77	339932	hypothetical protein [AF0072]
MJ0376	2183	+	367	341048	hypothetical protein region (ygcB [E. coli])
MJ0377	509	+	12	343243	hypothetical protein [AF2436]
MJ0378	965	+	169	343921	hypothetical protein, 22% basic aa [AF2435]
MJ0379	611	-	3	345500	transcriptional regulatory protein [AsnC/Lrp Helix-Turn-Helix family]
MJ0380	329	+	145	345645	
MJ0381	959	+	3	345977	hypothetical protein [APE1235]
MJ0382	728	+	19	346955	
MJ0383	1841	+	-6	347677	ATP-binding-probable helicase
MJ0384	725	+	16	349534	hypothetical protein, HD domain, L. Aravind and E.V. Koonin, Trends Biochem. Sci. 23, 469 (1998)
MJ0385	1052	+	-7	350252	
MJ0386	260	-	3	351567	hypothetical protein [AF2434]
SR3_12	28	-	131	351726	repeat
SR3_10	28	-	40	351794	repeat

SR3_11	28	-	41	351863	repeat
SR3_9	28	-	38	351929	repeat
SR3_8	28	-	42	351999	repeat
SR3_7	28	-	38	352065	repeat
SR3_6	28	-	39	352132	repeat
SR3_5	28	-	41	352201	repeat
SR3_4	28	-	39	352268	repeat
SR3_3	28	-	43	352339	repeat
SR3_2	28	-	38	352405	repeat
SR3_1	28	-	38	352471	repeat
MJ0387	296	-	378	353145	DNA-directed RNA polymerase (EC 2.7.7.6), subunit L (rpoL)
MJ0388	620	-	-28	353737	hypothetical protein, *Methanococcus vannielii*
MJ0389	917	+	91	353828	tyrosine--tRNA ligase (EC 6.1.1.1) (tyrosyl-tRNA synthetase)
MJ0390	389	-	15	355149	
MJ0391	539	-	10	355698	precorrin-8W decarboxylase (methyltransferase) (cbiT [Salmonella typhimurium])
MJ0392	1016	+	91	355789	hypothetical protein with zinc-binding domain of metallopeptidase [Synechocystis sp.] and 2 CBS domains
MJ0393	179	-	8	356992	SSU ribosomal protein S27AE
MJ0394	302	-	3	357297	SSU ribosomal protein S24E
MJ0395	473	-	17	357787	hypothetical protein [AF1115]
MJ0396	176	-	14	357977	DNA-directed RNA polymerase (EC 2.7.7.6), subunit E"
MJ0397	560	-	13	358550	DNA-directed RNA polymerase (EC 2.7.7.6), subunit E' (rpoE')
tRNA	74	+	221	358771	Arg (TCT)
tRNA	7	4 +	27	358872	_Glu (TTC) 2
MJ0398	812	+	165	359111	similar to rRNA intron-encoded homing endonuclease [PH0309, MJ0314]
tRNA_	75	-	52	360050	Meti (CAT)
MJ0399	1331	-	107	361488	phosphomannomutase (EC 5.4.2.8) (PMM)
MJ0400	818	+	105	361593	hypothetical aldolase protein (YneB [*E. coli*])
MJ0401	197	-	109	362717	
MJ0402	317	-	12	363046	13.8% gly, 22% basic aa
MJ0403	860	+	179	363225	hypothetical protein [Schizosaccharomyces pombe PID:g1184017]

MJ0404	449	-	270	364804	hypothetical protein, 7.8% pro [MTH1868]
MJ0405	383	-	198	365385	hypothetical protein [MTH1867]
MJ0406	893	+	176	365561	ribokinase (EC 2.7.1.15)
MJ0407	914	+	28	366482	quinolinate synthetase, subunit A
MJ0408	362	+	122	367518	hypothetical protein [AF2043]
MJ0409	2108	+	66	367946	
MJ0410	791	+	20	370074	cell division inhibitor /family with MJ0547/MJ0169/MJ0579
MJ0411	815	+	45	370910	imidazoleglycerol-phosphate synthase (cyclase) (HisF [E. coli])
MJ0412	764	-	28	372517	nitrate transport ATP-binding protein
MJ0413	797	-	18	373332	nitrate transport permease protein
MJ0414	1169	-	87	374588	hypothetical protein [AF0849]
MJ0415	428	+	181	374769	
MJ0416	569	+	25	375222	zinc-finger protein, 6.1% cys
MJ0417	710	-	9	376510	hypothetical protein [AF0505]
MJ0418	770	+	108	376618	hypothetical protein [PH0087]
MJ0419	1037	+	5	377393	hypothetical protein [PAB0054]
MJ0420	1058	+	45	378475	hypothetical protein, 22% basic aa [RSA00746]
MJ0421	1079	+	107	379640	hypothetical protein [AF2226]
MJ0422	821	-	201	381741	dihydrodipicolinate reductase (EC 1.3.1.26)
MJ0423	842	+	129	381870	pseudogene including MJ0423-MJ0425, similar to C-terminus of MJECL04 (aa 300-end)
MJ0426	539	-	-8	383243	
MJ0427	467	-	0	383710	28% basic aa
MJ0428	1280	-	-4	384986	UDP-N-acetyl-D-mannosaminuronic acid dehydrogenase (EC 1.1.1.-) (UDP-mannaca dehydrogenase
MJ0429	1184	+	168	385154	argininosuccinate synthase (EC 6.3.4.5)
MJ0430	611	-	14	386963	deoxycytidine triphosphate deaminase (EC 3.5.4.13) (dCTP deaminase)
MJ0431	215	-	172	387350	hypothetical protein [MJ1400]
MJ0432	275	-	502	388127	hypothetical helix-turn-helix transcriptional regulatory protein (c02008 [Sulfolobus solfataricus])
MJ0433	524	-	12	388663	
MJ0434	665	-	14	389342	hypothetical protein, 20% acidic aa [MJ0125]
MJ0435	287	-	0	389629	hypothetical protein [MJ0126]
MJ0436	1649	-	0	391278	paralog of queuine tRNA ribosyltransferase (EC 2.4.2.29); may be archaeosine tRNA ribosyltransferase.

MJ0436.5	179	-	138	391595	
MJ0437	236	-	72	391903	energy converting hydrogenase, *ehbD* [MTH1248] A. Tersteegen and R. Hedderich, Eur. J. Biochem. 264, 930 (1999)
MJ0438	1142	+	188	392091	hypothetical methyltransferase protein [AF2178]
MJ0439	1046	-	1	394280	ATPase
MJ0440	800	+	212	394492	hypothetical protein [Pyrococcus woesei]
MJ0441	788	+	156	395448	hypothetical transmembrane protein
MJ0442	662	-	97	396995	ATP-binding putative nickel incorporation protein (HypB [*Rhizobium leguminosarum*])
MJ0443	668	-	18	397681	RNA-binding protein (Rev interacting protein [*Homo sapiens*])
MJ0444	851	-	59	398591	hypothetical protein kinase
MJ0445	305	-	287	399183	translation initiation factor, eIF-1A
MJ0446	1082	+	312	399495	hypothetical protein similar to biotin synthase
MJ0447	482	-	-22	401037	
MJ0448	767	+	131	401168	metallo-beta-lactamase superfamily protein
MJ0449	848	+	13	401948	hypothetical membrane transporter protein (yeaB [B. subtilis])
tRNA	73	-	178	403047	Val (GAC)
MJ0450	557	+	230	403277	hypothetical protein with 2 CBS domains
MJ0451	665	+	17	403851	N-(5'-phosphoribosyl)anthranilate isomerase (EC 5.3.1.24)
MJ0452	440	-	3	404959	hypothetical protein [MTH777]
MJ0453	320	-	8	405287	hypothetical protein [MTH776]
MJ0454	986	-	9	406282	translation initiation factor aIF-2B, subunit I
MJ0455	578	-	3	406863	5.5% cys
MJ0456	923	+	157	407020	membrane protein possibly involved in tyrosine transport
MJ0457	1229	+	13	407956	succinyl-diaminopimelate desuccinylase (EC 3.5.1.18)
MJ0458	647	+	11	409196	hypothetical protein with similarity to kinases
MJ0458.5	170	+	53	409896	zinc finger protein ZNF141 [Homo sapiens]
MJ0459	266	+	22	410088	translation elongation factor EF-1, subunit beta
MJ0460	461	+	457	410811	LSU ribosomal protein L22P
MJ0461	623	+	29	411301	SSU ribosomal protein S3P
MJ0462	209	+	16	411940	LSU ribosomal protein L29P
MJ0463	305	+	88	412237	belongs to the SUI1 family of translation factors
MJ0464	284	+	100	412642	ribonuclease P, protein subunit p29 (Rpp29 [Homo sapiens])
MJ0465	350	+	202	413128	SSU ribosomal protein S17P

MJ0466	395	+	36	413514	LSU ribosomal protein L14P
MJ0467	359	+	31	413940	LSU ribosomal protein L24P
MJ0468	731	+	13	414312	SSU ribosomal protein S4E
MJ0469	569	+	26	415069	LSU ribosomal protein L5P
MJ0469.5	158	+	17	415655	SSU ribosomal protein S14P
MJ0470	389	+	46	415859	SSU ribosomal protein S8P
MJ0471	545	+	42	416290	LSU ribosomal protein L6P
MJ0472	437	+	26	416861	LSU ribosomal protein L32E
MJ0473	443	+	61	417359	LSU ribosomal protein L19E
MJ0474	584	+	36	417838	LSU ribosomal protein L18P
MJ0475	650	+	14	418436	SSU ribosomal protein S5P
MJ0476	461	+	69	419155	LSU ribosomal protein L30P
MJ0477	428	+	15	419631	LSU ribosomal protein L15P
MJ0478	1319	+	305	420364	protein transport protein (SecY/SEC61-alpha)
MJ0479	575	+	155	421838	adenylate kinase (EC 2.7.4.3), Ferber et al., Gene 185, 239 (1997)
MJ0480	593	+	57	422470	hypothetical protein [AF1901]
MJ0482	620	-	11	423694	hypothetical protein [AF1062]
MJ0481	293	+	98	423792	zinc finger protein, 9.9% cys
MJ0483	968	-	21	425074	hypothetical protein [PAB0746]
MJ0484	1460	+	505	425579	cobyric acid synthase
MJ0485	1010	+	55	427094	hypothetical protein [family with MJ1157,MJ1478]
LR1_10	423	+	76	428180	repeat
SR1_41	27	+	-27	428576	repeat
SR1_40	27	+	44	428647	repeat
SR1_39	27	+	42	428716	repeat
SR1_38	27	+	43	428786	repeat
SR1_37	27	+	39	428852	repeat
SR1_36	27	+	41	428920	repeat
SR1_35	27	+	48	428995	repeat
SR1_34	27	+	40	429062	repeat
SR1_33	27	+	39	429128	repeat
SR1_32	27	+	49	429204	repeat
SR1_31	27	+	46	429277	repeat
SR1_30	27	+	41	429345	repeat
SR1_29	27	+	41	429413	repeat
SR1_28	27	+	39	429479	repeat

SR1_27	27	+	41	429547	repeat
MJ0486	962	-	108	430644	hypothetical protein [AQ632]
MJ0487	1439	+	286	430930	phenylalanine--tRNA ligase (EC 6.1.1.20), beta chain (phenylalanyl-tRNA synthetase) (eukaryotic naming)
MJ0488	464	+	23	432392	hypothetical protein [MTH1144]
MJ0489	788	+	33	432889	hypothetical protein [MTH1145]
MJ0490	938	+	51	433728	malate dehydrogenase (NAD(P)+ dependent) (EC 1.1.1.82) (mdhII [*M. thermoautotrophicum*]), H. Thompson et al., Arch. Microbiol. 170, 38 (1998)
MJ0491	425	+	17	434683	hypothetical protein [MTH1197]
MJ0492	284	-	-5	435387	
MJ0493	848	-	137	436372	nicotinate-nucleotide pyrophosphorylase (carboxylating) (EC 2.4.2.19) (quinolinate phosphoribosyltransferase)
MJ0494	392	+	129	436501	hypothetical protein [AF0489]
MJ0495	1388	+	38	436931	translation elongation factor SelB
MJ0496	419	+	87	438406	
MJ0497	398	-	-2	439221	Holliday junction resolvase (Hjc [Pyrococcus furiosus])
MJ0498	464	-	-7	439678	hypothetical protein [MTH1866]
MJ0499	1271	-	155	441104	3-isopropylmalate dehydratase (EC 4.2.1.33), isomerase subunit
MJ0500	767	-	435	442306	hypothetical protein [MTH925]
MJ0501	596	-	90	442992	hypothetical protein [MTH1635]
MJ0502	1280	+	176	443168	3-phosphoshikimate 1-carboxyvinyltransferase (EC 2.5.1.19) (5-enolpyruvylshikimate-3-phosphate synthase)
MJ0503	1193	+	112	444560	2-oxosuberate (alpha-ketosuberate) synthase (AksA), D.M. Howell et al., Biochemistry 37, 10108 (1998)
MJ0504	587	+	34	445787	hypothetical protein [AF0103]
MJ0505	749	+	-4	446370	hypothetical protein [PH1993]
MJ0506	587	-	7	447713	amidotransferase hisH (EC 2.4.2.-)
MJ0507	539	-	19	448271	transcription initiation TATA-binding protein (TBP)
MJ0508	305	-	155	448731	LSU ribosomal protein L12A; homolog (or analog) of bacterial L7/12P
MJ0509	1013	-	130	449874	LSU ribosomal protein L10E
MJ0510	656	-	29	450559	LSU ribosomal protein L1P
MJ0511	665	-	1729	452953	deoxyuridylate hydroxymethyltransferase (EC 2.1.1.-) (deoxyuridylate hydroxymethylase)

MJ0512	701	-	341	453995	
MJ0513	719	-	156	454870	
MJ0514	749	-	270	455889	polyferredoxin [MTH405]
MJ0514.4	488	-	47	456424	energy converting hydrogenase, partial *ehaQ* [MTH401]A. Tersteegen and R. Hedderich, Eur. J. Biochem. 264, 930 (1999)
MJ0514.6	1223	-	-19	457628	energy converting hydrogenase, *ehaP* [MTH399]
MJ0515	1127	-	0	458755	energy converting hydrogenase, *ehaO* [MTH398]
MJ0516	443	-	8	459206	energy converting hydrogenase, *ehaN* [MTH397]
MJ0517	410	-	117	459733	energy converting hydrogenase, *ehaM* [MTH396]
MJ0518	281	-	6	460020	energy converting hydrogenase, *ehaL* [MTH395]
MJ0519	242	-	15	460277	energy converting hydrogenase, *ehaK* [MTH394]
MJ0520	875	-	194	461346	energy converting hydrogenase, *ehaJ* [MTH393]
MJ0521	206	-	205	461757	energy converting hydrogenase, *ehaI* [MTH392]
MJ0522	653	-	14	462424	energy converting hydrogenase, *ehaH* [MTH391]
MJ0523	692	-	112	463228	energy converting hydrogenase, *ehaG* [MTH390]
MJ0524	458	-	13	463699	energy converting hydrogenase, *ehaF* [MTH389]
MJ0525	158	-	142	463999	energy converting hydrogenase, *ehaE* [MTH388]
MJ0526	275	-	126	464400	energy converting hydrogenase, *ehaD* [MTH387]
MJ0526.5	245	-	12	464657	energy converting hydrogenase, *ehaC* [MTH386]
MJ0527	491	-	0	465148	energy converting hydrogenase, *ehaB* [MTH385]
MJ0528	260	-	3	465411	energy converting hydrogenase, *ehaA* [MTH384]
MJ0529	635	-	111	466157	HTH DNA-binding protein /family with MJ1641/MJ0272/MJ0173/MJ1243
MJ0530	593	+	193	466350	thioredoxin-2 (trx2)
MJ0531	437	+	224	467167	factor-dependent ATPase [MJ0577]
MJ0532	1172	-	9	468785	geranylgeranyl hydrogenase (bchP [Rhodobacter capsulatus]) related protein
MJ0533	167	-	3	468955	ferredoxin 2[4Fe-4S] homolog
MJ0534	1160	+	176	469131	FMN-containing flavoprotein [MTH1350]
MJ0535	1016	-	41	471348	acetylpolyamine aminohydolase (EC 3.5.1.-)
MJ0536	539	-	1035	472922	2-oxoglutarate ferredoxin oxidoreductase, gamma subunit

MJ0537	809	-	17	473748	2-oxoglutarate ferredoxin oxidoreductase, beta subunit
MJ0538	779	-	280	474807	hypothetical protein [PH1392] 21% acidic aa
MJ0539	1589	-	28	476424	lysine--tRNA ligase (Class I) (EC 6.1.1.6), M. Ibba et al., Science 278, 1119 (1997)
MJ0540	254	-	271	476949	21% acidic aa
MJ0541	503	-	24	477476	nicotinamide mononucleotide (NMN) adenylyltransferase (EC 2.7.7.1), N. Raffaelli et al., J. Bacteriol. 179, 7718 (1997)
McjPEPsyn	1235	-	1191	479902	intein
MJ0542	3530	- -	2333	481099	pyruvate, water dikinase (EC 2.7.9.2) (phosphoenolpyruvate synthase) (PEP synthase)
MJ0543	521	-	164	481784	LSU ribosomal protein L16P/L10E (QM protein) (GRC5 [Saccharomyces cerevisiae])
MJ0544	686	-	136	482606	dolichyl-phosphate beta-D-mannosyltransferase (EC 2.4.1.83)
MJ0545	740	-	107	483453	
MJ0546	167	-	5	483625	hypothetical protein [AF0697]
MJ0547	779	-	6	484410	cell division inhibitor (septum site-determining protein) (minD [Escherichia
MJ0548	878	-	181	485469	hypothetical protein [MTH752]
MJ0549	422	+	112	485581	hypothetical protein [Helicobacter pylori]
MJ0550	986	-	11	487000	hypothetical protein [MTH1751]
MJ0551	812	-	108	487920	similar to nitrite or sulfite oxidoreductase subunit
MJ0552	746	-	26	488692	cobalamin biosynthesis protein cbiJ [Salmonella typhimurium]
MJ0553	458	-	235	489385	similar to acylphosphatase
MJ0554	455	-	527	490367	hypothetical protein [MJ0129]
MJ0554.5	104	+	69	490436	
MJ0555	1049	+	245	490785	aminopeptidase
MJ0556	554	-	43	492431	hypothetical protein with 2 CBS domains [MJ0100, MJ0188]
MJ0557	614	-	143	493188	hypothetical protein [PAB1852], T.I. Zarembinski et al., Proc.Natl.Acad.Sci.USA 95, 15189 (1998)
MJ0558	782	-	16	493986	hypothetical archaeal protein [AF0998]
MJ0559	797	+	90	494076	survival protein surE [Escherichia coli]
MJ0560	404	-	20	495297	21% acidic aa
MJ0561	1016	+	309	495606	adenylosuccinate synthetase (EC 6.3.4.4)
MJ0562	212	-	71	496905	
MJ0563	929	-	149	497983	DNA-methyltransferase (C-5 cytosine-specific) (EC 2.1.1.73)

MJ0564	2675	-	231	500889	alanine--tRNA ligase (EC 6.1.1.7) (alanyl-tRNA synthetase)
LR1_9	426	+	208	501097	repeat
SR2_48	28	+	-28	501495	repeat
SR2_47	28	+	39	501562	repeat
SR2_46	28	+	44	501634	repeat
SR2_45	28	+	38	501700	repeat
SR2_44	28	+	31	501759	repeat
SR2_43	28	+	39	501826	repeat
SR2_42	28	+	40	501894	repeat
SR2_41	28	+	42	501964	repeat
MJ0565	440	-	56	502488	
MJ0566	2003	-	18	504509	ferrous iron transporter (GTP-binding) (FeoB [E. coli)]
MJ0567	245	-	-10	504744	similar to iron-binding transcriptional repressor, SH3-domain subunit
MJ0568	374	+	105	504849	similar to C-terminus of iron-binding transcriptional repressor protein (dtxR [Corynebacterium diphtheria])
MJ0569	875	-	3	506101	porphobilinogen deaminase (PBG) (EC 4.3.1.8) (Hem3 [Bacillus subtilis])
MJ0570	668	-	13	506782	hypothetical protein [PH1257]
LR1_8	425	+	88	506870	repeat
SR1_26	27	+	-27	507268	repeat
SR1_25	27	+	51	507346	repeat
SR1_24	27	+	38	507411	repeat
SR1_23	27	+	41	507479	repeat
SR1_22	27	+	35	507541	repeat
SR1_21	27	+	42	507610	repeat
SR1_20	27	+	38	507675	repeat
SR1_19	27	+	46	507748	repeat
SR1_18	27	+	45	507820	repeat
SR1_17	27	+	41	507888	repeat
SR1_16	27	+	41	507956	repeat
MJ0571	1418	+	418	508401	aspartokinase (EC 2.7.2.4) (aspartate kinase)
MJ0572	257	+	43	509862	hypothetical protein with TPR domain-like repeats
MJ0573	566	+	145	510264	
MJ0574	278	+	37	510867	
MJ0575	686	+	-22	511123	hypothetical protein [MTH1356]
MJ0576	1040	+	115	511924	malic acid transport protein
MJ0577	476	+	20	512984	factor-dependent ATPase

MJ0578	806	-	3	514269	ferredoxin
MJ0579	794	-	6	515069	cell division inhibitor (septum site-determining protein) (minD [Escherichia
MJ0580	326	-	8	515403	hypothetical protein [MTH1175]
MJ0581	245	+	291	515694	hypothetical protein [Anabaena sp. gi|1613872]
MJ0582	383	+	3	515942	hypothetical protein, 7.9% cys [AF1756]
MJ0583	170	+	70	516395	zinc finger protein ZNF134 [Homo sapiens]
MJ0584	1088	+	6	516571	hypothetical protein [TM0997]
MJ0585	614	+	23	517682	
MJ0585.5	167	-	11	518474	
MJ0586	494	+	91	518565	hypothetical archaeal protein [MTH729/AF1977/PAB0842]
MJ0587	458	-	479	519996	hypothetical protein [MJ0129]
MJ0588	1289	-	124	521409	hypothetical protein [PH1538]
MJ0589	305	+	56	521465	
MJ0589.5	140	-	9	521919	
MJ0590	2111	+	111	522030	acetyl-coenzyme A synthetase (ADP-forming), alpha and beta subunits, M. Musfeldt et al., J. Bacteriol. 181, 5885 (1999)
MJ0591	782	+	579	524720	proteasome (EC 3.4.99.46), alpha subunit
MJ0592	701	+	156	525658	hypothetical protein [AF0491/PAB0418/MTH685]
MJ0593	275	+	46	526405	LSU ribosomal protein L37AE
MJ0593.5	137	+	66	526746	DNA-directed RNA polymerase (EC 2.7.7.6), subunit P (rpoP), T.J. Darcy et al., J. Bacteriol. 181, 4424 (1999)
MJ0594	506	+	5	526888	hypothetical protein, 21% basic aa [MTH680]
MJ0595	227	+	269	527663	LSU ribosomal protein LXA (aka LX) (archae specific)
MJ0596	422	+	165	528055	hypothetical protein [PH0063]
MJ0597	1049	+	88	528565	hypothetical protein [fusion of MTH1279/MTH1278]
MJ0598	866	+	23	529637	site-specific DNA methyltransferase (N6-adenine) (EC 2.1.1.72) (MjaIII)
MJ0599	596	+	23	530526	
MJ0600	869	+	57	531179	type II DNA restriction enzyme (EC 3.1.21.4) (MjaIII)
MJ0601	788	+	127	532175	thiazole biosynthetic enzyme (THI4 [Saccharomyces cerevisiae])
MJ0602	782	-	9	533754	21% basic aa
MJ0603	1277	-	43	535074	glutamate-1-semialdehyde 2,1-aminomutase (EC 5.4.3.8) [MTH228, AF1241]
MJ0604	299	-	72	535445	hypothetical protein [RPH01192]

MJ0605	392	-	0	535837	hypothetical protein [AF0298]
MJ0606	272	-	87	536196	hypothetical protein [MTH1908]
MJ0607	236	-	5	536437	
MJ0607.5	125	-	23	536585	
MJ0608	920	-	28	537533	inorganic pyrophosphatase (EC 3.6.1.1), T.W. Young et al., Microbiol. 144, 2563 (1998); T. Shintani et al. FEBS Lett. 439, 263 (1998)
MJ0609	1304	-	243	539080	amino acid transporter
MJ0610	1301	-	15	540396	hypothetical protein, *H. sapiens* tRNP48 gene
MJ0610.5	404	+	290	540686	hypothetical protein [AF1011]
MJ0611	572	+	22	541112	membrane protein
MJ0611.5	512	+	9	541693	
MJ0612	1337	+	98	542303	prephenate dehydrogenase (EC 1.3.1.12) (PDH)
MJ0613	1031	-	299	544970	deoxyribonuclease (pyrimidine dimer) (EC 3.1.25.1) (exonuclease III)
MJ0614	353	-	93	545416	hypothetical protein [MTH952]
MJ0615	647	-	284	546347	ATP synthase (EC 3.6.1.34), subunit D (atpD [Sulfolobus acidocaldarius])
MJ0616	470	-	75	546892	phosphoribosylaminoimidazole carboxylase (EC 4.1.1.21)
MJ0617	584	+	109	547001	fumarate hydratase (EC 4.2.1.2), beta subunit (Class I)
MJ0618	293	-	1	547879	hypothetical protein [AF2072]
MJ0619	1517	-	-16	549380	similar to NARA protein
MJ0620	872	+	312	549692	ribosomal protein S6 modification protein
MJ0621	515	-	11	551090	hypothetical protein [AF0643]
MJ0622	1208	-	105	552403	cell division ftsZ protein [Escherichia coli]
7S_RNA	478	+	15	552418	7S RNA
MJ0623	476	+	-8	552888	hypothetical protein [MTH1774]
MJ0624	173	+	18	553382	ferredoxin 2[4Fe=4S]
MJ0625	1007	+	53	553608	ATPase
MJ0626	605	+	117	554732	hypothetical protein [MTH1020]
MJ0627	350	+	34	555371	
MJ0628	488	+	-4	555717	hypothetical protein in the TPR family; see MJ0572
MJ0629	626	+	20	556225	hypothetical protein, 23% acidic aa [AF2149]
MJ0630	890	+	24	556875	sodium-dependent phosphate transporter
MJ0631	506	+	3	557768	hydrogenase membrane anchor protein (hydD [Wolinella succinogenes])
MJ0632	1088	+	20	558294	ATPase (DNAA)
MJ0633	2816	+	150	559532	leucine--tRNA ligase (EC 6.1.1.4) (leucyl-tRNA synthetase)

MJ0634	1859	+	360	562708	hypothetical protein [DRA0027]
MJ0635	839	+	232	564799	
MJ0636	1172	+	261	565899	dihydrolipoamide dehydrogenase (EC 1.8.1.4)
MJ0637	815	-	28	567914	prephenate dehydratase (EC 4.2.1.51)
MJ0638	674	-	0	568588	hypothetical protein [MJ1252]
MJ0639	284	-	0	568872	
MJ0640	971	-	33	569876	hypothetical protein [MTH1864]
MJ0641	1250	-	162	571288	3-phosphoglycerate kinase (EC 2.7.2.3)
MJ0642	845	+	320	571608	hypothetical protein [MJ0872]
MJ0643	977	+	23	572476	porphobilinogen synthetase (EC 4.2.1.24) (delta-aminolevulinic acid dehydratase)
MJ0644	623	+	64	573517	S-adenosylmethionine:2-demethylmenaquinone methyltransferase related protein (MenG [E. coli])
MJ0645	572	+	33	574173	hypothetical protein [AF1294]
MJ0646	491	+	14	574759	hypothetical protein [MTH812]
MJ0647	212	-	48	575510	
MJ0648	440	-	-67	575883	hypothetical protein, 20% acidic aa [gi|2983340 Aquifex]
MJ0649	1343	+	218	576101	NAD(P)H:rubredoxin oxidoreductase (NROR)
MJ0650	2000	+	79	577523	
MJ0651	932	-	-2	580453	protease IV
MJ0652	389	-	20	580862	hypothetical protein (YqeI [Bacillus subtilis])
IS_B8	362	+	119	580981	repeat
MJ0653	581	+	80	581423	hypothetical protein with 2 CBS domains
MJ0654	905	-	75	582984	dihydroorotate dehydrogenase (EC 1.3.3.1)
MJ0655	266	+	182	583166	LSU ribosomal protein L34E
MJ0656	533	+	92	583524	cytidylate kinase (EC 2.7.4.14)
MJ0657	239	+	90	584147	LSU ribosomal protein L14E
MJ0658	953	+	138	584524	tungsten formylmethanofuran dehydrogenase (EC 1.2.99.5), subunit C related protein
MJ0659	413	+	151	585628	hypothetical protein in the TPR family; see MJ0572
MJ0660	230	-	97	586368	hypothetical archaeal protein [MTH1263/AF0280]
MJ0661	518	-	130	587016	hypothetical protein [PH1693]
MJ0662	650	-	-7	587659	
MJ0663	1481	+	94	587753	TPP-binding protein related to acetolactate synthase, T.L. Bowen et al., Gene 188, 77 (1997)
MJ0664	872	+	59	589293	Zn finger protein
MJ0665	1022	-	17	591204	hypothetical protein [MTH781]

MJ0666	1193	-	216	592613	molybdopterin biosynthesis protein (MoeA [E. coli])
MJ0667	1508	+	92	592705	thymidine phosphorylase (EC 2.4.2.4)
MJ0668	218	-	103	594534	hypothetical protein [MJ1390]
MJ0669	1085	-	192	595811	putative ATP-dependent RNA helicase, eIF-4A family
MJ0670	1067	-	78	596956	hypothetical protein [MTH175]
MJ0671	659	-	11	597626	riboflavin-specific deaminase (EC 3.5.4.-) (ribD) (ribG)
MJ0672	1295	-	72	598993	Na+ transporter
MJ0673	386	+	264	599257	SSU ribosomal protein S8E
MJ0674	1013	+	84	599727	pyruvate formate lyase activating apoprotein (EC 1.97.1.4), related
MJ0675	1172	-	15	601927	hypothetical protein [MJ1273]
MJ0676	1004	-	160	603091	hydrogenase expression/formation protein
MJ0677	989	-	112	604192	pyridoxone biosynthesis protein (SOR1 [Cercospora nicotianae])
MJ0678	959	-	73	605224	hypothetical protein [AF0785/MJ1221]
MJ0679	947	-	66	606237	transketolase (EC 2.2.1.1), carboxy-terminal half
MJ0680	701	-	43	606981	pentose-5-phosphate-3-epimerase (EC 5.1.3.1)
MJ0681	821	-	27	607829	transketolase (EC 2.2.1.1), amino-terminal half
MJ0682	2903	-	160	610892	hypothetical protein (rtcB [E. coli])
McjEc_hyp	1463	- -	1754	610601	intein
MJ0683	776	-	321	611698	hypothetical protein [AF1589/TM0117]
MJ0684	1109	-	33	612840	aminotransferase (subgroup I), [MJ0001, MTH1894, AF2366, AF2129, AF1623]
MJ0685	758	+	69	612909	nickel insertion enzyme (CooC [Rhodospirillum rubrum])
MJ0686	1739	-	3	615409	hypothetical protein (SC4G2.12c [Streptomyces coelicolor])
MJ0687	1004	-	71	616484	hypothetical protein [AF1312]
MJ0688	410	+	218	616702	hypothetical protein [PH0209]
MJ0689	152	-	101	617365	LSU ribosomal protein L39E
MJ0690	590	-	12	617967	hypothetical protein [MTH1614]
MJ0691	326	-	9	618302	hypothetical archaeal/eukaryal protein, 29% basic aa (TFAR19 [Homo sapiens])
MJ0692	443	-	15	618760	SSU ribosomal protein S19E
tRNA	71	-	406	619237	Cys (GCA)
MJ0693	614	-	67	619918	hypothetical protein [MTH1147]
MJ0694	1241	+	208	620126	RNA 2'-O-methyl modification protein (NOP5/NOP56)
MJ0695	323	-	121	621811	hypothetical protein [MJ0803]
MJ0696	419	-	124	622354	hypothetical protein [MJ0803]

MJ0697	689	+		195	622549	fibrillarin-like pre-rRNA processing protein (NOP1 [S. cerevisiae])
MJ0698	557	+		204	623442	imidazoleglycerol-phosphate dehydratase (EC 4.2.1.19)
SR2_40	28	-		159	624186	repeat
SR2_39	28	-		41	624255	repeat
LR1_7	425	-		-28	624652	repeat
MJ0699	1127	-		48	625827	atrazine chlorohydrolase relative [Pseudomonas gi|1493840]
MJ0700	845	+		152	625979	hypothetical protein [C-terminus of HP0284]
MJ0701	1241	+		0	626824	
MJ0702	1781	+		-13	628052	DNA-directed DNA polymerase II (EC 2.7.7.7), DP1 subunit, Y. Ishino et al., J. Bacteriol. 180, 2232 (1998)
MJ0703	677	+		28	629861	similar to phosphoribosylformimino-5-aminoimidazole carboxamide ribotide isomerase
MJ0703.5	305	+		129	630667	
MJ0704	737	+		492	631464	8-oxoguanine DNA glycosylase, A. Gogos and N.D. Clarke, J. Biol. Chem. 274, 30447 (1999)
MJ0705	1214	+		21	632222	3-hydroxy-3-methylglutaryl CoA reductase (NADPH) (EC 1.1.1.34)
MJ0706	611	+		36	633472	hypothetical protein [MJ0793]
MJ0707	140	-		51	634274	LSU ribosomal protein L40E
MJ0708	506	-		153	634933	hypothetical protein [MTH554]
MJ0709	1349	+		85	635018	hypothetical protein [AF1471]
MJ0710	1052	+		30	636397	hypothetical methyltransferase [MTH724]
tRNA	73	+		137	637586	Phe (GAA)
tRNA	72	+		11	637670	Asn (GTT)
tRNA	74	+		33	637775	Ile (CAT)
tRNA	7	4	+	22	637871	Glu (TTC) 1
tRNA	84	+		40	637985	Leu_(TAG)
tRNA	72	+		15	638084	His (GTG)
rRNA	1477	+		299	638455	SSU 2
tRNA	7	2	+	66	639998	Ala (TGC) 2
rRNA	2948	+		208	640278	LSU 2
rRNA	81	+		110	643336	5S rRNA 2
RNA	275	+		86	643503	Rnase P
MJ0711	965	+		219	643997	
MJ0712	1004	-		3	645969	glycerol-1-phosphate dehydrogenase (EC 1.1.1.6), M. Nishihara et al., J. Bacteriol. 181, 1330 (1999)

MJ0713	2270	-	116	648355	hydrogenase transcriptional regulatory (hypF [Rhizobium leguminosarum])
MJ0714	341	+	186	648541	hypothetical protein [MJ0310]
MJ0715	1013	+	45	648927	H2-forming N5,N10-methylenetetrahydromethanopterin dehydrogenase related protein
MJ0716	257	+	75	650015	hypothetical protein [MJ1642]
MJ0717	356	-	189	650817	molybdenum cofactor biosynthesis protein E (moaE gene)
MJ0718	1103	-	23	651943	chromate resistance protein A, D.H. Nies et al., J. Bacteriol. 180, 5799 (1998)
MJ0719	1763	-	131	653837	ABC transporter, ATP-binding subunit
MJ0720	998	-	95	654930	3-isopropylmalate dehydrogenase (EC 1.1.1.85)
MJ0721	1193	+	195	655125	aminotransferase (subgroup II) [MTH1337, AF0080]
MJ0722	230	-	7	656555	ferredoxin
MJ0723	419	+	199	656754	transcriptional regulatory protein [AsnC/Lrp Helix-Turn-Helix family], D. Charlier et al., Gene 201, 63 (1997)
MJ0724	617	-	18	657808	8-oxoguanine DNA glycosylase (mjOgg) (oxoG)
MJ0725	782	-	102	658692	coenzyme F420-reducing hydrogenase (EC 1.12.99.1), beta subunit
MJ0726	647	-	172	659511	coenzyme F420-reducing hydrogenase (EC 1.12.99.1), gamma subunit
MJ0727	893	-	127	660531	coenzyme F420-reducing hydrogenase (EC 1.12.99.1), alpha subunit
MJ0728	1871	-	49	662451	carbon monoxide dehydrogenase/acetyl-CoA synthase (EC 1.2.99.2), beta subunit [Clostridium thermoaceticum]
MJ0729	362	+	229	662680	hypothetical protein with 2 CBS domains
MJ0730	557	-	8	663607	hypothetical archaeal [4Fe-4S] cluster protein
MJ0731	593	-	15	664215	iron-sulfur flavoprotein (Isf [Methanosarcina thermophila])
MJ0732	1169	-	48	665432	FMN-containing flavoprotein [MTH1350]
MJ0733	284	-	91	665807	
MJ0734	584	-	117	666508	rubredoxin electron transport protein
MJ0735	239	-	70	666817	rubredoxin (RD)
MJ0736	665	-	93	667575	alkyl hydroperoxide reductase
MJ0737	182	-	79	667836	zinc finger protein (similar to rubrerythrin)
MJ0738	266	-	43	668145	hypothetical protein [MTH865]
MJ0739	452	-	32	668629	hypothetical protein [MJ0745]
MJ0740	155	-	75	668859	rubredoxin (RD)
MJ0741	347	-	184	669390	superoxide reductase [*Pyrococcus furiosus*], F.E.

					Jenney, Jr., et al., Science 286, 306 (1999)
MJ0742	311	-	108	669809	hypothetical protein [MTH691]
MJ0743	881	-	132	670822	heterodisulfide reductase, subunit B
MJ0744	581	-	52	671455	heterodisulfide reductase, subunit C
MJ0745	533	-	222	672210	hypothetical protein [MJ0739]
MJ0746	422	-	218	672850	hypothetical protein [MTH1351], 20% acidic aa
MJ0747	455	-	113	673418	DcrH chemoreceptor [Desulfovibrio vulgaris]
MJ0748	1196	-	272	674886	FMN-containing flavoprotein [MTH1350]
MJ0749	680	-	267	675833	ferredoxin-type protein NAPH homologue
MJ0750	713	-	166	676712	similar to ferredoxin-type protein NAPH 6.7% cys
MJ0751	833	-	85	677630	hypothetical protein 1 (insertion sequence ISH1.8)
MJ0752	224	-	87	677941	
MJ0753	620	-	207	678768	
MJ0754	554	-	9	679331	hypothetical protein [PAB2327]
MJ0755	1025	-	290	680646	probable amidase
MJ0756	407	-	245	681298	
MJ0757	731	-	128	682157	thymidylate synthase (EC 2.1.1.45), Xu et al., Nature Structural. Biol. 6, 750 (1999)
MJ0758	440	-	58	682655	
MJ0759	329	-	47	683031	hypothetical archaeal protein [MTH1416]
MJ0760	815	-	18	683864	hypothetical protein [MTH1417]
MJ0761	752	-	210	684826	hypothetical protein [AF1765]
MJ0762	1025	-	13	685864	malic acid transport protein
MJ0763	362	-	27	686253	
MJ0764	347	-	13	686613	hypothetical protein [MTH352]
MJ0765	1643	-	201	688457	[6Fe-6S] prismane-containing protein
MJ0766	197	-	274	688928	hypothetical protein [AF1756]
MJ0767	404	-	174	689506	hypothetical protein [MJ0549/PH0601/AF0743/MTH739]
MJ0768	746	+	85	689591	hypothetical protein [AF2256]
MJ0769	506	-	146	690989	
MJ0770	569	-	-4	691554	
MJ0771	689	+	188	691742	precorrin-2 C20-methyltransferase (EC 2.1.1.130) (cbiL [S. typhimurium])
MJ0772	1058	+	0	692431	
MJ0773	524	-	529	694542	
MJ0774	1226	+	688	695230	hypothetical protein [PH0212]
MJ0775	938	+	-13	696443	hypothetical protein [AF1179]

MJ0776	1094	+	50	697431	hypothetical protein [MTH947]
MJ0777	554	+	-31	698494	hypothetical zinc finger transcription protein
MJ0778	506	+	51	699099	hypothetical protein, HD domain, L. Aravind and E.V. Koonin, Trends Biochem. Sci. 23, 469 (1998)
MJ0779	896	-	10	700511	hypothetical protein [PH0654]
MJ0780	1004	-	24	701539	21% basic aa
MJ0781	2096	-	26	703661	member conjugal transfer group defined by klbA/VirB11 (family in MJ and AF)
MJ0782	2006	-	110	705777	transcription initiation factor IIB (TFB)
McjTFB	1004	- -	1289	705492	intein
MJ0782.5	242	-	307	706041	hypothetical archaeal protein [MTH886/PAB3084]
MJ0783	557	+	141	706182	hypothetical protein [PAB0997/MTH796]
MJ0784	1073	+	279	707018	H2-forming N5,N10-methylene-tetrahydromethanopterin dehydrogenase (EC 1.5.99.9)
MJ0785	1076	+	273	708364	biotin synthetase (EC 2.8.1.-) [Arabidopsis thaliana]
MJ0785.5	542	+	17	709457	
MJ0786	542	+	81	710080	
MJ0787	1511	+	153	710775	hypothetical archaeal protein [MTH1137]
MJ0788	224	+	31	712317	
MJ0789	347	+	86	712627	
MJ0790	686	+	38	713012	hypothetical protein [MTH1252], 22% basic aa
MJ0791	1451	-	28	715177	argininosuccinate lyase (EC 4.3.2.1), R. Cohen-Kupiec et al., FEMS Microbiol. Lett. 173, 231 (1999)
MJ0792	266	+	336	715513	
MJ0792.5	485	+	72	715851	
MJ0793	533	+	64	716400	hypothetical protein [PH0870]
MJ0794	413	+	61	716994	hypothetical protein [AF0793]
MJ0795	1511	+	83	717490	ABC transporter, substrate binding protein
MJ0795.5	509	+	8	719009	hypothetical protein [MJ1249.1]
MJ0796	704	+	23	719541	ABC transporter ATP-binding protein
MJ0796.5	407	+	-7	720238	
MJ0797	1100	+	16	720661	hypothetical membrane protein [AF1470]
MJ0798	1001	+	20	721781	protein containing 2 or more adjacent copies of the (34 aa) TPR motif
MJ0799	881	+	6	722788	phosphoribosylformylglycinamidine synthase II related protein (purL [Mycobacterium

					tuberculosis])
MJ0800	1235	+	107	723776	activator of (R)-2-hydroxyglutaryl-CoA dehydratase
MJ0801	1136	+	28	725039	ATPase
MJ0802	563	+	225	726400	hypothetical protein [MTH1870]
MJ0803	515	+	23	726986	hypothetical protein [MJ0210.5]
MJ0804	857	+	31	727532	hypothetical protein [AF0918/PAB1151]
MJ0805	599	+	8	728397	hypothetical protein [PH0223]
MJ0806	1016	-	3	730015	XAA-PRO aminopeptidase (EC 3.4.11.9)
MJ0807	521	+	136	730151	hypothetical protein [AF1396]
MJ0808	998	+	136	730808	pyruvate formate-lyase activating enzyme related protein
MJ0808.5	413	-	138	732357	
MJ0809	488	+	682	733039	similar to acylphosphatase
MJ0810	602	+	128	733655	hypothetical protein [PH0010/AF1969/MTH857]
MJ0811	1316	-	104	735677	hypothetical protein [MJ0077]
MJ0812	1118	-	7	736802	moxR gene product /family with MJ0079
MJ0813	746	-	14	737562	precorrin-3B C17-methyltransferase (EC 2.1.1.131) (cbiH [Salmonella typhimurium])
MJ0814	1109	-	8	738679	deoxyhypusine synthase (EC 1.5.1.-)
MJ0815	875	-	20	739574	hypothetical protein [AF1061]
MJ0816	872	-	80	740526	hypothetical protein [PH0471]
MJ0817	617	-	-22	741121	archaetidylserine decarboxylase
MJ0818	608	-	6	741735	
MJ0819	963	-	69	742767	pseudogene with similarity to ATPase
MJ0821	167	-	-167	742767	ATPase
MJ0822	1673	-	74	744514	surface layer protein (S-layer) (Slp)
MJ0823	770	+	318	744832	nickel insertion accessory protein (CooC [Rhodospirillum rubrum])
MJ0824	893	-	-1	746494	molybdenum cofactor biosynthesis moaA protein
MJ0825	692	-	104	747290	peptidyl-prolyl cis-trans isomerase (EC 5.2.1.8) (rotamase) (cyclophilin)
MJ0826	413	+	174	747464	hypothetical protein, 12.3 % gly [PH0471]
MJ0827	581	+	5	747882	erythrocyte membrane protein 7.2 [Homo sapiens]
MJ0828	197	+	35	748498	hypothetical protein, *Methanococcus voltae*
MJ0829	1262	+	146	748841	eukaryotic peptide chain release factor, subunit 1 (eRF)
MJ0830	746	-	0	750849	hypothetical protein [AF1556]

MJ0831	1295	+	103	750952	hypothetical phosphoesterase
McjRNR-1	1358	+	1274	753521	intein
MJ0832	5249	+ -	2369	752510	anaerobic ribonucleoside-triphosphate reductase (EC 1.17.4.2)
McjRNR-2	1598	+ -	2075	755684	intein
MJ0832.5	425	-	515	758222	
MJ0833	779	-	19	759020	
MJ0834	707	-	65	759792	
MJ0835	1130	-	32	760954	hypothetical protein [PH0648]
MJ0835.5	389	-	58	761401	hypothetical protein, *Methanococcus maripaludis*
MJ0835.6	1016	-	7	762424	hypothetical protein [PH0446]
MJ0836	356	-	8	762788	hypothetical protein, 8.9% cys [AF1105]
MJ0837	746	+	74	762862	hypothetical protein, HD domain, L. Aravind and E.V. Koonin, Trends Biochem. Sci. 23, 469 (1998)
tRNA	73	-	164	763845	Thr (GGT)
tRNA	72	+	179	764024	Gln (TTG)
MJ0838	668	+	54	764150	hypothetical protein [MTH1356]
MJ0839	1049	-	41	765908	DNA primase (EC 2.7.7.-), G. Desogus et al., Nucl. Acids Res. 27, 4444 (1999)
MJ0840	980	+	86	765994	hypothetical archaeal protein [MTH834/AF0479]
MJ0841	1244	-	9	768227	hypothetical archaeal protein [MTH1170]
MJ0842	1331	+	229	768456	methyl coenzyme M reductase, beta subunit
MJ0843	485	+	22	769809	methyl coenzyme M reductase operon, protein D
MJ0844	599	+	5	770299	methyl coenzyme M reductase operon, protein C
MJ0845	779	+	13	770911	methyl coenzyme M reductase, gamma subunit
MJ0846	1658	+	135	771825	methyl coenzyme M reductase, subunit alpha
MJ0847	908	+	223	773706	N5-methyltetrahydromethanopterin--coenzyme M methyltransferase (EC 2.1.1.86), subunit E
MJ0848	689	+	17	774631	N5-methyltetrahydromethanopterin--coenzyme M methyltransferase (EC 2.1.1.86), subunit D
MJ0849	794	+	33	775353	N5-methyltetrahydromethanopterin--coenzyme M methyltransferase (EC 2.1.1.86), subunit C
MJ0850	308	+	13	776160	N5-methyltetrahydromethanopterin--coenzyme M methyltransferase (EC 2.1.1.86), subunit B
MJ0851	722	+	35	776503	N5-methyltetrahydromethanopterin--coenzyme M methyltransferase (EC 2.1.1.86), subunit A
MJ0852	203	+	21	777246	N5-methyltetrahydromethanopterin--coenzyme M methyltransferase (EC 2.1.1.86), subunit F
MJ0853	251	+	24	777473	N5-methyltetrahydromethanopterin--coenzyme M methyltransferase (EC 2.1.1.86), subunit G

MJ0854	956	+	32	777756	methyltetrahydromethanopterin:cob(I)alamin methyltransferase (MtrH)
MJ0855	683	+	482	779194	hypothetical protein [fusion of MJ0668/MJ0716]
MJ0856	524	-	121	780522	hypothetical protein [MTH1881]
MJ0856.5	326	-	100	780948	hypothetical protein [C-terminus of PAB1452]
MJ0857	785	-	153	781886	
MJ0858	338	-	96	782320	hypothetical protein [MTH1880]
MJ0859	482	-	37	782839	hypothetical protein [PAB1584]
MJ0860	950	-	25	783814	bifunctional short chain isoprenyl diphosphate synthase (EC 2.5.1.1/EC 2.5.1
MJ0861	1343	-	17	785174	hypothetical nucleic acid-binding protein
MJ0862	1076	-	135	786385	carotenoid biosynthesis related protein [Erwinia herbicola]
MJ0863	884	-	180	787449	heterodisulfide reductase, subunit B
MJ0864	557	-	74	788080	heterodisulfide reductase, subunit C
MJ0865	1319	+	188	788268	hypothetical methyltransferase protein [MTH1171]
MJ0866	386	+	23	789610	hypothetical zinc-binding nucleotidyl beta-phosphatase (HIT1 [Saccharomyces cerevisiae])
MJ0867	1247	-	-4	791239	N6-(4-hydroxyisopentenyl)-2-methylthioadenosine (ms2io6A37) synthesis protein (miaB [E. coli])
MJ0868	380	-	124	791743	hypothetical protein with one CBS domain
MJ0869	959	+	216	791959	homolog of RAD51 repair protein, E.M. Seitz et al., Genes & Dev. 12, 1248 (1998)
MJ0870	1859	-	73	794850	coenzyme F420-reducing hydrogenase (EC 1.12.99.1), beta subunit
MJ0871	941	+	186	795036	putative permease
MJ0872	1142	-	47	797166	hypothetical protein [MJ0642]
MJ0873	869	+	205	797371	ferric enterobactin transport ATP-binding protein
MJ0874	257	+	-4	798236	22% acidic aa
MJ0875	2243	+	120	798613	hypothetical protein with TPR repeats
MJ0876	983	-	9	801848	iron(III) dicitrate transport system permease protein
MJ0877	993	-	73	802914	pseudogene with similarity to hemin permease
MJ0878	1241	+	235	803149	hypothetical protein [MJ0642]
MJ0879	836	+	153	804543	nitrogenase iron protein (nitrogenase reductase)
MJ0880	923	+	25	805404	hypothetical protein [MJ0871]
MJ0881	914	-	191	807432	ornithine carbamoyltransferase, (EC 2.1.3.3),

					subunit F
MJ0882	590	+	349	807781	hypothetical (N-6 adenine) DNA methyltransferase (YhhF [E. coli])
MJ0883	1007	+	28	808399	hypothetical methyltransferase protein [PAB0505/MTH1873/AF1973]
MJ0883.5	272	+	108	809514	hypothetical protein [MJ0023]
MJ0884	1547	+	61	809847	replication factor C, large subunit
Mcjpol-2	1427	-	837	813658	intein
MJ0885	4889	- -	2255	816292	DNA-dependent DNA polymerase family B (EC 2.7.7.7)
Mcjpol-1	1106	- -	2369	815029	intein
MJ0886	1859	-	1317	818205	molybdopterin biosynthesis moeA protein [Escherichia coli]
MJ0887	653	-	6	818864	hypothetical protein [AF2061]
MJ0888	617	-	26	819507	hypothetical protein [AF1748]
MJ0889	1403	+	92	819599	hypothetical protein [BB0634]
MJ0890	362	-	19	821383	hypothetical protein [AF2370]
MJ0891	650	+	325	821708	flagellin B1 precursor
MJ0892	650	+	86	822444	flagellin B2
MJ0893	647	+	106	823200	flagellin B3
MJ0894	422	+	219	824066	hypothetical flagellar biosynthesis protein, 25% acidic aa (*flpC* [*Methanococcus voltae*])
MJ0895	986	+	20	824508	hypothetical flagellar biosynthesis protein (*flpD* [*Methanococcus voltae*])
MJ0896	410	+	51	825545	hypothetical flagellar biosynthesis protein [*flpE* [*Methanococcus voltae*])
MJ0897	416	+	-7	825948	hypothetical flagellar biosynthesis protein (*flpF* [*Methanococcus voltae*])
MJ0898	461	+	109	826473	hypothetical flagellar biosynthesis protein (*flpG* [*Methanococcus voltae*])
MJ0899	698	+	13	826947	flagellar biosynthesis protein (similar to Sms [*E. coli*])
MJ0900	1640	+	25	827670	type II protein secretion ATPase (flagellar biosynthesis)
MJ0901	1676	+	14	829324	flagellar accessory protein, *Methanococcus voltae*
MJ0902	701	+	30	831030	hypothetical flagellar biosynthesis protein (similar to pilD [*P. aeruginosa*])
MJ0903	1913	+	213	831944	hypothetical protein [MJ1292]
MJ0904	584	+	108	833965	
MJ0905	332	+	75	834624	hypothetical protein [PAB1376], 21 % basic aa
MJ0906	1097	+	5	834961	hypothetical protein [PH0545]

MJ0907	854	-	16	836928	regulatory protein, *ydeM, E. coli*
MJ0908	3683	+	105	837033	cobalamin biosynthesis protein cobN [Pseudomonas denitrificans]
MJ0908.5	182	+	52	840768	nitrogenase iron protein
MJ0909	272	+	0	840950	hypothetical protein [MJ1103]
MJ0909.5	119	+	25	841247	
MJ0910	521	-	69	841956	magnesium chelatase
MJ0911	1091	-	435	843482	magnesium-chelatase, 38 kDa subunit
MJ0912	716	+	220	843702	putative Ser/Thr phosphatase
MJ0912.5	158	-	8	844584	
MJ0913	1193	-	0	845777	pantothenate metabolism flavoprotein (dfp [Escherichia coli])
MJ0914	398	-	8	846183	hypothetical protein (VapC [Dichelobacter nodosus])
MJ0914.5	191	-	-37	846337	
MJ0915	800	-	372	847509	hypothetical protein [AF1838]
MJ0916	266	-	102	847877	hypothetical protein [AF1532]
MJ0917	1712	+	84	847961	inositol phosphatase related bifunctional protein
MJ0918	797	-	8	850478	indole-3-glycerol phosphate synthase (EC 4.1.1.48) (IGPS)
MJ0919	446	-	74	850998	hypothetical protein [PH1781], 24 % acidic aa
MJ0920	554	-	8	851560	GTP-binding protein
MJ0921	899	-	13	852472	hypothetical protein [MTH1391]
MJ0922	413	-	293	853178	hypothetical protein with 2 CBS domains
MJ0923	890	+	192	853370	
MJ0924	767	-	68	855095	capsular polysaccharide biosynthesis protein capB [Staphylococcus aureus]
MJ0925	302	-	119	855516	
MJ0925.5	152	+	369	855885	
MJ0926	260	+	343	856380	23% acidic aa
MJ0927	731	-	29	857400	hypothetical protein, *Streptococcus mutans*
MJ0928	590	-	6	857996	N-6 adenine RNA methyltransferase (EC 2.1.1.48)
SR2_38	28	-	186	858210	repeat
SR2_37	28	-	39	858277	repeat
SR2_36	28	-	41	858346	repeat
LR1_6	425	-	-28	858743	repeat
MJ0929	1349	+	136	858879	adenylosuccinate lyase (EC 4.3.2.2)
MJ0930	638	+	67	860295	precorrin-8X methylmutase (precorrin

					isomerase) (CbiC [Salmonella typhimurium])
MJ0931	1118	+	14	860947	NTP-utilizing enzyme (GuaA family)
MJ0932	200	+	145	862210	archaeal histone
tRNA	71	+	183	862593	Arg_(TCG)
MJ0933	698	+	30	862694	hypothetical protein [AF0706]
tRNA	73	+	90	863482	Thr (TGT)
tRNA	74	+	18	863573	Pro (TGG)
tRNA	73	+	12	863659	Tyr (GTA)
tRNA	73	+	87	863819	Lys (TTT)
rRNA	81	+	14	863906	5S_rRNA_1
tRNA	74	+	80	864067	Asp (GTC)
MJ0934	617	+	84	864225	polyferredoxin
MJ0935	623	+	-16	864826	hypothetical protein [PAB2059]
MJ0936	497	+	98	865547	hypothetical protein [PH1530]
MJ0937	1331	+	67	866111	phosphoribosylamine--glycine ligase (EC 6.3.4.13) (glycinamide ribonucleotide synthetase)
MJ0938	728	-	33	868203	hypothetical protein [AQ1210]
MJ0939	818	+	83	868286	
MJ0940	953	+	377	869481	hypothetical protein with TPR domain-like repeats
MJ0941	968	+	38	870472	hypothetical protein with TPR domain-like repeats
MJ0942	1952	+	55	871495	probable ATP-dependent helicase
tRNA	74	-	245	873766	Ile (GAT)
MJ0943	1193	-	106	875065	hypothetical_exported_protein
MJ0944	359	+	237	875302	hypothetical protein [MJ0361]
MJ0945	626	-	28	876315	hypothetical protein [TM0778]
MJ0946	1121	-	8	877444	tRNA(guanine-26,N2-N2) methyltransferase (EC 2.1.1.32) (TRM1), F. Constantinesco et al., Nucleic Acids Res. 26, 3753 (1998)
MJ0947	3116	-	102	880662	isoleucine--tRNA ligase (EC 6.1.1.5) (isoleucyl-tRNA synthetase)
MJ0948	563	-	8	881233	20% acidic aa
MJ0949	368	-	38	881639	hypothetical archaeal protein [MTH730/PH1612]
MJ0950	686	-	47	882372	hypothetical protein [AF1936]
MJ0951	1046	-	200	883618	hypothetical protein [MTH1126]
tRNA	73	+	66	883684	Val (TAC)
MJ0952	416	-	127	884300	chaperone MtGimC, α-subunit, M.R. Leroux et al., EMBO J. 18, 6730 (1999)

MJ0953	299	+	190	884490	energy converting hydrogenase, *ehbA* [MTH1251]A. Tersteegen and R. Hedderich, Eur. J. Biochem. 264, 930 (1999)
MJ0954	1256	-	15	886060	N-terminus similar to S-layer protein MJ0822. Family with MJ1282/MJ0954/MJ0822/MJECS12
MJ0955	1106	+	124	886184	histidinol-phosphate aminotransferase (EC 2.6.1.9)
MJ0956	779	+	149	887439	erythrocyte binding protein
MJ0957	1049	+	3	888221	hypothetical protein [PH0796]
MJ0958	1277	+	8	889278	
MJ0959	1154	-	30	891739	aspartate aminotransferase (EC 2.6.1.1) [MTH1601, AF1417]
MJ0960	650	-	48	892437	transaldolase (EC 2.2.1.2)
MJ0961	2285	+	217	892654	DNA replication initiation protein (MCM family)
MJ0962	383	+	0	894939	hypothetical archaeal protein [PH1601/MTH1618/AF0109], 22% basic aa
MJ0963	1688	-	-8	897002	N-methylhydantoinase (EC 3.5.2.14)
MJ0964	2033	-	16	899051	N-methylhydantoinase
MJ0965	725	+	78	899129	uroporphyrin-III C-methyltransferase (EC 2.1.1.107) (cobA [S. typhimurium]) (cysG)
MJ0966	1322	+	23	899877	hypothetical methyltransferase protein [MTH1895]
MJ0967	530	-	129	901858	intracellular protease (PfpI [Pyrococcus furiosus])
MJ0968	818	+	140	901998	cation transporting ATPase (family E1-E2, superfamily IB)
MJ0969	809	-	312	903937	hypothetical protein [AF1646]
MJ0970	428	-	264	904629	hypothetical protein, 27% basic aa [Synechocystis sp.]
MJ0971	1088	+	129	904758	hypothetical protein [AF1322]
MJ0972	635	+	9	905855	hypothetical protein (YchE [Escherichia coli])
MJ0973	1232	-	8	907730	similar to phosphoadenylylsulfate reductase
MJ0974	431	-	13	908174	hypothetical protein [AF1091]
MJ0975	191	-	-10	908355	
MJ0976	1097	+	110	908465	
MJ0977	1406	+	34	909596	hypothetical phosphoesterase [AF0735], L. Aravind and E.V. Koonin, Trends Biochem. Sci. 23, 17 (1998)
MJ0978	440	+	248	911250	
MJ0979	590	-	31	912311	

MJ0980	665	-	626	913602	SSU ribosomal protein S3AE
MJ0981	605	-	41	914248	hypothetical archaeal protein [PAB0161/MTH828/AF1966]
MJ0982	665	-	37	914950	SSU ribosomal protein S2P
MJ0983	581	-	123	915654	LSU ribosomal protein L15E
MJ0984	653	+	147	915801	type II site-specific deoxyribonuclease (EC 3.1.21.4) (MjaI)
MJ0985	905	-	13	917372	site-specific DNA methyltransferase (N4-cytosine specific) (EC 2.1.1.113)
MJ0986	194	-	3	917569	hypothetical protein [APES038]
MJ0987	338	+	342	917911	chaperone MtGimC, β-subunit, M.R. Leroux et al., EMBO J. 18, 6730 (1999)
MJ0988	986	+	114	918363	hypothetical phosphoesterase, L. Aravind and E.V. Koonin, Trends Biochem. Sci. 23, 17 (1998)
MJ0989	341	-	3	919693	hypothetical protein, *Methanopyrus kandleri*
MJ0990	212	-	12	919917	hypothetical protein, *yedF, E. coli*
MJ0991	428	+	265	920182	22% acidic aa
MJ0992	530	+	4	920614	hypothetical protein [MJ0628]
MJ0993	1073	-	694	922911	hydrogenase expression/formation protein (hypD [Rhizobium leguminosarum])
MJ0994	719	-	8	923638	uroporphyrinogen-III synthase (EC 4.2.1.75)
MJ0995	665	-	0	924303	
MJ0996	1352	+	73	924376	hypothetical protein (TldD [E. coli])
MJ0998	413	-	286	926427	
MJ0999	1625	-	465	928517	thermosome, chaperonin, J.M. Kowalski et al., System. Appl. Microbiol. 21, 173 (1998)
MJ1000	1238	+	212	928729	histidine--tRNA ligase (EC 6.1.1.21) (histidyl-tRNA synthetase)
MJ1001	863	+	46	930013	ribosomal protein S6 modification protein II
MJ1002	926	+	91	930967	hypothetical protein [AF1759]
MJ1003	1259	+	105	931998	3-isopropylmalate dehydratase (EC 4.2.1.33), isomerase subunit
MJ1004	641	+	94	933351	hypothetical protein with two CBS domains
MJ1005	392	+	4	933996	
MJ1006	1025	+	26	934414	ATPase
MJ1007	2633	+	105	935544	valine--tRNA ligase (EC 6.1.1.9) (valyl-tRNA synthetase)
MJ1008	863	-	54	939094	aminotransferase (subgroup III) [MTH1430, AF0933]
MJ1009	707	-	104	939905	phosphate ABC transporter, regulatory protein

MJ1010	1121	-	55	941081	ATPase
MJ1011	389	-	392	941862	phosphate ABC transporter, regulatory protein
MJ1012	755	-	70	942687	phosphate ABC transporter, ATP-binding protein
MJ1013	842	-	34	943563	phosphate ABC transporter, permease protein A
MJ1014	935	-	-10	944488	phosphate ABC transporter, permease protein C
MJ1015	1130	-	177	945795	phosphate ABC transporter, periplasmic phosphate binding protein
MJ1016	881	+	267	946062	hypothetical ATP-binding protein [PH1608]
MJ1017	608	+	-7	946936	
MJ1018	1571	-	13	949128	D-3-phosphoglycerate dehydrogenase (serA) (EC 1.1.1.95)
MJ1019	1061	+	133	949261	carbamoyl-phosphate synthase (glutamine-hydrolyzing) (EC 6.3.5.5), small subunit
MJ1020	764	+	110	950432	hypothetical protein [APE1666]
MJ1021	488	-	50	951734	23% acidic aa
MJ1022	1706	-	236	953676	hypothetical protein [AF0587]
MJ1023	785	+	81	953757	ABC transporter ATP-binding protein
MJ1024	1190	+	14	954556	ABC transporter, permease subunit
MJ1025	1151	-	7	956904	hypothetical protein, *E. coli yhaN/yhaM*
MJ1026	1277	-	19	958200	thiamine biosynthesis protein thiC [Escherichia coli] (=thiA in Bacillus subtilis)
MJ1027	1121	+	171	958371	energy converting hydrogenase, *ehbN* [MTH1238] A. Tersteegen and R. Hedderich, Eur. J. Biochem. 264, 930 (1999)
MJ1028	1967	+	154	959646	Type VI topoisomerase (subunit B) (TOPVIB [*Sulfolobus shibatae*])
MJ1029	824	+	51	961664	dimethyladenosine transferase (EC 2.1.1.-) (DIM1 [*S. cerevisiae*])
MJ1030	440	+	6	962494	nucleotidyl-transferase
MJ1031	914	+	9	962943	putative transport protein
MJ1032	1097	+	130	963987	hypothetical protein [AF2058]
MJ1033	950	+	5	965089	magnesium, nickel and cobalt transport protein (CorA [*E. coli*]), R.L. Smith et al., J. Bacteriol. 180, 2788 (1998)
MJ1034	260	+	13	966052	signal recognition particle, subunit SRP19 (*H. sapiens*)
MJ1035	830	-	29	967171	F420-dependent N5,N10-methylenetetrahydromethanopterin dehydrogenase
MJ1036	689	+	418	967589	hypothetical protein [PH0632]

MJ1037	1202	+	75	968353	tryptophan synthase (EC 4.2.1.20), subunit beta
MJ1038	836	+	195	969750	tryptophan synthase (EC 4.2.1.20), subunit alpha
MJ1039	233	+	178	970764	DNA-directed RNA polymerase (EC 2.7.7.6), subunit H (*rpoH*)
MJ1040	1484	+	20	971017	DNA-directed RNA polymerase (EC 2.7.7.6), subunit B" (*rpoB"*)
MJ1041	1898	+	98	972599	DNA-directed RNA polymerase (EC 2.7.7.6), subunit B' (*rpoB'*)
MJ1042	4022	+	112	974609	DNA-directed RNA polymerase (EC 2.7.7.6), subunit A' (*rpoA'*)
McjRpolA'	1355	+ -	2645	975986	intein
McjRpolA"	1412	+	1593	978934	intein
MJ1043	2576	+ -	1637	978709	DNA-directed RNA polymerase (EC 2.7.7.6), subunit A" (*rpoA"*)
MJ1044	311	+	368	981653	LSU ribosomal protein L30E
MJ1045	548	+	3	981967	transcription termination-antitermination factor (*nusA* [*E. coli*])
MJ1046	443	+	145	982660	SSU ribosomal protein S12P
MJ1047	572	+	61	983164	SSU ribosomal protein S7P
MJ1048	2177	+	114	983850	translation elongation factor, EF-2
IS_B3	360	-	429	986816	repeat
MJ1049	482	+	71	986887	24% acidic aa
MJ1050	530	+	71	987440	hypothetical ATP/GTP-binding protein
MJ1051	1538	-	-18	989490	hypothetical O-carbamoyltransferase (nolO [Bradyrhizobium japonicum])
MJ1052	290	-	15	989795	hypothetical protein [MTH1187/AQ2067/PAB1240]
MJ1053	488	-	68	990351	hypothetical protein, 22% acidic aa [PH0283]
IS_B5	363	+	84	990435	repeat
MJ1054	2684	+	169	990967	UDP-hexose dehydrogenase
McjGDPM6DH	1361	+ -	1904	991747	intein
MJ1055	977	+	543	993651	capsular polysaccharide biosynthesis protein capI [Staphylococcus aureus]
MJ1056.5	1889	+	20	994648	pseudogene
MJ1057	857	+	-12	996525	glycosyl transferase
MJ1058	1826	+	14	997396	hypothetical O-carbamoyltransferase (nolO [Bradyrhizobium japonicum])
MJ1059	1217	+	20	999242	capsular polysaccharide biosynthesis protein capM [Staphylococcus aureus]
MJ1060	1370	+	240	1000699	
MJ1061	998	-	1	1003068	capsular polysaccharide biosynthesis protein

					capD [*Staphylococcus aureus*]
MJ1062	1451	-	16	1004535	spore coat polysaccharide biosynthesis protein spsG [*Bacillus subtilis*]
MJ1063	728	-	0	1005263	spore coat polysaccharide biosynthesis protein spsF [*Bacillus subtilis*]
MJ1064	635	-	5	1005903	galactoside acetyltransferase (EC 2.3.1.18) (Thiogalactoside acetyltransferase)
MJ1065	1010	-	6	1006919	sialic acid synthase (neuB [Escherichia coli])
MJ1066	1157	-	14	1008090	spore coat polysaccharide biosynthesis protein spsC [*Bacillus subtilis*]
MJ1068	1520	-	8	1009618	capsular polysaccharide biosynthesis membrane protein (CapF [*Staphylococcus aureus*])
MJ1069	1175	-	14	1010807	LPS biosynthesis protein (hexose transferase)
MJ1070	470	-	124	1011401	
MJ1071	938	-	0	1012339	hypothetical protein [MJ0977]
MJ1072	326	-	25	1012690	hypothetical protein [MJ0023]
MJ1073	809	-	191	1013690	hypothetical protein kinase
MJ1074	335	-	112	1014137	hypothetical protein [MJ0023]
MJ1075	1421	-	68	1015626	anthranilate synthase (EC 4.1.3.27), component I
MJ1076	1010	-	12	1016648	ATPase
MJ1077	1562	-	37	1018247	serine--tRNA ligase (EC 6.1.1.11) (seryl-tRNA synthetase)
MJ1078	701	-	93	1019041	hypothetical protein [AF2231]
MJ1079	1190	-	277	1020508	hypothetical archaeal protein [AF2040/MTH1700/PAB0115]
MJ1080	404	-	181	1021093	
MJ1081	398	+	146	1021239	hypothetical protein [TM1872]
MJ1082	359	+	22	1021659	autoregulatory transcription regulation factor (Lrs14 [*Sulfolobus solfataricus*])
MJ1083	578	+	73	1022091	iron-sulfur flavoprotein (Isf [*Methanosarcina thermophila*])
MJ1084	845	+	92	1022761	shikimate 5-dehydrogenase (EC 1.1.1.25)
MJ1085	1526	+	29	1023635	
MJ1086	1019	+	0	1025161	hypothetical protein [AF1967]
MJ1087	935	+	17	1026197	mevalonate kinase (EC 2.7.1.36)
MJ1088	836	-	8	1027976	cobalt transport ATP-binding protein cbiO [Salmonella typhimurium]
MJ1089	794	-	13	1028783	cobalt transport protein Q
MJ1090	284	-	53	1029120	cobalt transport protein N
MJ1091	698	-	4	1029822	cobalt transporter (ABC-type), periplasmic binding protein (CbiM [S.typhimurium])
MJ1092	635	+	288	1030110	putative nickel incorporation membrane protein (UreH [*B. subtilis*])

MJ1093	899	+	135	1030880	*nifB* protein [*Anabaena* sp.]
MJ1094	1154	-	120	1033053	hypothetical protein [MTH987]
MJ1095	1259	-	37	1034349	hypothetical protein [AF2259]
LR1_5	425	+	79	1034428	repeat
SR2_35	28	+	-28	1034825	repeat
SR2_34	28	+	46	1034899	repeat
SR2_33	28	+	40	1034967	repeat
SR2_32	28	+	39	1035034	repeat
SR2_31	28	+	43	1035105	repeat
SR2_30	28	+	40	1035173	repeat
SR2_29	28	+	45	1035246	repeat
SR2_28	28	+	39	1035313	repeat
SR2_27	28	+	38	1035379	repeat
SR2_26	28	+	42	1035449	repeat
SR2_25	28	+	49	1035526	repeat
SR2_24	28	+	41	1035595	repeat
SR1_15	27	+	39	1035662	repeat
SR1_14	27	+	40	1035729	repeat
MJ1096	1022	-	99	1036877	N-acetyl-gamma-glutamyl-phosphate reductase (EC 1.2.1.38)
MJ1097	1304	-	327	1038508	diaminopimelate decarboxylase (EC 4.1.1.20)
tRNA	76	-	40	1038624	Arg (GCG)
MJ1098	638	-	5	1039267	putative protein tyrosine phosphatase, 22% basic aa
MJ1099	704	-	354	1040325	hypothetical protein [AF2196]
MJ1100	1343	+	206	1040531	phosphohexosemutase related protein
MJ1101	1223	+	31	1041905	glucose-1-phosphate thymidylyltransferase (EC 2.7.7.24) (dTDP-glucose synthase)
MJ1102	467	+	120	1043248	deoxycytidine triphosphate deaminase (EC 3.5.4.14)
MJ1103	263	-	14	1043992	hypothetical protein [MJ0909]
MJ1103.5	158	-	0	1044150	
MJ1104	887	-	85	1045122	homoserine kinase (EC 2.7.1.39) (HK)
MJ1105	653	-	24	1045799	TRK system potassium uptake protein A
MJ1106	554	-	255	1046608	hypothetical protein [AF0426]
MJ1107	422	-	21	1047051	hypothetical protein [MJ0023]
MJ1108	1643	-	114	1048808	phenylalanine--tRNA ligase (EC 6.1.1.20), alpha chain (phenylalanyl-tRNA syntase) (eukaryotic naming)
LR1_4	396	+	179	1048987	repeat
SR1_13	23	+	-1	1049382	repeat
SR1_12	27	+	46	1049451	repeat

SR1_11	27	+	46	1049524	repeat
SR1_10	27	+	42	1049593	repeat
SR1_9	27	+	39	1049659	repeat
SR1_8	27	+	44	1049730	repeat
SR1_7	27	+	40	1049797	repeat
SR1_6	27	+	49	1049873	repeat
SR1_5	27	+	44	1049944	repeat
SR1_4	27	+	39	1050010	repeat
SR1_3	27	+	48	1050085	repeat
SR2_23	20	-	46	1050178	repeat
MJ1109	527	+	310	1050488	orotate phosphoribosyltransferase (EC 2.4.2.10) (uridine 5'-monophosphate synthase), L. Watrin et al., Mol. Gen. Genet. 262, 378 (1999)
MJ1110	1457	-	104	1052576	hypothetical protein [MTH588]
MJ1111	1121	-	-34	1053663	tRNA nucleotidyltransferase (EC 2.7.7.25) (CCA [Sulfolobus shibatae]), C-terminus
MJ1112	233	-	-16	1053880	tRNA nucleotidyltransferase (EC 2.7.7.25) (CCA [Sulfolobus shibatae]), N-terminus
MJ1113	902	-	131	1054913	UDP-N-acetylglucosamine--dolichyl-phosphate N-acetylglucosaminephosphotransferase (EC 2.7.8.15)
MJ1114	575	-	309	1055797	hypothetical protein [N-terminus of HP0141]
MJ1115	773	-	87	1056657	hypothetical protein (sgcQ [E. coli])
MJ1116	1622	-	27	1058306	asparagine synthetase (glutamine-hydrolysing) (EC 6.3.5.4)
MJ1117	587	+	146	1058452	hypothetical nucleotidyltransferase protein [MTH1152]
MJ1118	266	+	28	1059067	hypothetical protein [AF0359]
MJ1119	884	+	117	1059450	diaminopimelate epimerase (EC 5.1.1.7)
MJ1120	836	+	7	1060341	putative transcription regulator, LysR family
MJ1121	281	-	76	1061534	hypothetical protein [PAB2079/AF1996]
MJ1122	221	-	-24	1061731	hypothetical protein [AF1997]
MJ1123	614	+	80	1061811	ubiquinone/menaquinone biosynthesis C-methyltransferase (EC 2.1.1.-) (UbiE [E.coli])
McjRNAhel	1502	+	1029	1063454	intein
MJ1124	3584	+ -	2513	1062443	RTP-binding RNA helicase
MJ1124.5	143	-	288	1066458	
MJ1125	179	-	-57	1066580	hypothetical protein [MTH923]
MJ1126	1205	+	357	1066937	GTP-binding protein *hflX* [*Mycobacterium leprae*]
MJ1127	839	+	64	1068206	hypothetical protein [AF0899]
MJ1128	914	-	7	1069966	hypothetical protein, *H. sapiens*
MJ1129	284	-	15	1070265	MRP PROTEIN HOMOLOG

MJ1130	1604	-	129	1071998	Bifunctional enzyme: N-terminal is O-sialoglycoprotein endopeptidase, C-terminal is similar to protein kinase
MJ1131	929	-	101	1073028	GMP synthetase (EC 6.3.5.2) (glutamine amidotransferase)
MJ1132	293	-	22	1073343	hypothetical protein [MJ1205]
MJ1133	1262	+	249	1073592	hypothetical protein [PH0963/MTH1394/AF1643/APE2078]
MJ1134	677	-	29	1075560	hypothetical protein, 24% acidic aa [AF1559]
MJ1135	455	-	22	1076037	molybdenum cofactor biosynthesis protein (moaC)
MJ1136	1610	+	230	1076267	hypothetical protein (YPL086c [Saccharomyces cerevisiae]) (Lpg22p)
MJ1137	572	-	160	1078609	hypothetical protein [MTH1373]
MJ1138	452	+	156	1078765	hypothetical protein [C-terminus of MJ0943]
MJ1139	695	-	121	1080033	hypothetical protein [AF2317]
MJ1140	671	-	18	1080722	hypothetical protein [TM0797/MTH1182/DR1400]
MJ1141	641	+	45	1080767	hypothetical protein [AF0204]
MJ1142	1046	+	24	1081432	arsenical pump-driving ATPase
MJ1143	1082	+	46	1082524	hypothetical protein [AF1546]
MJ1144	968	-	3	1084577	cobalamin biosynthesis protein cbiG [Salmonella typhimurium]
MJ1145	194	-	343	1085114	HTH DNA-binding protein
MJ1146	1028	-	194	1086336	glyceraldehyde 3-phosphate dehydrogenase (EC 1.2.1.12)
MJ1147	1355	+	97	1086433	
MJ1148	323	-	8	1088119	DNA-directed RNA polymerase associated zinc-binding factor (RPA12/RPB9 family [Saccharomyces cerevisiae])
MJ1149	506	-	-4	1088621	ADP-ribose hydrolase (ADP-ribose pyrophosphatase), S. Sheikh et al., J. Biol. Chem. 273, 20924 (1998)
MJ1150	542	+	69	1088690	hypothetical protein [AF1433]
MJ1151	329	+	122	1089354	hypothetical protein [MJ1152/MJ1153]
MJ1152	209	+	12	1089695	hypothetical protein [MJ1151/MJ1153]
MJ1153	185	+	0	1089904	hypothetical protein [MJ1151/MJ1152]
MJ1154	1352	-	159	1091600	hypothetical protein, HD domain, L. Aravind and E.V. Koonin, Trends Biochem. Sci. 23, 469 (1998)
MJ1155	1808	-	278	1093686	hypothetical bifunctional protein. Serially similar to *E.coli* orfs f325 & f413
MJ1155.4	329	-	15	1094030	hypothetical protein [AF1754]
MJ1155.6	218	-	3	1094251	hypothetical protein [C-terminus of AF1755]

MJ1156	2708	-	165	1097124	probable CDC48 homolog (belongs to AAA family of ATPases)
MJ1158	356	-	123	1097603	zinc finger protein, 6.2% cys
MJ1157	911	+	124	1097727	hypothetical protein, 5.1% cys [family with MJ0485,MJ1478]
MJ1159	1934	+	40	1098678	ssDNA binding protein (RPA70/SSB family), T.J. Kelly et al., Proc.Natl.Acad.Sci.USA 95, 14634 (1998); F. Chedin et al., Trends Biochem. Sci. 23, 273 (1998)
MJ1160	1301	+	60	1100672	glutamyl-tRNA(Gln) amidotransferase, subunit A (gatA [B. subtilis])
MJ1161	431	+	227	1102200	
MJ1161.5	218	-	340	1103189	hypothetical protein [MJ1223.5]
MJ1162	641	-	114	1103944	hypothetical protein, *Methanothermus fervidus*
MJ1163	650	+	155	1104099	hypothetical archaeal protein [MTH1902]
MJ1163.5	98	+	117	1104866	
MJ1164	944	-	133	1106041	hypothetical archaeal protein [PH1808/MTH967/AF1787/APE0778]
MJ1165	437	+	248	1106289	tungsten formylmethanofuran dehydrogenase (EC 1.2.99.5), subunit E
MJ1166	1064	+	6	1106732	tungsten formylmethanofuran dehydrogenase (EC 1.2.99.5), subunit F
MJ1167	245	+	51	1107847	tungsten formylmethanofuran dehydrogenase (EC 1.2.99.5), subunit G
MJ1168	398	+	21	1108113	tungsten formylmethanofuran dehydrogenase (EC 1.2.99.5), subunit D
MJ1169	1700	+	30	1108541	tungsten formylmethanofuran dehydrogenase (EC 1.2.99.5), subunit A
MJ1170	257	+	45	1110286	hypothetical protein [PH1682]
MJ1171	818	+	29	1110572	tungsten formylmethanofuran dehydrogenase (EC 1.2.99.5), subunit C
MJ1172	251	+	142	1111532	hypothetical regulatory protein, similar to Cop6 protein of *S. aureus* plasmid pE194
MJ1173	281	+	4	1111787	hypothetical protein [AF1093]
MJ1174	1610	-	16	1113694	CTP synthase (EC 6.3.4.2)
MJ1175	1121	-	97	1114912	chorismate synthase (EC 4.6.1.4) (5-enolpyruvylshikimate-3-phosphate phospholyase)
MJ1176	1289	-	81	1116282	proteosome-activating nucleotidase (PAN), P. Zwickl et al., J. Biol. Chem. 274, 26008 (1999)
MJ1178	1043	-	125	1117450	capsular polysaccharide biosynthesis protein capM [Staphylococcus aureus]
MJ1177	1001	+	18	1117468	hypothetical protein [MTH1211/AF1800/PAB0658/AQ740]

MJ1179	446	+	372	1118841	glycerol-3-phosphate cytidylyltransrerase (EC 2.7.7.39)
MJ1180	392	+	302	1119589	
MJ1181	596	+	102	1120083	hypothetical transport protein [N-terminus of AF0640]
MJ1182	1097	+	410	1121089	hypothetical protein [AF0648]
MJ1183	470	+	16	1122202	hypothetical protein [MTH862]
MJ1184	467	+	23	1122695	riboflavin synthase (EC 2.5.1.9), S. Eberhardt et al., J. Bacteriol. 179, 2938 (1997)
MJ1185	1865	-	7	1125034	glyceraldehyde-3-phosphate:ferredoxin oxidoreductase (*gor* [*Pyrococcus furiosus*]), J. van der Oost et al., J. Biol. Chem. 273, 28149 (1998)
MJ1186	1037	+	162	1125196	molybdate ABC transporter, substrate-binding subunit
MJ1188	806	-	7	1127046	endonuclease IV related protein (end4 [*Mycoplasma pneumoniae*])
MJ1187	902	+	80	1127126	ADP-ribosylglycohydrolase (EC 3.2.-.-)
MJ1189	848	-	34	1128910	hypothetical protein [MJ0233]
MJ1190	1970	+	480	1129390	heterodisulfide reductase, subunit A; contains a SeC residue
MJ1190.5	401	+	236	1131596	coenzyme F420-non-reducing hydrogenase (VhuD [*Methanococcus voltae*]); contains 2 SeC residues)
MJ1191	854	+	127	1132124	coenzyme F420-non-reducing hydrogenase, subunit gamma (VhuG [*Methanococcus voltae*])
MJ1192	1253	+	37	1133015	coenzyme F420-non-reducing hydrogenase, subunit alpha (VhuA [*Methanococcus voltae*])
MJ1192.5	149	+	66	1134334	coenzyme F420-non-reducing hydrogenase (VhuU [*Methanococcus voltae*]); contains SeC residue
MJ1193	1172	+	117	1134600	polyferredoxin
MJ1194	1304	+	27	1135799	tungsten containing formylmethanofuran dehydrogenase (EC 1.2.99.5), subunit B; contains SeC residue
MJ1195	1553	-	257	1138913	2-isopropylmalate synthase (EC 4.1.3.12)
MJ1196	1097	-	139	1140149	cationic amino acid transporter MCAT-2
MJ1197	1859	+	98	1140247	threonine--tRNA ligase (EC 6.1.1.3) (threonyl-tRNA synthetase)
MJ1198	2282	+	219	1142325	hypothetical phosphesterase, L. Aravind and E.V. Koonin, Trends Biochem. Sci. 23, 17 (1998)
MJ1199	428	-	26	1145061	hypothetical protein [APE2001]

MJ1200	1085	-	54	1146200	DNA-methyltransferase (C-5 cytosine-specific) (EC 2.1.1.73)
MJ1201	209	-	85	1146494	LSU ribosomal protein L24E
MJ1202	224	-	61	1146779	SSU ribosomal protein S28E
MJ1203	350	-	124	1147253	SSU ribosomal protein L7AE
MJ1204	863	-	232	1148348	ATP phosphoribosyltransferase (EC 2.4.2.17)
MJ1205	308	-	25	1148681	hypothetical protein, 22% basic aa [AF2405]
MJ1206	1262	-	-4	1149939	hypothetical protein [AF1899]
tRNA	72	+	211	1150150	Gly (GCC)
tRNA	72	+	38	1150260	Gly (TCC)
MJ1207	677	+	247	1150579	acetyltransferase
MJ1208	1217	+	124	1151380	S-adenosylmethionine synthetase (EC 2.5.1.6) (MAT)
MJ1208.4	653	-	18	1153268	site-specific DNA-methyltransferase (adenine-specific) (EC 2.1.1.72) (HpaI)
MJ1208.6	167	-	-6	1153429	site-specific DNA-methyltransferase (adenine-specific) (EC 2.1.1.72), C-terminus
MJ1209	566	-	39	1154034	site-specific DNA-methyltransferase (N-4 cytosine specific) (EC 2.1.1.-)
MJ1210	764	-	116	1154914	hypothetical protein [AF0525]
MJ1211	347	-	31	1155292	hypothetical protein [PH1307]
MJ1212	602	-	285	1156179	CDP-archaeol--serine O-archaetidyltransferase (EC 2.7.8.8), Z. Ge and D.E. Taylor, J. Bacteriol. 179, 4970 (1997)
MJ1213	329	-	14	1156522	hypothetical protein, 22% basic aa [MJ0123]
MJ1214	3044	-	13	1159579	type I site-specific deoxyribonuclease (EC 3.1.21.3), subunit R
MJ1215	245	-	62	1159886	hypothetical protein [MJ0126]
MJ1216	362	-	-13	1160235	hypothetical protein [MJ0125]
MJ1217	293	-	14	1160542	hypothetical protein [MJ0126]
MJ1218	1262	-	45	1161849	type I site-specific deoxyribonuclease (EC 3.1.21.3), subunit S
MJ1219	302	-	28	1162179	
MJ1220	1733	-	82	1163994	site-specific DNA-methyltransferase (adenine-specific) (EC 2.1.1.72)
MJ1221	878	+	88	1164082	hypothetical protein [Streptomyces albus gi|416205]
MJ1222	719	-	26	1165705	dolichyl-phosphate beta-glucosyltransferase (EC 2.4.1.117)
MJ1223	275	-	-22	1165958	energy converting hydrogenase, *ehbB* [MTH1250]A. Tersteegen and R. Hedderich, Eur. J. Biochem. 264, 930 (1999)
MJ1223.5	233	-	220	1166411	hypothetical protein [MJ1161.5]

MJ1224	395	-	191	1166997	hypothetical archaeal protein [MTH949/PH0358/AF0460]
MJ1225	833	+	108	1167105	inosine-5'-monophosphate dehydrogenase related protein
MJ1226	2399	+	38	1167976	H+-transporting ATPase (EC 3.6.1.35)
MJ1227	719	+	12	1170387	pyruvate formate-lyase activating enzyme
MJ1228	395	-	21	1171522	translation initiation factor, eIF-5a (hypusine-containing)
MJ1229	1502	-	197	1173221	pyruvate carboxylase, subunit A, B. Mukhopadhyay et al., J. Biol. Chem. 273, 5155 (1998)
MJ1230	215	-	16	1173452	zinc finger protein
MJ1231	1700	-	38	1175190	pyruvate carboxylase, subunit B, B. Mukhopadhyay et al., J. Biol. Chem. 273, 5155 (1998)
MJ1232	872	-	259	1176321	hypothetical protein with 2 CBS domains
MJ1233	836	+	156	1176477	hypothetical methyltransferase protein [PH1028]
MJ1234	722	-	636	1178671	hypothetical protein [AF0072]
MJ1235	1274	-	19	1179964	ribulose 1,5-bisphosphate carboxylase/oxygenase (EC 4.1.1.39) (RubisCO), G.M.F. Watson et al., J. Bacteriol. 181, 1569 (1999)
MJ1236	1901	-	288	1182153	mRNA cleavage and polyadenylation specificity factor related, N-terminal 70%
MJ1237	671	-	243	1183067	proteasome (EC 3.4.99.46), subunit beta
MJ1238	1364	-	124	1184555	proline--tRNA ligase (EC 6.1.1.15) (prolyl-tRNA synthetase)
MJ1239	674	+	91	1184646	
MJ1240	290	-	9	1185619	
MJ1241	209	-	27	1185855	
MJ1242	1598	-	170	1187623	methyl-CoM reductase system component A2 (ABC transporter ATP-binding protein
MJ1243	368	-	3	1187994	HTH DNA-binding protein /family with MJ0529/MJ0272/MJ0173/MJ1641
MJ1244	323	-	95	1188412	hypothetical protein [MTH1110]
MJ1245	335	-	15	1188762	hypothetical protein [MTH1110]
MJ1246	863	-	152	1189777	succinate--CoA ligase (ADP-forming) (EC 6.2.1.5), alpha subunit
tRNA	107	+	173	1189950	Trp_(CCA)
MJINTR001	32	+	-67	1189990	intron
MJ1247	530	+	134	1190156	hypothetical protein [AF1796/PH1938]
MJ1248	455	-	39	1191180	
MJ1249	1082	+	189	1191369	hypothetical protein [AF0229]

MJ1249.5	263	+	244	1192695	hypothetical protein [MJ0795.5/MJ0785.5]
MJ1250	758	+	17	1192975	hypothetical archaeal protein [PAB0735/MTH552/APE0621]
MJ1252	698	-	21	1194452	ubiquinone/menaquinone biosynthesis C-methyltransferase (EC 2.1.1.-) (UbiE [E.coli])
MJ1251	500	+	19	1194471	hypothetical protein [MTH1148]
MJ1253	845	-	15	1195831	protein-export membrane protein *secF* [*E. coli*]
MJ1254	1253	+	318	1196149	
MJ1255	1181	+	30	1197432	similar to glycosyltransferase
MJ1256	923	+	9	1198622	hypothetical protein [MTH1018/AF0917]
MJ1257	1046	+	0	1199545	hypothetical protein [MJ0421]
MJ1258	200	+	86	1200677	archaeal histone
MJ1259	677	+	266	1201143	uridylate kinase (EC 2.7.4.-)
MJ1260	392	+	468	1202288	SSU ribosomal protein S6E
MJ1261	1232	+	287	1202967	translation initiation factor, eIF-2, subunit gamma
MJ1262	1088	+	245	1204444	hypothetical protein [MTH306]
MJ1263	1952	+	134	1205666	methionine--tRNA ligase (EC 6.1.1.10) (methionyl-tRNA synthetase)
MJ1264	2198	-	31	1209847	phosphoribosylformylglycinamidine synthase (EC 6.3.5.3), subunit II
MJ1265	419	-	80	1210346	nucleoside diphosphate kinase (EC 2.7.4.6)
MJ1266	1214	+	226	1210572	similar to branched-chain amino acid transport protein, LivJ
MJ1267	758	+	112	1211898	high-affinity branched-chain amino acid transport ATP-binding protein (livG [Escherichia coli])
MJ1268	485	+	249	1212905	high-affinity branched-chain amino acid transport ATP-binding protein livF [E
MJ1269	908	+	72	1213462	high-affinity branched-chain amino acid transport permease livH [Escherichia
MJ1270	1034	+	25	1214395	high-affinity branched-chain amino acid transporter, permease protein (livM [E. coli])
MJ1271	509	+	209	1215638	homoaconitase related protein
MJ1272	488	+	0	1216147	hypothetical protein [AF0440]
MJ1273	1049	+	3	1216638	hypothetical protein [AF1611]
MJ1274	770	-	23	1218480	diphthine synthase (EC 2.1.1.98)
MJ1275	1163	+	123	1218603	NA(+)/H(+) antiporter
SR2_22	28	-	84	1219878	repeat
SR2_21	28	-	48	1219954	repeat
SR2_20	28	-	45	1220027	repeat
SR2_19	28	-	40	1220095	repeat
SR2_18	28	-	46	1220169	repeat

MJ1276	1658	-	484	1222311	dihydroxy-acid dehydratase (EC 4.2.1.9)
MJ1277	503	-	245	1223059	3-isopropylmalate dehydratase (EC 4.2.1.33), small subunit
MJ1278	536	-	127	1223722	hypothetical protein [MJ1279]
MJ1279	542	-	4	1224268	hypothetical protein [MJ1278]
MJ1280	470	+	194	1224462	22% acidic aa
MJ1281	3044	+	20	1224952	
IS_B7	360	+	350	1228346	repeat
MJ1282	1055	+	10	1228716	similar to S-layer protein MJ0822. Family with MJECL39/MJ0954/MJ0822
MJ1282.5	761	-	-17	1230515	hypothetical protein [MJ0902]
MJ1282.6	536	-	16	1231067	
MJ1283	659	-	-48	1231678	
MJ1284	362	-	-9	1232031	
MJ1285	551	-	0	1232582	
MJ1286	1727	-	7	1234316	hypothetical protein [AF0995]
MJ1287	809	-	5	1235130	member conjugal transfer group defined by klbA/VirB11 (family in MJ and AF)
MJ1288	746	-	27	1235903	member conjugal transfer group defined by klbA/VirB11 (family in MJ and AF)
MJ1288.5	338	-	9	1236250	
MJ1289	494	-	36	1236780	
MJ1290	935	-	0	1237715	hypothetical protein [MJ0345]
MJ1291	719	-	16	1238450	hypothetical protein [MJ1403]
MJ1292	2462	+	214	1238664	hypothetical protein [MJ0903]
MJ1293	692	+	50	1241176	
MJ1294	854	-	114	1242836	fumarate hydratase (EC 4.2.1.2), alpha subunit (Class I)
MJ1295	653	-	13	1243502	hypothetical protein [AF1550]
MJ1296	1064	-	18	1244584	biotin synthase (EC 2.8.1.-) (bioB [E. coli])
MJ1297	698	+	60	1244644	6-carboxyhexanoate--CoA ligase (EC 6.2.1.14)
MJ1298	1115	+	16	1245358	8-amino-7-oxononanoate synthase (EC 2.3.1.47)
MJ1299	641	+	149	1246622	dethiobiotin synthetase (EC 6.3.3.3)
MJ1300	1382	+	9	1247272	adenosylmethionine-8-amino-7-oxononanoate aminotransferase (EC 2.6.1.62)
MJ1301	1178	-	269	1250101	ATPase (DNAA)
MJ1302	491	-	50	1250642	energy converting hydrogenase, *ehbL* [MTH1240] A. Tersteegen and R. Hedderich, Eur. J. Biochem. 264, 930 (1999)
MJ1303	1487	-	20	1252149	energy converting hydrogenase, *ehbK* [MTH1241]
MJ1304	455	-	15	1252619	hypothetical protein [APE1272/PH0690]

MJ1305	440	-	-21	1253038	hypothetical protein [AF0299]
MJ1306	248	-	16	1253302	energy converting hydrogenase, *ehbJ* [MTH1242]
MJ1307	785	-	25	1254112	energy converting hydrogenase, *ehbH* and *ehbI* [MTH1243 and MTH1244]
MJ1308	311	-	5	1254428	
MJ1309	1442	-	33	1255903	energy converting hydrogenase, *ehbF* [MTH1246]
MJ1310	353	-	41	1256297	energy converting hydrogenase, *ehbE* [MTH1247]
MJ1311	830	+	162	1256459	endonuclease IV related protein (end4 [*Mycoplasma pneumoniae*])
MJ1312	962	+	34	1257323	hypothetical protein [PH0981]
MJ1313	1208	+	105	1258390	hypothetical protein [AF2333]
MJ1314	920	+	3	1259601	cobalamin biosynthesis protein (cbiB [Salmonella typhimurium] (aminopropanol addition to cobyric acid))
MJ1315	1079	+	-9	1260512	hypothetical ATP/GTP-binding protein [AF2049]
MJ1316	227	+	17	1261608	hypothetical archaeal protein [PAB3259/MTH1281]
MJ1317	1181	-	-11	1263005	hypothetical protein [PAB0874]
MJ1318	1784	-	60	1264849	ATP-dependent protease LA (EC 3.4.21.53)
MJ1319	1466	+	122	1264971	sodium-dependent transporter
LR1_3	425	+	220	1266657	repeat
SR2_17	28	+	-28	1267054	repeat
SR2_16	28	+	42	1267124	repeat
SR2_15	28	+	38	1267190	repeat
SR2_14	28	+	51	1267269	repeat
SR2_13	28	+	39	1267336	repeat
SR2_12	28	+	49	1267413	repeat
SR2_11	28	+	39	1267480	repeat
SR2_10	28	+	48	1267556	repeat
SR2_9	28	+	39	1267623	repeat
SR2_8	28	+	41	1267692	repeat
MJ1320	392	-	84	1268196	hypothetical protein [AF0591]
MJ1321	2138	-	24	1270358	hypothetical protein [AQ1447]
MJ1322	3008	-	22	1273388	hypothetical myosin-related protein, C. Elie et al., J. Mol. Evol. 45, 107 (1997)
MJ1323	1097	-	6	1274491	RAD32 DNA repair protein (probable exonuclease)
MJ1324	701	+	95	1274586	molybdopterin-guanine dinucleotide biosynthesis protein B (mobB [E. coli])

MJ1325	266	+	143	1275430	transcriptional regulatory protein (ArsR Helix-Turn-Helix family)
MJ1326	1100	+	275	1275971	GTP-binding protein
MJ1327	722	+	24	1277095	type II DNA restriction enzyme (MjaIV)
MJ1327.5	830	+	-10	1277807	site-specific DNA-methyltransferase (adenine-specific) (EC 2.1.1.72) (MjaIV)
MJ1328	788	+	-49	1278588	site-specific DNA-methyltransferase (adenine) (MjaIV)
MJ1329	881	+	8	1279384	methionyl aminopeptidase (EC 3.4.11.18)
MJ1330	380	+	149	1280414	hypothetical protein [PH0374]
MJ1331	419	-	9	1281222	similar to acylphosphatase
MJ1332	1178	-	24	1282424	GTP-binding protein
MJ1333	251	+	93	1282517	
MJ1333.5	410	-	65	1283243	hypothetical protein [AQ1741]
MJ1334	848	-	36	1284127	UTP--glucose-1-phosphate uridylyltransferase (EC 2.7.7.9) (UDP-glucose pyrophosphorylase) (UDPGP)
MJ1335	428	+	215	1284342	phosphoheptose isomerase
MJ1336	443	+	71	1284841	similar to ADP-heptose synthase
MJ1337	692	+	242	1285526	
MJ1338	1025	+	113	1286331	H2-forming N5,N10-methylenetetrahydromethanopterin dehydrogenase related prot
MJ1339	461	+	35	1287391	translation initiation factor IF-2 related protein
MJ1340	341	+	75	1287927	hypothetical protein [MJ0310]
MJ1341	935	-	20	1289223	hypothetical protein [MTH721]
MJ1342	338	+	239	1289462	hypothetical protein [MJ0207], 6.7 % cys
MJ1343	1259	-	44	1291103	ammonium transport protein AMT1
MJ1344	335	-	31	1291469	nitrogen regulatory protein P-II
MJ1345	923	+	451	1291920	hypothetical protein with TPR domain-like repeats
MJ1346	1343	+	109	1292952	glutamate--ammonia ligase (EC 6.3.1.2) (glutamine synthetase), R. Cohen-Kupiec et al., J. Bacteriol. 181, 256 (1999)
MJ1347	695	+	93	1294388	hypothetical protein (YbaX [Escherichia coli])
MJ1348	1007	+	38	1295121	hypothetical protein with TPR repeats
MJ1349	1070	-	13	1297211	Ni/Fe hydrogenase, beta subunit [MTH193]
MJ1350	773	-	245	1298229	formylmethanofuran dehydrogenase (EC 1.2.99.5), subunit C-related
MJ1351	1514	-	127	1299870	glutamate synthase (NADPH) (EC 1.4.1.13), subunit alpha
MJ1351.5	1085	-	64	1301019	hypothetical protein [TM0398/AF0952]
MJ1352	758	-	181	1301958	NH(3)-dependent NAD+ synthetase (EC 6.3.1.5)

MJ1353	2018	-	61	1304037	formate dehydrogenase (EC 1.2.1.2), subunit alpha; contains a SeC residue
MJ1354	434	+	303	1304340	
MJ1355	1712	+	47	1304821	hypothetical protein [AQ752/TM1014]
MJ1356	566	+	198	1306731	hypothetical protein, 20% acidic aa [MJ1461]
MJ1357	1028	-	1	1308326	putative potassium channel protein
MJ1358	392	-	324	1309042	
MJ1359	725	-	124	1309891	hypothetical GTP-binding protein
MJ1360	284	-	64	1310239	
MJ1361	875	+	118	1310357	hypothetical protein [MTH1670]
MJ1362	965	-	76	1312273	energy converting hydrogenase, *ehbO* [MTH1237] A. Tersteegen and R. Hedderich, Eur. J. Biochem. 264, 930 (1999)
MJ1363	443	-	72	1312788	energy converting hydrogenase, *ehbM* [MTH1239]
tRNA	84	+	384	1313172	Ser (GCT)
MJ1364	1187	+	178	1313434	hypothetical protein [PH1538]
MJ1365	1178	+	106	1314727	transcriptional regulatory protein (TraB [*Enterococcus faecalis*])
MJ1366	851	-	27	1316783	ribose-phosphate pyrophosphokinase (EC 2.7.6.1)
MJ1367	890	-	21	1317694	molybdate ABC transporter, ATP-binding subunit (ModC [E. coli])
MJ1368	746	-	3	1318443	molybdate ABC transporter, permease protein (ModB [E. coli])
MJ1369	449	+	138	1318581	hypothetical protein [AF2349]
MJ1370	983	+	33	1319063	hypothetical protein [MTH1441]
MJ1371	722	+	9	1320055	methyltransferase, *Methanosarcina barkeri*
MJ1372	830	+	-4	1320773	undecaprenyl diphosphate synthase
MJ1373	485	+	0	1321603	
MJ1374	791	+	77	1322165	hypothetical protein [PAB1219/MTH751]
MJ1375	1244	-	-1	1324199	hypothetical protein [MTH379]
MJ1376	734	+	87	1324286	cell division protein (*ftsJ*)
MJ1377	1658	-	18	1326696	glutamate--tRNA ligase (EC 6.1.1.17) (glutamyl-tRNA synthetase)
MJ1378	1445	+	185	1326881	carbamoyl-phosphate synthase (glutamine-hydrolyzing) (EC 6.3.5.5), first half of large subunit
MJ1378.5	158	+	13	1328339	
MJ1379	299	+	29	1328526	hypothetical protein [MJ0126]
MJ1380	233	+	-4	1328821	hypothetical protein [MJ0125]
MJ1381	1823	+	158	1329212	carbamoyl-phosphate synthase (glutamine-hydrolyzing) (EC 6.3.5.5), second

					half of large subunit
MJ1382	437	-	3	1331475	hypothetical protein [AF2356]
MJ1383	767	-	124	1332366	hypothetical protein [AF1765]
MJ1384	581	-	232	1333179	Flavin-containing NADH dehydrogenase (EC 1.6.99.3) (nox [Thermus thermophilus])
MJ1385	536	-	28	1333743	hypothetical protein [AF0751]
MJ1386	125	+	142	1333885	
MJ1387	1136	-	289	1335435	hypothetical protein [AF2097]
MJ1388	1244	+	375	1335810	S-adenosylhomocysteine hydrolase (EC 3.3.1.1) (adenosylhomocysteinase)
MJ1389	401	-	360	1337815	hypothetical protein [AF1500/MJ0311]
MJ1390	317	-	198	1338330	hypothetical protein [MJ0668]
MJ1391	1253	-	200	1339783	aminotransferase (subgroup I) [AF0409]
MJ1392	1472	+	475	1340258	(R)-citramalate synthase (cimA), D.M. Howell et al., J. Bacteriol. 181, 331 (1999)
MJ1393	1823	+	251	1341981	hypothetical protein [AF2028]
MJ1394	2894	+	156	1343960	hypothetical protein [AF2028]
MJ1394.5	251	+	19	1346873	
MJ1395	395	+	54	1347178	hypothetical protein [AF2027]
MJ1396	8621	+	196	1347769	hypothetical protein [PH0954]
MJ1397	1448	+	69	1356459	hypothetical protein [AF2026]
MJ1398	1172	+	278	1358185	hypothetical archaeal protein [C-terminus similar to AF1564]
MJ1399	584	-	-16	1359925	hypothetical NTP-binding protein [PAB1725/MTH434/AF1395]
MJ1400	200	-	19	1360144	hypothetical protein [MJ0431]
MJ1401	2423	+	117	1360261	RTP-binding RNA helicase
IS_B6	361	-	95	1363140	repeat
MJ1402	1037	-	182	1364359	hypothetical protein [MJ1403]
MJ1403	1070	-	316	1365745	hypothetical protein [MJ1402]
MJ1404	1253	+	368	1366113	hypothetical protein with 4 CBS domains
MJ1405	212	+	63	1367429	acylphosphatase (EC 3.6.1.7)
MJ1406	437	+	102	1367743	aspartate carbamoyltransferase (EC 2.1.3.2), regulatory subunit
MJ1407	386	+	230	1368410	hypothetical protein, 15.9 % gly [AF0055]
MJ1408	1049	+	88	1368884	GTP-binding protein, GTP1/OBG-family
MJ1409	794	-	8	1370735	23% acidic aa
MJ1410	476	-	101	1371312	hypothetical protein [PH0281]
SR1_2	27	-	203	1371542	repeat
LR1_2	169	-	-27	1371684	repeat
MJ1411	1388	+	124	1371808	glycine betaine aldehyde dehydrogenase (GbsB [Bacillus subtilis])

MJ1412	1493	+	16	1373212	hypothetical protein [MTH1865]
MJ1413	299	-	29	1375033	hypothetical protein [PH1682]
MJ1414	713	-	63	1375809	
MJ1415	1100	+	85	1375894	tryptophan--tRNA ligase (EC 6.1.1.2) (tryptophanyl-tRNA synthetase)
MJ1416	1352	-	3	1378349	similar to hydrogenase expression/formation protein
MJ1417	1946	+	136	1378485	ATP-dependent endopeptidase La (EC 3.4.21.53)
MJ1417.5	431	+	-43	1380388	
MJ1418	542	+	333	1381152	L-fuculose-phosphate aldolase (EC 4.1.2.17) (L-fuculose-1-phosphate aldolase)
MJ1419	278	-	22	1381994	hypothetical protein [PHs058]
McjGF6PAT	1496	+	347	1382341	intein
MJ1420	3296	+ -	1709	1382128	glucosamine--fructose-6-phosphate aminotransferase (isomerizing) (EC 2.6.1.16) (L-glutamine-D-fructose-6-phosphate amidotransferase) [MTH171, PH0243]
IS_B2	329	-	97	1385850	repeat
MJ1421	1322	-	161	1387333	cobyrinic acid a,c-diamide synthase
McjRFC-3	1628	-	843	1389804	intein
MJ1422	5528	- -	2168	1393164	replication factor C, small subunit
McjRFC-2	1307	- -	3173	1391298	intein
McjRFC-1	1643	-	76	1393017	intein
MJ1423	1055	-	193	1394265	vanadium nitrogenase associated protein N
MJ1424	521	+	218	1394483	tRNA intron endonuclease [H. volcanii]
MJ1425	1031	+	29	1395033	malate dehydrogenase (NAD+ dependent) (EC 1.1.1.37) (mdhI [*M. thermoautotrophicum*]), H. Thompson et al., Arch. Microbiol. 170, 38 (1998)
MJ1426	461	+	26	1396090	hypothetical protein with 2 CBS domains
MJ1427	983	+	101	1396652	hypothetical protein [MTH830]
MJ1428	1700	+	10	1397645	protein containing 2 or more adjacent copies of the (34 aa) TPR motif
MJ1429	1499	+	0	1399345	hypothetical protein [PH0932]
MJ1430	380	+	92	1400936	phosphoribosyl-AMP cyclohydrolase (EC 3.5.4.19)
MJ1431	1076	+	8	1401324	similar to biotin synthetase
MJ1432	485	+	25	1402425	hypothetical protein [MTH650/PH0734/AF1405]
MJ1433	740	+	6	1402916	hypothetical protein, [MTH145] 26% basic aa
MJ1434	659	+	0	1403656	methylpurine DNA glycosylase (MpgII class of endonuclease III)

MJ1435	212	+	89	1404404	hypothetical protein [TM0526]
MJ1436	290	+	144	1404760	energy converting hydrogenase, *ehbP* [MTH1236] A. Tersteegen and R. Hedderich, Eur. J. Biochem. 264, 930 (1999)
MJ1437	683	+	7	1405057	L-2-haloalkanoate dehalogenase related protein
MJ1438	761	+	29	1405769	cobalamin (5'-phosphate) synthase (cobS [E. coli])
MJ1439	602	+	23	1406553	thermonuclease precursor
MJ1440	845	+	135	1407290	putative shikimate kinase
MJ1441	3677	+	148	1408283	magnesium chelatase subunit
MJ1442	563	+	214	1412174	
MJ1443	320	+	49	1412786	hypothetical protein [AF0736]
MJ1444	977	+	228	1413334	structure specific endonuclease (flap endonuclease-1, FEN1 family), D.J. Hosfield et al., J. Biol. Chem. 273, 27154 (1998)
MJ1445	398	+	151	1414462	hypothetical protein [AQ1666]
MJ1446	770	+	87	1414947	cytochrome-c3 hydrogenase (EC 1.12.2.1), subunit gamma
MJ1447	1142	+	125	1415842	Bifunctional protein: C-terminus is 3-hexulose-6-phosphate synthase (HumS [Methylomonas aminofaciens])
MJ1448	1589	+	0	1416984	site-specific DNA-methyltransferase (N4-cytosine specific) (EC 2.1.1.113)
MJ1449	1109	+	6	1418579	type II site-specific deoxyribonuclease (EC 3.1.21.4) (MjaII)
MJ1450	1112	+	13	1419701	hypothetical protein [MTH853]
MJ1451	1451	+	58	1420871	hypothetical protein [MTH1153]
MJ1452	776	+	296	1422618	protein arginine methyltransferase (EC 2.1.1.-) (HNRNP [Saccharomyces cerevisiae])
MJ1453	575	+	6	1423400	hypothetical archaeal protein [MTH863/PAB1814]
MJ1454	659	-	-2	1424632	3-dehydroquinate dehydratase (EC 4.2.1.10)
MJ1455	914	-	99	1425645	
MJ1456	1274	+	90	1425735	histidinol dehydrogenase (EC 1.1.1.23)
MJ1457	395	+	20	1427029	
MJ1458	653	+	65	1427489	energy converting hydrogenase, *ehbQ* [MTH1235] A. Tersteegen and R. Hedderich, Eur. J. Biochem. 264, 930 (1999)
MJ1459	1667	+	141	1428283	adenine deaminase (EC 3.5.4.2)
MJ1460	476	-	-5	1430421	hypothetical protein [MJ0739]
MJ1461	548	-	141	1431110	hypothetical protein [MJ1356]
MJ1462	245	-	150	1431505	
MJ1463	698	-	27	1432230	hypothetical protein [AQ756/PAB1806]

MJ1464	1103	+	172	1432402	hypothetical GTP-binding protein [PH0645]
MJ1465	1214	-	701	1435420	threonine synthase (EC 4.2.99.2)
MJ1466	641	-	338	1436399	transposase
IS_A2	666	-	-651	1436414	repeat
MJ1467	377	+	219	1436633	
MJ1468	3026	+	21	1437031	similar to surface antigens
MJ1469	224	+	0	1440057	hypothetical protein [MJ0431]
MJ1469.5	461	+	-7	1440274	hypothetical protein [MJ0027]
MJ1470	1871	+	14	1440749	hypothetical protein [MJ1472]
MJ1471	533	+	0	1442620	
MJ1472	1631	+	14	1443167	hypothetical protein [MJ1470]
MJ1473	923	-	7	1445728	5-methyl-tetrahydromethanopterin-homocysteine-S-methyltransferase (EC 2.1.1.13) (cobalamin-independent methionine synthase)
MJ1474	581	+	0	1445728	hypothetical protein [MTH1862/PH0709]
MJ1475	374	+	140	1446449	hypothetical protein [PAB2413/MTH589]
MJ1476	680	+	9	1446832	hypothetical RNA ribosyl 2'-O-methyltransferase (LasT [E. coli])
MJ1477	1007	+	20	1447532	hypothetical protein [TM1410]
MJ1478	896	-	3	1449438	hypothetical protein [potential ortholog of PH0300, family with MJ0485,MJ1157]
MJ1479	1295	+	272	1449710	aminotransferase (subgroup I)
MJ1479.5	311	-	3	1451319	
MJ1480	1268	+	135	1451454	hypothetical protein [MTH867/AF1278]
MJ1481	638	+	15	1452737	21% acidic aa
MJ1482	926	+	18	1453393	2-phosphoglycerate kinase (EC 2.7.2.-) [Methanothermus fervidus]
MJ1483	446	+	20	1454339	
MJ1484	449	+	-15	1454770	
MJ1485	1421	+	10	1455229	TRK system potassium uptake protein
SR1_1	27	-	147	1456824	repeat
SR2_7	28	-	50	1456902	repeat
SR2_6	28	-	51	1456981	repeat
SR2_5	28	-	43	1457052	repeat
SR2_4	28	-	44	1457124	repeat
SR2_3	28	-	49	1457201	repeat
SR2_2	28	-	43	1457272	repeat
SR2_1	28	-	41	1457341	repeat
LR1_1	394	-	-28	1457707	repeat
MJ1486	1166	-	23	1458896	phosphoribosylglycinamide formyltransferase 2 (EC 2.1.2.2) (GART 2)
MJ1487	1277	+	122	1459018	hypothetical protein [PH1701]

MJ1488	1178	+	22	1460317	hypothetical protein [RPF00845]
MJ1489	2018	+	129	1461624	DNA replication initiation protein (MCM family)
MJ1489.5	416	+	116	1463758	hypothetical protein [MJECS05]
MJ1490	1268	+	157	1464331	dihydroorotase (EC 3.5.2.3)
MJ1491	371	+	87	1465686	15.2 % gly
MJ1492	269	+	210	1466267	hypothetical protein [MTH518]
MJ1493	683	+	18	1466554	
MJ1494	1112	-	3	1468352	ATP-dependent 26S protease regulatory subunit 8
MJ1495	833	+	194	1468546	hypothetical membrane protein [MJ0137]
MJ1496	341	+	-7	1469372	hypothetical protein [PH1112]
MJ1497	1028	+	9	1469722	similar to succinyl-diaminopimelate desuccinylase
MJ1498	875	+	23	1470773	site-specific DNA methyltransferase (N4-cytosine) (EC 2.1.1.113) (Mja V)
MJ1499	479	-	3	1472130	
MJ1500	689	-	103	1472922	type II DNA restriction enzyme (MjaV)
MJ1501	668	-	27	1473617	coenzyme F420-dependent NADP reductase, H. Berk and R.K. Thauer, Arch. Microbiol. 168, 396 (1997)
MJ1502	941	-	8	1474566	arylsulfatase related protein (AtsA [*Pseudoalteromonas carrageenovora*])
MJ1503	395	-	23	1474984	autoregulatory transcription regulation factor (Lrs14 [*Sulfolobus solfataricus*])
MJ1504	1097	-	167	1476248	lipopolysaccharide biosynthesis protein (bplD)
MJ1505	2333	-	6	1478587	putative ATP-dependent RNA helicase, eIF-4A family
MJ1506	1181	-	182	1479950	ABC transporter, substrate binding protein [MJ1561]
MJ1507	1184	+	82	1480032	hypothetical membrane protein [MJ0797]
MJ1508	671	+	6	1481222	ABC transporter, probable ATP-binding subunit
MJ1509	458	+	133	1482026	DNA-binding protein(some similarity to HMVA)
MJ1510	578	-	24	1483086	hypothetical protein [PH1069]
MJ1511	320	+	168	1483254	hypothetical protein [MJ0742]
MJ1512	4838	+	185	1483759	reverse gyrase (EC 5.99.1.3)
Mcjr-gyr	1481	+ -	2240	1486357	intein
MJ1513	995	-	770	1489603	
MJ1514	359	+	118	1489721	hypothetical protein (PH0828/E. coli ytfP)
MJ1515	1064	+	6	1490086	hypothetical protein [MTH1466]
MJ1516	287	+	31	1491181	
MJ1516.5	464	-	10	1491942	

MJ1517	821	+	102	1492044	
MJ1518	959	+	153	1493018	
MJ1519	3506	+	137	1494114	exodeoxyribonuclease V (EC 3.1.11.5), subunit alpha
MJ1520	1160	-	38	1498818	bacteriochlorophyll synthase, 43kDa subunit related protein [Rhodobacter cap
MJ1521	1265	+	89	1498907	Na+/H+ exchanger (NAH1 [Bos taurus])
MJ1522	632	+	150	1500322	precorrin-6Y C5,15-methyltransferase [decarboxylating] (EC 2.1.1.132) (cbiE [Salmonella typhimurium])
MJ1523	371	+	69	1501023	hypothetical protein, CRCB family 12.3% gly
MJ1524	323	+	12	1501406	hypothetical protein [Thermococcus celer]
MJ1525	2777	+	-4	1501725	oligosaccharyl transferase STT3 subunit
MJ1526	731	-	27	1505260	hypothetical protein [MTH727/AF0679]
MJ1527	326	-	23	1505609	hypothetical protein, 22% basic aa [AF0736]
MJ1528	656	-	141	1506406	triosephosphate isomerase (EC 5.3.1.1)
MJ1529	500	+	20	1506426	methylated-DNA--protein-cysteine S-methyltransferase (EC 2.1.1.63)
MJ1530	467	+	182	1507108	N-terminal acetyltransferase complex, subunit ARD1
MJ1531	1262	+	9	1507584	type I restriction enzyme CfrI (EC 3.1.21.3), specificity subunit
MJ1532	710	+	38	1508884	phosphoribosylformimino-5-aminoimidazole carboxamide ribotide isomerase (EC 5.3.1.16) (PRAC ribotide isomerase)
MJ1533	1910	-	18	1511522	hypothetical nucleotide binding protein [Methanococcus vannielii]
MJ1534	992	+	291	1511813	F420-dependent N5,N10-methylenetetrahydromethanopterin reductase
MJ1535	896	+	67	1512872	
MJ1536	902	-	38	1514708	thioredoxin reductase (NADPH) (EC 1.6.4.5)
MJ1537	1004	-	8	1515720	
MJ1538	743	+	185	1515905	hypothetical protein (Kti12p [Saccharomyces cerevisiae])
MJ1539	314	+	82	1516730	23% acidic aa
MJ1540	368	+	56	1517100	
MJ1541	1259	-	3	1518730	atrazine chlorohydrolase related protein (AtzA [Pseudomonas sp. (strain ADP)]
MJ1542	2411	-	18	1521159	
MJ1543	989	+	260	1521419	ketol-acid reductoisomerase (EC 1.1.1.86)
MJ1544	1280	-	64	1523752	hypothetical nucleotidyl triphosphate binding protein [Haemophilus influenzae]

MJ1545	692	+	150	1523902	hypothetical archaeal protein [AF2414]
MJ1546	1034	+	14	1524608	acyl carrier protein synthase
MJ1547	365	+	0	1525642	nucleotidyl-transferase
MJ1548	308	+	114	1526121	hypothetical protein [AF0947]
MJ1549	1175	+	0	1526429	acetyl--CoA ligase (EC 6.2.1.1) (acetyl CoA synthetase)
MJ1550	182	+	247	1527851	
MJ1551	170	+	15	1528048	hypothetical protein [AF1733]
MJ1552	392	+	137	1528355	hypothetical zinc finger protein [PH0675]
MJ1553	470	+	25	1528772	transcriptional regulatory protein (ArsR Helix-Turn-Helix family)
MJ1554	1865	+	86	1529328	hypothetical protein [AF0817]
MJ1555	1301	+	166	1531359	aspartate--tRNA ligase (EC 6.1.1.12) (aspartyl-tRNA synthetase)
MJ1556	935	+	43	1532703	hypothetical protein [AF2276]
MJ1557	746	+	8	1533646	hypothetical methyltransferase protein [MTH1200/PAB1538/AF1973]
MJ1558	269	-	7	1534668	hypothetical protein [AF2138]
MJ1559	509	+	87	1534755	hypothetical GTP-binding protein [PAB1537/TM0036/AF0814/MTH1068]
MJ1560	1157	-	13	1536434	quinolone resistance norA protein protein
MJ1561	1658	-	77	1538169	ABC transporter, substrate binding protein [MJ1506]
MJ1562	1151	-	0	1539320	transport protein
MJ1563	446	-	26	1539792	hypothetical protein [Bacillus subtilis]
MJ1564	509	+	395	1540187	hypothetical protein, 21 % acidic aa [AF1395]
MJ1565	1538	+	4	1540700	hypothetical nucleotidyl binding protein (Family - MJ1429/MJECL08)
MJ1566	1340	-	-5	1543573	
MJ1567	515	-	-15	1544073	
MJ1568	518	-	6	1544597	2'-5' RNA ligase (EC 6.5.1.-) (LigT [E. coli])
MJ1569	665	+	376	1544973	cobalt transporter (ABC-type), periplasmic binding protein (CbiM [S.typhimurium])
MJ1570	344	+	0	1545638	hypothetical protein [MTH1706]
MJ1571	830	+	175	1546157	cobalt transport protein (cbiQ)
MJ1572	833	+	0	1546987	ABC transporter ATP-binding protein
MJ1572.5	335	-	3	1548158	similar to C-terminus of helicases
MJ1573	182	-	113	1548453	
MJ1574	2051	-	-42	1550462	putative ATP-dependent RNA helicase, eIF-4A family
MJ1575	563	+	490	1550952	GMP synthase (glutamine-hydrolyzing) (EC 6.3.5.2) (glutamine amidotransferase)
MJ1576	566	+	84	1551599	hypothetical protein [TM0961]

MJ1577	1793	+	33	1552198	
MJ1578	776	+	70	1554061	precorrin-4 C11-methyltransferase (EC 2.1.1.133) (cbiF [Salmonella typhimurium])
MJ1579	209	-	101	1555147	hypothetical protein [RPH00978]
MJ1580	341	-	-19	1555469	hypothetical protein, 21% basic aa [MTH137]
MJ1581	899	+	309	1555778	aspartate carbamoyltransferase (EC 2.1.3.2), catalytic subunit
MJ1582	746	+	6	1556683	hypothetical universal protein [MJ1582]
MJ1583	377	+	3	1557432	hypothetical protein [PH1056]
MJ1584	425	-	8	1558242	
MJ1585	917	-	14	1559173	hypothetical protein [APE0011/PAB0049/MTH579]
MJ1586	1433	+	87	1559260	hypothetical protein (yjeF [E. coli])
MJ1587	512	+	61	1560754	hypothetical protein [HI1400]
MJ1588	335	+	20	1561286	
MJ1589	722	+	37	1561658	hypothetical protein [MJ0871]
MJ1590	314	+	391	1562771	
MJ1591	1046	+	135	1563220	selenium donor protein (selD [Escherichia coli]); contains SeC residue.
MJ1592	725	+	338	1564604	phosphoribosylaminoimidazolesuccinocarboxamide synthase (EC 6.3.2.6) (PRAD succinocarboxamide Sase)
MJ1593	248	+	21	1565350	hypothetical protein [MTH169/AF1941/PAB3433]
MJ1594	632	+	69	1565667	phosphoserine phosphatase (EC 3.1.3.3) (O-phosphoserine phosphohydrolase)
MJ1595	1004	-	34	1567337	hypothetical protein [AF2231]
MJ1596	1040	-	3	1568380	erythro-3-methyl-D-malate dehydrogenase
MJ1597	1286	+	227	1568607	serine hydroxymethyltransferase (glyA [E.coli]) (EC 2.1.2.1)
LR1_18	425	+	92	1569985	repeat
SR2_58	28	+	-28	1570382	repeat
SR2_57	28	+	47	1570457	repeat
SR2_56	28	+	41	1570526	repeat
SR2_55	28	+	38	1570592	repeat
SR2_54	28	+	43	1570663	repeat
MJ1597.5	383	-	-170	1570904	
SR2_53	28	+	-172	1570732	repeat
SR1_121	27	+	114	1570874	repeat
MJ1598	1049	-	126	1572076	hypothetical protein [MTH1426]
MJ1599	830	-	19	1572925	ATP-binding protein [PH1968]
MJ1600	530	+	78	1573003	hypothetical protein, 12.1 % gly [AF1740]

MJ1601	479	+	7	1573540	hypothetical protein, 22% acidic aa [AF0934]
MJ1602	1007	+	23	1574042	homoserine dehydrogenase (EC 1.1.1.3) (HDH)
LR1_17	424	+	81	1575130	repeat
SR1_120	27	+	-27	1575527	repeat
SR1_119	27	+	41	1575595	repeat
SR1_118	27	+	38	1575660	repeat
SR1_117	27	+	46	1575733	repeat
SR1_116	27	+	39	1575799	repeat
SR1_115	27	+	42	1575868	repeat
SR1_114	27	+	46	1575941	repeat
SR1_113	27	+	46	1576014	repeat
SR1_112	27	+	38	1576079	repeat
SR1_111	27	+	45	1576151	repeat
SR1_110	27	+	44	1576222	repeat
SR1_109	27	+	41	1576290	repeat
MJ1603	677	+	297	1576614	ribose-5-phosphate isomerase A (EC 5.3.1.6)
MJ1604	1373	-	18	1578682	ADP-dependent 6-phosphofructokinase (ADP-PFK) [*Pyrococcus furiosus*], J.E. Tuininga et al., J. Biol. Chem. 274, 21023 (1999)
MJ1605	1202	-	59	1579943	glucose-6-phosphate isomerase (EC 5.3.1.9)
MJ1606	1562	-	41	1581546	glycogen (starch) synthase (EC 2.4.1.11)
MJ1607	1169	+	168	1581714	LPS biosynthesis protein (hexose transferase)
MJ1608	209	+	35	1582918	
MJ1609	1121	+	42	1583169	ATPase
MJ1610	1835	+	246	1584536	glucan 1,4-alpha-glucosidase (EC 3.2.1.3) (glucoamylase)
MJ1611	1400	+	25	1586396	alpha-amylase (EC 3.2.1.1) [Pyrococcus furiosus]
MJ1612	1232	+	13	1587809	similar to phosphoglycerate mutase [MTH1591]; M.Y. Galperin et al., Protein Sci. 7, 1829 (1998)
MJ1613	764	-	18	1589823	hypothetical protein [AF1400]
MJ1614	752	-	8	1590583	endonuclease IV related protein (end4 [Mycoplasma pneumoniae])
MJ1615	764	-	4	1591351	hypothetical protein [PH0494]
MJ1616	1487	+	229	1591580	inosine-5'-monophosphate dehydrogenase (EC 1.1.1.205) (IMPDH)
MJ1617	272	+	43	1593110	hypothetical protein [AF2279]
MJ1618	365	-	16	1593763	hypothetical polyketide biosynthesis protein (tcmJ [Streptomyces glaucescens])
MJ1619	710	-	23	1594496	biotin operon repressor/biotin--acetyl-CoA carboxylase synthetase (EC 6.3.4.15) (birA [Escherichia coli]), B. Mukhopadhyay et al., J. Biol. Chem. 273, 5155 (1998)

MJ1620	1514	+	75	1594571	hypothetical protein [MTH1171]
MJ1621	170	-	43	1596298	duplication of a part of MJ1068
MJ1621.5	374	+	1048	1597346	ATPase 2
MJ1623	1535	+	220	1597940	
MJ1624	383	-	128	1599986	hypothetical protein, 21% basic aa [AF0905]
MJ1626	2069	-	102	1602157	
MJ1625	2012	+	56	1602213	hypothetical protein (YPL009c [Saccharomyces cerevisiae])
MJ1627	452	-	7	1604684	hypothetical protein, 5.0% cys [AF2154]
MJ1628	1343	-	101	1606128	putative ABC transporter/ATP-binding protein
MJ1629	875	-	291	1607294	hypothetical protein [AF0504]
MJ1630	3407	-	37	1610738	DNA-directed DNA polymerase II (EC 2.7.7.7), DP2 subunit, Y. Ishino et al., J. Bacteriol. 180, 2232 (1998)
MJ1631	1556	+	447	1611185	glycogen phosphorylase (EC 2.4.1.1)
MJ1632	686	+	20	1612761	similar to coenzyme PQQ synthesis protein III
MJ1633	1424	-	21	1614892	hypothetical NADP phosphatase, L. Aravind and E.V. Koonin, Trends Biochem. Sci. 23, 17 (1998)
MJ1634	722	-	120	1615734	hypothetical protein [AF0878]
MJ1635	1241	+	200	1615934	putative transposase [PF0736]
MJ1636	968	+	67	1617242	N5,N10-methenyltetrahydromethanopterin cyclohydrolase (EC 3.5.4.27)
MJ1637	1346	+	131	1618341	hypothetical ATP/GTP binding protein [PH0922]
MJ1638	779	-	-8	1620458	ATP-binding protein
MJ1639	431	+	148	1620606	
MJ1640	614	-	21	1621672	hypothetical protein [AF1056]
MJ1641	842	-	133	1622647	putative HTH DNA-binding protein, sigma 70 family, 22% basic aa
MJ1642	482	+	386	1623033	hypothetical protein [MJ0150]
MJ1643	3506	+	-34	1623481	P115 protein (eukaryotic chromosome segregation protein homolog)
MJ1644	521	+	160	1627147	hypothetical protein [PH1319/APE1545/TM1675]
MJ1645	728	+	19	1627687	hypothetical protein [PH0635/MTH1227]
MJ1646	620	+	40	1628455	orotate phosphoribosyltransferase related protein
MJ1647	287	+	69	1629144	dna binding protein (hmvA)
MJ1648	689	+	147	1629578	phosphoribosylformylglycinamidine synthase (EC 6.3.5.3), subunit I
MJ1649	1148	-	3	1631418	hypothetical methyltransferase protein [MJ1653]
MJ1650	1106	-	18	1632542	hypothetical protein [AF1972]
MJ1651	776	-	90	1633408	hypothetical protein [AF0461]

MJ1652	2270	+	101	1633509	DNA topoisomerase, Type I (EC 5.9.1.2)
MJ1653	1154	+	19	1635798	hypothetical methyltransferase protein [MJ1653]
MJ1654	596	+	146	1637098	hypothetical protein, 21% acidic aa [AF0779]
MJ1655	548	+	21	1637715	hypoxanthine (guanine) phosphoribosyltransferase (EC 2.4.2.8) [MTH1320], J. Sauer and P. Nygaard, J. Bacteriol. 181, 1958 (1999)
MJ1656	710	+	400	1638663	5-carboxymethyl-2-oxo-hex-3-ene-1,7-dioate decarboxylase (EC 4.1.1.68) (2-hydroxyhepta-2,4-diene-1,7-dioate isomerase)
MJ1657	995	+	60	1639433	hypothetical protein [MJ0090]
MJ1658	356	+	0	1640428	
MJ1659	1070	+	17	1640801	hypothetical ATP/GTP binding protein [PH0922]
MJ1660	1616	+	17	1641888	phenylalanyl-tRNA synthetase, alpha chain related protein
MJ1661	557	-	25	1644086	hypothetical pyridoxine biosynthetic glutamine amidotransferase (YMR095c [Saccharomyces cerevisiae])
MJ1662	1664	-	51	1645801	methyl coenzyme M reductase system, component A2
MJ1663	611	+	95	1645896	molybdopterin-guanine dinucleotide biosynthesis protein A
MJ1664	677	+	-4	1646503	hypothetical protein, *cagS Helicobacter pylori*
MJ1665	1373	-	3	1648556	hypothetical protein [APE2446]
MJ1666	1394	-	131	1650081	hypothetical protein [PH0168]
MJ1667	1118	-	3	1651202	hypothetical protein [PH0167]
MJ1668	1127	-	-7	1652322	hypothetical protein [PH0166]
MJ1669	743	-	55	1653120	hypothetical protein [*Mycobacterium tuberculosis*]
MJ1670	398	-	30	1653548	hypothetical protein [MTH1081/TM1810]
MJ1671	134	-	3	1653685	
MJ1672	2399	-	123	1656207	hypothetical protein, HD domain, L. Aravind and E.V. Koonin, Trends Biochem. Sci. 23, 469 (1998)
MJ1673	386	-	38	1656631	
MJ1674	1898	-	8	1658537	hypothetical protein [TM1812]
MJ1675	776	+	117	1658654	tRNA-pseudouridine synthase I (EC 5.4.99.12)
MJ1676	713	+	192	1659622	hypothetical protein [MTH564]
MJ1677	614	+	13	1660348	hypothetical protein (YchE [*E. coli*])
MJ1678	1142	+	23	1660985	pyridoxal phosphate-bound protein
MJ1679	290	+	16	1662143	hypothetical protein, 25% acidic aa [PH0468]
MJ1680	464	+	-30	1662403	hypothetical protein [PH0469]

MJ1681	1139	-	-4	1664002	hypothetic Zn finger protein
MJ1682	851	-	16	1664869	heat shock protein X
Contig: ECES					
MJECS01	827	-	432	1259	
MJECS02	3542	-	13	4814	similar to type IIS restriction enzyme
MJECS03	444	-	37	5295	pseudogene with similarity to a hypothetical protein (with frameshift) [Archeoglobus]
MJECS04	404	-	164	5863	
MJECS05	431	+	520	6383	hypothetical protein [MJ1489.5]
MJECS06	419	-	195	7428	hypothetical protein [MJECL27]
MJECS07	1214	-	0	8642	
MJECS08	3191	-	96	11929	hypothetical protein, AA xcpc
MJECS09	710	-	-4	12635	hypothetical protein [MJECS08]
MJECS10	833	-	546	14014	
MJECS11	551	+	465	14479	hypothetical protein, E. coli yfbU
MJECS11.5	479	-	381	15890	
Contig: ECEL					
MJECL01	746	+	302	302	hypothetical protein [PH1064, C-terminus]
MJECL02	347	-	37	1432	
MJECL03	323	-	-55	1700	20% acidic aa
MJECL04	1316	+	234	1934	hypothetical protein [PH0977]
MJECL05	185	+	15	3265	
MJECL06	347	-	337	4134	23% basic aa
MJECL07	1235	-	427	5796	hypothetical protein [AF1033/PH0928/MJ1262]
MJECL08	1556	-	36	7388	hypothetical nucelotidyl binding protein (Family - MJ1565/MJ1429)
MJECL08.5	206	+	143	7531	
MJECL09	314	+	52	7789	
MJECL10	1001	+	55	8158	hypothetical protein [PH0155], C-terminus
MJECL12	722	+	6	9165	hypothetical protein [PH0156]
MJECL13	1796	+	800	10687	DNA replication initiation protein (MCM family)
MJECL13.5	1235	-	221	13939	
MJECL13.6	209	+	387	14326	
MJECL14	896	+	-4	14531	ATPase
MJECL15	1094	+	20	15447	ATPase
MJECL16	209	+	61	16602	hypothetical protein [MJ1072]
MJECL17	119	-	3078	20008	archaeal histone
MJECL18	623	+	874	20882	hypothetical protein in the TPR family; see MJ0572
MJECL19	521	+	-7	21498	hypothetical protein in the TPR family; see MJ0572

MJECL20	440	+	831	22850	hypothetical protein tyrosine phosphatase [MJ0215]
MJECL21	1277	-	8	24575	
MJECL22	236	-	279	25090	hypothetical protein [MJECL23]
MJECL23	2474	-	46	27610	hypothetical protein ser/thr kinase [MJECL22]
MJECL24	776	+	459	28069	chromosome maintenance/partitioning protein (ParA family)
MJECL25	329	+	-7	28838	ATPase
MJECL26	1016	-	11	30194	ATPase
MJECL27	485	-	377	31056	hypothetical protein [MJECS06]
MJECL28	3782	-	478	35316	
MJECL29	200	-	904	36420	archaeal histone
MJECL30	458	-	731	37609	hypothetical protein [MJ0974]
MJECL31	194	-	-10	37793	hypothetical protein [MJ0975]
MJECL32	1313	-	1035	40141	
MJECL33	1223	-	-16	41348	
MJECL34	875	-	883	43106	
MJECL35	1889	-	9	45004	ATP binding protein
MJECL36	431	-	390	45825	hypothetical protein [MJ1382]
MJECL37	641	+	399	46224	
MJECL38	599	-	332	47796	
MJECL39	503	-	533	48832	hypothetical protein. Family with MJ1282/MJ0954/MJ0822/MJECS12
MJECL40	3092	-	624	52548	type I site-specific deoxyribonuclease (EC 3.1.21.3)
MJECL41	1286	-	65	53899	similar to restriction modification enzyme /family with MJ0130/MJ1531/MJ1218
MJECL42	1673	-	372	55944	type I site-specific deoxyribonuclease (EC 3.1.21.3), modification subunit
MJECL43	1151	-	243	57338	
MJECL44	992	-	3	58333	hypothetical protein, bacteriophage φPVL orf41

Acknowledgments

The authors thank C. R. Woese and Gary Olsen for many helpful discussions. D.E.G. and N.K. were supported in part by NASA Grant NAG5-8479 (to C. R. Woese and G. Olsen). D.E.G. was also supported by a NIH Cellular and Molecular Training Grant. N.K. was also supported by a DOE grant (DEFG C02-95ER61963) to C. R. Woese. I.J.A. was supported by a DOE grant (DEFG05-97ER20269) to W.B.W. The authors also thank M. D'Souza, G. D. Pusch, N. Maltsev, K. Farahi, W. Kim, W. Lin, W. Gardner, M. Furlong, and D. Singleton for helpful discussions.

[6] Genome of *Pyrococcus horikoshii* OT3

By Yutaka Kawarabayasi

Introduction

Pyrococcus horikoshii OT3, a member of the hyperthermophilic euryarchaeota isolated in 1992 from a hydrothermal vent at a depth of 1395 m
in the Okinawa trough, is an anaerobic microorganism whose growth is
greatly enhanced in the presence of sulfur.[1] This strain is obligately heterotrophic, and tryptophan is required for growth, in contrast to closely related
strains.[1] The organism grows over a pH range of 5–8 (optimum pH 7.0)
and a NaCl concentration range of 1–5% (optimum 2.4%). This archaeon
grows at temperatures ranging from 90° to 105° and optimally at 98°, which
is highest among all members of Archaea for which the entire genomic
sequence has been determined.[2–4] Thus, it was expected that the entire
genomic nucleotide sequence of this archaeon would provide important
information about evolution, the origin of life, the thermostability of its
gene products, and the industrial applications of its proteins. A genome
project focusing on this strain was supported by Ministry of International
Trade and Industry and was completed in April 1998.

Determination of Entire Genomic Sequence

To determine the entire genomic sequence of *P. horikoshii* OT3, a
genomic library was first constructed using the single-copy fosmid vector,
pBAC108L.[5] The fosmid clones in this library were ordered and mapped
on the physical map of the *P. horikoshii* OT3 genome. To determine the
nucleotide sequence of each fosmid clone, a shotgun library was constructed
by cloning fragments generated by sonication of each fosmid DNA, after
purification by CsCl ultracentrifugation. After determination of the nucleotide sequence of each fosmid clone, the remaining gaps, where the corresponding fosmid clones were not found, were filled by sequencing the

[1] J. M. González, Y. Masuchi, F. T. Robb, J. W. Ammerman, D. L. Maeder, M. Yanagibayashi,
J. Tamaoka, and C. Kato, *Extremophiles* **2**, 123 (1998).
[2] C. J. Bult, O. White, G. J. Olsen *et al.*, *Science,* **273**, 1058 (1996).
[3] D. R. Smith, L. A. Doucette-Stamm, C. Deloughery *et al.*, *J. Bacteriol.* **179**, 7135
(1997).
[4] H. P. Klenk, R. A. Clayton, J. F. Tomb *et al.*, *Nature* **390**, 364 (1997).
[5] H. Shizuya, B. Birren, U. Kim *et al.*, *Proc Natl. Acad. Sci. U.S.A.* **89**, 8794 (1992).

TABLE I
FEATURES OF *P. horikoshii* OT3 GENOME

Parameter	Value		
Genome size	1,738,505 bp		
G + C content	41.88%		
Genome coverage by coding region	90.72%		
Potential coding regions		Total: 2061	
	Long ORFs: 2002		Short ORFs: 59
Function assigned	537 (26.06%)		22 (1.07%)
Function unknown	429 (20.82%)		22 (1.07%)
With protein motifs	148 (7.18%)		15 (0.73%)
No similarity and no motifs	888 (43.09%)		

shotgun clones constructed from long polymerase chain reaction (PCR) products directly amplified from genomic DNA. To confirm the authenticity of the genomic sequence, the restriction patterns of the long PCR products amplified from the fosmid and genomic DNA were compared with those deduced from the determined nucleotide sequence. The nucleotide sequences of 60 fosmid clones and 8 long PCR products together produced the 1,738,505-bp-long entire nucleotide sequence of the *P. horikoshii* OT3 genome with high accuracy.[6] The G + C content of this genome is 41.9%.

Extraction of Information from Genomic Sequence

The most important information extracted from the nucleotide sequence of the *P. horikoshii* OT3 genome was the assignment of potential protein coding regions.[6] At first, open reading frames (ORFs) consisting of 100 codons or longer were designated as potential coding regions. The conditions under which this assignment was made were as follows. All of the longest ORFs, between an ATG or GTG and a stop codon, with 100 codons or longer were taken to be potential protein coding regions (long ORFs), even though overlapping ORFs may be present in the same or the opposite strand. At the second step, focusing on the regions where ORFs over 100 codons long were not identified, ORFs consisting of 50 to 99 codons and showing similarity to any data in the protein database or containing identifiable protein motifs were designated as short ORFs. As shown in Table I, the total number of long and short ORFs were 2002 and 59, respectively, and in total these ORFs occupied 90.7% of the entire genome.

[6] Y. Kawarabayasi, M. Sawada, H. Horikawa *et al.*, *DNA Res.* **5**, 55 and 147 (1998).

For the assignment of gene function, a similarity search comparing data obtained to sequences in the public protein database was performed. During the similarity search, in many cases it was found that some ORF in *P. horikoshii* OT3 had 30% identity to some genes in other archaea, but the ORF had only 10% identity to the original genes experimentally identified, which had 30% identity to genes in other archaea and which were used for annotation of genes in other archaea. If the gene products or functions were assigned according to the similarity to unconfirmed genes deduced from the genomic sequence and, moreover, if these gene names were registered in the public database, serious problems would be created for all database users. Thus, gene functions or gene names were deemed to be acceptable only when these were similar to genes with experimentally confirmed functions. The total ORFs were classified into four classes: ORFs with known functions, ORFs with unknown functions, ORFs with just protein motifs, and ORFs with neither. The relative proportions of the ORFs of these four categories are shown in Table I.

All of the information and data indicated in this section are available to the public and an ftp service for sequences and a homology search service have been made available (URL; http://www.bio.nite.go.jp/ot3db_index.html).

Identification of Enzymes Encoded on the Genome of *P. horikoshii* OT3

As mentioned in the previous section, there were 559 ORFs for which a function could be assigned, including those encoding enzymes, structural proteins, regulatory factors, and other functional proteins. Among all of these function-assigned ORFs, those coding for proteins that catalyze some reaction were selected; a total of 203 of such ORFs and 154 kinds of enzymes were identified. Among these 203 ORFs, 49 ORFs were found to code for the same enzyne as those encoded by other ORFs or the subunits of one enzyme. The ORFs coding for enzymes are summarized in Table II.

Approximately 154 enzymes have been classified according to functional categories. Through the functional analysis of these enzymes, the total number of enzyme species and ORFs involved in carbohydrate metabolism, energy metabolism, lipid metabolism, nucleotide metabolism, amino acid metabolism, cofactor metabolism, tRNA acylation, and other functions was estimated and these findings were summarized in Table III. The number of enzymes involved in amino acid, nucleotide, carbohydrate, and energy metabolism is 90, 48, 52, and 45, respectively. These values are slightly larger than those for other categories. The identified enzymes occupy just small parts of entire pathways.

TABLE II

ENZYMES IDENTIFIED IN ANALYSIS OF *P. horikoshii* OT3 GENOME

EC No.	ORF ID	EC No.	ORF ID	EC No.	ORF ID	EC No.	ORF ID
1.1.1.1333	PH0417	2.1.2.10	PH1146	3.1.26.4	PH1650	4.3.2.2	PH0852
1.1.1.205	PH0307	2.1.2.11	PH0951	3.1.3.3	PH1885	4.4.1.1	PH1093
1.1.1.3	PH1075	2.1.3.3	PH0726	3.1.31.1	PH1212	4.6.1.3	PH0013
1.1.1.34	PH1805	2.3.1.16	PH0676	3.2.1.4	PH1527	5.1.3.13	PH0416
1.1.1.37	PH1277	2.4.2.10	PH1128	3.2.2.21	PH0784	5.1.3.2	PH0378
1.1.1.38	PH1275	2.4.2.14	PH0240	3.3.1.1	PH0540		PH1742
1.1.1.85	PH1722	2.4.2.19	PH0011	3.4.11.18	PH0628	5.2.1.8	PH1399
1.1.1.95	PH1387	2.4.2.28	PH0125	3.4.13.9	PH1149	5..3.1.1	PH1884
1.17.4.1	PH0363		PH1143	3.4.21.92	PH0201	5.3.1.6	PH1375
1.2.1.11	PH1088	2.4.2.4	PH1598	3.4.24.-	PH0003	5.4.2.1	PH0037
1.2.1.12	PH1830	2.5.1.1	PH1072		PH0246	5.4.2.2	PH1210
1.2.1.2	PH1353	2.5.1.16	PH0211	3.4.24.57	PH1987	5.4.2.8	PH0923
	PH1434	2.5.1.3	PH1156	3.4.99.46	PH1402	5.4.99.2	PH1306
	PH1437		PH1357		PH1553	5.99.1.2	PH0622
1.2.1.38	PH1720	2.6.1.1	PH0771	3.5.1.1	PH0066	5.99.1.3	PH0800
1.2.7.1	PH0678		PH1308		PH0232		PH1011
	PH0679		PH1371		PH1463	6.1.1.1	PH1011
	Ph0680	2.6.1.11	PH1716	3.5.1.19	PH0999	6.1.1.10	PH0993
	PH0681	2.6.1.16	PH0243	3.5.2.3	PH1963	6.1.1.11	PH0710
	PH0682					6.1.1.12	PH1020
	PH0684	2.6.1.19	PH0138	3.5.3.11	PH0083		
	PH0685		PH1423	3.5.4.12	PH1589	6.1.1.14	PH1614
	PH1660	2.7.1.15	PH1845	3.5.4.13	PH1997	6.1.1.15	PH1006
	PH1661	2.7.1.36	PH1625	3.6.1.-		6.1.1.16	PH0636
	PH1662	2.7.1.39	PH1087	3.6.1.1	PH1907	6.1.1.17	PH1686
	PH1663				PH1972		
	PH1665	2.7.2.2	PH1282	3.6.1.34	PH1974	6.1.1.19	PH1478
	PH1666	2.7.2.3	PH1218		PH1975	6.1.1.2	PH1921
1.3.3.1	PH1516				PH1976	6.1.1.20	PH0657
		2.7.3.-	PH0484		PH1977		PH0658
		2.7.4.14	PH1265		PH1977	6.1.1.21	PH0290
1.4.1.4	PH1593	2.7.4.3	PH1117		PH1981		
1.4.3.16	PH0015		PH1753	4.1.1.-	PH1014	6.1.1.22	PH0241
1.4.4.2	PH1994	2.7.4.6	PH0698	4.1.1.21	PH0320		
	PH1995	2.7.4.7	PH1155	4.1.1.3	PH0834	6.1.1.3	PH0699
1.5.1.-	PH1397	2.7.4.9	PH1695	4.1.1.39	PH0939	6.1.1.5	PH1065
1.5.3.1	PH1364	2.7.6.1	PH1923	4.1.1.41	PH1283	6.1.1.6	PH0224
1.6.-.-	PH0572	2.7.7.10	PH0365		PH1284	6.1.1.7	PH0297
	PH1509	2.7.7.13	PH1697		PH1287	6.1.1.9	PH0314
1.6.4.-	PH1217	2.7.7.22	PH0925	4.1.2.17	PH0191	6.3.1.2	PH0359
1.6.4.5	PH1426	2.7.7.24	PH0413	4.1.3.12	PH1727	6.3.2.6	PH0239
1.6.5.3	PH1447		PH1925	4.2.1.11	PH1942	6.3.3.1	PH0316
	PH1449	2.7.7.6	PH1544	4.2.1.20	PH1583	6.3.4.13	PH0323
2.1.1.-	PH1823		PH1545	4.2.1.32	PH1683	6.3.4.15	PH0147
2.1.1.14	PH1089		PH1546		PH1684	6.3.4.2	PH1792
2.1.1.32	PH1829		PH1632	4.2.1.33	PH1724	6.3.4.4	PH0438
2.1.1.63	PH1835		PH1637		PH1726	6.3.5.1	PH0182
2.1.1.72	PH1032		PH1908			6.3.5.2	PH1346
2.1.1.73	PH0039		PHS044	4.2.1.46	PH0414		PH1347
2.1.1.77	PH1886	2.7.7.7	PH0112	4.2.1.52	PH0847	6.3.5.3	PH1953
2.1.1.80	PH0481		PH1947	4.2.1.70	PH1242		PH1955
2.1.1.98	PH0725	2.7.9.2	PH0092	4.2.99.18	PH1498	6.3.5.4	PH1102
2.1.2.-	PH0318	3.1.1.61	PH0483	4.2.99.2	PH0857	6.5.1.2	PH1622
2.1.2.1	PH1654				PH1406	6.5.1.4	PH1529

TABLE III
CLASSIFICATION OF ENZYME SPECIES AND ORFs

Category of function	Number of ORFs	Number of enzyme species
Carbohydrate metabolism	52	33
Energy metabolism	45	21
Lipid metabolism	6	6
Nucleotide metabolism	48	36
Amino acid metabolism	90	75
Cofactor metabolism	18	15
tRNA acylation	20	19
Other functions	32	32

The presence of inteins was examined in the all ORFs estimated. The intein is a self-splicing portion of a polypeptide that can also act as a homing endonuclease,[7,8] and the presence of 33 inteins has been reported in archaeal genes.[2,3,9–13] For example, *Methanococcus jannaschii* is known to be a species that possesses 15 genes with inteins. Intein portions were originally identified in 10 ORFs in *P. horikoshii* OT3 and now 14 inteins have been found in 13 genes (see InBase; http://www.neb.com/neb/frame_tech.html). Among these, three ORFs, coding for DNA repair protein (PH0263), cell division control protein (PH0606), and ATP synthetase (PH1975), were identified as the first cases of intein-containing genes in each of these gene families.

Discovery of New Enzymes from Genomic Data

On examining gene organization and gene function, this information can be used to identify new thermostable enzymes. As mentioned previously, the genomic sequence provides information allowing the identification of enzymes and their functions, which in many cases are easily determined by means of a database similarity search. However, the remaining unidentified ORFs comprise more than 50% of the total ORFs.

[7] R. Hirata, Y. Ohsumi, A. Nakano, H. Kawasaki, K. Suzuki, and Y. Anraku, *J. Biol. Chem.* **265**, 6726 (1990).
[8] P. M. Kane, C. T. Yamashiro, D. F. Wolczyk, N. Neff, M. Goebl, and T. H. Stevens, *Science* **250**, 651 (1990).
[9] J. Riera, F. T. Robb, R. Weiss, and M. Fontecave, *Proc. Natl. Acad. Sci. U.S.A.* **94**, 475 (1990).
[10] M. Xu, M. W. Southworth, F. B. Mersha, L. J. Hornstra, and F. B. Peler, *Cell* **75**, 1371 (1993).
[11] M. Takagi, M. Nishioka, H. Kakihara *et al.*, *Appl. Environ. Microbiol.* **63**, 4504 (1997).
[12] F. B. Perler, D. G. Comb, W. E. Jack *et al.*, *Proc. Natl. Acad. Sci. U.S.A.* **89**, 5577 (1992).
[13] F. Niehaus, B. Frey, and G. Antranikian, *Gene* **204**, 153 (1997).

Identification of Essential Enzymes from Genomic Information

Through genomic sequence analysis, genes that may be absent in the organism can be identified. However, it is very difficult to confirm the absence of a particular enzyme. In analysis of the information extracted from the *P. horikoshii* OT3 genomic sequence, some enzymes thought to be necessary for cell viability were not identified in the similarity search. This does not necessarily mean that the genes encoding these enzymes are absent from the *P. horikoshii* OT3 genome. In some cases, the apparent absence of necessary enzymes results from the sequence diversity of the genes of *P. horikoshii* OT3 in relation to the genes with the same functions in other organisms or results from the absence of the metabolic pathway itself.

For example, direct acylation of tRNA^{Gln} with glutamine has been observed in *Escherichia coli*[14] and in the cytoplasm of mammalian cells.[15,16] However, a transamidation pathway in which Gln-tRNA^{Gln} is formed by the amidation of amino acid misacylated Glu-tRNA^{Gln} is operative in the archaea, gram-positive eubacteria, mitochondria, and chloroplasts.[17-19] This amidation reaction is catalyzed by a tRNA-specific glutamine amidotransferase, Glu-tRNA^{Gln} amidotransferase. In the genome of *P. horikoshii* OT3, the gene encoding glutamyl-tRNA synthetase was identified easily by means of a similarity search, although a gene encoding glutaminyl-tRNA synthetase was not identified. This indicates that the Glu-tRNA^{Gln} amidotransferase activity is necessary for this strain and likely for other archaea. To identify the three subunits of this amidotransferase, a similarity search was performed based on the amino acid sequence of the subunits of this enzyme of *Archaeoglobus fulgidus* and *Bacillus subtilis*.[20] Results indicated that the subunits B and C of the Glu-tRNA^{Gln} amidotransferase were encoded by genes present in the *P. horikoshii* OT3 genome, but a gene coding for the subunit A was not apparent.

It has been reported that both Gln-tRNA and Asn-tRNA transamidation activity were exhibited by a single enzyme encoded by orthologous ORFs of three Glu-tRNA^{Gln} amidotransferase subunits in *Deinococcus radiodurans*.[21] In many archea, asparaginyl-tRNA synthetase, which catalyzes

[14] R. A. Lazzarini and A. H. Mehler, *Biochemistry* **3**, 1445 (1964).

[15] J. M. Ravel, S. F. Wang, C. Heinemeyer, and W. Schive, *J. Biol. Chem.* **240**, 432 (1965).

[16] M. P. Deutscher, *J. Biol. Chem.* **242**, 1123 (1967).

[17] A. Schon, S. Kannangara, S. Gough, and D. Soll, *Nature* **331**, 187 (1988).

[18] A. Schon, H. Hottinger, and D. Söll, *Biochimie* **70**, 391 (1988).

[19] A. W. Curnow, M. Ibba, and D. Söll, *Nature* **382**, 589 (1996).

[20] A. W. Curnow, K. Hong, R. Yuan, O. Martins, W. Winkler, T. M. Henkin, and D. Soll, *Proc. Natl. Acad. Sci. U.S.A.* **94**, 11819 (1997).

[21] A. W. Curnow, D. L. Tumbula, J. T. Pelaschier, B. Min, and D. Soll, *Proc. Natl. Acad. Sci. U.S.A.* **95**, 12838 (1998).

the direct acylation of tRNA^{Asn} with asparagine, has not been identified. In other archaea, Glu-tRNA^{Gln} amidotransferase shows two distinct activities, catalyzing the amidation of glutamate or aspartate bound to a tRNA molecule. However, the ORF encoding asparaginyl-tRNA synthetase was identified easily on the genome of *P. horikoshii* OT3. Thus, the amidotransferase in *P. horikoshii* OT3 needs not possess aspartate amidotransferase activity. It would seem that this difference in activity should reflect a difference in amino acid sequence when compared to other archaea. These observations indicate that the Glu-tRNA^{Gln} amidotransferase, including a subunit A, should be present in *P. horikoshii* OT3, although this enzyme is expected to have a different amino acid sequence and different activity as compared to other archaeal amidotransferases. Thus it is very important to use experimental approaches to identify apparently essential enzymes from genomic information.

New Enzyme Identification from Pathway Analysis

As shown in Table III, the enzymatic functions related to 203 ORFs (9.8% of the total ORFs, 36% of the total ORFs with assigned functions) encoded on the genome were determined through similarity searches. Most of these gene products serve to perform reactions in metabolic pathways. However, as in the case of other archaea, it seems likely that these identified enzymes are not enough to comprise all reactions in all metabolic pathways. Among the enzymes involved in the tryptophan biosynthesis pathway, some were not found in *P. horikoshii* OT3, but this strain needs tryptophan for its viability and growth. It is likely in the case of this pathway and most other metabolic pathways in *P. horikoshii* OT3 that function-assigned enzymes are not sufficient to construct these pathways. The findings that essential enzymes in a metabolic pathway are missing or lacking can be explained by two possibilities. One possibility is that the gene product has the correct enzymatic activity but possesses an amino acid sequence strikingly different from that of enzymes identified and analyzed previously in other species. Especially in the case of a pathway in which all enzymes are assigned except for one enzyme, the missing enzyme may be identified through experimental analysis. Such an experimental approach includes analysis of the expression of enzymatic activity, characterization of the enzymes, complementation of the defect in a mutant strain, and structural analysis.

The other possibility is that such a pathway is actually not active in this organism. In such a case, the identification of such enzymes in this unidentified pathway requires much experimentation. However, all of these enzymes should be encoded by ORFs deduced from the genomic sequence.

Thus, utilization of the genomic sequence information should make identification of new pathways easier than by traditional biological analysis.

Identification of Unexpected Enzymatic Activities

Gene function was assigned on the basis of similarity to genes that are already deposited in public databases and for which the function has been identified experimentally. However, the similarity shown in this type of analysis is relatively weak even among species in the archaea. Most of the function-assigned genes have only 30% identity to known genes. This weak similarity of putative gene products is due to the diversity of archaea, so it is unclear whether this similarity is a reliable basis for the identification of gene activity and function. However, this low level of similarity also may mean that such genes may have multiple functions.

The β-glucosidase produced by *Pyrococcus furiosus* is known to be an enzyme with a wide substrate range.[22] The gene encoding this enzyme has been cloned and sequenced,[23] and it was found to show highest similarity with that of β-galactosidase in *Sulfolobus solfataricus*.[24,25] These findings indicate that genes with only weak similarity may encode products with different enzymatic activity(ies). Fortunately, in the case of *P. horikoshii* OT3, genes homologous to the β-glucosidase gene of *P. furiosus* have been identified (PH501 and PH366). However, the similarities between the genes in *P. furiosus* and *P. horikoshii* OT3 are rather weak (38% identity for PH501, 30% identity for PH366). In addition, these genes in *P. horikoshii* OT3 genes show the same degree of similarity to other genes: β-mannosidase in *P. furiosus*[26] and β-glycosidase in *S. solfataricus*[27] and *S. shibatae*.[28] This similarity is lower than expected for archaea belonging to the same genus. These observations also suggest the possibility that these two enzymes in *P. horikoshii* OT3 may possess other activities in addition to those of the enzymes already isolated from *P. furiosus*.

The gene (PH1371) identified as an aspartate aminotransferase gene on the basis of homology has been expressed in *E. coli,* and the enzymatic activity of its gene products has been analyzed. Indeed, the results of

[22] S. W. M. Kengen, E. J. Luesink, J. M. Stams, and A. J. B. Zehnder, *Eur. J. Biochem.* **213,** 305 (1993).

[23] W. G. B. Voorhorst, R. I. L. Eggen, E. J. Luesink, and W. M. de Vos, *J. Bacteriol.* **177,** 7105 (1995).

[24] M. V. Cubellis, C. Rozzo, P. Montecucchi, and M. Rossi, *Gene* **94,** 89 (1990).

[25] S. Little, P. Cartwright, C. Campbell, A. Prenneta, J. McChesney, A. Mountain, and M. Robinson, *Nucleic Acids Res.* **17,** 7980 (1989).

[26] M. W. Bauer, E. J. Bylina, R. V. Swanson, and R. M. Kelly, *J. Biol. Chem.* **271,** 23749 (1996).

[27] L. H. Pearl, A. M. Hemmings, R. Nucci, and M. Rossi, *J. Mol. Biol.* **229,** 561 (1993).

[28] H. Connaris and B. M. Charalambous, unpublished data.

analysis indicated that this enzyme could use aromatic amino acids as a substrate.[29] It certainly seems likely that the genes encoding enzymes with multiple specificities may be present in this genome.

Identification from Gene Dosage

Similarity searches comparing all ORFs deduced from the entire genomic sequence with each other serve to provide information about gene dosage. Analysis of gene dosage data from entire genomes provides us with information on the gene families showing conserved gene dosage and gene families showing diversed gene dosage. Information on gene dosage was obtained through analysis of the genomic sequences of four archaea: *M. jannaschii, Methanobacterium thermoautotrophicum, A. fulgidus,* and *P. horikoshii* OT3. A total of 224 gene families were identified as single genes in one species. However, the number of gene families composed of two genes in each of the four species was found to be only three, and no gene family composed of over three genes was identified. Among the three gene families composed of two genes in each of the four species, one family was very interesting, as the two genes in every species are located adjacently and the direction of these two genes is conserved in each of the genomes (PH0982 and 0983; AF2380 and 2381, MJ0578 and 0579, MTH1173 and 1174). This conservation of location and direction suggests that these genes may have an important function.

However, gene families of different gene dosage in the four species were also identified. Both function-assigned genes and hypothetical genes were included in these gene families, especially those that include a large number of genes in some species but none or less in most other species. Those genes included in gene families showing conserved gene dosage and those showing diverse gene dosage are interesting candidates for functional analysis through experimentation.

Confirmation of Thermostable Enzyme Activity

To confirm the enzymatic activity of the candidate enzymes encoded by genes in the *P. horikoshii* OT3 genome, expression of the candidate proteins in *E. coli,* analysis of enzymatic activity, crystallization, and three-dimensional structure of these proteins need to be perused. Indeed, 15 proteins have been expressed successfully in *E. coli* using pET vectors

---

[29] I. Matsui, personal communication.

TABLE IV
ANALYZED GENE PRODUCTS OF *P. horikoshii* OT3

Gene product	Expression	Detection of activity	Crystallization	Three-dimensional structure
Ribose-5-phosphate Isomerase	Yes	Yes	Yes	Solved
Aromatic amino acid aminotransferase	Yes	Yes	Yes	Solved
N-Methyltransferase	Yes	Yes		
Carboxypeptidase + aminoacylase	Yes	Yes	Yes	
Acylamino acid-releasing enzyme	Yes	Yes	Yes	
Aminopeptidase	Yes	Yes	Yes	
Proteasome β subunit	Yes	Yes		
L-aspartate oxidase	Yes	Yes		
β-Glycosidase	Yes	Yes		
Flap endonuclease	Yes	Yes		
DNA polymerase I	Yes	Yes		
NADPH dehydrogenase	Yes	Yes		
Threonine dehydrogenase	Yes	Yes	Yes	
Initiation factor 5A	Yes			
Signal recognition protein 54	Yes	Yes		

(Novagen, Madison, WI) and these results are summarized in Table IV. Six proteins have already been crystallized and the tertiary structure of two enzymes has been solved.[29–31]

Preparation of Materials for Postgenomic Period

The next step in the analysis of genomic data will focus on functional and structural genomics. In the case of *B. subtilis, S. cerevisiae,* and *E coli,* disruptant mutant strains for every one of the genes estimated from the entire sequence have been constructed. However, it is difficult to use this approach for hyperthermophiles because most of these organisms cannot grow on plates and no methods are available for transformation with DNA molecules. Thus, to identify or confirm gene function, in some instances it is necessary to express the gene products in transformable bacteria in order to assess the enzymatic activity, to analyze the thermostability of gene products, and to define the tertiary structure of the gene products. For these analyses, the construction of clones showing gene expression is neces-

[30] K. Ishikawa, H. Ishida, Y. Koyama *et al., J. Biol. Chem.* **273,** 17726 (1998).
[31] S. Ando, K. Ishikawa, H. Ishida, Y. Kawarabayasi, H. Kikuchi, and Y. Kosugi, *FEBS Lett.* **447,** 25 (1999).

sary. To support this step, plasmid clones each containing a *P. horikoshii* OT3 genomic fragment appropriately 2 kb in size have been constructed. The sequences of ORFs and genomic regions contained in these plasmid clones have been determined by sequencing both ends of the 2-kb insert. These clones are available (kyutaka@nite.go.jp).

[7] Genomic Sequence of Hyperthermophile, *Pyrococcus furiosus:* Implications for Physiology and Enzymology

By FRANK T. ROBB, DENNIS L. MAEDER, JAMES R. BROWN,
JOCELYNE DIRUGGIERO, MARK D. STUMP, RAYMOND K. YEH,
ROBERT B. WEISS, and DIANNE M. DUNN

Introduction

Microorganisms that are able to grow at temperatures above 90° are defined as hyperthermophiles.[1] They form a diverse group consisting of autotrophic and heterotrophic prokaryotes, including several bacteria, although the majority of hyperthermophiles are Archaea.[2] Most of the conventional tools of genetic and physiological analysis are either not effective or very difficult to apply to these microorganisms due to their unusual growth conditions. As a result, relatively slow progress has characterized the field since its inception. A new paradigm has characterized the field recently, with the availability of complete genome sequences of five hyperthermophiles. Comprehensive inventories of coding sequences are a unique resource that may, to some degree, compensate for the lack of genetic systems in the hyperthermophilic Archaea and are an invaluable aid during the interpretation of physiological studies.

A unique resource for comparative studies of hyperthermophiles, namely the complete genomic sequences of three species in the genus *Pyrococcus,* is now accessible[3,4,5] (Forterre, personal communication, 1998).

[1] K. O. Stetter, G. Fiala, R. Huber, and A. Segerer, *FEMS Microbiol. Rev.* **75,** 117 (1990).
[2] C. R. Woese, O. Kandler, and M. L. Wheelis, *Proc. Natl. Acad. Sci. U.S.A.* **87,** 4576 (1990).
[3] D. L. Maeder, R. Weiss, D. Dunn, J. Cherry, J. M. Gonzalez, J. DiRuggiero, and F. T. Robb, *Genetics* **152,** 1299 (1999).
[4] Y. Kawarabayasi, M. Sawasa, H. Horikawa, Y. Haikawa, Y. Hino, S. Yamamoto, M. Sekine, S.-I. Baba, H. Kosugi, and A. Hosoyama *DNA Res.* **5,** 55 (1998).
[5] G. Erauso, A. L. Reysenbach, A. Godfroy, J. R. Meunier, B. Crump, F. Partensky, J. A. Baross, V. Marteinsson, G. Barbier, N. R. Pace, and D. Prieur, *Arch. Microbiol.* **160,** 338 (1993).

Although these strains are quite similar in their fermentative, sulfur-reducing growth physiology and optimal growth temperatures, which are in the range 98–100°, significant issues of genome divergence are emerging from the ongoing study of their genomic sequences.

Pyrococcus furiosus, an Archaeon Growing Optimally at 100°

Pyrococcus furiosus DSM 3838 was isolated from a shallow marine solfatara at Vulcano Island off southern Italy. The cells are fermentative, sulfur-reducing, flagellated cocci with a maximal growth temperature of 103° and optimal growth at 100°.[6] Generation times of 35 min and cell densities of more than 10^{10} cells/ml have been reported.[7,8]

Pyrococcus furiosus is unusual in that it was the first hyperthermophile growing optimally at 100° to be described thoroughly.[6] It is able to grow to high cell densities[7] and in the absence of elemental sulfur.[8] These factors have contributed to its widespread adoption as a source of highly thermostable enzymes. Research with this strain has been increasing steadily since the early 1990s, resulting in 259 publications relating to enzymes purified from *P. furiosus,* or expressed from cloned genes. More than 50 papers on the physiology or biochemistry have appeared per year for both 1998 and 1999. *P. abyssi* and *P. horikoshii* are represented by 6 and 11 publications, respectively, on biochemical or physiological topics. This reflects the more recent description of these strains (*P. abyssi* was described in 1993[5] and *P. horikoshii* in 1997[9]) and their inability to grow to high cell densities without elemental sulfur. Without doubt, the availability of their complete genome sequences will stimulate further research on *P. abyssi* and *P. horikoshii*. However, for the time being, the *P. furiosus* genome project is significant because it provides the most comprehensive resource for both genomics and proteomics available for the study of hyperthermophilic Archaea.

Sequencing Strategy and Methods

The determination of the 1,908,255-bp genomic sequence of *P. furiosus* was accomplished in three phases. The first phase included pulsed-field gel electrophoresis to determine the size of the genome and a sequencing

[6] G. Fiala and K. O. Stetter, *Arch. Microbiol.* **145,** 56 (1986).

[7] N. Raven, N. Ladwa, D. Cossar, and R. Sharp, *Appl. Microbiol. Biotechnol.* **38,** 263 (1992).

[8] M. W. Adams, *in* "Protocols for Archaebacterial Research" (F. T. Robb, E. M. Fleischmann, A. R. Place, S. DasSarma, and H. J. Schreier, eds.). Univ. of Maryland Press, College Park, MD, 1991.

[9] J. M. Gonzalez, Y. Masuchi, F. T. Robb, J. W. Ammerman, D. L. Maeder, M. Yanagibayashi, J. Tamaoka, and C. Kato, *Extremophiles* **2,** 123 (1998).

survey of several thousand individual, random reads, which were placed on record in the GST database at GenBank.[10] The genome was completed by the University of Utah Genome Center using the following strategy. Complete genomic sequencing was accomplished using a combination of multiplex sequencing and BigDye terminator reactions as follows.

Three subclone libraries (two medium-insert plasmid and one cosmid library) were constructed to obtain saturating physical coverage of the genome. The vectors used are based on modified plasmid R1 replication origins and copy number controls.[11] The sequence of the vectors used in the sequencing of the *P. furiosus* genome can be found in GenBank (accession numbers AF 129072, AF 129504). The vector system uses a modified plasmid R1 origin that allows stable propagation of inserts at low copy number, and yields high-quality DNA preparations, after a brief induction of a runaway replication system.

Plasmid Libraries

Genomic DNA is prepared from 1-liter cultures of *P. furiosus* grown at 95° using previously described methods.[10] The resuspended cells are mixed with 1 ml of 1.0% molten agarose (FMC SeaPlaque GTG grade) in TE buffer (10 mM Tris, 1 mM EDTA, pH 7.5), and the molten suspension is drawn up into 1-ml plastic syringes and allowed to set. Hardened plugs are extruded and incubated in 6.0 ml of 1.0 mg/ml proteinase K, 100 mM EDTA, pH 8.0, 1% (w/v) N-laurylsarcosine, 0.2% sodium deoxycholate for 20 hr at 42° with gentle shaking. The plugs are washed six times in 20 mM Tris–HCl, pH 8, 50 mM EDTA, incubated for 1 hr in 20 mM Tris–HCl, pH 8, 50 mM EDTA, 1 mM phenylmethylsulfonyl fluoride (PMSF), and washed twice in 20 mM Tris–HCl, pH 8, 50 mM EDTA and twice in 2 mM Tris–HCl, pH 8, 5 mM EDTA. The plugs are stored at 4°.

Agarose plugs (~200 μl) of *P. furiosus* genomic DNA are equilibrated in restriction buffer, heated to 65° for 5 min, and brought to 37°. Partial digestions with either *Tsp*509I (1 unit/μg DNA, 15 min at 65°) or *Alu*I (1 unit/μg DNA, 20 min at 37°) restriction enzymes are immediately loaded onto a 0.6% Seakem Gold agarose gel. Fragments ranging from 5 to 15 kb are excised from the gel and eluted into 0.5× TBE using a S&S electroelution trap (Schleicher and Schuell, Keene, NH). Buffer is removed, and the DNA is concentrated by centrifugation in a Centricon-100 (Amersham Pharmacia Biotech, Piscataway, NJ). Oligonucleotide adapters (15-mers)

[10] K. M. Borges, S. R. Brummet, A. P. Bogert, M. C. Davis, K. M. Hujer, S. T. Domke, J. Szasz, J. Ravel, J. DiRuggiero, C. Fuller, J. W. Chase, and F. T. Robb, *Genome Sci. Technol.* **1,** 37 (1996).

[11] J. E. K. G. Larsen, J. Light, and S. Molin, *Gene* **28,** 45 (1984).

are ligated onto the genomic ends in a 100 : 1 (adapters : genomic DNA) ratio by incubation with T4 DNA ligase. Excess adapters are removed by a second agarose gel purification, followed by electroelution and concentration as stated earlier. The cosmid library is generated from a *Sau*3A partial digest, using similar conditions to those just described. The inserts are ligated to dephosphorylated vector arms of the cosmid vector pWDYcos1 (GenBank accession AF129504). The ligation mix is packaged with Gigapack III Gold packaging extract (Stratagene, LaJolla, CA).

Vector DNA (10 µg) is linearized by complete digestion with *Pme*I restriction endonuclease. A series of 20 vectors are used for the plasmid library, each vector differing only by a multiplex identifier, which comprises the 18 nucleotides flanking each side of the *Pme*I cloning sites. The multiplex identifier sequences are used as target sequences for hybridization probes in the multiplex mapping and sequencing phases. Oligonucleotide adapters (15-mers) are ligated onto the blunt ends in a 100 : 1 (adapters : vector DNA) ratio for 18 hr at room temperature (same conditions as described earlier). Excess adapters are removed by agarose gel purification followed by electroelution and concentration as stated earlier.

Mixtures of 50 fmol of vector DNA + 100 fmol of genomic DNA are incubated with T4 DNA ligase in 1× ligase buffer in a total volume of 20 µl for 3 hr at room temperature. Aliquots of the ligation mix are electroporated into *Escherichia coli* DH10B cells. Cells are incubated in LB for 30 min at 37° for expression, and aliquots are plated on LB agar plates containing 100 µg/ml ampicillin/2 m*M* isopropylthiogalactoside (IPTG)/ 40 µg/ml X-Gal. The plates are incubated at 37° for 24 hr to form single colonies. For each of the vectors, 288 white colonies are picked into three microtiter dishes containing 1.5 ml of Terrific Broth (TB) + 100 µg/ml ampicillin, and cultures are grown for 18 hr at 30°. Saturated cultures (80 µl) are transferred into Corning (Cambridge, MA) Costar (V-bottom) microtiter dishes containing 10 µl of 22.5% glycerol for long-term storage at −70°. The remaining culture is diluted 1 : 2 with fresh broth and grown at 42° to generate high yields of plasmid DNA by inactivation of the copy number control system.

The plasmid libraries contain a total of 2440 clones with a size range of 3–15 kb and an average insert size of 5 kb. The total length of the plasmid libraries represents 6.4× physical coverage of the genome. The cosmid library contains a total of 238 clones with an average insert size of 35 kb, representing 4.4× physical coverage of the genome. Libraries constructed in these vectors are the starting material for an automated multiplex mapping procedure using sequence tags at the vector : insert junction both to size the subclone and to map sites of randomly inserted sequencing transposons. A restriction digest, using a rare-cutting endonuclease, followed by agarose gel electrophoresis, transfer to a nylon membrane, and hybridization to labeled probes, sizes the inserts and maps the sequencing transposons. The vector sequence tags are multiplexed, such that 20 different fragments are sized per gel lane.

The overall sequencing strategy used is a modified whole genome shotgun strategy that proceeds through three phases. In the first phase, paired-end sequences are generated from each clone in the plasmid and cosmid libraries. The approximate size of each subclone is determined. In the second phase, selected plasmid inserts are fully sequenced using mapped transposon inserts as priming sites for dideoxy sequencing. Pairwise sequence comparison between the complete plasmid sequences and the library of pair-end sequences allows identification of overlapping clones. Optimally overlapped plasmid clones are then chosen for a subsequent round of transposon mapping and sequencing, and this process is iterated until all detectable overlaps are exhausted.

At this point, 1,812,100 bp are completed, covering 95% of the genome, and the sequence is in 39 contigs. In the third phase, templates spanning the gaps are recovered from the cosmid library and then from genomic polymerase chain reaction (PCR) products. Cosmid clones span 12 gaps. Physical templates for the remaining 27 gaps are recovered using PCR from *P. furiosus* genomic DNA, with primers from the ends of the sequence contigs. PCR fragments spanning all the gaps are identified, initially using combinatorial mixtures of primers and, finally, directed PCR.

Multiplex Transposon Mapping and Sequencing: $\gamma\delta$ Transposition

For anchor contig assembly, 150 random 8- to 10-kb plasmid clones are chosen for complete sequencing, generating large contiguous pieces of DNA for scaffolding the remaining clones in the plasmid library. Subsequent bridging and extending clones are chosen by sequence comparison of end sequence reads. The transposon-mapping phase involves the *in vivo* insertion of a $\gamma\delta$ transposon that carries a *Not*I site for mapping the location of the transposon and primer sites for generating bidirectional sequence ladders from the transposon into flanking DNA.

Plasmid DNA from library clones are transformed into *E. coli* DH5α containing the tetracycline-resistant plasmid pCTA4401 (R. B. Weiss, unpublished), which carries the gene for $\gamma\delta$ transposase. A series of matings are first transferred in a chloramphenicol-resistant F factor carrying a mini 480 nucleotide (nt) $\gamma\delta$ transposon (R. B. Weiss, unpublished, 1998) and are then selected for $\gamma\delta$ transposition into the target plasmid by a second mating into a streptomycin-resistant donor, coselecting for the plasmid and F factor resistances. Typical matings yield 10,000 cells/ml containing plasmids with $\gamma\delta$ transposon inserts.

The contents of three 96-well microtiter plates (288 colonies) are picked for each transposed plasmid using an automated colony picker. Typically, 5 to 10 different plasmid inserts are processed in parallel; each insert is carried on a plasmid with different multiplex identifier tags. Microtiter plates, containing 200 μl LB medium, streptomycin (80 μg/ml), and ampicillin (100 μg/ml), are grown at 30° overnight in a humidified incubator. Sample multiplexing consists of duplicate 15-μl transfers, from each of the 10 culture plate wells, into a deep-well titer dish each containing 400 μl of TB plus streptomycin and ampicillin. Each well of the original culture plates is amended with 60 μl of 50% dimethyl sulfoxide, and the plates are sealed with aluminum foil sealers and stored at $-70°$, awaiting clone retrieval. Runaway replication of the plasmid is induced by a 2-hr incubation at 42°. After induction, the cells are pelleted by centrifugation. Pelleted cells are the input to the DNA preparation process.

96-Well Plate Plasmid Purification

Plasmid DNA is isolated by alkaline lysis using semiautomated 96-well minipreparation methods. Induced cultures are centrifuged for 20 min at 2000g at room temperature to pellet the cells. The medium is decanted, and the cells are resuspended in 180 μl of 10 mM Tris–HCl, pH 8, 10 mM EDTA, 50 mM glucose, and 100 μg/ml RNase and transferred to a polyvinylidene difluoride (PVDF) Millipore filter dish; 18 μl of 0.2 N NaOH/0.1% (w/v) sodium dodecyl sulfate (SDS) is then added, and the cells are lysed gently by pipetting 4× with a Robbins Hydra located on a carousel. An aliquot (180 μl) of 3 M potassium acetate pH 4.5, is added, the filter sealed with a Beckman aluminum sheet, and the mixture shaken vigorously for 40 sec. The filter is nested on top of a Dynatech (Chantilly, VA) polystyrene microtiter dish and centrifuged for 10 min at 2000g. The cleared lysate is transferred to a Beckman (Fullerton, CA) deep-well microtiter dish containing 100 mg of diatomaceous earth resuspended in 80 μl of 6 M guanidine thiocyanate. The solution is mixed thoroughly, transferred (three successive transfers of 35 μl) to a PVDF Millipore filter, washed 2× with 400 μl of 8.5 mM Tris–HCl, pH 7.5, 80 mM potassium acetate, 40 mM EDTA, and 55% ethanol, and washed 1× with 70% ethanol by vacuum filtration. Purified DNA is eluted in 80 μl of sterile distilled water at 65° by centrifugation at 2000g for 10 min into a Dynatech microtiter dish. The method produces DNA yields of 300–500 fmol/well using 12- to 20-kb plasmids.

The mapping reactions use a rare-cutting restriction endonuclease, *Not*I, to release the insert from the vector and to cut at the $\gamma\delta$ transposon. During mapping, fragments are sized on a 25-cm 1.0% agarose gel containing three tiers of 96 wells/comb. The gel is electroblotted onto a 28- × 30-cm mem-

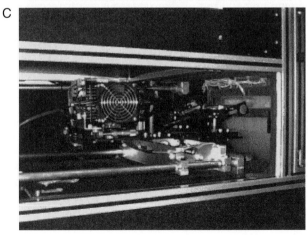

brane. This accounts for 288 lane sets per membrane, and six membranes can be batch processed in a custom-built automated hybridization instrument (AHII, University of Utah Genome Center), which uses fluorogenic probes to detect membrane-bound DNA fragments. Figure 1 depicts one automated hybridization instrument used to process membranes through successive rounds of hybridization and detection. A typical batch process is composed of 21 hybridization/imaging cycles, which provides sizing information for 17,280 transposed plasmids. This occurs over the course of 7 days of unattended AHII operation.

A typical hybridization/detection cycle uses 100 pM of biotinylated oligonucleotide probe, which is injected in 120 ml of 100 mM NaPO$_4$, pH 7.2, 125 mM NaCl (PBS) + 1% SDS, and hybridized for 1 hr at 42°. The unbound probe is removed by five washes with 180 ml PBS + 1% SDS. Streptavidin–alkaline phosphate (SAAP, Boehringer-Mannheim, Frankfurt, Germany, 1000 units/ml), diluted to 1 unit per 120 ml PBS + 1% SDS, is added to the chamber and incubated for 20 min. Residual SAAP is removed by three washes with 120 ml PBS + 1% SDS, followed by five washes with 120 ml PBS and five washes with 120 ml of 0.1 M diethanolamine, pH 9.0. Attophos (150 ml of a 3-mg/ml solution in 0.1 M diethanolamine,

---

FIG. 1. Nylon membrane processing and the automated hybridization and imaging instrument (AHII). (A) The AHII comprises six subsystems: (1) the hybridization chamber, (2) the fluid delivery system, (3) the computer control system, (4) the imaging system, (5) the electrical system, and (6) the chassis and packaging system. The hybridization chamber is composed of eccentric acrylic drums. Nylon membranes are secured to the exterior surface of the rotating inner drum, which continually bathes the membranes in a fluid puddle that is formed in the bottom 5-mm gap between the two drums. The membrane-carrying drum has 6000 cm^2 of available surface area, which is sufficient for six membranes that, in single lane density form, carry a total of 1728 mapping lane sets. Through dedicated software, the user is able to construct, save, and execute desired hybridization protocols. The control software orchestrates the additions of buffering solutions, membrane temperature control, drum speeds, timers, and precision deliveries of concentrated solutions such as probes, enzymes, fluorogenic substrates, and image acquisition. Concentrated solutions are stored in refrigerated vials and by an on-board pipetting robot. Electric resistance heaters secured to the drum surfaces provide membrane temperature control. (B) Concentrated oligonucleotide probe and enzyme conjugate solutions are stored in refrigerated vials and by an on-board pipetting robot. (C) The imaging system consists of a Peltier-cooled Dalsa 2048 × 96 pixel CCD camera, a custom-designed SCSI camera controller (SCC), and a xenon arc lamp coupled to a fiber-optic light line for fluorescent excitation. The camera is operated in a time delay and integration mode (TDI), in which pixel shifts are synchronized with the surface motion of the rotating drum. At 150 dpi, 2048 pixels capture ~34 cm of drum width, and TDI permits scan lengths that are limited only by available SCC memory. The SCSI camera controller (SCC) performs 12-bit analog-to-digital conversions at a 15-MHz sampling rate and is capable of acquiring 254 megabytes of image data.

pH 9.0) is added, incubated for 8 min at room temperature, and drained. The membranes are then incubated for 30 to 120 min to allow a linear increase in the fluorescent signal. Digital images are captured by the on-board charge-coupled device (CCD) camera and processed using custom software. Figure 2 is an example of a section of a mapping membrane depicting a gray scale image of the digital output from the scanning of the probed membrane-bound DNA fragments and the resultant pixel traces from 20 mapping lanes.

The mapping of 288 lanes/plasmid generates dense sets of transposons that are then used for sequencing the insert. A typical transposon map is shown in Fig. 3, where a predicted, optimally spaced set of transposons is shown overlying the total mapped set. In phase two of *P. furiosus* sequencing, ~187,000 transposons are mapped across ~650 plasmid inserts. The optimal subset of ~11,500 transposons is retrieved iteratively from the frozen microtiter culture plates and sequenced. Collection of 1.5 Mb of a contiguous sequence at high coverage is accomplished using multiplex sequencing and the AHII system, with a nonradioactive version of cycle sequencing chemistry as described previously.[12] The remainder of the genome sequence is completed using BigDye terminator chemistry on the ABI 377 sequencing instrument.

Genome Closure and Finishing

After the plasmid library has been sequenced to completion, 39 gaps remain in a total database of 1,812,000 bp. Primers are chosen on the 5' and 3' ends on each of the 39 sequence contigs. BlastN analysis of the cosmid end sequences against the plasmid sequence contigs render information as to order and orientation, as well as gap sizes for 12 gaps. PCR is performed on the cosmid templates using the appropriate primers. The 3 gaps that are larger than 2 kb are subcloned into plasmid vectors transposed with $\gamma\delta$ to yield deletions and sequenced to close the gaps. Nine gaps spanned by cosmids are <1.5 kb and are closed by genomic PCR and primer walking. The remaining 27 gaps are spanned successively by combinatorial and directed PCR experiments. The short 16 gaps of <1.5 kb, 16 in number, are closed by genomic PCR and primer walking. Large gaps between 2 and 10 kb do not have cosmid bridging templates. These are closed by genomic PCR and subcloning into one of the plasmid vectors used to build the original subclone libraries. There are 23 regions that have single clone coverage. Sequence assembly of these areas is verified by genomic PCR. All regions of low-quality sequence or ambiguity are checked by resequencing

[12] M. D. Stump, J. L. Cherry, and R. B. Weiss., *Nucleic Acids Res.* **27**, 4642 (1999).

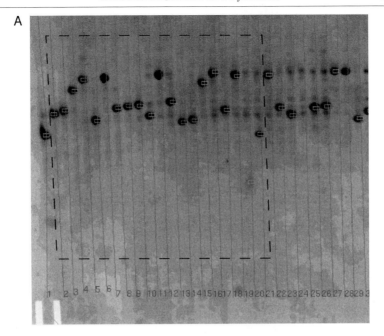

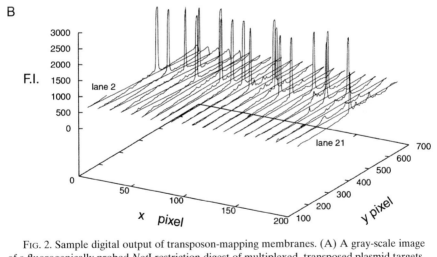

FIG. 2. Sample digital output of transposon-mapping membranes. (A) A gray-scale image of a fluorogenically probed *Not*I restriction digest of multiplexed, transposed plasmid targets. The first 30 lanes of a 288-lane agarose gel blot are shown. (B) Raw CCD pixel value output for the boxed lane set shown in (A).

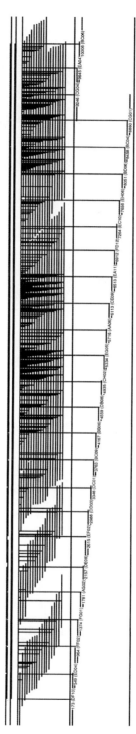

FIG. 3. Sample mapping set of transposon insertions. The AHII system was used to map 288 independent transposon insertions into a plasmid carrying a 10.2-kb insert. The location of each insertion is displayed as an inverted T, where the trunk of the T indicates the predicted location and the bottom arms of the T are the lengths of bidirectional 40-bp sequence ladders. A subset of 26 insertions is numbered with their fragment coordinates and microtiter plate-well location. This subset is the optimal set for sequencing this fragment.

templates covering that region. The final physical coverage of the genome from the plasmid and cosmid library is shown in Fig. 4.

The complete genome sequence of *P. furiosus* is available from the following website—http://www.genome.utah.edu/sequence.html—and is in preparation for submission.

Nucleotide Preference, Codon Usage, and Aminoacyl-tRNA Synthetases

In order to apply the available coding sequences to significant problems in the biochemical and structural characterization of hyperstable proteins, it is necessary to carry out recombinant expression of these genes, normally in *E. coli*. Because of the significant differences in coding between Bacteria and the hyperthermophilic Archaea, this poses problems on occasion. Observations on the characteristics of translation in *P. furiosus* are discussed later in order to provide insights into these problems and their solutions.

As shown in Table I, the nucleotide usage of the *P. furiosus* genome, which is very similar to the other *Pyrococcus* genomes, is heavily biased in favor of A and G, at the expense of C and G, in the open reading frames (ORFs). The CG dinucleotide is strongly avoided, whereas GA is highly used as shown in Table II. The significance of this bias in *P. furiosus* and other hyperthermophiles, including hyperthermophilic bacteria, may be that at optimal growth temperature approaching 100°, pervasive cytosine deamidation occurs as a result of ongoing hydrolytic damage to DNA,[13] which causes the formation of U in the chromosome. When DNA repair fails to remove these U's, a C-to-T transition occurs. The mutational event will be the substitution of Leu for Arg, a radical change that results in significant loss of protein function. For this reason, selection against C within codons may have resulted in the bias observed on a genomic scale. It is interesting to note that some hyperthermophiles have high G + C genomes, including *Methanopyrus kandleri* and *Pyrodictium* spp., both of which have GC contents of 60 mol%. Genomic studies on these strains are needed to examine the genetic basis for CG bias.

Implications for Recombinant Expression of P. furiosus Genes

The major impact of dinucleotide bias is that usage of codons containing CG is avoided: AGA and AGG codons for Arg are strongly favored at the expense of the four alternative codons containing CG dinucleotides, namely CGT, CGA, CGG, and CGC (see Table III). This causes very

---

[13] T. Lindahl, *Nature* **362,** 709 (1993).

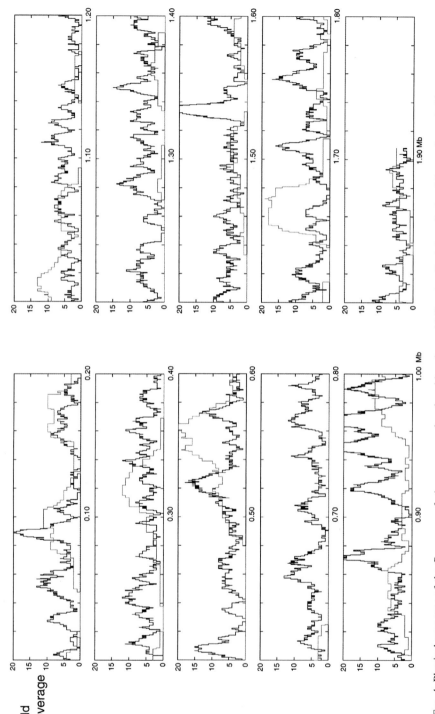

FIG. 4. Physical coverage of the *Pyrococcus furiosus* genome by plasmid and cosmid libraries. Each box depicts 0.2 megabases (Mb) of the *P. furiosus* genome; the ordinate indicates fold coverage and the abscissa indicates genome coordinate. The dark solid line is plasmid coverage, and the light dotted line is cosmid.

TABLE I
NUCLEOTIDE OCCURRENCE AND CODON USAGE IN *P. furiosus*[a]

Nucleotide	Genome		ORFs	
	nt/strand	Relative	nt/strand	Relative
A	565,154	0.30	586,283	0.33
	388,630	0.20	297,845	0.17
G	389,364	0.20	429,610	0.24
T	565,105	0.30	457,945	0.26
Total	1,908,253	1.00	1,771,683	1.00

[a] A and G are preferred in ORFs relative to global distribution.

significant problems in the expression of many genes from *P. furiosus* in *E. coli* and other hyperthermophiles that have a similar codon bias. *Escherichia coli* has the reverse bias for Arg, in which almost no AGA or AGG Arg codons occur, and the cognate tRNAs are extremely rare. A similar, although less severe, effect is seen in the case of the rare Leu and Ile codons. The solution to this problem has been the use of the plasmid pSJS,[14] kindly supplied by Dr. Steven J. Sandler, University of Massachusetts. This construct expresses the Arg tRNA genes encoding AGA and AGG, and the Ile tRNA AUA. The major ColE replicon-based vectors may be used in conjunction with pSJS based on the replicon from plasmid pACYC, which is compatible with ColE plasmids.[14] In our experience, the genes encoding a 500-kDa reverse gyrase and a 16.5-kDa small heat shock protein from *P. furiosus* could not be expressed at detectable levels before repeating the experiment in a strain of *E. coli* BL21 cotransformed with pSJS; this resulted in satisfactory expression of soluble, active proteins (W. Kanoksilapathan and F. Robb, unpublished, 1999). Major bands of recombinant protein were detected following induction.

Aminoacyl-tRNA Synthetase Structure and Function

Aminoacyl-tRNA synthetases fulfill a critical role in the interpretation of the genetic code. Therefore, the characterization of genes encoding these enzymes in a complete genome sequence can provide important insights into the process of protein synthesis within a given organism. [Herein, specific aminoacyl-tRNA synthetases (RS) will be referred to by the corre-

[14] R. Kim, S. Sandler, S. Goldman, H. Yokota, J. Clark, and S.-H. Kim, *Biotech. Lett.* **207** (1998).

TABLE II
NORMALIZED DINUCLEOTIDE USAGE IN
P. furiosus ORFs

First	Second			
	A	C	G	T
A	0.93	1.15	1.15	1.26
C	0.91	0.69	0.55	0.86
G	1.52	1.13	1.05	1.22
T	1.03	0.68	0.76	0.89

sponding three letter amino acid code; i.e., glutamyl-tRNA synthetase would be GluRS.]

Aminoacyl-tRNA synthetases catalyze the esterification or charging of a tRNA molecule with a specific amino acid via a two-step reaction. First, the amino acid is activated in the presence of ATP and Mg^{2+}, which results in the attachment of an aminoacyl adenylate (amino acid-AMP) to the synthetase and the release of PP_i. Second, the synthetase aminoacylates

TABLE III
RELATIVE CODON FREQUENCY IN *P. furiosus* ORFs

First base	Second base								Third base
		t		c		a		g	
T	F	1.66	S	0.58	Y	1.37	C	0.24	t
T	F	1.17	S	0.40	Y	1.18	C	0.14	c
T	L	1.25	S	0.66	*	0.08	*	0.11	a
T	L	0.90	S	0.16	*	0.05	W	0.79	g
C	L	1.55	P	0.68	H	0.45	R	0.07	t
C	L	1.09	P	0.59	H	0.51	R	0.07	c
C	L	1.14	P	1.27	Q	0.61	R	0.08	a
C	L	0.53	P	0.19	Q	0.51	R	0.05	g
A	I	2.09	T	0.95	N	1.14	S	0.66	t
A	I	0.87	T	0.49	N	1.08	S	0.69	c
A	I	2.59	T	1.02	K	2.44	R	1.78	a
A	M	1.41	T	0.37	K	2.73	R	1.35	g
G	V	2.25	A	1.35	D	1.76	G	0.79	t
G	V	0.68	A	0.88	D	1.02	G	0.49	c
G	V	1.22	A	1.57	E	3.09	G	2.36	a
G	V	0.89	A	0.38	E	2.58	G	0.90	g

TABLE IV

COMPLETE (CLASS I OR II) AND PARTIAL (CLASS X) AMINOACYL-tRNA SYNTHETASE (-tRS)
GENES PRESENT IN GENOMES OF *Pyrococcus furiosus, P. abyssi,* AND *P. horikoshii*

Synthetase	Class	Estimated length (amino acids)
Arginyl-tRS	I	629
Cysteinyl-tRS	I	483
Glutamyl-tRS	I	580
Isoleucyl-tRS	I	1066
Leucyl-tRS	I	968
Lysyl-tRS	I	523
Methionyl-tRS	I	724
Tryptophanyl-tRS	I	385
Tyrosyl-tRS	I	376
Valyl-tRS	I	891
Alanyl-tRS	II	914
Asparaginyl-tRS	II	437
Aspartyl-tRS	II	439
Glycyl-tRS	II	568
Histidyl-tRS	II	432
Phenylalanyl-tRS α and β subunits	II	499, 556
Prolyl-tRS	II	479
Seryl-tRS	II	455
Threonyl-tRS	II	626
AlaX157	X	149–157
AlaX213	X	213
AlaX405	X	404–405
MetX	X	109–112

the tRNA and then releases the amino acid–tRNA complex and AMP.[15,16] For each of the 20 amino acids, a specific aminoacyl-tRNA synthetase exists, although there are significant exceptions, which will be discussed later. The 20 different synthetases are split evenly among two families called class I and class II, which differ in terms of primary sequences, crystallographic structures, and tRNA charging sites[16] (Table IV). Class I synthetases share the amino acid motifs HIGH and KMSKS, which are located in a nucleotide-binding fold of alternating α helices and parallel β sheets called the Rossmann fold.[16,17] Class II synthetases lack these class I motifs and, instead, have three different amino acid motifs, which are conserved more loosely. The class II binding site is also different, being

[15] G. Eriani, J. Cavarelli, F. Martin, L. Ador, B. Rees, J.-C. Thierry, J. Gangloff, and D. Moras, *J. Mol. Evol.* **40,** 499 (1995).
[16] G. Eriani, M. Delarue, O. Poch, J. Gangloff, and D. Moras, *Nature* **347,** 203 (1990).
[17] D. Moras, *Trends Biochem. Sci.* **17,** 159 (1992).

composed of six antiparallel β sheets and two long α helices. The two classes of synthetases acylate the tRNA molecule at different sites: class I attaches the activated amino acid to the 2'-OH whereas class II charges the 3'-OH.

Class I and II synthetases are widely considered to be of independent origin. The two aminoacyl-tRNA synthetase subdivisions are not concordant with amino acid biochemistry, anticodon assignments in the genetic code, or tRNA sequences.[18] However, within each class, specific aminoacyl-tRNA synthetases appear to be related phylogenetically.[19,20] Class I and class II type synthetases likely diverged prior to the separate evolution of prokaryotic and eukaryotic lineages. However, there are many probable examples of interspecies transfer of aminoacyl-tRNA synthetase genes both within and between the three domains of life: Archaea, Bacteria, and Eukarya.[21,22]

Having a complete complement of aminoacyl-tRNA synthetases was long viewed as being essential for every cellular organism, and indeed, the first few completed bacterial genome sequences by and large supported that supposition. Therefore, it was a major surprise when several aminoacyl-tRNA synthetases were absent in the first completed archaeal genomes. *Methanococcus jannaschii*,[23] *Methanobacterium thermoautotrophicum*,[24] and *Archaeoglobus fulgidus*[25] all lacked recognizable AsnRS GlnRS, and LysRS, while the methanogens were also missing CysRS.

Reasonable biochemistry could be proposed to account for the absence of AsnRS and GlnRS. Two alternative biosynthetic pathways for Gln-tRNAGln were known to exist from early biochemical studies. "Crown eukaryotes" (animals, plants, and fungi) and *E. coli* have specific GlnRS and GluRS enzymes.[26] GlnRS has been demonstrated biochemically not to exist in several bacteria, organelles, and an archaeal halophile.[27–30] In

[18] J. Nicholas and W. H. McClain, *J. Mol. Evol.* **40**, 482 (1995).

[19] G. M. Nagel and R. F. Doolittle, *Proc. Natl. Acad. Sci. U.S.A.* **88**, 8121 (1991).

[20] G. M. Nagel and R. F. Doolittle, *J. Mol. Evol.* **40**, 487 (1995).

[21] J. R. Brown, *in* Thermophiles. The Keys to Molecular Evolution and The origin of Life? (J. Wiegeland and M. Adams, eds.), p. 217. Taylor and Francis Group, 1998.

[22] Y. I. Wolf, L. Aravind, N. V. Grishin, and E. V. Koonin, *Genome. Res.* **9**, 689 (1999).

[23] C. J. Bult, O. White, G. J. Olson, L. Zhou, R. D. Fleishmann, G. G. Sutton, J. A. Blake, L. M. FitzGerald, R. A. Clayton, and J. D. Gocayne, *Science* **273**, 1058 (1996).

[24] D. R. Smith, L. A. Doucette-Stamm, C. Deloughery, H. Lee, J. Dubois, T. Aldredge, R. Bashirzadeh, D. Blakely, R. Cook, K. Gilbert *et al.*, *J. Bacteriol.* **179**, 7135 (1997).

[25] H. P. Klenk, R. A. Clayton, J. F. Tomb, O. White, K. E. Nelson, K. A. Ketchum, R. J. Dodson, M. Gwinn, E. K. Hickey, J. D. Peterson *et al.*, *Nature* **390**, 364 (1997).

[26] R. A. Lazzarini and A. H. Mehler, *Biochemistry* **3**, 1445 (1964).

[27] J. Lapointe, L. Duplain, and M. Proulx, *J. Bacteriol.* **165**, 88 (1986).

[28] R. Breton, D. Watson, M. Yaguchi, and J. Lapointe, *J. Biol. Chem.* **265**, 18248 (1990).

these organisms, Gln-tRNAGln is synthesized via a two-step process where tRNAGln is first misacylated with glutamate by GluRS and then glutamate is converted to glutamine by a specific amidotransferase. *Bacillus subtilis* Glu-tRNAGln amidotransferase has been identified as a heterotrimeric enzyme where the three subunits are encoded by a single operon, gatCAB.[31] Orthologus *gat* genes have been found in most organisms lacking GlnRS, including the three archaeal genomes listed earlier.

In a halophilic archaeon, evidence was found for the synthesis of Asn-tRNAAsn via transamidation of Asp-tRNAAsn formed by AspRS.[31] This alternative pathway might also exist in certain bacteria, namely *Helicobacter pylori*[32] and *Aquifex aoelicus*,[33] as well as the archaea, as these organisms all lack a recognizable AsnRS gene.

The incorporation of lysine in archaeal proteins was solved by the discovery of an entirely novel aminoacyl-tRNA synthetase. LysRS of the typical class II family had been previously found in all eukaryotes and bacteria and in at least one archaeal species, *Sulfolobus solfataricus,* a member of the divergent kingdom Crenarchaeota. However, a novel class I-like synthetase was discovered to catalyze the formation of Lys-tRNALys in four species from the other archaeal kingdom, the Euryarchaeota,[34] and one group of bacteria, the Spirochaetes.[35] Homologs to class I LysRS have been found in several archaeal species where it has been suggested to have originated and subsequently acquired by spirochaetes through horizontal gene transfer. The enzyme or pathway furnishing the last "missing" archaeal synthetase, CysRS, has yet to be found in methanogens, although CysRS was readily identified in *A. fulgidus.*

Pyrococcus Aminoacyl-tRNA Synthetases

It is interesting to contrast the aminoacyl-tRNA synthetase complements of *P. furiosus, P. horikoshii,* and *P. abyssi* with that of earlier se-

[29] Y. Gagnon, L. Lacoste, N. Champagne, and J. Lapointe, *J. Biol. Chem.* **271,** 4856 (1996).

[30] A. Schön, C. G. Kannangara, S. Gough, and D. Söll, *Nature* **331,** 187 (1988).

[31] A. W. Curnow, M. Ibba, and D. Söll, *Nature* **382,** 589 (1996).

[32] J. F. Tomb, O. White, A. R. Kerlavage, R. A. Clayton, G. G. Sutton, R. D. Fleischmann, K. A. Ketchum, H.-P. Klenk, S. Gill, B. A. Dougherty, K. Nelson, J. Quackenbush, L. Zhou, E. F. Kirkness, S. Peterson, B. Loftus, D. Richardson, R. Dodson, H. G. Khalak, A. Glodek, K. McKenney, L. M. FitzGerald, N. Lee, M. D. Adams, and J. C. Venter, *Nature* **388,** 539 (1997).

[33] G. Deckert *et al., Nature* **392,** 353 (1998).

[34] M. Ibba. S. Morgan, A. W. Curnow, D. R. Pridmore, U. C. Vothknecht, W. Gardner, W. Lin, C. R. Woese, and D. Söll, *Science* **278,** 1119 (1997).

[35] M. Ibba, H. C. Losey, Y. Kawarabayasi, H. Kikuchi, S. Bunjun, and D. Soll, *Proc. Natl. Acad. Sci. U.S.A.* **96,** 418 (1999).

quenced archaeal genomes. All three *Pyrococcus* species have the full regiment of synthetases, with the exception of GlnRS, as well as several genes that appear to code for additional, albeit truncated, aminoacyl-tRNA synthetase-like proteins (Table IV).

Recognizable genes coding for CysRS, AsnRS, and class I LysRS are found in the three *Pyrococcus* species. PheRS is unique among the amino-acyl-tRNA synthetases because it is composed of distinct α and β subunits. In most bacteria, the two subunits are encoded by a two-gene operon, pheST. The same transcriptional unit is found in *Pyrococcus* but not in other Archaea. Phylogenetic analysis of PheRS subunits suggests that Spiro-chaetes may have been gene transfer recipients of this synthetase (J. R. Brown, unpublished, 1999), as well as class I LysRS from some *Pyrococcus*-related species.[34]

The absence of GlnRS is congruent with other studies and lends further support to the scenario that GlnRS evolved in eukaryotes from a gene duplication of GluRS.[36,37] Thus *Pyrococcus* species should employ the alter-native Gln-tRNAGln pathway, using an enzyme homologous to the *Bacillus subtilis* heterotrimeric Glu-tRNAGln amidotransferase (GAT). However, only a single weakly similar homolog to the gatB subunit (also called yeast Pet112 gene product) could be found in all three *Pyrococcus* species and convincing matches to either gatA or gatC subunits could not be found. The presence of AsnRS suggests that Asn-tRNAAsn formation is direct. Therefore, further experiments are needed to verify or determine the mode of Gln-tRNAGln synthesis in *Pyrococcus* species.

In all three *Pyrococcus* genomes, there are several short-length ORFs showing strong homology to specific domains of either AlaRS or MetRS (Fig. 5). It is clear that these ORFs are derived from synthetase fragments as they can be fully aligned within a specific region of the parent protein. Given that the true biochemical function of these ORFs is presently un-known, homologs to AlaRS and MetRS will be referred to here as AlaX and MetX, respectively.

AlaX is actually a family of proteins falling into three different amino acid residue size categories: 149–157 residues (named here AlaX 157 fam-ily), 213 residues (AlaX213 family), and 404–405 residues (AlaX405 family). The three AlaX families show overlapping homology to a similar region of AlaRS (Fig. 1). Both AlaX216 and AlaX405 families begin at D515 in *P. furiosus* AlaRS, whereas AlaX157 begins at or near W593, depending on the species. All three AlaX families have two highly conserved regions centered about three histidines (H613, H617, and H721) and a single glycine

[36] M. Siatecka, M. Rozek, J. Barciszewski, and M. Mirande, *Eur. J. Biochem.* **256,** 80 (1998).
[37] J. R. Brown and W. F. Doolittle, *J. Mol. Evol.* **49,** 485 (1999).

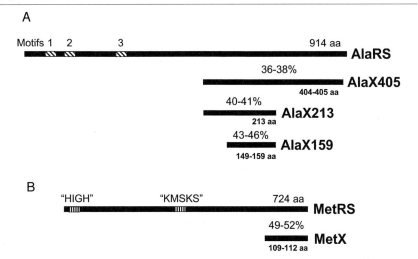

Fig. 5. Regions of overlap between partial (A) AlaX and (B) MetX proteins with the parent synthetase. The locations of class I (MetRS) and class II (AlaRS) defining motifs are shown. Percentages are the range of amino acid identity shared between the smaller ORF and the related aminoacyl-tRNA synthetase. Small numbers indicate the range of lengths found across all three *Pyrococcus* species. MetRS and AlaRS amino acid lengths are given for *P. furiosus* only. The drawing is not to scale.

(G729). The three-dimensional structure of AlaRS has not been determined; however, the major functional domains of human and *E. coli* AlaRS are known.[38] The major class II determinants and regions involved in tRNA acceptor helix recognition are located in the N terminus portion of AlaRS and are not included in the regions of *Pyrococcus* AlaX-AlaRS homology. The corresponding region of AlaX homology in *E. coli* AlaRS is dispensable for aminoacylation and is thought to play a role in stabilizing oligomeric enzyme complexes.

A putative single copy gene encodes the MetX protein, which varies in length from 109 amino acids in *P. furiosus* and *P. abyssi* to 112 residues in *P. horikoshii*. MetX shows strong homology to the extreme COOH terminus of *P. furiosus* MetRS beginning around M617. Again, this region is outside of the N-terminal catalytic core of this class I synthetase. However, MetX is highly concordant with the catalytic domain of the yeast protein Arc1p, which is implicated in facilitating tRNA binding to multifunctional complexes of MetRS and GluRS. Other homologies of MetX and Arc1p can

---

[38] K. Shiba, N. Suzuki, R. Nichols, P. Plotz, T. Noda, and P. Schimmel, *Biochemistry* **34,** 10340 (1995).

be found in the N-terminal domain of bacterial PheRS β subunit, an uncharacterized *E. coli* ORF, and several human, mouse, and rice EST fragments.[39]

All proteins within the particular MetX or AlaX protein types are highly related, although the two most closely related *Pyrococcus* species vary from family to family. It is difficult to derive statistically robust phylogenetic analysis on short proteins. However, preliminary trees suggest that neither AlaX nor MetX families arose from recent gene duplications of AlaRS or MetRS within the *Pyrococcus* lineage. Instead these partial aminoacyl-tRNA synthetases may have diverged very early in the evolution of the archaeal lineage and perhaps "predate" the Crenarchaeota–Euryarchaeota split.

Similarly deep divergence events may be postulated for the evolution of other "synthetase fragments" evolved from HisRS and PheRS (J. R. Brown, in preparation). PheX family proteins are limited to gram-positive cocci. While the precise cellular role of PheX is unknown, gene knockout experiments suggest that PheX is not essential for growth in culture (J. R. Brown, unpublished result). The HisX or HisZ proteins occur in a number of bacterial and archaeal species and are correlated with the presence of a shortened version of the *hisG* gene.

Interestingly, single AlaX and MetX proteins were also found in the Crenarchaeota, as they occur in the genome sequence of *Aeropyrum pernix.* These sequences do not cluster with those of *Pyrococcus* spp., which suggests an independent or parallel evolutionary recruitment event in a different species lineage. Knowing the functions of AlaX and MetX proteins will, perhaps, add important new insights into how and why evolution recruits particular pieces of existing genes into new cellular services.

Insights into Metabolic Functions in *P. furiosus*

The availability of a comprehensive gene list provides insights into the presence and absence of pathways and metabolic capabilities, despite the relatively high level of unannotated sequences. Table V indicates the assignments of genes to functional classes, with the exception of 1111 genes that are in the unknown or conserved hypothetical categories. The following initial observations on metabolic functions in *P. furiosus* may be inferred from the gene list.

Glycolysis and Gluconeogenesis

The discovery of two novel ADP-dependent kinases in *P. furiosus,* without significant homology to other kinases or previously recognized

[39] G. Simos, A. Segref, F. Fasiolo, K. Hellmuth, A. Shevchenko, M. Mann, and E. C. Hurt, *EMBO J.* **15,** 5437 (1996).

TABLE V
ASSIGNMENTS OF ORFs IN *P. furiosus* GENOME INTO
FUNCTIONAL CLASSES

Functional class	Count
Amino acid biosynthesis	90
Biosynthesis of cofactors, prosthetic groups, and carriers	47
Cell envelope	43
Cellular processes	54
Central intermediate metabolism	50
DNA metabolism	35
Energy metabolism	117
Fatty acid and phospholipid metabolism	11
Purines, pyrimidines, nucleosides, and nucleotides	44
Regulatory functions	44
Transcription and RNA processing	31
Translation	146
Transport and binding proteins	121
Transposon related functions	25
Unknown general	41
Conserved hypothetical	1060

glycolytic enzymes, illustrates the need for rigorous biochemical studies.[40] In conjunction with genomic sequence data, will almost certainly result in the discovery of novel enzyme functions in the unannotated genes of the *P. furiosus* genome. The fact that ADP appears to be utilized in these major steps of glycolysis, rather than ATP, has significant implications for the energy balance of the hyperthermophiles. The higher thermostability of ADP over ATP may possibly result in greater metabolic efficiency in cells growing at or near 100°. The novel hexokinase and phosphofructo-kinase appear to function in glycolysis rather than gluconeogenesis. The latter function is accomplished by an NAD(P)-dependent GAP dehydrogenase that is strictly phosphate dependent[40] and is repressed under glycolytic growth conditions, e.g., during growth on sugars.

van der Oost and colleagues[41] addressed the reciprocal regulation by repression and kinetic parameters of glyceraldehyde-3-phosphate oxidoreductase, a tungsten-containing enzyme, and GAP dehydrogenase, apparently leading to mid-pathway regulation during the switch between gluconogenic and glycolytic growth conditions.

The TCA cycle in *P. furiosus* is apparently incomplete due to the absence

[40] P. Zwickl, S. Fabry, C. Bogedain, A. Haas, and R. Hensel, *J. Bacteriol.* **172,** 4329 (1990).
[41] J. van der Oost, G. Schut, S. W. Kengen, W. R. Hagen, M. Thomm, and W. M. de Vos, *J. Biol. Chem.* **273,** 28149 (1998).

TABLE VI
FUNCTIONAL PATHWAYS IN *P. furiosus* ABSENT IN *P. horikoshii*

Amino acid biosynthesis
   Trp biosynthesis
   His biosynthesis
   Thr biosynthesis
   Val, Leu, Ile biosynthesis
   Arg, Pro biosynthesis
   Cys biosynthesis
   Glu biosynthesis
Carbohydrate uptake and utilization
   malEFG transport system
   18 unknown ABC transporters
   Cation efflux system
   Uracil permease
   Anion permease
   Cation-transport ATPase
   Ferritin
   Chitinase, two-ORF cluster
   β-Mannosidase
   2-amylopullulanase
   Endo-1,4-β-glucanase
   Endo-β-1,3-glucanase
   Cyclomaltodextrin glucanotransferase precursor
   Two α-amylases
   β-Glucosidase
Cell wall components, proteases, and amino acid utilization
   UDP-glucose dehydrogenase
   Putative acetyltransferase
   dTDP-glucose 4,6-dehydratase
   UTP-gluycose-1-phosphate uridylyltransferase
   Proline permease
   Three aminopeptidases
   Pyrolysin
   Neutral protease
Energy metabolism
   Hydrogenase cluster, five-ORF cluster
   3-Hydroxyisobutyrate dehydrogenase, NAD(P)H-flavin oxidoreductase
   3-Oxoacyl-[acyl-carrier-protein] reductase
   Isocitrate dehydrogenase, citrate synthase, two-ORF cluster
   nifS protein
   Glycerol-3-phosphate dehydrogenase
   Glycerol kinase

of succinyl-CoA hydrolase (EC 3.1.2.3) and 2-oxoglutarate synthase (EC 1.2.7.3). The absence of malate dehydrogenase (EC 1.1.3.7) from the genome is not necessarily conclusive because a copy of a putative malate oxidoreductase has been annotated in the genome of *P. furiosus* (R. Weiss, unpublished result). Interestingly, the TCA cycle in *P. horikoshii* is reduced significantly due to the absence of a three-ORF cluster found in *P. furiosus* encoding isocitrate dehydrogenase, aconitate hydratase, and citrate synthase.

The *P. furiosus* genome is distinguished from those of the two other *Pyrococcus* species by a significantly more complete inventory of amino acid biosynthetic pathways and by the addition of several pathways for the uptake and catabolism of the carbohydrates cellobiose, maltose, trehalose, laminarin, and chitin. In contrast, the genome of *P. horikoshii* lacks several amino acid biosynthetic pathways (see Table VI).

This may have the following implications. Possibly the abyssal milieus of the Pacific species of *Pyrococcus* rarely contain available sugars, whereas the onshore isolation locale of *P. furiosus* is characterized by high-sugar, low-protein conditions. *P. furiosus* can use many carbohydrate sources efficiently, leading to the potential for the absence of amino acids in the growth medium. This may have led to the selection of the amino acid biosynthetic pathways that are absent in *P. horikoshii.*

Recent Mergers

The possible acquisition of His, Trp, Aro, Arg, and Ile/Val, as well as several carbohydrate utilization pathways by *P. furiosus,* implies that the histories of these pathways are more recent than the history of the genome at large. In fact, the phylogeny of the genes making up these blocks presents a phylogenetic picture that is at variance with the Archaeal phylogeny, based on a preliminary inspection of the blast analysis of the genome[10] (D. L. Maeder and F. T. Robb, unpublished result). In agreement with this, the *trp* operon of *M. thermoautotrophicum* and the *his* operon of *S. solfataricus* show mixed phylogeny.[42]

Acknowledgments

The authors gratefully acknowledge support from the Microbial Genome Program of the U. S. Department of Energy, the National Science Foundation, the Knut and Alice Wallenberg Foundation, and the U.S. Department of Commerce. This is contribution number 521 from the Center of Marine Biotechnology.

[42] W. F. Doolittle, *Nature* **392,** 15 (1998).

[8] Genome of *Aquifex aeolicus*

By RONALD V. SWANSON

Introduction

Aquifex aeolicus is the first hyperthermophilic bacterium to have its genome sequence completely determined.[1] It is also the first obligate chemo-lithoautotrophic bacterium to be completely sequenced. Prior to the genome sequence, little else was known about *A. aeolicus* other than that it is a motile, flagellated, hydrogen-oxidizing, microaerophilic, gram-negative bacterium. The known phenotypic properties of *A. aeolicus* led to the prediction of specific genes in the genome; for instance, flagellar and chemo-taxis genes were expected to be present because the organism is motile. In addition, a large number of genes were predicted to be present because they represent core functions present in all cells, or in all bacteria. Both hyperthermophily and chemoautotrophy were expected to have an observable effect on the content of the genome. *A. aeolicus* has a relatively small genome for a free-living bacterium, 1.55 million base pairs. This suggested that there is likely to be minimal noncoding DNA and few unexpected or apparently extraneous genes. Although these predictions were largely fulfilled, in some cases genes that were expected to be present were not found. In many cases, these discrepancies between prediction and observation can lead to interesting insights and avenues of research.

Prior to the genome project, no gene sequences from *A. aeolicus* were available in the public domain. This is currently an atypical situation but may become the rule rather than the exception as the efficiency of having the genome sequence as a starting point for studying an organism becomes fully recognized. The genome sequence and the catalog of encoded functions identified by similarity directly provide ready answers to many questions. The sequence also provides the template for the rapid subcloning and expression of genes in heterologous hosts for structural and biochemical characterization,[2–5] but the sequence also raises its own questions. Princi-

---

[1] G. Deckert, P. V. Warren, T. Gaasterland, W. G. Young, A. L. Lenox, D. E. Graham, R. Overbeek, M. A. Snead, M. Keller, M. Aujay, R. Huber, R. A. Feldman, J. M. Short, G. J. Olsen, and R. V. Swanson, *Nature* **392,** 353 (1998).

[2] H. S. Duewel, G. Y. Sheflyan, and R. W. Woodard, *Biochem. Biophys. Res. Commun.* **263,** 346 (1999).

[3] M. S. Finnin, J. R. Donigian, A. Cohen, V. M. Richon, R. A. Rifkind, P. A. Marks, R. Breslow, and N. P. Pavletich, *Nature* **401,** 188 (1999).

pally, these arise from our inability to find or recognize a particular function predicted to be present. The discovery of analogous replacements, i.e., nonhomologous enzymes that catalyze the same reaction, novel pathways, or even entirely different strategies, is likely to rapidly expand our understanding of biochemistry and microbial physiology.

The actual genome sequencing was performed by a small group of three to four researchers working part-time on the genome over a period of about 18 months. The sequence was analyzed by two independent groups: Diversa Corporation and researchers affiliated with the University of Illinois, Champaign–Urbana. The published annotation represents the merging and reconciliation of their results. It is significant that the project was not carried out within a dedicated genome-sequencing environment. Laboratories working with unusual microbes may do well to consider the investment in an upfront sequencing effort. Many researchers may object to such an effort on the basis that it is not hypothesis-driven or otherwise outside their purview. However, the genome sequence provides an organizing principle and investigational resource whose value is difficult to overstate. One need only consider the time and effort spent by individual researchers on cloning and sequencing particular genes to see that a concentrated effort up front would rapidly pay for itself. Viewing the sequence in its entirety fosters the formation of intelligent hypotheses. The value of a definitive yes or no answer as to whether a recognizable homolog of a particular gene is present rapidly rules out many false hypotheses. In addition, the recognition that a close paralog of a gene of interest is, or is not, present can profoundly influence the interpretation of results.

Sequencing and Assembly

The overall strategy used to efficiently sequence the *A. aeolicus* genome was the whole genome random approach.[6] The first phase of the project consisted of shotgun sequencing, resulting in 10,500 successful sequencing reads. The sequence fragments were assembled on an Apple Power Macintosh computer with the commercially available user-friendly assembly and editing program Sequencher (Gene Codes, Ann Arbor, MI). Assembly was typically performed in batches of approximately 200–400 sequences followed by inspection and editing of the assemblies. All sequences in

[4] C. Chatelet, J. Gaillard, Y. Petillot, M. Louwagie, and J. Meyer, *Biochem. Biophys. Res. Commun.* **261,** 885 (1999).
[5] A. J. Morales, M. A. Swairjo, and P. Schimmel, *EMBO J.* **18,** 3475 (1999).
[6] E. S. Lander and M. S. Waterman, *Genomics* **2,** 231 1988.

the set were compared to all others through this process. The assembled sequences comprised approximately 750 assemblies of contiguous sequences (contigs) at the end of the random phase.

Sequences for the shotgun phase were generated from a small insert plasmid excision library constructed in Lambda ZAP II vectors.[7,8] The average insert length was determined to be 2.9 kb pairs by sequencing. Two different methods were used for sequencing. In the first, dye-primer M13 (−21) and M13 reverse primer ABI (Foster City, CA) Prism CS+ ready reaction kits were analyzed on 48-cm 4% polyacrylamide gels. The second method utilized ABI Prism FS+ dye-terminator reactions and one of two pBluescript-specific primers. These reactions were analyzed on 36-cm 5% Long-Ranger (J. T. Baker, Phillipsburg, NJ) gels.

Following assembly of the shotgun sequences, a closure phase is necessary where directed techniques are used to close gaps and complete the assembly.[6] Closure is initiated by sequencing of the small insert templates with primers directed outward from the ends of contigs. It was rapidly recognized that closure by sequencing of a fosmid[9,10] library was preferable as it provided an independent check of the assembly and was logistically easier to manage. Sequences were obtained from both ends of approximately 200 randomly chosen clones from a library constructed in the fosmid vector pFos.

These fosmid end sequences were then assembled with consensus sequences of the contigs from the random phase sequences, again using Sequencher. Contigs were joined by directly sequencing fosmids whose ends were contained on separate contigs. This strategy requires no explicit "ordering" of the contigs prior to gap closure nor does it require any mapping. The positions of end sequences of fosmid clones not directly used in gap closure in the final map are used to independently confirm the final assembly. One consequence of reducing the number of sequences in the random phase is the larger number of gaps that remain to be closed in the directed phase. A second consequence is that overall coverage is reduced. In order to ensure that reduced coverage did not compromise accuracy, ~200 oligonucleotide primers were synthesized to resequence regions of ambiguity identified by visual inspection of the entire assembly. These reactions were performed primarily on fosmid templates. The final complete

[7] J. M. Short, J. M. Fernandez, J. A. Sorge, and W. D. Huse, *Nucleic Acids Res.* **16,** 7583 (1988).

[8] M. A. Alting-Mees and J. M. Short, *Nucleic Acids Res.* **17,** 9494 (1989).

[9] H. Shizuya, B. Birren, U. J. Kim, V, Mancino, T. Slepak, Y. Tachiiri, and M. Simon, *Proc. Natl. Acad. Sci. U.S.A.* **89,** 8794 (1992).

[10] U.-J. Kim, H. Shizuya, P. J. de Jong, B. Birren, and M. I. Simon, *Nucleic Acids Res.* **20,** 1083 (1992).

assembly is composed of 13,785 sequences with an average edited read length of 557 bp.[1]

Fosmid Sequencing

Sequences for the closure phase were generated from fosmid clones and polymerase chain reaction (PCR) products. Fosmid sequencing was performed using one full-dried Autogen 740 minipreparation from a 3-ml overnight culture grown in Luria broth resuspended in 16 μl sterile H_2O combined with 8 μl of primer (1.4 pmol/μl) and 16 μl Perkin-Elmer (Norwalk, CT) dye terminator cycle sequencing ready reaction mix, with AmpliTaq DNA Polymerase, FS+. The reaction mix was transferred to a PCR tube and subjected to the following program in a thermocycler: Hold, 96°, 3 min/25 cycles, 96°, 20 sec/50°, 20 sec/60°, 4 min/hold 60°, 5 min. Hold, 4°, indefinite. Purification: add 150 μl 70% ethanol containing 0.5 mM $MgCl_2$, centrifuge 15 min, and allow pellet to air dry. Loading (48-well comb): add 1.75 μl loading buffer to each sample. Load 1.3 μl.

PCR Amplification and Sequencing

The final eight gaps in the sequence were not covered by the fosmid clones and were closed by direct sequencing of PCR products generated with the TaqPlus Long PCR System (Stratagene Cloning Systems, La Jolla, CA). Gap sequences were amplified with the following reaction: *A. aeolicus* genomic DNA (100 ng/μl), 2.0 μl; dNTPs (1.25 mM each), 8.0 μl; primer 1 (10 pmol/μl), 5.0 μl; primer 2 (10 pmol/μl), 5.0 μl; low salt buffer (supplied with Taq Plus Long), 5.0 μl; H_2O, 24.5 μl; Stratagene Taq Plus Long solution, 0.5 μl; 50 μl total reaction volume. The reaction was performed in a thermocycler using the following PCR cycling scheme: 94° 1.5 min/25 cycles of 94° 10 sec/58° 30 sec/72° 2.5 min/72° 30 min/4° hold.

Following amplification, the PCR products are sequenced directly in a dye terminator reaction. The product is purified from the 50-μl reaction PCR reaction using a Wizard preparation column (Promega, Madison, WI) and resuspended in 35 μl sterile water. The sequencing reaction contains 6 μl of the purified product; 4 μl primer (1.4 pmol/μl); 2 μl water; and 8 μl dye terminator cycle sequencing ready reaction mix. Following thermocycling and precipitation using the conditions specified previously, the reaction is resuspended in 8 μl loading buffer; 1.75 μl of this is loaded per lane.

Gene (ORF + RNA) Identification and Functional
Assignment Approaches

Open reading frames (ORFs) of greater than 300 bp were analyzed by a variety of methods, and the predicted gene product was categorized as (i) similar to a protein of known function, (ii) similar to a protein of unknown function (hypothetical), (iii) unknown function (putative), or noncoding. Assignments to the first category, which represent functional identifications, were based primarily on FASTA[11] alignments. Secondary information used in some cases for final assignments included BLASTP,[12] PROSITE[13] analysis, and, in rare instances, contextual information. Detailed annotation available from the Swiss-Prot database[14] was particularly useful for reconciling assignments. Known bacterial proteins of less than 100 amino acids were searched against the *A. aeolicus* sequence to identify protein coding genes of less than 300 bp. tRNA genes were identified with the program tRNAscan-SE.[15]

ORFs were assigned as similar to a protein of unknown function (or hypothetical) based on FASTA alignments. ORFs were assigned as unknown function (or putative) based on length, purine/pyrimidine ratio of the coding strand,[16] and position relative to other ORFs. For instance, an ORF, with no database similarity that largely overlaps an ORF with functional assignment would be designated noncoding. The relatively low G + C content of the genome (average 43.4%) results in random open reading frames of approximately 50 bp on average. We used 300 bp as a cutoff, and the average ORF designated as putative was 762 bp. ORFs not designated as coding in one of these three categories are assumed to be noncoding. BLASTX[17] was used to examine all presumptive "intergenic regions" for similarities to known protein sequences. In some cases, based on these results, the original base calls were reinspected.

Using these methods, 849 genes were assigned as similar to proteins of known function, 256 as similar to proteins of unknown function, and 407 as coding regions of unknown function. Similarity alone is an imperfect guide to function and, for more divergent sequences and functions, the assignments are even more tentative. It is also important to note that these

[11] W. R. Pearson and D. J. Lipman, *Proc. Natl. Acad. Sci. U.S.A.* **85,** 2444 (1988).
[12] S. F. Altschul, W. Gish, W. Miller, E. W. Myers, and D. J. Lipman, *J. Mol. Biol.* **215,** 403 (1990).
[13] K. Hofmann, P. Bucher, L. Falquet, and A. Bairoch, *Nucleic Acids Res.* **27,** 215 (1999).
[14] A. Bairoch and R. Apeweiler, *Nucleic Acids Res.* **28,** 45 (2000).
[15] T. M. Lowe and S. R. Eddy, *Nucleic Acids Res.* **25,** 955 (1997).
[16] A. L. Mazin and B. L. Alkin, *Molek. Biol.* **9,** 716 (1975) [In Russian].
[17] W. Gish and D. J. States, *Nature Genet.* **3,** 266 (1993).

assignments are based on the information in databases at the time of the analyses and are dependent on the annotations within those databases. Therefore, it is important to use the functional assignments only as an initial guide to in-depth analyses.

Genome Structure

The most striking feature of the *A. aeolicus* genome is the organizational relationship of the genes relative to one another. The majority of genes in *A. aeolicus* appear to be organized within operons. This in itself is not surprising for a bacterial genome; however, the composition of many of the operons is unexpected. While some genes are clustered in operons of common function, such as flagellar proteins, oxidoreductases, transporters, and ribosomal components, many pathway components are scattered throughout the genome. Two examples are explored here; the *his* and *trp* genes responsible for histidine and tryptophan biosynthesis, respectively. Both pathways are found as operons in a number of organisms, including Archaea, and have been studied extensively. In *A. aeolicus,* there are eight *his* genes: *hisA,B,C,D,IE,F,G,* and *H*. None of these genes are cotranscribed with one another, although five of the eight are apparently in operons. At this point, it should be noted that none of the operons of *A. aeolicus* have been experimentally confirmed; they are deduced based solely on the proximity of the initiation and stop codons of adjacent genes. It is difficult to determine where transcripts end, as *A. aeolicus* apparently does not rely on hairpins.[18] For the following gene arrangements, the numbers in parentheses represent the distance between the genes; a negative number indicates that the genes overlap.

Genomic Context of his Genes in A. aeolicus

his*A:* dicistronic operon with pyrDB (−1 bp)
his*B:* dicistronic operon with 426 amino acid putative protein (+15 bp)
his*C:* monocistronic
his*D:* monocistronic
his*IE:* dicistronic operon with DNA polymerase (−1 bp)
his*F:* monocistronic
his*G:* dicistronic operon with oxaloacetate decarboxylase (−4 bp)
his*H:* tricistronic with cytochrome *c* biogenesis protein and pyruvate
    formate lyase activating protein (0 bp)

[18] T. Washio, J. Sasayama, and M. Tomita, *Nucleic Acids Res.* **26,** 5456 (1998).

Similarly, the *trp* genes are dispersed throghout the genome. Paralogs are present for two of genes *trpB* and *trpD,* but none of the nine genes are transcribed together.

Genomic Context of trp Genes in A. aeolicus

trpA: dicistronic operon with a hypothetical protein (−11 bp)

trpB1: dicistronic operon with a hypothetical protein (+16 bp)

trpB2: dicistronic operon with a 361 amino acid putative protein (+5 bp)

trpC: dicistronic operon with a hypothetical protein (+46 bp)

trpD1: dicistronic operon with a 400 amino acid putative protein (0 bp)

trpD2: tricistronic operon, *trpD2* lies between *cobA* (+5 bp) and nitrite reductase (+16 bp)

trpE: monocistronic

trpF: monocistronic

trpG: tricistronic operon, *trpG* lies between NADH dehydrogenase chain 7 (−1 bp) and a hypothetical protein (−1 bp)

The observation that many of the genes overlap is common in *A. aeolicus.* It is also common to find overlaps between genes that are convergently transcribed. In some cases, the overlaps are tens of base pairs, and in some cases, only the stop codons overlap, as in the sequence below, where two TAA stops on opposite strands are underlined.

$$5'\text{-TT}\underline{\text{AA}}\text{-}3'$$
$$3'\text{-}\underline{\text{AA}}\text{TT-}5'$$

Interestingly, not only are pathways dispersed throughout the genome, but so are genes that encode subunits of the same enzyme. *trpA* and *trpB* from the previous illustration are an example; these genes encode the α and β subunits of tryptophan synthase and can be expected to be required in stoichiometric amounts. This particular example is complicated by the fact that there are apparently two versions of the *trpB* gene present. However, there are many other examples where enzyme subunits are encoded separately: *gltB* and *gltD,* the large and small subunits of glutamate synthase; *ribH* and *ribC,* the α and β subunits of riboflavin synthase; *nrdA* and *nrdF,* the α and β chains of ribonucleotide reductase; and *pheS* and *pheT,* the α and β subunits of phenylalanyl-tRNA synthetase.

It is intriguing to hypothesize that the lack of cocistronic arrangements for biosynthetic genes is due to the obligate chemosynthetic metabolism of *A. aeolicus.* Typically, heterotrophic organisms appear to cotranscribe biosynthetic pathways while autotrophs appear less likely to do so. This allows the heterotrophs to take advantage when these compounds appear in the environment. This pattern appears to hold in both the Bacteria and the Archaea. It is more difficult to rationalize why subunits of enzymes would not be cotranscribed. This possibly reflects a very limited range of gene expression responses, either biosynthesis is "on" or it is "off." Alternatively, the *A. aeolicus* versions of these proteins may not be "subunits" but may function as monomers. Whatever the ultimate reason may be, the sequence has raised an intriguing question.

"Missing" Genes

Catalase. The *A. aeolicus* genome encodes three superoxide dismutase genes: two of the Cu/Zn family and one of the Fe/Mn family. Conspicuously absent from the list of genes involved in responses to oxidative stress, however, is catalase. The absence of an identifiable catalase gene was even more unexpected in light of the fact that *A. aeolicus* cells and colonies react positively with H_2O_2, resulting in the rapid evolution of gas bubbles. Initial speculation concerning the inability to identify a catalase gene centered on the idea that the catalase activity observed in *A. aeolicus* and *A. pyrophilus*[19] may result from a novel enzyme that cannot yet be identified by sequence similarity. Targeted searches using all the known catalase sequences and motifs, as well as the sequence of pseudocatalase, a manganese-containing nonheme enzyme found in lactic acid bacteria,[20] failed to produce any candidate genes. A pseudocatalase sequence was reported from *Thermus* sp.[21] This gene is homologous to the gene from *Lactobacillus* and is apparently unrelated to any genes in *A. aeolicus*. An *A. aeolicus* library in a *katG katE Escherichia coli* host (P. Kretz, Diversa Corp., personal communication, 1997) was screened at elevated temperatures in an attempt to identify the gene functionally by expression cloning. Despite extensive screening, however, no catalase-positive clones were identified (J. Lopez-Dee, A. M. Ramirez and R. V. Swanson, unpublished results, 1997).

With the entire genome sequence in hand, only a short N-terminal or peptide-derived amino acid sequence is necessary to identify a gene. Therefore, following the negative results of the bioinformatics and expression cloning approaches, an attempt was made to purify the catalase activity (K. Ma and M. W. W. Adams, unpublished results, 1997). *A. aeolicus* cells were obtained from K. Stetter and R. Huber (University of Regensburg, Germany), and cell lysates were prepared. The catalase activity was not

[19] R. Huber, T. Wilharm, D. Huber, A. Trincone, S. Burggraf, H. Konig, R. Rachel, I. Rockinger, H. Fricke, and K. O. Stetter, *Arch. Microbiol.* **15,** 340 (1992).

[20] T. Igarashi, Y. Kono, and T. Kiyoshi, *J. Biol. Chem.* **271,** 29521 (1996).

[21] M. Kagawa, N. Murakoshi, Y. Nishikawa, G. Matsumoto, Y. Kurata, T. Mizobata, Y. Kawata, and J. Nagai, *Arch. Biochem. Biophys.* **362,** 346 (1999).

present in the soluble fraction of these extracts, but was associated with the pellets after centrifugation. Acid treatment of this pellet did not destroy the activity. The activity remained with the insoluble fraction following extraction with organic solvents such as ethanol, butanol, acetone, or hexane. Inductively coupled plasma atomic emission spectrometric analyses revealed that the remaining precipitate consisted of a mixture of phosphate salts containing $P:Fe:Ca:Zn:Cr:Ni:Cu:Si:Mn$ $(71:30:28:7:6:3:2:1:1, w/w)$. Among all components, Fe alone has strong catalase activity, and Cu and Mn have very little activity. Because Fe is the major component of phosphate salts, it is considered to be responsible for the observed *A. aeolicus* "catalase" activity (K. Ma and M. W. W. Adams, unpublished observations, 1997). It currently remains unknown whether this inorganic complex is an artifact of the culture medium or if specific genes encoded by the genome play a role in the precipitation of salts to provide the protective catalase function. If the latter is the case, then the absence of a gene predicted to be present in the genome on the basis of physiology and catalytic activity led to the discovery of an unsuspected alternative microbial solution to the problem. Obviously, further work would be required to substantiate this conclusion.

Ribonuclease P. RNase P is the endoribonuclease responsible for removal of the precursor-tRNA leaders to generate the mature 5' ends of tRNA.[22] Bacterial RNase P's are ribonucleoproteins that contain both a protein and an RNA component. The RNA subunit of bacterial RNase P is catalytically active *in vitro* in the absence of the protein subunit.[23] Genes encoding neither the protein nor the RNA component were identified during the annotation of the *A. aeolicus* genome. As in the case of catalase, an extensive search was carried out using known sequences and motifs derived from consensus sequences. Searches for both components yielded negative results.

The fact that one of the genes is expected to encode a stable RNA allowed us to develop an additional search strategy. This strategy was based on the observation that while the average G + C content of the genome is 43.4% and relatively constant, the high growth temperature of the organism leads to a G + C content of the readily identified stable RNA genes that is significantly higher: tRNA genes (average G + C, 68.5%) and rRNA genes (average G + C, 65.0%). Therefore, the genome was scanned to identify high G + C segments that could harbor the rnpB gene. One such segment was identified by its high G + C content. BLASTN searches of this sequence against GenBank revealed similarity to tmRNA. This is a

[22] D. N. Frank and N. R. Pace, *Annu. Rev. Biochem.* **67**, 153 (1998).

[23] C. Guerrier-Takada, K. Gardiner, T. Marsh, N. Pace, and S. Altman, *Cell* **35**, 849 (1983).

stable RNA whose name derives from its characteristics as both a tRNA and an mRNA. TmRNA is involved in a *trans*-translation process to add a C-terminal peptide tag to the protein product of an incomplete mRNA. The tmRNA-directed tag targets the partial protein for proteolysis.[24] The identification of the *A. aeolicus ssrA* gene encoding tmRNA validated the G + C scan search strategy, as this gene was unidentified previous to the search. The *A. aeolicus* tmRNA is 347 bp long and the G + C content is 66.9%. In contrast, however, the search did not reveal candidates for the *rnpB* gene. It is also notable that the search did not reveal candidates that might encode novel catalytic RNAs of unknown function. At this point, it appears likely that the RNase P activity will need to be purified in order to identify the biochemical machinery responsible for this function in *A. aeolicus.*

Chemotaxis. *A. aeolicus* is motile and possesses monopolar polytrichous flagella. More than 25 genes involved in the flagellar structure and biosynthesis are present in the *A. aeolicus* genome. However, no homologs of the bacterial chemotaxis system were identified. In enteric bacteria, flagellar rotation, and consequently swimming behavior, is controlled by a phosphotransfer network triggered by ligand binding to membrane-bound receptors.[25] None of these receptors were identified in *A. aeolicus* either. Chemotaxis systems, homologous to those found in enteric bacteria, are present in the Archaea *Halobacterium salinarum*[26] and *Pyrococcus horikoshii*[27] and *Archaeoglobus fulgidus,*[28] despite the fact that bacterial and archeal flagellar apparati are not homologous.[29] The *M. jannaschii* genome also lacks homologs of known chemotaxis genes. Thus, motility in *A. aeolicus* and *M. jannaschii* is either undirected or the input for controlling taxis is through another unidentified system. In *E. coli,* the flagellar switch is essential for flagellar structure, function, and coupling of chemotaxis signals. However, the *A. aeolicus* genome contains homologs of only two of the three *E. coli* proteins that comprise the switch: FliG and FliN. Biochemical and genetic

[24] K. C. Keiler, P. R. Waller, and R. T. Sauer, *Science* **271,** 990 (1996).

[25] R. B., Bourret, K. A. Borkovich, and M. I. Simon, *Annu. Rev. Biochem.* **60,** 401 (1991).

[26] J. Rudolph, N. Tolliday, C. Schmitt, S. C. Schuster, and D. Oesterhelt, *EMBO J.* **14,** 4249 (1995).

[27] Y. Kawarabayasi, M. Sawada, H., Horikawa, Y. Haikawa, Y. Hino, S. Yamamoto, M. Sekine, S. Baba, H. Kosugi, A. Hosoyama, Y. Nagai, M. Sakai, K. Ogura, R. Otsuka, H. Nakazawa, M. Takamiya, Y. Ohfuku, T. Funahashi, T. Tanaka, Y. Kudoh, J. Yamazaki, N. Kushida, A. Oguchi, K. Aoki, and H. Kikuchi, *DNA Res.* **5,** 55 (1998).

[28] H. P. Klenk, R. A. Clayton, J. F. Tomb, O. White, K. E. Nelson, K. A. Ketchum, R. J. Dodson, M. Gwinn, E. K. Hickey, J. D. Peterson, D. L. Richardson, A. R. Kerlavage, D. E. Graham, N. C. Kyrpides, R. D. Fleischmann, J. Quackenbush, N. H. Lee, G. G. Sutton, S. Gill, E. F. Kirkness, B. A. Dougherty, K. McKenney, M. D. Adams, B. Loftus, and J. C. Venter, *Nature* **390,** 364 (1997).

[29] K. F. Jarrell, D. P. Bayley, and A. S. Kostyukova, *J. Bacteriol.* **178,** 5057 (1996).

studies implicate the other switch protein, FliM, as the receptor for phospho-Che Y, the switch signal. The absence of both FliM and the chemotaxis proteins in *A. aeolicus* is consistent with this identification of FliM as the phospho-Che Y receptor in *E. coli*. This result also argues against a direct role for FliM in torque generation in *E. coli,* despite the observation that *fliM* mutants are defective in torque generation. The absence of FliM in *A. aeolicus* offers insight into the extensively studied *E. coli* chemotaxis system. The most extensively studied chemotaxis systems respond to sugars and amino acids, although in some cases, responses to other inputs (e.g., metals, redox potential, and light) are also mediated. In contrast to all the organisms known to possess the classical chemotactic signal transduction pathways, both *A. aeolicus* and *M. jannaschii* are obligate chemoautotrophs. Chemoautotrophs may respond to a different set of factors, such as the concentrations of dissolved gases, such as CO_2, H_2, or O_2, or another critical parameter, such as temperature. If this trend in chemotaxis system type, known for heterotrophs but uncharacterized in chemoauto-trophs, continues as more genomes are sequenced, it may offer an important physiological marker within the genome. As more data are accumulated and in combination with other genome characteristics, such as the operon arrangements of biosynthetic genes, which also appear to diverge along heterotrophic/chemotrophic lines, the predictive value of this type of obser-vation will grow. These types of correlations will become increasingly impor-tant as we move into sequencing uncultured organisms with a primary goal of predicting their physiology from the genome sequence alone.[30,31]

Concluding Comments

A relatively small research group can carry out a microbial genome-sequencing project in a reasonable amount of time—approximately the same amount of time that the same size group of researchers could clone and sequence several dozen genes based on similarities or limited protein sequence data. Genome data and associated similarity-based annotation provide immediate answers to many of the fundamental questions about the characteristics and potentials of an organism. In addition, data provided allow researchers to target interesting additional questions. The genome sequence is a powerful resource in combination with traditional biochemis-try and should accelerate the discovery of novel enzymes rapidly. Complete data make comprehensive comparative studies possible and expose the

[30] J. L. Stein, T. L. Marsh, K. Y. Wu, H. Shizuya, and E. F. DeLong, *J. Bacteriol.* **178,** 591 (1996).

[31] C. Schleper, E. F. DeLong, C. M. Preston, R. A. Feldman, K. Y. Wu, and R. V. Swanson, *J. Bacteriol.* **180,** 5003 (1998).

weakness of extrapolating from limited data sets. Eventually we will be able to look at a genome and predict the physiology and role of an organism in the environment with confidence, even if that organism has never been studied before or is uncultured in the laboratory.

[9] Genome of *Thermotoga maritima* MSB8

By Karen E. Nelson, Jonathan A. Eisen, and Claire M. Fraser

Introduction

The non-spore-forming, rod-shaped bacterium *Thermotoga maritima* was isolated from geothermal heated marine sediments at Vulcano, Italy, in 1986.[1] *T. maritima* has a temperature optimum for growth of 80° and can metabolize many simple and complex carbohydrates, including glucose, sucrose, starch, xylan, and cellulose.[1,2] Both xylan and cellulose are complex plant polymers; xylan represents the most abundant noncellulosic polysaccharide in angiosperms, where it accounts for 20–30% of the dry weight of wood tissues. Cellulose is the most abundant biopolymer occurring in nature, estimated to account for 75×10^9 tons of dry plant biomass annually.[3] Both cellulose and xylan, through conversion to fuels (e.g., H_2), have major potential as renewable carbon and energy sources.

Most other *Thermotoga* strains have been isolated in continental oil reservoirs and oil-producing wells.[4] All, however, demonstrate the characteristic toga-like sheath structure that surrounds the cells and an ability to ferment a range of carbohydrate substrates. From an evolutionary standpoint, the Thermotogales are significant, as small subunit ribosomal RNA (SSU rRNA) phylogenetic analysis of *T. maritima* placed the bacterium as the deepest and most slowly evolving lineage in the Eubacteria.[5] Later analysis of the SSU rRNA from *Aquifex pyrophilus* (isolated and character-

---

[1] R. Huber, T. A. Langworthy, H. Konig, M. Thomm, C. R. Woese, U. B. Sleytr and K. O. Stetter, *Arch. Microbiol.* **144**, 324 (1986).

[2] R. Huber and K. O. Stetter, *in* "The Prokaryotes" (A. Balows, *et al.,* eds.), p. 3809. Springer, Berlin, 1992.

[3] N. S. Thompson, Wood and Agricultural Residues, p. 101. Academic Press, New York, 1983.

[4] C. Jeanthon, A. L. Reysenbach, S. L'Haridon, A. Gambacorta, N. R. Pace, P. Glenat, and D. Prieur, *Arch. Microbiol.* **164**(2), 91 (1995).

[5] L. Achenbach-Richter, R. Gupta, K. O. Stetter, and C. R. Woese, *Syst. Appl. Microbiol.* **9**, 34 (1987).

METHODS IN ENZYMOLOGY, VOL. 330

ized in 1992[6]) suggested that the Aquificales represent a branch more ancient than the Thermotogales. Numerous phylogenetic studies[7,8] have, however, cast doubt on this proposed branching order, particularly as relating to the particular enzyme or gene being characterized.

In an attempt to further elucidate the phylogeny and physiology of this organism, the genome of the type strain *T. maritima* MSB8 was sequenced at The Institute for Genomic Research (TIGR), in Rockville, Maryland. The genome was sequenced by the whole genome random sequencing method that has been described previously for other genomes sequenced at TIGR[9,10] (see later).

Genome Sequencing and Annotation Procedure

Random Sequencing, Closure, and Annotation

Thermotoga maritima MSB8 was grown from a culture derived from a single cell isolated by optical tweezers and provided by Robert Huber (University of Regensburg). For random sequencing, one small-insert plasmid library (1.5–2.5 kb) was generated by random mechanical shearing of genomic DNA, and one large-insert lambda library was generated by partial Tsp5091 digestion and ligation to λ-DASHII/*Eco*RI vector (Stratagene). In the initial random sequencing phase, an approximate sevenfold sequence coverage was achieved with 27,789 sequences from plasmid clones (average

[6] R. Huber, T. Wilharn, D. Huber, A. Trincone, S. Burggraf, H. Konig, R. Rachel, I. Rockinger, H. Fricke, and K. O. Stetter, *System. Appl. Microbiol.* **15,** 340 (1992).

[7] S. L. Baldauf, J. D. Palmer, and W. F. Doolittle, *Proc. Natl. Acad. Sci. U.S.A.* **93,** 7749 (1996).

[8] J. R. Brown and W. F. Doolittle, *Microbiol. Mol. Biol. Rev.* **61,** 456 (1997).

[9] R. D. Fleischmann, M. D. Adams, O. White, R. A. Clayton, E. F. Kirkness, A. R. Kerlavage, C. J. Bult, J.-F. Tomb, B. A. Dougherty, J. M. Merrick, K. McKenney, G. Sutton, W. FitzHugh, C. Fields, J. D. Gocayne, J. Scott, R. Shirley, L.-I. Liu, A. Glodek, J. M. Kelley, J. F. Weidman, C. A. Phillips, T. Spriggs, E. Hedblom, M. D. Cotton, T. R. Utterback, M. C. Hanna, D. T. Nguyen, D. M. Saudek, R. C. Brandon, L. D. Fine, J. L. Fritchman, J. L. Fuhrmann, N. S. M. Geoghagen, C. L. Gnehm, L. A. McDonald, K. V. Small, C. M. Fraser, H. O. Smith, and J. C. Venter, *Science* **269,** 496 (1995).

[10] H.-P. Klenk, R. A. Clayton, J.-F. Tomb, O. White, K. E. Nelson, K. A. Ketchum, R. J. Dodson, M. Gwinn, E. K. Hickey, J. D. Peterson, D. L. Richardson, A. R. Kerlavage, D. E. Graham, N. C. Kyrpides, R. D. Fleischmann, J. Quackenbush, N. H. Lee, G. G. Sutton, S. Gill, E. F. Kirkness, B. A. Dougherty, K. McKenney, M. D. Adams, B. Loftus, S. Peterson, C. I. Reich, L. McNeil, J. H. Badger, A. Glodek, L. Zhou, R. Overbeek, J. D. Gocayne, J. F. Weidman, L. McDonald, T. Utterback, M. D. Cotton, T. Spriggs, P. Artiach, B. P. Kaine, S. M. Sykes, P. W. Sadow, K. P. D'Andrea, C. Bowman, C. Fujii, S. A. Garland, T. M. Mason, G. J. Olsen, C. M. Fraser, H. O. Smith, C. R. Woese, and J. C. Venter, *Nature* **390,** 364 (1997).

read length 531 bases). In addition, sequences from 546 λ clones served as a genome scaffold. Plasmid and λ sequences were jointly assembled using TIGR-Assembler,[11] and the remaining sequence gaps were closed by editing the ends of sequence traces and/or primer walking on plasmid clones. Physical gaps were closed by direct sequencing off genomic DNA or combinatorial polymerase chain reaction (PCR) followed by sequencing of the PCR product.

Open Reading Frame (ORF) Prediction and Gene Family Identification

An initial set of ORFs likely to encode proteins was identified by GLIMMER,[12] and those shorter than 30 codons were eliminated. ORFs that overlapped were inspected visually and, in some cases, removed. ORFs were searched against a nonredundant protein database as described previously.[9] Frameshifts and point mutations were detected and corrected where appropriate.[13] Remaining frameshifts and point mutations are considered authentic, and corresponding regions were annotated as "authentic frameshift" or "authentic point mutation," respectively. Two sets of hidden Markov models (HMMs) were used to determine ORF membership in families and superfamilies. These included 527 HMMs from pfam v2.0 and 199 HMMs from the TIGR ortholog resource. TopPred[14] was used to identify membrane-spanning domains in proteins.

Comparative Genomics and Phylogenetics

Paralogous gene families were constructed by searching the ORFs against themselves using BLASTX,[15] identifying pairwise matches above $P \leq 10^{-5}$ over 60% of the query search length, and subsequently clustering these matches into multigene families. Multiple sequence alignments for

[11] G. G. Sutton, O. White, M. D. Adams, and A. R. Kerlavage, *Genome Sequence Technol.* **1,** 9 (1995).

[12] S. L. Salzberg, A. L. Delcher, S. Kasif, and O. White, *Nucleic Acids Res.* **15,** 544 (1998).

[13] J. F. Tomb, O. White, A. R. Kerlavage, R. A. Clayton, G. G. Sutton, R. D. Fleischmann, K. A. Ketchum, H.-P. Klenk, S. Gill, B. A. Dougherty, K. E. Nelson, J. Quackenbush, L. Zhou, E. F. Kirkness, S. Peterson, B. Loftus, D. Richardson, R. Dodson, H. G. Khalak, A. Glodek, K. McKenney, L. M. FitzGerald, N. Lee, M. D. Adams, E. K. Hickey, D. E. Berg, J. D. Gocayne, T. R. Utterback, J. D. Peterson, J. M. Kelley, M. D. Cotton, J. M. Weidman, C. Fujii, C. Bowman, L. Watthey, E. Wallin, W. S. Hayes, M. Borodovsky, P. Karp, H. O. Smith, C. M. Fraser, and J. C. Venter, *Nature* **388,** 539 (1997).

[14] M. G. Claros and G. von Heijne, *Comput. Appl. Biosci.* **10,** 685 (1994).

[15] S. F. Altschul, W. Gish, W. Miller, E. W. Myers, and D. J. Lipman, *J. Mol. Biol.* **215,** 403 (1990).

these protein families were generated using msa,[16] an annealing algorithm, and the alignments were scrutinized.

The *T. maritima* ORFs were added to a set of all ORFs from 16 published microbial genomes.[9,10,17–29] Pairwise matches were identified by searching

[16] G. G. Sutton and T. Bussey, Abstract presented at the Fifth International Genome Sequencing and Analysis Conference. Hilton Head Island, South Carolina, 1993.

[17] C. J. Bult, O. White, G. J. Olsen, L. Zhou, R. D. Fleischmann, G. G. Sutton, J. A. Blake, L. M. Fitzgerald, R. A. Clayton, J. D. Gocayne, A. R. Kerlavage, B. A. Dougherty, J.-F. Tomb, M. D. Adams, C. I. Reich, R. Overbeek, E. F. Kirkness, K. G. Weinstock, J. M. Merrick, A. Glodek, J. L. Scott, N. S. M. Geoghagen, J. F. Weidman, J. L. Fuhrman, D. Nguyen, T. R. Utterback, J. M. Kelley, J. D. Peterson, P. W. Sadow, M. C. Hanna, M. D. Cotton, K. M. Roberts, M. A. Hurst, B. P. Kaine, M. Borodovsky, H.-P. Klenk, C. M. Fraser, H. O. Smith, C. R. Woese, and J. C. Venter, *Science* **273,** 1058 (1996).

[18] G. Deckert, P. V. Warren, T. Gaasterland, W. G. Young, A. L. Lenox, D. E. Graham, R. Overbeek, M. A. Snead, M. Keller, M. Aujay, R. Huber, R. A. Feldman, J. M. Short, G. J. Olsen, and R. V. Swanson, *Nature* **392,** 353 (1998).

[19] Y. Kawarabayasi, M. Sawada, H. Horikawa, Y. Haikawa, Y. Hino, S. Yamamoto, M. Sekine, S. Baba, H. Kosugi, A. Hosoyama, Y. Nagai, M. Sakai, K. Ogura, R. Otsuka, H. Nakazawa, M. Takamiya, Y. Ohfuku, T. Funahashi, T. Tanaka, Y. Kudoh, Y. Yamazaki, N. Kushida, A. Oguchi, K. Aoki, and H. Kikuchi, *DNA Res.* **5,** 55 (1998).

[20] D. R. Smith, L. A. Doucette-Stamm, C. Deloughery, H. Lee, J. Dubois, T. Aldredge, R. Bashirzadeh, D. Blakely, R. Cook, K. Gilbert, D. Harrison, L. Hoang, P. Keagle, W. Lumm, B. Pothier, D. Qiu, R. Spadafora, R. Vicaire, Y. Wang, J. Wierzbowski, R. Gibson, N. Jiwani, A. Caruso, D. Bush, and J. N. Reeve, *J. Bacteriol.* **179,** 7135 (1997).

[21] F. Kunst, N. Ogasawara, I. Moszer, A. M. Albertini, G. Alloni, V. Azevedo, M. G. Bertero, P. Bessieres, A. Bolotin, S. Borchert, R. Borriss, L. Boursier, A. Brans, M. Braun, S. C. Brignell, S. Bron, S. Brouillet, C. V. Bruschi, B. Caldwell, V. Capuano, N. M. Carter, S. K. Choi, J. J. Codani, I. F. Connerton, and A. Danchin, *Nature* **390,** 249 (1997).

[22] C. M. Fraser, S. Casjens, W. M. Huang, G. G. Sutton, R. Clayton, R. Lathigra, O. White, K. A. Ketchum, R. Dodson, E. K. Hickey, M. Gwinn, B. Dougherty, J. F. Tomb, R. D. Fleischmann, D. Richardson, J. Peterson, A. R. Kerlavage, J. Quackenbush, S. Salzberg, M. Hanson, R. van Vugt, N. Palmer, M. D. Adams, J. Gocayne, and J. C. Venter, *Nature* **390,** 580 (1997).

[23] F. R. Blattner, G. Plunkett, C. A. Bloch, N. T. Perna, V. Burland, M. Riley, J. Collado-Vides, J. D. Glasner, C. K. Rode, G. F. Mayhew, J. Gregor, N. W. Davis, H. A. Kirkpatrick, M. A. Goeden, D. J. Rose, B. Mau, and Y. Shao, *Science* **277,** 1453 (1997).

[24] S. T. Cole, R. Brosch, J. Parkhill, T. Garnier, C. Churcher, D. Harris, S. V. Gordon, K. Eiglmeier, S. Gas, C. E. Barry III, F. Tekaia, K. Badcock, D. Basham, D. Brown, T. Chillingworth, R. Connor, R. Davies, K. Devlin, T. Feltwell, S. Gentles, N. Hamlin, S. Holroyd, T. Hornsby, K. Jagels, and B. G. Barrell, *Nature* **11,** 537 (1998).

[25] C. M. Fraser, J. D. Gocayne, O. White, M. D. Adams, R. A. Clayton, R. D. Fleischmann, C. J. Bult, A. R. Kerlavage, G. Sutton, J. M. Kelley, J. L. Fritchman, J. F. Weidman, K. V. Small, M. Sandusky, J. Fuhrmann, D. Nguyen, T. R. Utterback, D. M. Saudek, C. A. Phillips, J. M. Merrick, J.-F. Tomb, B. A. Dougherty, K. F. Bott, P.-C. Hu, T. S. Lucier, S. N. Peterson, H. O. Smith, C. A. Hutchison III, and J. C. Venter, *Science* **270,** 397 (1995).

[26] R. Himmelreich, H. Hilbert, H. Plagens, E. Pirkl, B. C. Li, and R. Herrmann, *Nucleic Acids Res.* **24,** 4420 (1996).

this dataset against itself using BLASTX,[15] and subsequently clustered, and converted to multiple sequence alignments as described earlier. From this initial set of alignments, a subset of 33 homologous gene families ($P \leq 10^{-5}$ over 60% of the query search length) were identified. These 33 homologous gene families included adenosine phosphoribosyltransferase (*apt*), arginyl-tRNA synthetase (*argS*), enolase (*eno*), cell division protein (*ftsZ*), a putative secreted metalloendopeptidase (*gcp*), glutamyl-tRNA synthetase (*gltX*), groES protein (*groES*), pseudouridylate synthase (*hisT*), phenylalanyl-tRNA synthetase (*pheS*), phosphoglycerate kinase (*pgk*), phosphoribosyl pyrophosphate synthetase (*prs*), the ribosomal proteins *rplA, rplB, rplC, rplE, rplF, rplK, rplN, rplR, rplV, rpsB, rpsC, rpsD, rpsE, rpsG, rpsH, rpsJ, rpsK, rpsL, rpsM, rpsQ,* and *rpsS,* and preprotein translocase (*secY*).

The sequence alignments were enhanced using the CLUSTAL X program, and regions that were ambiguous, hypervariable, or contained large alignment gaps were excluded from subsequent phylogenetic analysis. Phylogenetic trees were generated from the curated alignments using the PAUP 4.0.0d64 and PHYLIP computer programs. Parsimony analysis was conducted using the *heuristic search* algorithm of PAUP and *protpars* of PHYLIP. Distance-based trees were generated using the neighbor-joining algorithm of Saitou and Nei.[30] One hundred bootstrap replicates were conducted for all trees. For all PAUP analyses, multiple step matrices were used in calculation of distances and in parsimony analysis. The whole genome set of pairwise BLASTX[15] search results was also used to determine "best hits" of *T. maritima* to other genomes.

Variation in nucleotide composition within the genome was analyzed by χ^2 analysis of trinucleotide content. The distribution of all 64 trinucleotides (3-mers) was computed in 2000-bp windows across the genome. Windows that overlapped by half their length, i.e., 1000 bp, were used to compute the χ^2 statistic on the difference between its 3-mer content and

[27] T. Kaneko, S. Sato, H. Kotani, A. Tanaka, E. Asamizu, Y. Nakamura, N. Miyajima, M. Hirosawa, M. Sugiura, S. Sasamoto, T. Kimura, T. Hosouchi, A. Matsuno, A. Muraki, N. Nakazaki, K. Naruo, S. Okumura, S. Shimpo, C. Takeuchi, T. Wada, A. Watanabe, M. Yamada, M. Yasuda, and S. Tabata, *DNA Res.* **3**, 185 (1996).

[28] C. M. Fraser, S. J. Norris, G. M. Weinstock, O. White, G. G. Sutton, R. Dodson, M. Gwinn, E. K. Hickey, R. Clayton, K. A. Ketchum, E. Sodergren, J. M. Hardham, M. P. McLeod, S. Salzberg, J. Peterson, H. Khalak, D. Richardson, J. K. Howell, M. Chidambaram, T. Utterback, L. McDonald, P. Artiach, C. Bowman, M. D. Cotton, K. Horst, K. Roberts, B. Hatch, H. O. Smith, and J. C. Venter, *Science* **281**, 375 (1998).

[29] A. Goffeau, B. G. Barrell, H. Bussey, R. W. Davis, B. Dujon, H. Feldmann, F. Galibert, J. D. Hoheisel, C. Jacq, M. Johnston, E. J. Louis, H. W. Mewes, Y. Murakami, P. Philippsen, H. Tettelin, and S. G. Oliver, *Science* **274**, 563 (1996).

[30] N. Saitou and M. Nei, *Mol. Biol. Evol.* **4**, 406 (1987).

that of the whole genome. A large value of this statistic means that the composition within the window is different from the rest of the genome; the probability values for these regions are based on the assumption that the DNA composition is relatively uniform throughout the genome.

Genome Sequence

Main Characteristics of Thermotoga maritima Genome

The *T. maritima* MSB8 genome is a single circular chromosome composed of 1,860,725 bp (Fig. 1, see color insert). The genome contains 1877 predicted coding regions, 54% of which have functional assignments.[31] Eight hundred sixty three (46%) are of unknown function. The genome has an average G + C content of 46%, with one region of significantly higher (62%) and one region of significantly lower (34%) G + C content. These two regions correspond to the single ribosomal RNA operon and lipopolysaccharide biosynthesis (LPS) proteins, respectively.

Evolutionary Analysis of Conserved Genes

To determine whether the genome sequence could resolve the phylogenetic positioning of *T. maritima* within the bacteria, molecular phylogenetic, analysis of a subset of 33 genes for which homologs were found in a set of complete genome sequences (see Comparative Genomics and Phylogenetics) was conducted. While there were some patterns that were consistently observed in the trees of the different genes, we found little support for the rRNA-based positions of *Aquifex* and *Thermotoga* (Table I). In particular, the position of *Thermotoga* within the bacteria varies greatly depending on which gene is used. Because these differences are in regions of high bootstrap support, it suggests that they are due to real differences in the histories of the different genes. Thus, based on single gene analysis, the phylogenetic position of *Aquifex* and *Thermotoga,* and the nature of the deepest branching eubacterial species, should be considered ambiguous.

Evolutionary Analysis of Whole Genome

The focus on single gene analyses, by definition, involves only a small portion of the genome of each species. As an alternative to single gene phylogenetic analysis, it was important to get a picture of the overall similarity of the whole genome to other species. This was done by characterizing the level of similarity of each *T. maritima* ORF to ORFs from other com-

[31] M. Riley, *Microbiol. Rev.* **57**, 862 (1993).

TABLE I
MOLECULAR PHYLOGENETIC ANALYSIS

Protein	Archaea monophyletic	*Thermotoga* with *Aquifex*	*Aquifex* deepest	*Thermotoga* deepest	Both as in 16S rRNA (3/33)
Adenine phosphoribosyl-transferase	−	−	−	−	−
Arginyl-tRNA synthetase	−	+	−	+	−
Enolase	−	+	−	−	−
Cell division protein	+	−	−	−	−
Secreted metalloendopeptidase	+	−	+	−	+
Glutamyl-tRNA synthetase	+	+	−	−	−
GroES protein	+	−	−	−	−
Pseudouridylate synthase I	+	−	−	−	−
Phenylalanyl-tRNA synthetase	−	−	−	−	−
Phosphoglycerate kinase	+	−	−	−	−
Phosphoribosyl pyrop. synth.	+	+	−	−	−
Ribosomal protein L1	+	−	−	−	−
Ribosomal protein L2	+	−	+	−	−
Ribosomal protein L3	+	−	+	−	−
Ribosomal protein L5	−	−	+	−	−
Ribosomal protein L6	+	−	+	−	−
Ribosomal protein L11	+	−	−	−	−
Ribosomal protein L14	+	−	−	−	−
Ribosomal protein L18	+	+	−	−	−
Ribosomal protein L22	+	−	+	−	−
Ribosomal protein S2	+	+	−	−	−
Ribosomal protein S3	+	+	−	−	−
Ribosomal protein S4	+	−	−	−	−
Ribosomal protein S5	+	−	+	−	−
Ribosomal protein S7	+	+	−	−	−
Ribosomal protein S8	+	−	+	−	+
Ribosomal protein S10	+	−	−	−	−
Ribosomal protein S11	+	−	−	−	−
Ribosomal protein S12	−	+	−	−	−
Ribosomal protein S13	−	−	−	+	−
Ribosomal protein S17	−	+	−	−	−
Ribosomal protein S19	−	−	+	−	+
Preprotein translocase SecY	+	−	−	−	−

pleted genomes. The initial focus was on the highest hits of each *T. maritima* gene and, in particular, the species that the highest hit came from. As expected, many of the genes from *T. maritima* had as their highest hit genes from other bacterial species. However, surprisingly, 24% (451 genes) of the genes had as their highest hit a gene from an archaeal species with more

TABLE II
ARCHAEAL GENES IN BACTERIAL GENOMES

Bacterial species	Best hits[a] to archaeal proteins
Thermotoga maritima	451 (24.0%)
Aquifex aeolicus	246 (16.0%)
Bacillus subtilis	286 (6.9%)
Synechocystis sp.	126 (4.0%)
Borrelia burgdorferi	45 (3.6%)
Escherichia coli	99 (2.3%)

[a] 10^{-5} over 60% of the length of the sequence.

than half of these being most similar to genes from the archaeon *Pyrococcus horikoshii*.[19] For convenience, these have been termed archaeal-like genes in the *T. maritima* genome. This percentage of archaeal-like genes is higher than for any other bacterial species [the next highest is the only other sequenced hyperthermophilic bacterium *A. aeolicus*[18] with 16% of its genes (241 total) being archaeal-like (see Table II)]. Thus, by whole genome similarity, *T. maritima* is the most archaeal-like of any of the sequenced Eubacteria.

There are numerous ways in which the observed similarities could have arisen, including convergence of the genomes, loss of genes in mesophiles, amplification of genes shared among thermophiles, and different rates of evolution in thermophiles and mesophiles. While each of these may explain some of the similarity, the most plausible explanation for much of the similarity between the bacterium *T. maritima* and members of the Archaea is extensive lateral gene transfer between these lineages in the past.

The possibility of lateral gene transfer is supported by multiple lines of evidence. The first line of evidence involves nucleotide composition heterogeneity within the genome. Based on the assumption that the DNA composition is relatively uniform throughout the genome, χ^2 analysis indicates that there are at least 51 regions in the chromosome that have a significantly different composition ($P \leq 1.9 \times 10^{-9}$) (Fig. 1, see color insert). Compositional bias as seen in χ^2 distributions have been used previously in support of lateral gene transfer because mutation and selection biases should push most parts of the genome to a uniform composition.[32] Genes that have been transferred from other species will require long periods of time before they "ameliorate" to their new hosts' composition.

A second line of evidence supporting lateral gene transfer is clustering

[32] J. G. Lawrence and H. Ochman, *J. Mol. Evol.* **44**, 383 (1997).

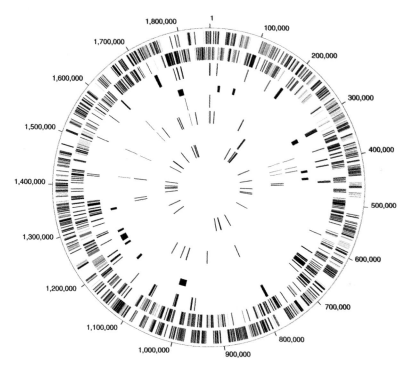

Fig. 1. Circular representation of the *T. maritima* MSB8 genome illustrating predicted-coding regions and other features. Outer circle, predicted protein-coding regions on the plus strand. Second circle, predicted protein-coding regions on the minus strand. Third circle, χ^2 composition in 2000-bp windows (bands correspond to χ^2 values with a probability smaller than 1.9×10^{-9}). Fourth circle, archaeal-like islands on the genome. Fifth circle, small repeats. Sixth circle, large repeats (black), large repeats associated with small repeats (red). Seventh and eighth circles, rRNAs and tRNAs, respectively. Reprinted by permission from *Nature* **399**, 323 (1999) Macmillan Magazines Ltd.

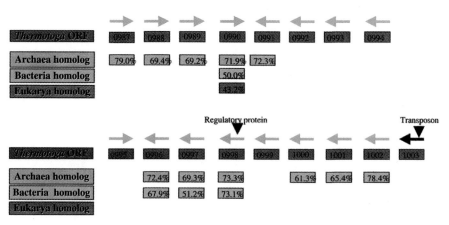

Fig. 2. Section of the *T. maritima* genome showing the levels of similarity to archaeal vs bacterial homologs. Top archaeal hits in this region were to *P. horikoshii* and *A. fulgidus*.

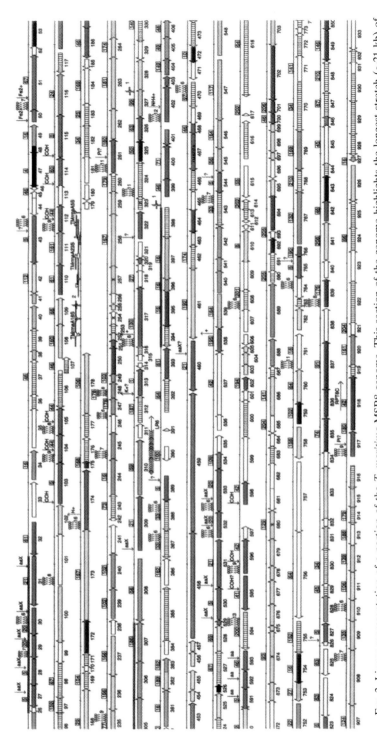

FIG. 3. Linear representation of a portion of the *T. maritima* MSB8 genome. This section of the genome highlights the longest stretch (~21 kb) of archael-like genes in the *T. maritima* genome. This particular region encompasses open reading frames TM0987 through TM1001, which are also presented in Fig. 2. Reprinted by permission from *Nature* **399**, 323 (1999) Macmillan Magazines Ltd.

🏠	Signal peptide	
	LP	Lipoprotein
	Fe+++ →	Transporter
	〰〰〰 9	GES region
	84	Paralogous gene family
	⟱	Authentic Frame Shift

├─┤ 1 kb

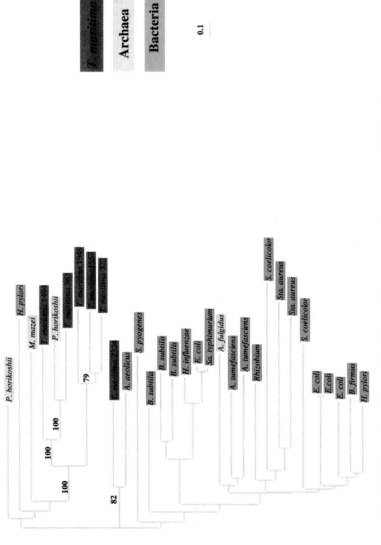

FIG. 5. Phylogenetic positions of six oligopeptide transporters (permease protein) from the genome sequence of *T. maritima. A tumefaciens, Agrobacterium tumefaciens; B. firmus, Bacillus firmus; H. influenzae, Haemophilus influenzae; H. pylori, Helicobacter pylori; M. mazei, Methanosarcina mazei; P. horikoshii, Pyrococcus horikoshii; Sa. typhimurium, Salmonella typhimurium; S. pyogenes, Streptococcus pyogenes; Sta. aureus, Staphylococcus aureus; and S. coelicolor, Streptomyces coelicolor.*

of the archaeal-like genes within the *T. maritima* genome. Eighty-one of the archaeal-like genes in the *T. maritima* genome are clustered in 15 regions of the chromosome that range in size from ∼4 to 21 kb. In all of these regions, there is some conservation of gene order between *T. maritima* and the archaeon with which it shares its top database hit for that region. The longest of these archaeal-like regions is composed of genes that are either hypothetical proteins or proteins that have highest levels of similarity to proteins from the Archaea *P. horikoshii*[19] and *Archaeoglobus fulgidus* (Fig. 2, see color insert) and is presented in one panel of the linear representation of the *T. maritima* genome[33] (Fig. 3, see color insert). This clustering and conservation of gene order is consistent with the occurrence of transfers of large genome pieces. In addition, many of these clusters correspond to regions of biased nucleotide composition identified in the χ^2 analysis. Forty-two of these include genes and repeat structures that have highest levels of similarity to regions on the chromosomes of other thermophiles, including the thermophilic Archaea and *Aquifex* (Fig. 1, see color insert). Interestingly, all of the 30-bp small repeat areas have a χ^2 composition that is substantially different from the rest of the genome.

The level of similarity between many of the *T. maritima* genes and genes from Archaea is striking, even in cases in which the gene is only found in *T. maritima* and in one species of Archaea. Alignments of two such *T. maritima* genes from the longest identified archaeal-like stretch on the genome sequence are presented in Fig. 4. This extensive similarity and the lack of homologs in other species suggest that there was a lateral transfer event in the recent past for these genes.

A final line of support for the occurrence of lateral transfers between Archaea and *T. maritima* comes from phylogenetic analysis. Phylogenetic analyses clearly show that some *T. maritima* and archaeal genes share a common ancestor, to the exclusion of other species, instead of with homologs from bacteria, as would be expected if there were no gene transfers (Fig. 5, see color insert).

Genes Associated with Thermophily

As an initial point of reference for understanding what contributes to the survival of thermophiles in their environment, 108 genes were identified on the *T. maritima* genome, which have orthologs only in the genomes of

[33] K. E. Nelson, R. A. Clayton, S. R. Gill, M. L. Gwinn, R. J. Dodson, D. H. Haft, E. K. Hickey, J. D. Peterson, W. C. Nelson, K. A. Ketchum, L. McDonald, T. R. Utterback, J. A. Malek, K. D. Linher, M. M. Garrett, A. M. Stewart, M. D. Cotton, M. S. Pratt, C. A. Phillips, D. Richardson, J. Heidelberg, G. G. Sutton, R. D. Fleischmann, O. White, S. L. Salzberg, H. O. Smith, J. C. Venter, and C. M. Fraser, *Nature* **399,** 323 (1999).

```
TM0989    MNSAIEYRISLKYVKKSTLPVKRTQKLAEFVPVFIAGKNGKEEQIPGKAE
PYRHO     MG--MKYLIKLKYIKPVQLAVNREQKLAEFYPIFEDGTLG-------KVE

TM0989    ARAKVFLPEFLNFAKKVGAIHDEQISLFQMENDRSRKLVLKVSDDYAYLR
PYRHO     AKARVFLPEYSDFAKGLG---------FEVKK-R-R---MTTENDYAYLR

TM0989    LMIYGVLMSVLKNPVEWGYLEDLVLSMEPLTLRFWGSKIKYTFWKAKNRR
PYRHO     AMIYGVVMGVLRNPGDWERLEERVLEMEPIELRYWASKFRNVYWKYKNMR

TM0989    VLNYLARRILEVERLGKVT
PYRHO     RLMFLARRFMEVEMI----


TM1002    ---------------------------------------------MSK
ARCFU     EKAREVSSAMLDMVFRENNIFYPTLKVLLSEGEWRAIREQEEIIGYYKLK

TM1002    PSSFRERVA----DEIDPVVSPETYEKLPDEIK-RVAGMLTPDTEYSLVR
ARCFU     PAEWKSDAKPVHPYEIDTTITAEQLISLPEEVQNLLKGQLLKPDSYEVVR

TM1002    DNDIKIESGYLSLEELNAIFKTLPFDITFVDKHGRVRFFSGGHRIFHRAP
ARCFU     EDDLRLDEGYLSQREINAIFKALPFDITFVDSNDRVRFFSGSDRIFARTP

TM1002    TVLGRPVQFCHPPRSVHIVNKILKAFKEGRKEPAEFWINMGDRKIHIRYF
ARCFU     SVIGRRVHLCHPPRSVHIVDKILKAFRNGERDFAEFWINMGGRIIHIRYF

TM1002    QVLDKEGNYLGTLEVVQDITRIKELEGEKRLLDWEN
ARCFU     AVRD-EGEYLGTVEVVQDVTEIKRLEGEKRLLGWK-
```

FIG. 4. Alignments of two protein sequences (TM0989 and TM1002) conserved between *T. maritima* and *P. horikoshii* (PYRHO) and *Archaeoglobus fulgidus* (ARCFU), respectively.

other thermophilic Eubacteria (*A. aeolicus*) and Archaea, 71 of which are shared only by *T. maritima* and the thermophilic Archaea. These 108 genes are primarily conserved hypotheticals, but also include *flgA, hndB, rps18,* rubrerythrin, two ABC transporters, glutamate synthase-related protein, putative glycerate kinase, a putative hydrogenase, putative NADH dehydrogenase, putative LPS biosynthesis protein, putative phosphonopyruvate decarboxylase, putative pyruvate formate lyase activating enzyme, alanine acetyltransferase-related protein, and the sensory box protein.

Metabolism

Routine genome analysis has identified pathways involved in the degradation of sugars and plant polysaccharides. Close to 7% of the predicted coding sequences in the *T. maritima* genome are involved in the metabolism of simple and complex sugars, a percentage more than twice that seen in other eubacterial and archaeal species sequenced to date. Several genes encoding proteins involved in the sequential degradation of xylan were identified. These include an endo-1,4-β-xylanase ($xynA/B$), acetylxylan esterase ($axeA$), α-glucuronidase ($aguA$), xylose isomerase ($xylA$), and α-xylosidase ($xloA$). Genes encoding endoglucanases ($celA/B$) and β-glucosidases ($bglB$) that are involved in cellulose degradation were identified, confirming the presence of the cellulolytic systems predicted by biochemical studies of this organism.[34] The *T. maritima* cellulosic system is evidently far less complicated than that described for the thermophilic bacterium *Clostridium thermocellum,* in which cellulose breakdown is dependent on a multienzyme complex composed of between 14 and 26 subunits.[35] The reduction in complexity of the cellulolytic system in *T. maritima* would make it easier for industrial-type applications for the utilization of cellulose as an energy source.

Industrial Applications and Relevance

One of the main interests/aspects for sequencing the *Thermotoga* genome was the potential that this organism has for providing a range of thermostable enzymes able to withstand the rigors of industrial processes. The bacterium also stands to reveal new pathways for the conversion of plant polymers, and the generation of a renewable energy source in the form of hydrogen gas.

Before the project was initiated in January of 1997, close to 143 nucleotide sequences from this organism were publicly available. The *T. maritima* genome has provided the complete 1877 open reading frames in the genome sequence. Many of these open reading frames correspond to hypothetical and conserved hypothetical proteins, which will likely provide additional interesting information on the biology of this organism. The sequences that are conserved across hyperthermophiles will likely give insight into thermophile biology. Already, there are numerous projects underway to characterize the proteins from these gene sequences. The majority of initial studies have shown *T. maritima* gene products to be among the most thermostable, and many can be readily expressed with *Escherichia coli.* Although there are many gene sequences available from thermophilic Archaea, difficulty in understanding the gene expression patterns of many of these species has limited the amount of characterization that has been conducted on proteins from these species. Clearly, *T. maritima,* which does

[34] K. Bronnenmeier, A. Kern, W. Liebl, and W. L. Staudenbauer, *Appl. Environ. Microbiol.* **61,** 1399 (1995).
[35] C. R. Felix and L. G. Ljungdahl, *Annu. Rev. Microbiol.* **47,** 791 (1993).

not demonstrate the same limitations, will contribute a tremendous amount of information to the understanding of thermophile biology. The complete genome sequence has allowed for a comprehensive analysis of the metabolic properties of *T. maritima,* and a description of the phylogenetic position that this species holds, but future analysis will reveal a wealth of novel thermostable enzymes.

Acknowledgment

The *Thermotoga maritima* MSB8 whole genome sequencing project was supported by DOE Grant DE-FC02-95ER61962.

Section II

Saccharolytic Enzymes

[10] Classification of Glycoside Hydrolases and Glycosyltransferases from Hyperthermophiles

By BERNARD HENRISSAT and PEDRO M. COUTINHO

Introduction

Glycoside hydrolases and glycosyltransferases are widespread groups of carbohydrate-active enzymes present in virtually all organisms and are involved, respectively, in the hydrolysis and in the biosynthesis of glycosidic bonds between carbohydrates or between a carbohydrate and a noncarbohydrate moiety. The wide diversity of carbohydrate structures is accompanied by an equal diversity of the enzymes responsible for both their synthesis and selective hydrolysis. Carbohydrate-active enzymes have long been classified according to their substrate specificity, which formed the basis of their Enzyme Commission (EC number) classification by the IUBMB. A more recent classification system was introduced in 1991 for glycoside hydrolases based on primary structure comparisons and the grouping of sequences in families of related sequences.[1] From 300 sequences in 35 families in 1991, the family classification of glycosidases has been updated regularly and has grown to almost 3000 sequences in 77 families in 1999.[2–4] This classification system, which received immediate acceptance by molecular and structural biologists, has now become a standard means for the classification of glycoside hydrolases. A similar classification system has been developed for glycosyltransferases,[5] the enzymes involved in the formation of glycosidic bonds from activated sugar species (nucleotide diphospho sugars, nucleotide monophospho sugars, or sugar 1-phosphates), that presently comprise 36 families. Classifications of these and other classes of carbohydrate-active enzymes,[4] like polysaccharide lyases and carbohydrate esterases, can be used to understand the carbohydrate flux in organisms made possible by the advent of genomics. This universal system can be used for the identification and classification of glycoside hydrolases and glycosyltransferases from hyperthermophiles.

---

[1] B. Henrissat, *Biochem. J.* **280,** 309 (1991).

[2] B. Henrissat and A. Bairoch, *Biochem. J.* **293,** 781 (1993).

[3] B. Henrissat and A. Bairoch, *Biochem. J.* **316,** 695 (1996).

[4] P. M. Coutinho and B. Henrissat, *in* "Recent Advances in Carbohydrate Bioengineering" (H. J. Gilbert, G. Davies, B. Henrissat, and B. Svensson, eds.). The Royal Society of Chemistry, Cambridge, p. 3.

[5] J. A. Campbell, G. J. Davies, V. Bulone, and B. Henrissat, *Biochem. J.* **326,** 929 (1997).

Classification of Carbohydrate-Active Enzymes
  from Hyperthermophiles

Family Definitions and Availability of Classification on the Web

In the sequence classification system, proteins that display amino acid sequence similarities are grouped in families. Consequently, members of different families display no significant sequence similarity. Sequences being grouped in families solely on the basis of sequence relatedness, one frequently finds enzymes of different substrate specificities in the same family.[1,2] In this case the family is called "polyspecific," even if each of the enzymes it contains has a strict specificity; about one-third of the families of glycoside hydrolases are polyspecific. When a sequence displays no similarity with any other sequence, it is left unclassified until a related sequence is found and a new family is defined. With the accumulation of sequence data, the number of glycoside hydrolase families has grown from 35 in 1991 to 77 in 1999 and that of glycosyltransferases from 27 in 1997 to 36 in 1999. All the families of carbohydrate-active enzymes can be accessed on the web through the carbohydrate-active enzymes server (CAZy: http://afmb.cnrs-mrs.fr/~pedro/CAZY/db.html).[4] In this server, a regularly updated listing of the members in each family is provided with direct links to the SwissProt and EMBL/GenBank data banks. Other links are provided such as the Taxonomy Browser of the National Center of Biological Information, PROSITE, ENZYME, and the Protein Data Bank. A document giving the classification of glycoside hydrolase families and the index of glycoside hydrolase entries in SwissProt is available on the ExPASy server at URL: http://www.expasy.ch/cgi-bin/lists?glycosid.txt.[3]

Assigning a New Sequence to a Family

Whether a sequence is from a hyperthermophile, a thermophile, a mesophile, or a psychrophile, the procedure is identical. The new sequence should be subjected to a similarity search using BLAST[6] or FASTA[7] on one of the many servers dedicated to similarity searches. Most SwissProt entries for glycoside hydrolases and polysaccharide lyases are annotated with the family number in the comment lines. If the search results contain SwissProt entries from these families, a simple examination of the comment lines for the entry will indicate the family to which it belongs. However, the best hits with the sequence under study may be found with entries in

[6] S. F. Altschul, T. L. Madden, A. A. Schäffer, J. Zhang, Z. Zhang, W. Miller, and D. J. Lipman, *Nucleic Acids Res.* **25,** 3389 (1997).
[7] W. R. Pearson, *Methods Enzymol.* **183,** 63 (1990).

nucleotide databanks (EMBL/GenBank/DDBJ) and not in SwissProt. In this case, to browsing through the CAZy server (URL: http://afmb.cnrs-mrs.fr/~pedro/CAZY/db.html) is recommended to find the family which contains the best hit sequences. Tools for a fast automatic family assignment are being prepared and will soon become available in CAZy.

Glycoside Hydrolases from Hyperthermophiles

The known glycoside hydrolases from hyperthermophilic organisms are classified according to families and are given in Table I. As of June 1999, entries from hyperthermophiles are found in 33 of the 77 known families of glycoside hydrolases (namely families 1–5, 8–13, 15, 16, 26, 28, 29, 31, 32, 35, 36, 38, 39, 42–44, 48, 51, 53, 57, 65, 67, 74, and 77). Hyperthermophilic glycosidases from other families are likely to be found; for example, two chitinases from family 18 have been reported in the genome of *Pyrococcus furiosus*.[8]

Glycosyltransferases from Hyperthermophiles

In hyperthermophiles, 9 out of 36 glycosyltransferases families are represented (families 2, 4, 5, 19, 26, 28, 30, 35, and 36) as of June 1999. Most have been identified by homology and arise from genomic analysis. Known glycosyltransferases from hyperthermophilic organisms classified in families are given in Table II.

Advantages of Family Classifications

The classification system developed for carbohydrate-active enzymes has several advantages in understanding structure–function relationships. These advantages are reflected at several different levels.

1. Structure: Unlike the traditional EC classification, the sequence-based classification reflects the structural features of the enzymes, with the members of a family sharing a similar three-dimensional (3-D) structure.[9,10] Once established for one member, the 3-D structure may prove useful for solving the structure of other members of the family by molecular replacement or for homology modeling.

2. Mechanisms: As the molecular mechanism is firmly dictated by the active site geometry, the sequence-based family classification has been

---

[8] L. E. Driskill, K. Kusy, M. W. Bauer, and R. M. Kelly, *Appl. Environ. Microbiol.* **65**, 893 (1999).

[9] G. Davies and B. Henrissat, *Structure* **3**, 853 (1995).

[10] B. Henrissat and G. Davies, *Curr. Opin. Struct. Biol.* **7**, 637 (1997).

TABLE I
GLYCOSIDE HYDROLASES, TRANSGLYCOSIDASES, AND RELATED PROTEINS OF HYPERTHERMOPHILES[a]

Organism	Protein or ORF name	Family	EC number	SwissProt	GenBank	Mechanism
Archae[a]						
Methanococcus jannaschii	ORF MJ1610	15	n.d.	Q59005	U67601	inverting
	ORF MJ1611	57	n.d.	Q59006	U67601	retaining
Pyrococcus						
P. furiosus	β-glucosidase CelB	1	3.2.1.21	Q51723	U37557	retaining
	β-mannosidase BmnA	1	3.2.1.25	Q51733	U60214	retaining
	endo-1,4-glucanase A	12	3.2.1.4	Q9V2T0	AF181032	retaining
	α-amylase	13	3.2.1.1	O08452	AF001268	retaining
	β-1,3-glucanase	16	3.2.1.39	O73951	AF013169	retaining
	α-amylase	57	3.2.1.1	P49067	L22346	retaining
	amylopullulanase	57	3.2.1.41	O30772	AF016588	retaining
P. horikoshii	ORF PH0366	1	n.d.	O58104	AP000002	retaining
	ORF PH0501	1	n.d.	O58237	AP000002	retaining
	ORF PH1171	5	n.d.	O58925	AP000005	retaining
	ORF PH0511	35	n.d.	O58247	AP000002	retaining
	ORF PH0835	38	n.d.	O58565	AP000003	retaining
	ORF PH0193	57	n.d.	O57932	AP000001	retaining
	ORF PH0746	65	n.d.	O58512	AP000003	retaining
P. kodakaraensis	β-glycosidase	1	3.2.1.-	Q9YGB8	AB028601	retaining
P. sp.	α-amylase	57	3.2.1.1.	O32450	D87907	retaining
P. sp. KOD1	α-amylase	13	3.2.1.1	O33476	D83793	retaining
P. woesei	β-galactosidase	1	3.2.1.23	O52629	AF043283	retaining
Sulfolobus						
S. acidocaldarius	α-amylase	13	3.2.1.1	Q53641	D83245	retaining
	maltooligosyltrehalose synthase	13	2.4.1.-	Q53688	D83245	retaining
	isoamylase	13	n.d.	O05152	D83245	retaining
S. shibatae	β-galactosidase	1	3.2.1.23	P50388	L47841	retaining

S. solfataricus	β-galactosidase	1	3.2.1.23	P14288	X15950	retaining
	β-galactosidase S (LacS)	1	3.2.1.23	P22498	AF133096	retaining
	α-amylase	13	3.2.1.1	P95869	Y08256	retaining
	α-amylase	13	3.2.1.1	Q55088	D64130	retaining
	α-amylase c06020	13	3.2.1.1	P95867	Y08256	retaining
	GlgX c06021	13	n.d.	P95868	Y08256	retaining
	glycosyltrehalose-producing enzyme	13	2.4.1.-	Q55262	D64128	retaining
	α-glucosidase	31	3.2.1.20	Q59645	AF042494	retaining
Thermococcus						
T. hydrothermalis	α-amylase	13	3.2.1.1	Q93647	AF068255	retaining
	amylopullulanase	57	3.2.1.41	Q9Y818	AF113969	retaining
T. litoralis	4-α-glucanotransferase	57	2.4.1.-	O32462	D88253	retaining
T. sp. 9ON-7	β-glucosidase	1	3.2.1.21	O08324	Z70242	retaining
T. sp. Rt3	α-amylase	13	3.2.1.1	O50200	AF017454	retaining
Bacteria						
Anaerocellum thermophilum	endo-1,4-glucanase CelB	5	3.2.1.4	P96310	Z86104	retaining
	endo-1,4-glucanase CelD	5	3.2.1.4	Q59154	Z77855	retaining
	CelA (N-terminal domain)	9	3.2.1.4	P96311	Z86105	inverting
	xylanase A	10	3.2.1.8	Q59150	Z69782	retaining
	CelA (C-terminal domain)	48	3.2.1.91	P96311	Z86105	inverting
Aquifex aeolicus	ORF CelY	8	n.d.	O67401	AE000738	inverting
	branching enzyme	13	2.4.1.18	O66936	AE000704	retaining
	ORF aq_720	57	n.d.	O66934	AE000704	retaining
	ORF MalM	77	n.d.	O66937	AE000704	retaining

(continued)

TABLE I (*continued*)

Organism	Protein or ORF name	Family	EC number	SwissProt	GenBank	Mechanism
Caldicellulosiruptor						
C. saccharolyticus	BglA	1	3.2.1.21	P10482	X12575	retaining
	ManA (N-terminal domain)	5	3.2.1.78	P22533	L01257	retaining
	CelB (C-terminal domain)	5	3.2.1.4	P10474	X13602	retaining
	CelA (N-terminal domain)	9	3.2.1.4	P22534	L32742	inverting
	CelB (N-terminal domain)	10	3.2.1.91	P10474	X13602	retaining
	ORF4	10	n.d.	P23557	M34459	retaining
	XynA	10	3.2.1.8	P23556	M34459	retaining
	XynE	10	3.2.1.8	O30427	AF005383	retaining
	XynI	10	3.2.1.8	O30421	AF005382	retaining
	pullulanase	13	3.2.1.41	Q59319	L39876	retaining
	β-mannanase	26	3.2.1.78	P77847	U39812	retaining
	XynB	39	3.2.1.37	P23552	M34459	retaining
	XynD	39	3.2.1.37	O30428	AF005383	retaining
	XynF (N-terminal domain)	43	3.2.1.37; 3.2.1.55	O30426	AF005383	inverting
	XynF (C-terminal domain)	43	3.2.1.37; 3.2.1.55	O30426	AF005383	inverting
	ManA (C-terminal domain)	44		P22533	L01257	inverting
	CelA (C-terminal domain)	48		P22534	L32742	inverting
C. sp. 14B	β-glucosidase BglS	1	3.2.1.21	Q9ZEN0	AJ131346	retaining
C. sp. Rt69B.1	XynB	10	3.2.1.8	O52373	AF036923	retaining
	XynC (N-terminal domain)	10	3.2.1.8	O52374	AF036924	retaining
	XynD	11	3.2.1.8	O52375	AF036925	retaining
	XynC (C-terminal domain)	43	3.2.1.8	O52374	AF036924	inverting

Organism / enzyme	Family	EC number	SwissProt	GenBank	Mechanism
C. sp. Tok7B.1					
CelB (multidomain protein)	5	n.d.	Q9X3P6	AF07837	retaining
CelA (N-terminal domain)	10	3.2.1.8; 3.2.1.55	Q9X3P5	AF078737	retaining
CelB (multidomain protein)	10	n.d.	Q9X3P6	AF078737	retaining
CelA (C-terminal domain)	43	3.2.1.8; 3.2.1.55	Q9X3P5	AF078737	inverting
Dictyoglomus thermophilum					
xylanase	10	3.2.1.8	Q12603	L39866	retaining
xylanase	11	3.2.1.8	P77853	U76545	retaining
α-amylase 2	13	3.2.1.1	P14898	X13199	retaining
α-amylase 3	13	3.2.1.1	P14899	X15948	retaining
b-mannanase	26	3.2.1.78	O30654	AF013989	retaining
α-amylase 1	57	3.2.1.1	P09961	X07896	retaining
Fervidobacterium pennivorans					
pullulanase	13	3.2.1.41	Q9XDB5	AF096862	retaining
Rhodothermus marinus					
xylanase	10	3.2.1.8	P96988	Y11564	retaining
cellulase A	12	3.2.1.4	O33897	U72637	retaining
laminarinase	16	3.2.1.39	O52754	AF047003	retaining
lichenase	16	3.2.1.73	P45798	U04836	retaining
β-mannanase	26	3.2.1.78	P49425	X90947	retaining
Thermoanaerobacter					
T. brockii					
β-glucosidase CglT	1	3.2.1.21	Q60026	Z56279	retaining
β-glucosidase/xylosidase XglS	3	3.2.1.21; 3.2.1.37	P96090	Z56279	retaining
T. ethanolicus					
β-galactosidase	2	3.2.1.23	P77989	Y08557	retaining
amylase-pullulanase	13	3.2.1.1; 3.2.1.41	P16950	M28471	retaining
amylase-pullulanase	13	3.2.1.1; 3.2.1.41	P38939	M97665	retaining
T. polysaccharolyticum					
cyclodextrinase	13	3.2.1.54	P29964	M88602	retaining
α-galactosidase	36	3.2.1.22	P77988	Y08557	retaining
β-mannanase A	5	3.2.1.78; 3.2.1.4	Q9ZA17	U82255	retaining

(continued)

TABLE I (*continued*)

Organism	Protein or ORF name	Family	EC number	SwissProt	GenBank	Mechanism
T. saccharolyticum	xylanase A	10	3.2.1.8	P36917	M97882	retaining
	amylase-pullulanase	13	3.2.1.1; 3.2.1.41	P36905	L07762	retaining
T. sp. ATCC53627	β-xylosidase XynB	39	3.2.1.37	P36906	M97883	retaining
	cyclodextrin glucanotransferase	13	2.4.1.19		Z35484	retaining
T. sp. JW/SL-YS 485	xylanase	10	3.2.1.8	Q60043	U27183	retaining
	β-xylosidase	39	3.2.1.37	Q30360	AF001926	retaining
T. thermosaccharolyticum	glucoamylase	15	3.2.1.3	O85672	AF071548	inverting
T. thermosulfurogenes	β-galactosidase	2	3.2.1.23	P26257	M57579	retaining
	xylanase A	10	3.2.1.8	Q60046	U50952	retaining
	amylase-pullulanase	13	3.2.1.1; 3.2.1.41	P38536	M57692	retaining
	cyclodextrin glucanotransferase	13	2.4.1.19	P26827	X54654	retaining
	polygalacturonase PglA	28	3.2.1.15	Q60045	U50951	inverting
Thermotoga						
T. maritima	β-glucosidase	1	3.2.1.21	Q08638	X74163	retaining
	β-galactosidase Z (TM1193)	2	3.2.1.23	Q56307	U08186	retaining
	ORF TM1624	2	n.d.	Q9X1V9	AE001806	retaining
	ORF TM1062	2	n.d.	Q9X0F2	AE001766	retaining
	ORF TM0809	3	n.d.	Q9WZR6	AE001748	retaining
	ORF TM0076	3	n.d.	Q9WXT1	AE001694	retaining
	ORF TM0025	3	n.d.	Q9WXN2	AE001690	retaining
	α-glucosidase (TM1834)	4	3.2.1.20	O33830	AJ001089	
	ORF TM1281	4	n.d.	Q9X108	AE001783	
	ORF TM1068	4	n.d.	Q9S5X4	AE001767	
	ORF TM0752	4	n.d.	Q9WZL1	AE001745	
	ORF TM0434	4	n.d.	Q9WYR5	AE001722	
	ORF TM1751	5	n.d.	Q9X273	AE001813	retaining
	ORF TM1752	5	n.d.	Q9X274	AE001813	retaining
	ORF TM1227	5	n.d.	Q9X0V4	AE001779	retaining

xylanase A (TM0061)	10	3.2.1.8	Q60037	Z46264	retaining
ORF TM0070	10	n.d.	Q9WXS5	AE001693	retaining
endo-1,4-glucanase A (TM1524)	12	3.2.1.4	Q60032	Z69341	retaining
endo-1,4,-glucanase B (TM1525)	12	3.2.1.4	Q60033	Z69341	retaining
4-α-glucanotransferase (TM0767)	13	2.4.1.25	Q33838	AJ001090	retaining
4-α-glucanotransferase (TM0364)	13	2.4.1.25	Q60035	Z50813	retaining
α-amylase (TM1840)	13	3.2.1.1	P96107	Y11359	retaining
pullulanase (TM1845)	13	3.2.1.41	O33840	AJ001087	retaining
ORF TM1835	13	n.d.	Q9X2F4	AE001820	retaining
ORF TM1650	13	n.d.	Q9X1Y3	AE001807	retaining
ORF TM0024	16	n.d.	Q9WXN1	AE001690	retaining
ORF TM0437	28	n.d.	Q9WYR8	AE001722	retaining
ORF TM0306	29	n.d.	Q9WYE2	AE001712	retaining
ORF TM0308	31	n.d.	Q9WYE4	AE001712	inverting
invertase (TM1414)	32	3.2.1.26	O33833	AJ001073	retaining
α-galactosidase (TM1192)	36	3.2.1.22	O33835	AJ001072	retaining
ORF TM1231	38	n.d.	Q9X0V8	AE001779	retaining
ORF TM1851	38	n.d.	Q9X2G6	AE001822	retaining
β-galactosidase (TM1195)	42	3.2.1.23	Q56306	U08186	retaining
ORF TM0310	42	n.d.	Q9WYE6	AE001713	retaining
ORF TM0281	51	n.d.	Q9WYB7	AE001710	retaining
ORF TM1201	53	n.d.	Q9X0S8	AE001777	retaining
α-glucuronidase	67	3.2.1.139	P96105	Y09510	retaining
ORF TM0305	74	n.d.	Q9WYE1	AE001712	retaining
T. neapolitana β-glucosidase	1	3.2.1.21	O33843	Z97212	retaining
β-galactosidase	2	3.2.1.23	O85250	AF055482	retaining
β-glucosidase	3	3.2.1.21	Q60038	Z77856	retaining
β-xylosidase	3	3.2.1.37	Q56322	U58632	retaining
α-glucosidase	4	3.2.1.20	O86960	AJ009832	retaining
xylanase A	10	3.2.1.8	Q60042	Z46945	retaining

(continued)

TABLE I (*continued*)

Organism	Protein or ORF name	Family	EC number	SwissProt	GenBank	Mechanism
	xylanase B	10	3.2.1.8	Q60041	Z49961	retaining
	endo-1,4-glucanase	12	3.2.1.4	O08428	U93354	retaining
	endo-1,4-glucanase B	12	3.2.1.4	P96492	U93354	retaining
	4-α-glucanotransferase	13	2.4.1.25	O86956	AJ009831	retaining
	cyclomaltodextrin glucanotransferase	13	3.2.1.54	O86959	AJ009832	retaining
	laminarinase A	16	3.2.1.39	Q60039	Z47974	retaining
	β-mannanase	26	3.2.1.78	O33841	U58632	retaining
	α-galactosidase	36	3.2.1.22	Q9R7H1	AF011400	retaining
	β-galactosidase bgalA	42	3.2.1.23	O85248	AF055482	retaining
T. sp. strain FjSS3-B.1	xylanase A	10	3.2.1.8	Q60044	U33060	retaining
	xylanase B	10	3.2.1.8	Q9R6T4	AF126689	retaining
	xylanase C	10	3.2.1.8	Q9WWJ9	AF126690	retaining
Thermus						
T. aquaticus	trehalose synthase	13	5.4.99.16	O06458	D86216	retaining
	amylomaltase	77	2.4.1.-	O87172 Q9X6C7	AB016244	retaining
T. brockianus ITI360	α-galactosidase	36	3.2.1.22		AF135398	retaining
T. sp. A4	β-galactosidase	42	3.2.1.23	O69315	D85027	retaining
T. sp. IM6501	maltogenic α-amylase	13	3.2.1.133	O69007	AF060204	retaining
	pullulanase	13	3.2.1.41	O69008	AF060205	retaining
T. sp. T2	α-galactosidase	36	3.2.1.22	Q9WXC1	AB018548	retaining
	β-galactosidase	42	3.2.1.23	O54315	Z93773	retaining
T. thermophilus	β-glycosidase	1	3.2.1.21	Q9X9D4 Q9RA61 Q9X6D2	Y16753	retaining
T. thermophilus TH125	β-glycosidase bglT	1	3.2.1.-		AF135400	retaining
	α-galactosidase	36	3.2.1.22		AF135400	

shown to correlate with the molecular mechanism of glycosidases.[11] Glycoside hydrolases operate via two general mechanisms leading to overall retention or inversion of the anomeric configuration.[12–14] In glycosyltransferases, catalysis also occurs via two mechanisms leading to either retention or inversion of the stereochemistry at the anomeric carbon.[13,15] Once established for one member of a family, the molecular mechanism can be safely extended to all members of the family. Similarly, once identified (in position and function) for one member of a family, the catalytic residues can be inferred easily for all members of the family.

3. Multiple enzyme classes: Although glycoside hydrolases and transglycosidases have similar 3-D structures and mechanisms, they are assigned to totally different groups in the EC classification, with the former being found under hydrolases (EC 3.2.1.x) and the later classified as glycosyltransferases (EC 2.4.1.x). The classification in families based on amino acid sequence similarities resolves this contradiction as several families contain the two types of enzymes: examples include glycoside hydrolase families 13, 16, 17, 32, 33, 57, 66, and 68.

4. Evolutionary relationships: Examination of the families of glycoside hydrolases (and transglycosidases) and of glycosyltransferases shows divergent evolution (different substrate specificities resulting in a polyspecific family; tuning of the reaction to favor hydrolysis or transglycosylation, favoring one substrate over a closely related one, etc.) as well as convergent evolution (identical substrate and reaction specificity in unrelated families).

5. Data mining: Because the new classification is based on amino acid sequence similarities, it is useful for mining genomes for carbohydrate-active enzymes, as well as for predicting the function of putative coding regions in genomes. For instance, one can safely predict if an open reading frame (ORF) encodes a carbohydrate-active enzyme and, if already known for a member of the family, the molecular mechanism and the 3-D fold. However, the precise substrate specificity can be harder to predict as a number of families are polyspecific. In a similar fashion, the classification can also assist enzyme discovery by designing primers corresponding to the conserved region(s) of a family.[16]

---

[11] J. Gebler, N. R. Gilkes, M. Claeyssens, D. B. Wilson, P. Béguin, W. W. Wakarchuk, D. G. Kilburn, R. C. Miller, Jr., R. A. Warren, and S. G. Withers, *J. Biol. Chem.* **267,** 12559 (1992).

[12] D. E. Koshland, *Biol. Rev. Cambridge Phil. Soc.* **28,** 416 (1953).

[13] M. L. Sinnott, *Chem. Rev.* **90,** 1171 (1990).

[14] J. D. McCarter and S. G. Withers, *Curr. Opin. Struct. Biol.* **4,** 885 (1994).

[15] I. Saxena, R. M. Brown, Jr., M. Fèvre, R. A. Geremia, and B. Henrissat, *J. Bacteriol.* **177,** 1419 (1995).

[16] H. Dalboge and L. Lange, *Trends Biotechnol.* **16,** 265 (1998).

TABLE II
GLYCOSYLTRANSFERASES AND RELATED PROTEINS OF HYPERTHERMOPHILES[a]

Organism	Protein or ORF name	Family	EC number	SwissProt	GenBank	Mechanism
Archae[a]						
Aeropyrum pernix	ORF APE1192	2	n.d.	Q9YCR9	AP000061	inverting
	ORF APE1184	2	n.d.	Q9YCS7	AP000061	inverting
	ORF APE0426	2	n.d.	Q9YF12	AP000059	inverting
	ORF APE2066	4	n.d.	Q9YA73	AP000063	retaining
	ORF APE1191	4	n.d.	Q9YCS0	AP000061	retaining
Archaeoglobus fulgidus	ORF AF0039	2	n.d.	O30196	AE001104	inverting
	ORF AF0321	2	n.d.	O29924	AE001082	inverting
	ORF AF0322	2	n.d.	O29923	AE001082	inverting
	ORF AF0328	2	n.d.	O29919	AE001082	inverting
	ORF AF0387	2	n.d.	O29860	AE001078	inverting
	ORF AF0581	2	n.d.	O29674	AE001064	inverting
	ORF AF0620	2	n.d.	O29635	AE001062	inverting
	ORF AF2115	2	n.d.	O28165	AE000958	inverting
	ORF AF0038	4	n.d.	O30197	AE001104	retaining
	ORF AF0043	4	n.d.	O30192	AE001104	retaining
	ORF AF0045 (N-terminal domain)	4	n.d.	O30191	AE001103	retaining
	ORF AF0045 (C-terminal domain)	4	n.d.	O30191	AE001103	retaining
	ORF AF0327	4	n.d.	O29920	AE001082	retaining
	ORF AF0617	4	n.d.	O29638	AE001062	retaining
	ORF AF1728	4	n.d.	O28546	AE000983	retaining

Organism	ORF					
Methanococcus jannaschii	ORF MJ0544	2	n.d.	Q57964	U67504	inverting
	ORF MJ1057	2	n.d.	Q58457	U67549	inverting
	ORF MJ1059	4	n.d.	Q58459	U67549	retaining
	ORF MJ1069	4	n.d.	Q58469	U67549	retaining
	ORF MJ1178	4	n.d.	Q58577	U67559	retaining
	ORF MJ1607	4	n.d.	Q59002	U67601	retaining
	ORF MJ1606	5	n.d.	Q59001	U67600	retaining
	ORF MJ1062	28	n.d.	Q58462	U67549	inverting
	ORF MJ1255	28	n.d.	Q58652	U67566	inverting
	ORF MJ1631	35	n.d.	Q59025	U67603	retaining
Pyrococcus horikoshii	ORF PH0051	2	n.d.	O57812	AP000001	inverting
	ORF PH0396	2	n.d.	O58133	AP000002	inverting
	ORF PH0401	2	n.d.	O58138	AP000002	inverting
	ORF PH0424	2	n.d.	O58161	AP000002	inverting
	ORF PH0430	2	n.d.	O58167	AP000002	inverting
	ORF PH0433	2	n.d.	O58170	AP000002	inverting
	ORF PH0456	2	n.d.	O58219	AP000002	inverting
	ORF PH1585	2	n.d.	O59263	AP000006	inverting
	ORF PH1596	2	n.d.	O59252	AP000006	inverting
	ORF PH1879	2	n.d.	O59540	AP000007	inverting
	ORF PH1844	4	n.d.	O59512	AP000007	retaining
	ORF PH0434	4	n.d.	O58171	AP000002	retaining
	ORF PH0069	5	n.d.	O57794	AP000001	retaining
	ORF PH1512	35	n.d.	O59181	AP000006	retaining
Thermococcus litoralis	maltodextrin phosphorylase (MalP)	35	2.4.1.1	Q9YGA7	AF115479	retaining

(continued)

TABLE II (*continued*)

Organism	Protein or ORF name	Family	EC number	SwissProt	GenBank	Mechanism
Bacteria						
Aquifex aeolicus	ORF Alg (Aq_1684)	2	n.d.	O67594	AE000751	inverting
	ORF BcsA (Aq_1407)		n.d.	O67406	AE000738	inverting
	ORF Dmt (Aq_1899)	2	n.d.	O67737	AE000762	inverting
	ORF LgtF (Aq_1742)	2	n.d.	O67628	AE000754	inverting
	ORF aq_1080	4	n.d.	O67173	AE000722	retaining
	ORF Cap (Aq_1641)	4	n.d.	O67559	AE000749	retaining
	ORF aq_572	4	n.d.	O66840	AE000696	retaining
	ORF MtfA (Aq_1641)	4	n.d.	O67183	AE000723	retaining
	ORF MtfB (Aq_515)	4	n.d.	O66801	AE000693	retaining
	ORF MtfC (Aq_516)	4	n.d.	O66802	AE000693	retaining
	ORF RfaG (Aq_2115)	4	n.d.	O67880	AE000773	retaining
	ORF GlgA (Aq_721)	5	n.d.	O66935	AE000704	retaining
	ORF aq_362	9	n.d.	O66687	AE000686	inverting
	ORF RfaCa (Aq_1543)	9	n.d.	O67498	AE000745	inverting
	ORF RfaC2 (Aq_145)	9	n.d.	O66538	AE000675	inverting
	ORF LpxB (Aq_1427)	19	n.d.	O67420	AE000740	inverting
	ORF MurG (Aq_1177)	28	n.d.	O67238	AE000727	inverting
	ORF kdtA (Aq_326)	30	n.d.	O66663	AE000684	inverting
	ORF GlgP (Aq_717)	35	n.d.	O66932	AE000704	retaining

Thermotoga maritima					
ORF TM0756	2	n.d.	Q9WZL5	AE001745	inverting
ORF TM1229	2	n.d.	Q9X0V6	AE001779	inverting
ORF TM0757	2	n.d.	Q9WZL6	AE001745	inverting
ORF TM0619	4	n.d.	Q9WZ88	AE001736	retaining
ORF TM0622	4	n.d.	Q9WZ90	AE001736	retaining
ORF TM0624	4	n.d.	Q9WZ92	AE001737	retaining
ORF TM0744	4	n.d.	Q9WZK3	AE001744	retaining
ORF TM0756	4	n.d.	Q9WZL5	AE001745	retaining
ORF TM1405	4	n.d.	Q9X1C5	AE001793	retaining
ORF TM0760	4	n.d.	Q9WZL9	AE001746	retaining
ORF TM0392	4	n.d.	Q9WYM3	AE001719	retaining
ORF TM0631	4	n.d.	Q9WZ99	AE001737	retaining
ORF TM0895	5	n.d.	Q9WZZ7	AE001754	retaining
ORF TM0818	26	n.d.	Q9WZS5	AE001749	inverting
ORF TM0232	28	n.d.	Q9WY74	AE001707	retaining
α-glucan phosphorylase (AgpA)	35	n.d.	O33831	AJ001088	inverting
ORF TM1848	36	n.d.	Q9X2G3	AE001822	inverting
Thermotoga neapolitana					
cellobiose-phosphorylase	36	2.4.1.20	O87964	Z99777	inverting

a Compiled June 1999.

6. Metabolic mapping: Finally, this system allows an immediate perception of the repertoire of the different classes of carbohydrate-active enzymes produced by any organism, and hyperthermophiles in particular.[8]

Classification and Genomics in Hyperthermophiles

With the advent of genomics, the exhaustive classification of glycoside hydrolases and glycosyltransferases into families of structurally related enzymes provides the necessary framework to understand glycosidic bond degradation and biosynthesis in different organisms. The study of the role of glycosidases in these organisms is, however, more advanced, as interest in thermostable glycosidases for the industrial degradation of saccharides has fueled research since the mid 1980s.[17] Many questions are now arising from the genomic analysis of these organisms, and we will try to illustrate the impact of carbohydrate-active enzyme family analysis.

Glycosidases and Transglycosidases

As mentioned earlier, glycosidases can be either inverting or retaining the anomeric configuration. Retaining glycosidases have a clear preference in hyperthermophiles (see Table I): in archae, all but one putative family 15 glycosidase from *Methanococcus jannaschii* are retaining while only 10% are inverting in bacteria. This bias toward retaining mechanism at high temperatures could arise from difficulties in controlling the inverting mechanism with the formation of undesired condensation products. Furthermore, high temperatures induce the nonenzymatic degradation of saccharides by the Maillard reaction for peptide-containing media and caramelization.[8] To minimize the effect of these side reactions, hyperthermophilic (and thermophilic) organisms may have developed strategies to reduce the exposure of the more labile substrate-reducing function in solution. These strategies could involve (1) extracellular degradation of saccharides to produce di- and oligosaccharides that are specifically transported into the cell were intracellular degradation will follow, rather than full extracellular degradation and transport of monosaccharides[8]; (2) transglycosylation reactions to yield cyclic compounds with family 13 and 77 transglycosidases as to prevent substrate degradation in the media; and (3) production of the nonreducing α,α-trehalose from maltooligosaccharides, as in *Sulfolobus,* by the combined use of maltosyl trehalose synthase and a maltooligosyl trehalose trehalohydrolase.[18] All these strategies preserve the integrity of the substrate.

[17] M. W. Bauer, S. B. Halio, and R. M. Kelly, *Adv. Prot. Chem.* **48,** 271 (1996).
[18] K. Maruta, H. Mitsuzumi, T. Nakada, M. Kubota, H. Chaen, S. Fukuda, T. Sugimoto, and M. Kurimoto, *Biochim. Biophys. Acta* **1291,** 177 (1996).

The most common glycosidases in hyperthermophilic archaea belong to families 1, 13, and 57 (see Table I), an indication of the use of both β-linked (family 1) and α-linked (families 13 and 57) substrates by most. The repertoire of glycoside hydrolases of hyperthermophilic archaea, however, is usually smaller than that of hyperthermophilic bacteria (e.g., xylanases have been found only in bacteria). This takes the most dramatic proportions in *Aeropyrum pernix* and *Archaeoglobus fulgidus,* two archaeons whose genomes have been completely sequenced[19,20] and appear totally devoid of known glycosidases. This observation is not restricted to hyperthermophiles, as the completely sequenced genome of *Methanobacterium thermoautotrophicum,*[21] a thermophilic, but not hyperthermophilic, archaeon also completely lacks glycoside hydrolases. Some sulfate-reducing and methanogenic archaea are known to grow in the absence of carbohydrates,[17] but many archaea are carbohydrate-utilizing.[22] In *Methanococcus,* the two putative glycosidases from families 15 and 57, together with a putative phosphorylase from the glycosyltransferase family 35, found in the *M. jannaschii* genome,[23] are likely to participate intracellularly to the glycogen metabo-

[19] Y. Kawarabayasi, Y. Hino, H. Horikawa, S. Yamazaki, Y. Haikawa, K. Jin-no, M. Takahashi, M. Sekine, S. Baba, A. Ankai, H. Kosugi, A. Hosoyama, S. Fukui, Y. Nagai, K. Nishijima, H. Nakazawa, M. Takamiya, S. Masuda, T. Funahashi, T. Tanaka, Y. Kudoh, J. Yamazaki, N. Kushida, A. Oguchi, K. Aoki, K. Kubota, Y. Nakamura, N. Nomura, Y. Sako, and H. Kikuchi, *DNA Res.* **6,** 83 (1999).

[20] H. P. Klenk, R. A. Clayton, J. F. Tomb, O. White, K. E. Nelson, K. A. Ketchum, R. J. Dodson, M. Gwinn, E. K. Hickey, J. D. Peterson, D. L. Richardson, A. R. Kerlavage, D. E. Graham, N. C. Kyrpides, R. D. Fleischmann, J. Quackenbush, N. H. Lee, G. G. Sutton, S. Gill, E. F. Kirkness, B. A. Dougherty, K. McKenney, M. D. Adams, B. Loftus, S. Peterson, C. I. Reich, L. K. McNeil, J. H. Badger, A. Glodek, L. Zhou, R. Overbeek, J. D. Gocayne, J. F. Weidman, L. McDonald, T. Utterback, M. D. Cotton, T. Spriggs, P. Artiach, B. P. Kaine, S. M. Sykes, P. W. Sadow, K. P. D'Andrea, C. Bowman, C. Fujii, S. A. Garland, T. M. Mason, G. J. Olsen, C. M. Fraser, H. O. Smith, C. R. Woese, and J. C. Venter, *Nature* **390,** 364 (1997).

[21] D. R. Smith, L. A. Doucette-Stamm, C. Deloughery, H. Lee, J. Dubois, T. Aldredge, R. Bashirzadeh, D. Blakely, R. Cook, K. Gilbert, D. Harrison, L. Hoang, P. Keagle, W. Lumm, B. Pothier, D. Qiu, R. Spadafora, R. Vicaire, Y. Wang, J. Wierzbowski, R. Gibson, N. Jiwani, A. Caruso, D. Bush, H. Safer, D. Patwell, S. Prabhakar, S. McDougall, G. Shimer, A. Goyal, S. Pietrovski, G. M. Church, C. J. Daniels, J.-i. Mao, P. Rice, J. Nolling, and J. N. Reeve, *J. Bacteriol.* **179,** 7135 (1997).

[22] S. W. M. Kengen, A. J. M. Stams, and W. M. de Vos, *FEMS Microbiol. Rev.* **18,** 119 (1996).

[23] C. J. Bult, O. White, G. J. Olsen, L. Zhou, R. D. Fleischmann, G. G. Sutton, J. A. Blake, L. M. FitzGerald, R. A. Clayton, J. D. Gocayne, A. R. Kerlavage, B. A. Dougherty, J. F. Tomb, M. D. Adams, C. I. Reich, R. Overbeek, E. F. Kirkness, K. G. Weinstock, J. M. Merrick, A. Glodek, J. L. Scott, N. S. M. Geoghagen, J. F. Weidman, J. L. Fuhrmann, D. T. Nguyen, T. Utterback, J. M. Kelley, J. D. Peterson, P. W. Sadow, M. C. Hanna, M. D. Cotton, M. A. Hurst, K. M. Roberts, B. B. Kaine, M. Borodovsky, H. P. Klenk, C. M. Fraser, H. O. Smith, C. R. Woese, and J. C. Venter, *Science* **273,** 1058 (1996).

lism described for *M. maripaludis,* a noncarbohydrate-utilizing organism.[24] The absence of glycoside hydrolase in the genome of *A. fulgidus* is intriguing as this organism is reported to have a starch/glucose fermentative metabolism.[25] The apparent lack of glycoside hydrolase in three of the five archaeal genomes sequenced to date indicates either that *A. pernix, A. fulgidus,* and *M. thermoautotrophicum* do not need any glycoside hydrolase or that they have glycoside hydrolases completely different from those already known and which remain to be discovered, as one would expect an organism to be able to hydrolyze the bonds it builds.

Glycoside hydrolases are frequently modular, with a catalytic module carrying one or several ancillary modules.[4] A frequent function of these ancillary modules is binding to carbohydrates, an essential feature for the efficient degradation of insoluble polysaccharides, but other functions, such as cell attachment and protein binding, can also be present. The enzymes from hyperthermophiles are no exception to this, particularly those from *Caldicellulosiruptor* sp. Modularity is apparently more frequent in bacteria than in archaea.

Glycosyltransferases versus Glycoside Hydrolases

Glycosyltransferases from hyperthermophiles are compiled in Table II. Virtually all these entries are putative proteins found during genome sequencing because of the greater difficulty in working with glycosyltransferases (frequent transmembrane regions, difficult and expensive assays for biochemical characterization, lack of straightforward screening assays for expression cloning, etc.). The classification in families here again proves useful as some general functions can be assigned to these putative coding regions: glycosyltransferases using a nucleotide diphospho sugar, a nucleotide monophospho sugar, or a sugar 1-phosphate as the sugar donor and the molecular mechanism if it is known. In addition, examination of the substrate and product specificity of well-characterized homologs can also give some clue to the substrate and product specificity, although—like for glycoside hydrolases—the glycosyltransferase families are often polyspecific. Despite the lack of detailed knowledge on substrate and product specificity for most of the entries, some features of the repertoire of glycosyltransferases of hyperthermophiles are immediately apparent from the sequence-based families. Hyperthermophilic glycosyltransferases are concentrated in families 2 and 4, which are responsible "at large" for the biosynthesis of most β- and α-linked exopolysaccharides, cellular wall components, and glycogen.

[24] J. P. Yu, J. Lapodapo, and W. B. Whitman, *J. Bacteriol.* **176,** 325 (1994).
[25] K. O. Stetter, *System. Appl. Microbiol.* **10,** 172 (1988).

Like other organisms, hyperthermophiles appear to have a larger repertoire of glycosyltransferases than of glycoside hydrolases. *Aeropyrum pernix*
and *Archaeoglobus fulgidus,* whose genomes are totally devoid of known
glycosidases, contain 5 and 15 potential glycosyltransferases (Table II),
respectively. Even if some of the putative glycosyltransferases discovered
in genomes turn out to be inactive or not even enzymes, the trend toward
glycosyltransferases will probably continue.

There is accumulating evidence for an important gene transfer between
archaeal and bacterial hyperthermophiles.[26,27] The most spectacular example is the recent bacterial genome of *Thermotoga maritima* bearing 24% of
genes most similar to archaeal ones.[27] Even if there is evidence suggesting
a lower degree of horizontal transfer from bacteria to archaea,[26] one would
expect the majority of glycosidases to be already "sampled" in both archaea
and bacteria. In fact, no glycosidase (and transglycosidase) families are
purely archaeal, but the reverse is not true. Therefore, the division of Tables
I and II into archaeal and bacterial enzymes does not necessarily reflect
the "real" origin of a given enzyme.

[26] L. Aravind, R. L. Tatusov, Y. I. Wolf, D. R. Walker, and E. V. Koonin, *Trends Genet.* **14,**
442 (1998).
[27] K. E. Nelson, R. A. Clayton, S. R. Gill, M. L. Gwinn, R. J. Dodson, D. H. Haft, E. K.
Hickey, J. D. Peterson, W. C. Nelson, K. A. Ketchum, L. McDonald, T. R. Utterback,
J. A. Malek, K. D. Linher, M. M. Garrett, A. M. Stewart, M. D. Cotton, M. S. Pratt,
C. A. Phillips, D. Richardson, J. Heidelberg, G. G. Sutton, R. D. Fleischmann, J. A. Eisen,
O. White, S. L. Salzenberg, H. O. Smith, J. C. Venter, and C. M. Fraser, *Nature* **399,**
323 (1999).

[11] β-Glycosidase from *Sulfolobus solfataricus*

By Marco Moracci, Maria Ciaramella, and Mose' Rossi

Introduction

The hydrolysis of glycosidic bonds is crucial for several cell functions,
such as energy uptake, cell wall expansion, and turnover of signaling molecules. Because the diversity of saccharides, there is a wide variety of glycosylhydrolases that have been classified in more than 74 families, based on
amino acid sequence similarities.[1] However, despite the vast assortment
of glycosylhydrolases, these enzymes follow only two mechanisms, both

[1] http://afmb.cnrs-mrs.fr/~pedro/CAZY/db.html

involving two carboxyl groups in the active site. One involves the *retention* and the other the *inversion* of the anomeric configuration of the substrate. Interest in the structure and mechanisms of these enzymes has increased considerably in recent years because of the potential use of such biocatalysts including stable glycosidases isolated from thermophilic organisms.[2-4] In particular, the β-glycosidase from *Sulfolobus solfataricus* (Ssβ-gly) has been studied extensively. This organism is an extremely thermoacidophilic archaeon that grows optimally in hot springs at 87° and pH 3.0; due to the absence of tools for molecular genetic studies in *S. solfataricus,* the physiological role of Ssβ-gly is currently unknown. Nevertheless, an increasing amount of structural and biochemical data is becoming available for this enzyme. This article provides a description of two purification methods for Ssβ-gly from recombinant *Escherichia coli,* the standard assay for its activity, as well as a summary of the functional and structural properties of the enzyme.

Assay

Assays for β-glycosidase activity are extremely simple and can be performed either in liquid solutions or *in situ;* a wide range of synthetic substrates, exploiting either chromogenic or fluorogenic methods of detection, can be used, allowing high sensitivity and precise definition of the substrate specificity. Moreover, the activity of this class of enzymes can be assayed easily on disaccharides by coupling other enzymatic reactions, available in commonly used diagnostic kits, for detection of the monosaccharides produced.

The standard assay for Ssβ-gly is performed at 65° in 1 ml of 5 mM 2- or 4-nitrophenyl-β-D-galactopyranoside (2-/4-Np-β-Gal) and 50 mM sodium phosphate buffer at pH 6.5. The pH value of the buffer is measured at 20°; although pH can change with temperature, sodium phosphate buffer is less affected by temperature than other systems. Assays are performed by following changes in absorbance with a spectrophotometer equipped with a circulating water bath; the temperature of the assay is measured accurately in the assay cuvette. Quartz cuvettes containing the standard assay mixtures are equilibrated at 65° and the temperature is kept constant during all activity measurements. After the addition of 5 μl of Ssβ-gly (0.1 mg/ml), the change at 405 nm is followed for 2 min. One unit of enzyme activity is defined as the amount of enzyme catalyzing the hydrolysis of 1 μmol of substrate over 1 min at 65°, under the conditions described. The molar

[2] A. Sunna, M. Moracci, M. Rossi, and G. Antranikian, *Extremophiles* **1**, 2 (1997).

[3] M. M. Bauer, L. E. Driskill, and R. M. Kelly, *Curr. Opin. Biotechnol.* **9**, 141 (1998).

[4] B. Henrissat and P. M. Coutinho, *Methods Enzymol.* **330** [10] (2001) (this volume).

extinction coefficients of 2- and 4-nitrophenol measured at 405 nm, 65° in 50 mM sodium phosphate buffer, pH 6.5, are 1711 M^{-1} cm^{-1} and 9340 M^{-1} cm^{-1}, respectively, and are used for calculating enzymatic activity. Most of the substrates used are stable to heat; however, a blank mixture containing all the reactants, except the enzyme, is always used to correct for spontaneous hydrolysis. This method of assay can be adapted to all the Ssβ-gly preparations used, including those from *Escherichia coli* and *S. solfataricus*. However, when Ssβ-gly is expressed from *E. coli*, crude extracts have to be heated for at least 30 min at 65° before the assay in order to avoid interference by the endogenous proteins that aggregate during the assay.

When Ssβ-gly is assayed at different temperatures or in differnt buffer systems, the extinction coefficients of 2- and 4-nitrophenol groups have to be measured accurately, as they are sensitive to pH and temperature. Alternatively, the methods generally used for β-glycosidase can be adapted.[5]

In order to analyze the activity of Ssβ-gly on disaccharides, such as lactose, cellobiose, and laminaribiose, or on glucooligosaccharides, such as cellotriose and cellotetraose, the amount of D-glucose produced in the reaction can be determined by using a coupled assay for its quantitation (i.e., the glucose oxidase/peroxidase-coupled assay). For this purpose, several commercial diagnostic kits can be used and all give precise and reproducible results.

Purification

β-Glycosidase is expressed in *S. solfataricus* cells as about 0.1% of the total cell protein. The purification procedure has been modified[6] from the one described earlier[7] and includes several chromatographic steps that allow one to obtain homogeneous protein (specific activity of about 180 U/mg when assayed on 4-Np-Gal at 75°) with good yields (44%). Both methods give samples of Ssβ-gly with a high degree of purity that can be crystallized successfully.[8] A complete description of the purification of Ssβ-gly from *S. solfataricus* has been reported previously.[6] However, the recovery from this source does not exceed 15 mg of pure protein per 500 g of frozen archaeal cells. For this reason, Ssβ-gly enzyme is routinely purified from *E. coli* cells, exploiting the fact that it is expressed efficiently as a soluble protein in

[5] S. Hestrin, D. S. Feingold, and M. Schramm, *Methods Enzymol.* **1**, 241.

[6] R. Nucci, M. Moracci, C. Vaccaro, N. Vespa, and M. Rossi, *Biotechnol. Appl. Biochem.* **17**, 239 (1993).

[7] F. M. Pisani, R. Rella, C. A. Raia, C. Rozzo, R. Nucci, A. Gambacorta, M. De Rosa, and M. Rossi, *Eur. J. Biochem.* **187**, 321 (1990).

[8] L. H. Pearl, A. M. Hemmings, R. Nucci, and M. Rossi, *J. Mol. Biol.* **229**, 558 (1993).

several mesophilic hosts while maintaining the features of the enzyme purified from the archaeon (see later).

Recombinant β-Glycosidase

Recombinant Ssβ-gly has been cloned and expressed in *E. coli, Saccharomyces cerevisiae,* and *Pichia pastoris.*[9–11] Ssβ-Gly proteins isolated from *E. coli* and *S. cerevisiae* are very similar in enzymatic activity, specificity, and stability to the enzyme purified from *S. solfataricus.*[10,12]

Bacterial Strains and Growth. We report here a modification of the expression/purification procedure of Ssβ-gly from *E. coli* described previously,[10] which improves both the plasmid stability and the expression level of the enzyme. In order to remove a noncoding region of about 700 bp located downstream of the Ssβ-gly gene (*lacS*), *Nde*I and *Nco*I sites are added to the 5′ and 3′ ends of *lacS,* respectively, by polymerase chain reaction (PCR) amplification. The resulting fragment is cloned in the *Nco*I–*Nde*I sites of pET29 (Novagen, Madison, WI), which carries a kanamycin resitance gene, producing plasmid pETGly29. In this new vector, *lacS* gene expression is driven by a T7 RNA polymerase-dependent promoter from its first natural codon. Expression is performed in *E. coli* strain BL21(DE3), which contains a chromosomal copy of the T7 RNA polymerase gene under the control of an isopropylthio-β-D-galactopyranoside (IPTG)-inducible *lacUV* promoter. In order to ensure the integrity of the pETGly29 plasmid in the recombination-competent BL21(DE3) strain, freshly transformed cells are always used instead of cryopreserved *E. coli* cells. For 2-liter scale preparations, one single colony selected on LB/kanamycin plates is inoculated in 50 ml of LB medium containing kanamycin (30 μg/ml) and is grown overnight at 37° with vigorous shaking. This culture is used to inoculate (at 1/100 dilution, v/v) four 2-liter flasks containing 500 ml each of the SB medium supplemented with kanamycin (30 μg/ml) and with 50 μl of antifoam A (Sigma, St. Louis, MO). Cultures are incubated at 37° with vigorous shaking until the OD$_{600}$ is 1.0–1.2, induced with 0.5 m*M*IPTG, and grown overnight. Kanamycin selection, instead of ampicillin, increases plasmid stability in large-scale cultures.

[9] M. Moracci, A. La Volpe, J. F. Pulitzer, M. Rossi, and M. Ciaramella, *J. Bacteriol.* **174,** 873 (1992).

[10] M. Moracci, R. Nucci, F. Febbraio, C. Vaccaro, N. Vespa, F. La Cara, and M. Rossi, *Enzyme Microb. Technol.* **17,** 992 (1995).

[11] S. Fusco, M. Moracci, I. Di Lernia, M. De Rosa, and M. Rossi, unpublished results (1998).

[12] S. D'Auria, A. Morana, F. Febbraio, C. Vaccaro, M. De Rosa, and R. Nucci, *Protein Exp. Purif.* **7,** 299 (1996).

Crude Extracts and Thermal Treatment. Cells collected by centrifugation at 4000g at 4° for 15 min and resuspended in 50 ml of 50 mM sodium phosphate, pH 6.8, are broken by three cycles of French Press (American Instrument Company, Travenol Laboratories Inc., Silver Spring, MD). The lysate is clarified by centrifugation at 17,000g at 4° for 30 min; DNA is removed by adding 0.5 mg of DNase I and MgSO₄ (0.1 mM final concentration) to the supernatant and incubating the sample at 37° for 30 min.

The extreme resistance to heat of Ssβ-gly allows one to perform a simple thermal treatment, which is crucial for both the speed and the yield of purification. To this aim, the sample is heated for 30 min at 75° and aggregated host proteins are eliminated by centrifugation at 27,000g at 4° for 30 min. Finally, the supernatant (90 ml) is filtered through 0.45-μm filters (Millipore, Bedford, MA). After this step, more than 5000 total units of Ssβ-gly at the specific activity of about 26 U/mg can be recovered, as assayed under standard conditions (4-Np-Gal substrate at 65°). Because of the interference produced in the standard assay by endogenous proteins in the crude extract, which aggregate at 65°, it is not possible to estimate the yields and the purification fold after this step.

Hydrophobic Chromatography. (NH₄)₂SO₄ (final concentration 1 M) is added to the sample obtained from the previous step, and the sample is loaded at a flow rate of 2 ml/min on a fast protein liquid chromatography (FPLC) Phenyl-Superose HR 10/10 column (Pharmacia Biotech, Uppsala, Sweden), equilibrated with 50 mM phosphate buffer, pH 6.8, 1 M (NH₄)₂SO₄ (running buffer). After washing, the column is eluted with a nonlinear gradient of running buffer and running buffer without (NH₄)₂SO₄ and then with 50 ml of 50 mM phosphate buffer, pH 6.8. Enzyme activity is eluted as a broad peak after the gradient during the elution with running buffer. The active pool (around 120 ml) is dialyzed against 3 liters of 20 mM phosphate buffer, pH 7.0, for 24 hr at 4° and then concentrated by ultrafiltration using a YM30 membrane (Amicon, Beverly, MA). This step usually produces high yields (>70%) and a purification fold of about 6.0 from the thermal treatment. The purity of the sample is checked by SDS–PAGE on continuous 7% polyacrylamide gels run at room temperature; at this step the enzyme is generally >95% homogeneous (specific activity above about 150 U/mg under the conditions described earlier). When Ssβ-gly is stored at 4° in 20 mM phosphate buffer, pH 7.0, and sodium azide 0.02%, no appreciable loss of activity can be detected even after several months. In contrast, the enzyme is stable at −20° only if supplemented with 50% (v/v) glycerol, as freezing could inactivate homogeneous samples rapidly.

Size-Exclusion Chromatography. When a higher state of purity is required, i.e., for crystallization studies, Ssβ-gly can be further purified by gel filtration using an FPLC Superdex 200 HG 26/60 column (Pharmacia

Biotech). In this case, the enzyme solution (<5 ml) is applied to the column equilibrated previously with 20 mM phosphate buffer, pH 7.0, at a flow rate of 2 ml/min. Fractions containing β-galactosidase activity are pooled and concentrated as described earlier.

Chimeric β-Glycosidase

Mutations of Ssβ-gly residues involved in the activity and/or stability of the enzyme could hamper the normal protocol of purification. For these reasons, a different expression system in *E. coli* has been developed in which wild-type and mutant Ssβ-gly are expressed as fusions with the glutathione *S*-transferase (GST) protein from *Schistosoma japonicum*. This system allows the easy purification of proteins by affinity chromatography on glutathione Sepharose 4B matrix, following GST activity. The fused polypeptides can then be separated by thrombin cleavage.

Plasmid vectors pGEXGly and pGEX-K-Gly are obtained by inserting the *lacS* gene into the *Bam*HI site of pGEX-2T and pGEX-2TK (Pharmacia Biotech), respectively. Both resulting plasmids contain an open reading frame (ORF) coding for a polypeptide of around 80 kDa, consisting of GST (26 kDa) at the N terminus, and of Ssβ-gly (56 kDa) at the C terminus. In pGEX-K-Gly, the recognition sequence for the catalytic subunit of cAMP-dependent protein kinase (PK) is inserted between the GST and the Ssβ-gly portions (see later). The sequences at the junction of the polypeptides deduced from the two plasmids are the following:

pGEXGly   -Leu-Val-Pro-<u>Arg-Gly</u>-Ser-**Met-Tyr-**

pGEX-K-Gly   -Leu-Val-Pro-<u>Arg-Gly</u>-Ser-*Arg-Arg-Ala-Ser-Val*-**Met-Tyr-**

The first two Ssβ-gly amino acids are in boldface type, the residues involved in the peptidic bond hydrolyzed by the thrombin protease are underlined, and the PK recognition sequence is in italics.

Bacterial Growth and Crude Extract. Expression is performed in *E. coli* strain BL21, harboring either pGEXGly or pGEX-K-Gly, starting from freshly transformed cells. For 2-liter scale preparations, one single colony selected on ampicillin is inoculated in 50 ml of LB medium containing ampicillin (50 μg/ml) and is grown overnight at 37° with vigorous shaking. This culture is used to inoculate (at 1/100 dilution, v/v) four 2-liter flasks containing LB medium (500 ml each), supplemented with ampicillin (50 μg/ml) and antifoam A (50 μl each). Cell cultures are incubated at 37° with vigorous shaking until the OD$_{600}$ is 0.6–0.8, induced by adding 0.5 mM IPTG, and grown overnight. Crude extract is prepared by collecting cells by centrifugation at 4000g at 4° for 15 min, resuspending the pellets in 50 ml of phosphate-buffered saline (PBS) (150 mM NaCl, 50 mM sodium phosphate buffer, pH 7.3) supplemented with Triton 0.1%, and by three cycles of French Press (American Instrument Company). The lysate is clarified by centrifugation at 17,000g for 30 min, diluted to 200 ml of PBS/Triton 0.1%, and filtered through a 0.45-μm filter (Millipore).

Glutathione Affinity Chromatography and Thrombin Treatment. The crude extract is slowly loaded overnight at 4° on a 2-ml bed volume of a gluthatione Sepharose 4B column (Pharmacia Biotech), equilibrated with PBS/Triton 0.1%, in order to allow the binding of GST to the matrix. The column is washed with 60 ml of PBS. The efficiency of binding is followed by a standard GST assay at 37° performed following the manufacturer's protocol (Pharmacia Biotech). The fusion protein is eluted from the column by the addition of 2 ml of 500 mM Tris–HCl, pH 8.0, supplemented with 10 mM reduced gluthatione, and incubation at room temperature (22–25°) for 10 min; the eluate is collected and assayed for GST activity at 37° and for Ssβ-gly activity at 65°. This procedure is repeated until all the fusion protein is eluted from the column; finally, active fractions are collected and stored at 4°. No accurate estimation of the yields and the purification fold from the crude extract can be performed after this step because of interference from aggregated *E. coli* proteins and because Ssβ-gly specific activity is reduced severely when the protein is fused to GST.

The active pool is then subjected to thrombin treatment in order to recover Ssβ-gly free of the fused GST polypeptide; to this aim, pooled fractions (about 50 ml) are incubated at 25° overnight with 50 units of thrombin solution prepared as suggested by the manifacturer (Pharmacia Biotech). The efficiency of thrombin cleavage is tested by SDS–PAGE.

Affinity chromatography samples obtained with pGEXGly vector, analyzed by SDS–PAGE, reveal the expected 80-kDa band corresponding to the fusion; however, the fused polypepeptide is inefficiently cleaved by thrombin, suggesting that the thrombin recognition site in this chimera is unaccessible to the protease.[13] In contrast, the chimeric protein expressed from the pGEX-K-Gly vector, which contains five extra amino acids of the PK recognition sequence between the two portions, is cleaved efficiently and specifically by thrombin, suggesting that the PK recognition sequence works as a spacer, exposing the protease site. After this purification step, 1500 total units of Ssβ-gly (specific activity about 20 U/mg) are usually recovered, as assayed under standard conditions (4-Np-Gal substrate at

[13] M. Moracci, L. Capalbo, M. De Rosa, R. La Montagna, A. Morana, R. Nucci, M. Ciaramella, and M. Rossi, *in* "Carbohydrate Bioengineering" (S. B. Petersen, B. Svenson, and S. Pedersen, eds.), p. 77. Elsevier, 1995.

65°). Only samples obtained from pGEX-K-Gly are, therefore, used for further purification steps.

Chromatography on Q-Sepharose. The sample obtained after thrombin treatment is loaded at a flow rate of 2 ml/min on an FPLC Q-Sepharose column (Pharmacia Biotech) (1.5 × 11 cm) equilibrated with 10 mM Tris–HCl buffer, pH 8.0. The column is washed with 30 ml of the same buffer, Ssβ-gly is eluted after the first 12 ml, indicating that the enzyme did not bind the column. All the contaminants are eluted with a nonlinear gradient of 0–0.3m NaCl. This step usually produces high yields (about 80%) and a purification fold of about 5.0 from the affinity chromatography. After this step the sample is about 90% homogeneous by SDS–PAGE (specific activity above 130 U/mg) and can be stored as described earlier.

Ssβ-gly purified by this method should contain seven extra amino acids at its N terminus (see earlier discussion). However, only four extra amino acid residues are observed by direct N-terminal sequencing of samples obtained with this method, suggesting that the tail produced by thrombin cleavage is further hydrolyzed by contaminating proteases. This "long" form of Ssβ-gly shows the same features of the "normal" form; therefore, this procedure is used routinely for the production of Ssβ-gly mutants severely affected in their properties (see later).

Table I reports the comparison of the two purification methods described; it is clear that the expression system pETGly29/BL21(DE3) is far more efficient; however, the pGEX-K-Gly/BL21 expression system, although less efficient, allows purification of inactive and/or unstable mutant proteins, which are very difficult to purify using traditional methods.

Enzyme Function

Molecular mass determined by SDS–PAGE under reducing and nonreducing conditions, and by gel filtration chromatography, showed that Ssβ-

TABLE I
Comparative Ssβ-Gly Expression in *E. coli*

Expression system	Antibiotic resistance	Culture volume (liter)	Cell pellet yield (g)	Yield per culture volume (units/liter)[a] (g of protein/liter)	Yield per cell (units/g) (g of protein/g of cells)
pETGly29/BL21(DE3)	Kanamicin	2	30	1912	127
				12.5	0.8
pGEX-K-GLy/BL21	Ampicillin	2	20	666	66.6
				5.2	0.52

[a] Assays were performed at standard conditions using 4-Np-β-D-Gal as substrate.

TABLE II
KINETIC CONSTANTS OF S. solfataricus β-GLYCOSIDASE[a]

Substrate	K_m (mM)	k_{cat} (sec^{-1})	k_{cat}/K_m (sec^{-1} mM^{-1})
4-Np-Fuc	1.0	994	994
4-Np-Glc	0.5	698	1396
2-Np-Glc	3.5	1898	542
4-Np-Gal	4.7	1313	279
2-Np-Gal	4.0	2146	536
4-Np-Xyl	4.0	366	41
Laminaribiose[b,c]	1.0	908	908
β-Gentiobiose[b,c]	100	1360	14
Cellobiose[b,c]	30	1333	44
Cellotriose[b]	3.0	197	66
Cellotetraose[b]	1.7	584	343
Cellopentaose[b]	2.4	270	112
4-Np-Cellobioside[b]	0.3	503	1676

[a] At 75°.
[b] Hydrolysis products were detected as glucose, and K_m values are shown as glucose equivalents.
[c] Kinetic values for these substrates were calculated taking into account that two glucose equivalents were produced for every glucosidic bond hydrolyzed.

gly is a homotetrameric protein (240 kDa) with 56-kDa subunits. The enzyme is thermostable (half-life 48 hr at 85°), resistant to detergents and organic solvents, and shows maximal activity above 95°. One of the most interesting properties of Ssβ-gly is its wide substrate specificity, its ability to hydrolyze oligosaccharides, and its efficient synthesis of α-D-glycosides by transglycosylation.[14] For these reasons, the catalytic properties and the reaction mechanism have been studied extensively by classical enzymological methods and by site-directed mutagenesis.

Substrate Specificity

Kinetic constants of Ssβ-gly for several substrates are summarized in Table II: the enzyme is active on arylglycosides of galactose, fucose, glucose, and xylose and on di- and oligosaccharides derived from glucose. No hydrolysis occurred with 4-Np-α-D-gluco- and galacto-pyranosides, 4-Np-β-D-N-acetylglucosamine and galactosamine, 4-Np-β- and α-L-fucopyranoside, 4-Np-β-D-glucuronic or galacturonic acid, or with maltose. Mannose,

[14] A. Trincone, R. Improta, R. Nucci, M. Rossi, and Gambacorta, *Biocatalysis* **10,** 195 (1994).

4-Np-α-L-arabinopyranoside, and 2-Np-β-D-galactopyranoside 6-phosphate were susceptible to hydrolysis but with only limited efficiency.[6] Ssβ-Gly hydrolyzes arylglycosides very efficiently, showing that the k_{cat}/K_m values increase in the order of 4-Np-β-D-Glc > Fuc > Gal. The hydrolysis of arylgalactosides is fast, but the k_{cat}/K_m values are smaller than those for arylglucosides. β-Fucosidase activity is comparable to β-glucosidase, whereas the hydrolysis of 4-Np-β-D-xylopyranoside is much slower. It is interesting that Ssβ-gly shows, at high temperature, both catalytic efficiency and turnover number comparable to those of other β-glycosidases from mesophilic sources.

The catalytic efficiency of hydrolysis of disaccharides and oligosaccharides is far lower than that of chromogenic substrates. Nevertheless, all three β-linked glucose dimers tested are substrates of the enzyme, with the order of preference being β1–3 > β1–4 > β1–6; laminaribiose is the disaccharide hydrolyzed with the highest efficiency.

Ssβ-Gly is able to hydrolyze oligosaccharides with good efficiency: k_{cat}/K_m values are such that cellotetraose > cellotriose > cellobiose, whereas the hydrolysis of cellopentaose is slower. When incubating the enzyme with cellotetraose, the production of glucose and cellotriose is observed by thin-layer chromatography. Trace amounts of cellobiose are detected only after 5 min of incubation, probably as a result of hydrolysis of the initial product cellotriose. These results indicate that Ssβ-gly exhibits exoglucosidase activity. This is further confirmed using 4-Np-β-D-cellobioside as substrate: the production of both 4-nitrophenol and glucose could be detected, but while glucose release increases linearly from the onset of the reaction, the presence of 4-nitrophenol could be detected only after several minutes. This indicated that Ssβ-gly exoglucosidase activity is specific toward the nonreducing end of the oligosaccharide substrate.

Analysis of Catalytic Machinery

Retaining glycosidases (Scheme 1) follow a double-displacement mechanism involving a glycosyl-enzyme intermediate. The carboxyl group in the active center functions as a general acid/base catalyst and the carboxylate as the nucleophile of the reaction.[15,16] The two carboxylic residues essential for catalysis in Ssβ-gly were determined from glutamic and aspartic acid residues conserved among all family 1 glycosylhydrolases members. The glutamic acid-206 and -387 residues of Ssβ-gly were found in two totally conserved motifs: Asn-**Glu**-Pro and **Glu**-Asn-Gly, corresponding to the general acid/base catalyst and the nucleophile, respectively, and identified

---

[15] D. E. Koshland, *Biol. Rev.* **28**, 416 (1953).
[16] A. White, and D. R. Rose, *Curr. Opin. Struct. Biol.* **7**, 645 (1997).

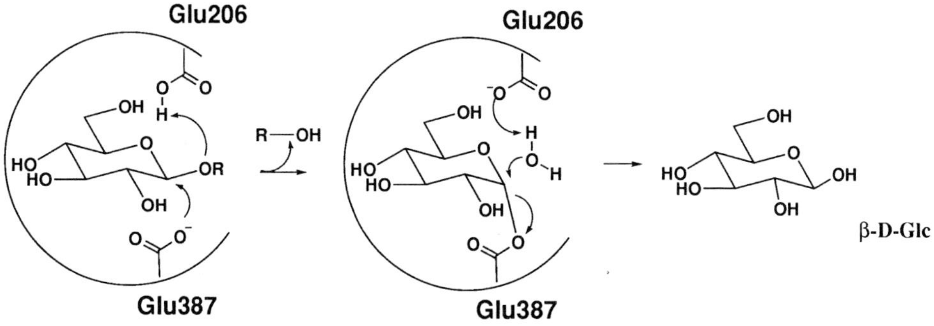

SCHEME 1

previously in the active site of *Agrobacterium* β-glucosidase.[17,18] These residues were changed to isosteric glutamine by site-directed mutagenesis; the resulting mutant proteins were expressed as fusions to GST as described previously.

The Glu387Gln mutant was completely inactive against all the substrates tested.[19] These data are consistent with the function of Glu387 as the attacking nucleophile. Moreover, this function has been further confirmed using active site-directed inhibitors[20] and inspection of the three-dimensional structure (see later discussion).

In contrast, the Glu206Gln mutation strongly affects, but does not completely eliminate, the activity on 2/4-Np-β-D-glycosides.[19] Kinetic constants determined for the Glu206Gln mutant at 65° are compared to the wild type in Table III. The mutant showed a 10- to 30-fold decrease in K_m values and a 60-fold decrease of the k_{cat} values on 2/4-Np-Glc, suggesting that the glucosyl-enzyme intermediate is accumulated during the reaction. k_{cat} on 2/4-Np-Gal and 4-NpFuc are less affected. These results suggest that 2/4-Np-glucoside and 2/4-Np-galactoside substrates elicit different interactions with the Glu206 residue in the Ssβ-gly active site. Moreover, the pH sensitivity of the reaction was affected by the mutation: whereas wild-type Ssβ-gly showed a typical bell-shaped curve, with maximum activity at pH 6.5, the Glu206Gln residual activity was apparently independent of pH.[19]

These results strongly suggest that glutamic acid-206 in Ssβ-gly is the

[17] S. G. Withers, R. A. J. Warren, I. P. Street, K. Rupitz, J. B. Kempton, and R. Aebersold, *J. Am. Chem. Soc.* **112,** 5887 (1990).

[18] Q. Wang, D. Trimbur, R. Graham, R. A. J. Warren, and S. G. Withers, *Biochemistry* **34,** 14554 (1995).

[19] M. Moracci, L. Capalbo, M. Ciaramella, and M. Rossi, *Protein Eng.* **9,** 1191 (1996).

[20] F. Febbraio, R. Barone, S. D'Auria, M. Rossi, R. Nucci, G. Piccialli, L. De Napoli, S. Orrù, and P. Pucci, *Biochemistry* **36,** 3068 (1997).

TABLE III
KINETIC CONSTANTS OF HYDROLYSIS BY WILD-TYPE AND MUTANT E206Q[a]

| | K_m (mM) | | k_{cat} (sec^{-1}) | |
Substrate	Wild type	Glu206Gln	Wild type	Glu206Gln
2-NpGal	0.95 ± 0.08	1.53 ± 0.1	537 ± 11	62 ± 1.4
4-NpGal	1.17 ± 0.06	4.79 ± 0.17	383 ± 5	43 ± 0.5
2-NpGlc	1.01 ± 0.24	0.03 ± 0.006	457 ± 22	8.1 ± 0.13
4-NpGlc	0.30 ± 0.04	0.03 ± 0.003	334 ± 10	6.1 ± 0.08
4-NpFuc	0.45 ± 0.11	0.09 ± 0.005	366 ± 17	49 ± 0.5

[a] At 65°.

general acid/base catalyst for the hydrolysis reaction. The studies reported so far on glycohydrolases mutated to influence general acid/base catalysis demonstrate that the kinetic parameters can be affected to different extents.[18,21,22] The reasons for these differences are not clear, it is possible that enzymes with different substrate specificities require the assistance of the general acid/base catalyst to different degrees.

Enzyme Structure

Ssβ-Gly has been crystallized in its native, tetrameric form using multiple isomorphous replacement and has been refined at 2.6 Å resolution.[23] The enzyme shows the classic $(\beta\alpha)8$ fold that has also been observed in other mesophilic members of the glycosylhydrolase family 1 crystallized so far: the cyanogenic β-glucosidase from *Trifolium repens*,[24] the phospho-β-glucosidase from *Lactococcus lactis*,[25] the β-glucosidase from *Bacillus polymixa*,[26] and the myrosinase from *Sinapis alba*.[27] Inspection of the Ssβ-gly three-demensional structure gave insights into both the architecture of the substrate binding site and the possible molecular determinants of thermal stability.

[21] E. Witt, R. Frank, and W. Hengstenberg, *Protein Eng.* **6,** 913 (1993).

[22] M. MacLeod, T. Lindhorst, S. G. Withers, and R. A. J. Warren, *Biochemistry* **33,** 6371 (1994).

[23] C. Aguilar, I. Sanderson, M. Moracci, M. Ciaramella, R. Nucci, M. Rossi, and L. H. Pearl, *J. Mol. Biol.* **271,** 789 (1997).

[24] T. Barret, C. G. Suaresh, S. P. Tolley, E. J. Dodson, and M. A. Hughes, *Structure* **3,** 951 (1995).

[25] C. Weisman, G. Best, W. Hengstenberg, and G. E. Schultz, *Structure* **3,** 961 (1995).

[26] J. Sanz-Aparicio, J. A. Hermoso, M. Martinez-Ripoll, J. L. Lequerica, and J. Polaina, *J. Mol. Biol.* **275,** 491 (1998).

[27] W. P. Burmeister, S. Cottaz, H. Driguez, R. Iori, S. Palmieri, and B. Henrissat, *Structure* **5,** 663 (1997).

Active Site and Substrate-Binding Tunnel

The Ssβ-gly (βα)8 fold is quite complex: the connections between β strands and α helices within βα repeats at the top of the barrel are substantially elaborated and contain several loops with extra secondary structural elements. In particular, a deep indentation at the top of the barrel is connected to the outside via a radial channel, about 30 Å in length, which emerges between the helices of the fifth and sixth βα units (Fig. 1). The opening is generated by a 45° kink in the helix of the fifth unit, occurring at an unusual sequence of aromatic residues (Trp-287, Trp-288, Phe-289, Phe-290). Because the tops of the barrel in the tetrameric structure are in contact, the loops provide a "roof" for the radial channel in each monomer,

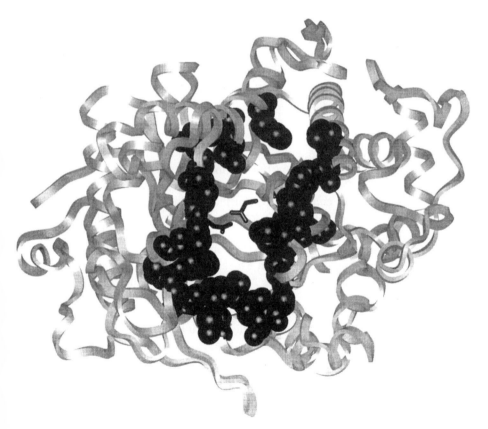

FIG. 1. Substrate binding site of Ssβ-gly. Residues in the radial channel are shown with the Van der Waals radius of their atoms. The side chains of the catalytic diad Glu-206 and Glu-387 are shown as sticks.

converting it into a tunnel. The radial channel in the monomer contains the catalytic residues Glu-206 and Glu-387 and a high concentration of other residues highly conserved in family 1 glycohydrolases and likely implicated in catalysis and/or substrate binding. Moreover, a cellohexaose chain could be modeled into the tunnel, with its nonreducing end and the scissile glycosidic bond in close proximity to the catalytic diad Glu-387–Glu 206.[23] These findings strongly suggest that this tunnel is the substrate binding site and support enzymological studies showing the exoglucosidase activity of Ssβ-gly.

The distance between the two carboxylate residues responsible of the catalytic activity is diagnostical for the assignment of the reaction mechanism in glycohydrolases: these two groups are, on average, 5.0 and 9.3 Å apart, in *retainers* and *inverters,* respectively. Analysis of the Ssβ-gly three-dimensional structure confirmed that this enzyme follows a *retaining* mechanism, as the distance between the Cδ groups of Glu387 and Glu206 is 4.6 Å.

Factors Producing Thermostability

The comparison of the Ssβ-gly amino acid sequence with those of thermophilic and mesophilic β-glycosidases revealed that the enzyme shows a lower content of alanine and isoleucine and a higher content of tryptophan and asparagine than the mean of the mesophiles.[23] This suggests that most of the rules proposed to explain the molecular origins of the enhanced stability of enzymes from hyperthermophiles do not hold for this protein. In contrast, the Ssβ-gly three-dimensional structure revealed that two structural features might contribute to thermal stabilization in this enzyme: a large number of both buried hydrophilic cavities and ionic groups involved in ion pairs. In Ssβ-gly, one buried solvent molecule on 11.4 amino acid residues is observed versus a 1 : 27 ratio in proteins in general. So far this feature has only been observed in Ssβ-gly, whereas the abundance of ion pairs arranged in large networks has been found in other enzymes from hyperthermophiles.[28–30] The Ssβ-gly tetramer contains 524 charged groups; the majority of these (58%) are involved in ion pair interactions and around 60% of these interactions occur as part of multiple ion pair networks involving at least three charged groups. The cyanogenic β-glycosidase from clover

[28] M. Hennig, B. Dairmont, R. Sterner, K. Kirschner, and J. N. Jansonius, *Structure* **3,** 1295 (1995).
[29] K. S. P. Yip, T. J. Stillman, K. L. Britton, P. J. Artymiuk, P. J. Baker, S. E. Sedelnikova, P. C. Engel, A. Pasquo, R. Chiaraluce, V. Consalvi, R. Scandurra, and D. W. Rice, *Structure* **3,** 1147 (1995).
[30] A. Goldman, *Structure* **3,** 1277 (1995).

has far less of its charged residues involved in ion pairs (41%) and the majority of them are isolated pairs.[24]

Most of the networks in Ssβ-gly occur in noncontiguous positions on the surface of the molecule and spanning different domains and subunits; hence, they could act as cross-linkers between folded structural elements. It is interesting that arginines occur frequently in ion pair networks, presumably for their ability to form multidentate interactions; this might explain the higher number of arginines found in enzymes from hyperthermophiles.[31,32]

These findings suggest that large networks of ion pairs are important for hyperthermostability in Ssβ-gly and provide a useful framework for direct experimental test, by site-directed mutagenesis, of the contribution of each interaction to the overall stability.

Acknowledgments

Work in the authors' laboratory was supported by EC project: "Biotechnology of Extremophiles" contract No. BIO2-CT93–0274 and by the EC FAIR project: "Enzymatic Lactose Valorization" contract No. 1048/96.

[31] P. Argos, M. Rossmann, U. Grau, H. Zuber, G. Frank, and J. Tratschin, *Biochemistry* **25,** 5698 (1979).
[32] L. Menéndez-Arias and P. Argos, *J. Mol. Biol.* **206,** 397 (1989).

[12] Xylose Isomerases from *Thermotoga*

By CLAIRE VIEILLE, DINLAKA SRIPRAPUNDH, ROBERT M. KELLY, and J. GREGORY ZEIKUS

Introduction

$$\text{D-Xylose} \rightleftharpoons \text{D-xylulose}$$
$$\text{D-Glucose} \rightleftharpoons \text{D-fructose}$$

Typically present in microorganisms that grow on xylose, xylose isomerase (XI) (EC 5.3.1.5) converts xylose to xylulose, which is then phosphorylated and enters the pentose–phosphate pathway. Because it also accepts glucose as substrate, XI is used extensively to isomerize glucose to fructose in the manufacture of high fructose corn syrup (HFCS). Performed at temperatures around 55–60°, this isomerization process requires thermostable enzymes. XIs have been characterized from a number of eubacterial sources and from rice and barley, but they have not been reported in fungi or archaea. XIs from the hyperthermophilic eubacteria *Thermotoga maritima* and *Thermotoga neapolitana* are the most thermostable yet characterized XIs.

Thermotoga maritima and *T. neapolitana* Growth Conditions

Thermotoga maritima (DSM 3109) and *T. neapolitana* (DSM 5068) are grown at 80° and 85°, respectively, in artificial sea water (per liter: 15.0 g NaCl, 2.0 g Na_2SO_4, 2.0 g $MgCl_2 \cdot 6H_2O$, 0.5 g $CaCl_2 \cdot 2H_2O$, 0.25 g $NaHCO_3$, 0.1 g K_2HPO_4, 50 mg KBr, 20 mg H_3BO_3, 20 mg KI, 15 mg $Fe[NH_4]_2[SO_4]_2[SO_4]_2$, 3 mg $Na_2WO_4 \cdot 2H_2O$, and 2 mg $NiCl_2 \cdot 6H_2O$) supplemented with 0.5% (w/v) tryptone, 0.1% (w/v) yeast extract, and 0.5% (w/v) xylose as the inducer. Xylose and phosphate are added after sterilization. Oxygen is removed from the medium prior to inoculation by incubating the medium at 98° for 30 min while sparging with N_2 and by adding Na_2S (final concentration, 0.05%).

Cloning of *T. neapolitana xylA* Gene in *Escherichia coli*

Thermotoga neapolitana chromosomal DNA is purified by the method of Goldberg and Ohman.[1] A library of 3–7 kb *Sau*3A partially digested *T. neapolitana* DNA fragments is constructed in *Bam*HI-digested pUC18.[2] *Escherichia coli* Sure strain (Stratagene, La Jolla, CA) is transformed with the ligation mixture by electroporation.[3] The *T. neapolitana xylA* gene is isolated by colony hybridization of the just described library with a homologous, polymerase chain reaction (PCR)-generated DNA probe. Primers used for the PCR reaction[2] encode ENYVFWG and EPKP (N,K)EP, conserved peptides in type II XIs. The template is *T. neapolitana* chromosomal DNA.

DNA Manipulations

Plasmid DNA purification, restriction analysis, PCR reaction, and colony and DNA hybridizations are performed by conventional techniques.[3,4]

[1] J. B. Goldberg and D. E. Ohman, *J. Bacteriol.* **158,** 1115 (1984).
[2] C. Vieille, J. M. Hess, R. M. Kelly, and J. G. Zeikus, *Appl. Environ. Microbiol.* **61,** 1867 (1995).
[3] F. M. Ausubel, R. Brent, R. E. Kingston, D. D. Moore, J. G. Seidman, J. A. Smith, and K. Struhl (eds.), "Current Protocols in Molecular Biology." Greene Publishing & Wiley-Interscience, New York, 1993.
[4] J. Sambrook, E. F. Fritsch, and T. Maniatis, "Molecular Cloning: A Laboratory Manual," 2nd Ed. Cold Spring Harbor Laboratory Press, Cold Spring Harbor, NY, 1989.

Sequences are determined using the Sequenase Version 2.0 kit (U.S. Bio-chemical Corp., Cleveland, OH).

T. maritima and T. neapolitana Xylose Isomerase Purification

All purification steps (see Table I) are performed at room temperature under aerobic conditions. *T. maritima* cells (250 g) are resuspended in 1 liter of 50 mM Tris–HCl (pH 8.0) (buffer A) containing 1 mM MgCl$_2$ and 0.02% NaN$_3$. A cell homogenate is prepared by twice passing the cell suspension through a French press cell at 12,000 lb/in^2. After centrifugation at 10,000g for 30 min, the supernatant is loaded onto a Q-Sepharose FF column (Pharmacia, Uppsala, Sweden) preequilibrated with buffer A. The column is washed with buffer A, and proteins are eluted with a 5-bed volume 0–1.0 M NaCl linear gradient in buffer A. Fractions with XI activity are pooled and (NH$_4$)$_2$SO$_4$ is added to 1.3 M final concentration. The solution is then loaded onto a Toyopearl Phenyl-650M hydrophobic interaction column (TosoHaas, Philadelphia, PA) equilibrated with 1.3 M (NH$_4$)$_2$SO$_4$ in buffer A. The column is washed with 2-bed volumes of the same solution, and proteins are eluted with a 10-bed volume 1.3–0.0 M (NH$_4$)$_2$SO$_4$ linear gradient in buffer A. Fractions with XI activity are pooled, concentrated in a stirred cell equipped with a 10,000 molecular weight cutoff membrane, and dialyzed at 4° against buffer A. The enzyme is then loaded onto a HiLoad Q-Sepharose HP anion-exchange column equilibrated with buffer A. The column is washed with 2-bed volumes of buffer A, and the enzyme is eluted with a 0.1–0.4 M NaCl linear gradient in buffer A. Active fractions are pooled, concentrated, and dialyzed against buffer A containing 0.15 M NaCl and 0.02% (w/v) NaN$_3$. Finally, the enzyme solution is loaded onto a HiLoad Superdex 200 prep grade gel-

TABLE I
PURIFICATION OF *T. maritima* XYLOSE ISOMERASE[a]

Purification	Total protein (mg)	Specific activity[b] (U/mg)	Overall purification (-fold)	Overall yield (%)
Cell-free extract	22,582	0.15	1.0	100
Q-Sepharose FF	4,225	0.47	3.1	60
Phenyl-650M	296	4.0	27.0	35
Q-Sepharose HP	180	5.0	33.3	27
Superdex 200	30	19.5	130.0	17

[a] Brown *et al.*[9]
[b] Determined using fructose as the substrate.

filtration column and eluted with buffer A containing 0.15 M NaCl and 0.02% NaN_3. The purified enzyme is dialyzed against buffer A containing 0.02% NaN_3.

The native *T. neapolitana* enzyme is purified at room temperature under aerobic conditions as follows: *T. neapolitana* cells (12.5 g wet weight) are resuspended in 100 mM sodium phosphate buffer (pH 7.4) (buffer B) containing 10 mM $MgSO_4$ and 1 mM $CoCl_2$. The cell extract obtained after passage through a French pressure cell (pressure drop, 18,000 psi) is centrifuged at 10,000g for 20 min and the supernatant is used as the crude enzyme. The crude extract is applied to a DEAE-Sepharose column equilibrated with buffer B. The enzyme is eluted with a 0.0–0.2 M NaCl linear gradient in buffer B. The active fractions are pooled, and NaCl is added to a final concentration of 4 M; this enzyme solution is then loaded into a Toyopearl Phenyl-650M hydrophobic interaction column equilibrated with 4 M NaCl in buffer B. The proteins are eluted using 4.0–2.8, 2.8–1.6, 1.6–0.8, and 0.8–0.0 M NaCl linear gradients, separated by 1 column volume plateaus. Active fractions are pooled, concentrated in a stirred cell concentrator, and the NaCl concentration is adjusted to 0.2 M. This enzyme solution is finally loaded onto a Superose 6 prep grade gel filtration column equilibrated with 0.2 M NaCl in buffer B and eluted with the same buffer. The purified enzyme is dialyzed overnight against 50 mM 4-morpholino propanesulfonic acid (MOPS) (pH 7.0 at room temperature) (buffer C) containing 5 mM $MgSO_4$, 0.5 mM $CoCl_2$, and 0.007% NaN_3.

Thermotoga neapolitana recombinant XI is purified from *Escherichia coli* HB101 carrying plasmid pTNE2-kan[2], as described by Lee *et al.*,[5] except that (i) the cells are grown in TB supplemented with kanamycin; (ii) buffer C containing 5 mM $MgSO_4$ plus 0.5 mM $CoCl_2$ is used as buffer; (iii) the cell extract is heat-treated at 90° for 2 hr and 30 min in a shaking oil bath; (iv) an $(NH_4)_2SO_4$ fractionation step is added after the heat treatment[6]; (v) ion-exchange chromatography is performed on Q-Sepharose Fast Flow, and proteins are eluted with a linear 0.0–0.3 M NaCl gradient in buffer C.

Protein concentrations are determined routinely by the method of Bradford[7] using bovine serum albumin as the standard. Enzyme fractions are analyzed by sodium dodecyl sulfate–12% polyacrylamide gel electrophoresis (SDS–PAGE) and are visualized by Coomassie blue staining.

Enzyme Assays

The effect of temperature on *T. neapolitana* XI specific activity is determined by incubating the enzyme (0.02–0.1 mg/ml) for 20 min in buffer C

[5] C. Y. Lee, M. Bagdasarian, M. H. Meng, and J. G. Zeikus, *J. Biol. Chem.* **265,** 19082 (1990).
[6] C. Y. Lee and J. G. Zeikus, *Biochem. J.* **273,** 565 (1991).
[7] M. M. Bradford, *Anal. Biochem.* **72,** 248 (1976).

containing 1 mM $CoCl_2$ and 0.8 M glucose at the temperatures of interest. The pH at 90° is calculated to be 6.3 (using a $\Delta pK_a/\Delta t$ of -0.011 for MOPS). The effect of temperature on *T. maritima* XI specific activity is determined by incubating the enzyme for 30 min in 100 mM piperazine-N,N'-bis[2-ethanesulfonic acid] (PIPES) (pH 7.0) containing 1 mM $CoCl_2$ and 1 M fructose. To determine TMXI and TNXI kinetic parameters, enzymatic assays are performed in 100 mM PIPES (pH 7.0 at room temperature) containing 1 mM $CoCl_2$ (plus 10 mM $MgSO_4$ for TNXI) and 1 M to 10 mM substrate (0.5 M to 1 mM for xylose). All reactions are stopped by placing the tubes in ice. Controls without enzyme are used in every experiment. The fructose and xylulose formed are measured by the cysteine–carbazole–sulfuric acid method.[8] Absorbances are measured at 560 nm (fructose) and 537 nm (xylulose). Glucose is quantified using Sigma kit 510A (Sigma, St. Louis, MO). One unit of XI activity is defined as the amount of enzyme that produces 1 μmol of product/min under the assay conditions.

Thermostability Studies

The enzyme is incubated at various temperatures in the presence of 100 mM MOPS (pH 7.0), 0.5 mM $CoCl_2$ for different periods of time. Thermoinactivation is stopped by cooling the tubes in a water bath equilibrated at room temperature. Residual XI activity is measured at 90° in the conditions described earlier.

pH Studies

The effect of pH on TMXI and TNXI activities is measured using the enzyme assays described earlier, except that MOPS and PIPES buffers are substituted by acetate 100 mM (pH 4.0 − 5.7), PIPES 100 mM (pH 6.0–7.5), or N-[2-hydroxyethyl]piperazine-N'-[3-propanesulfonic acid] (EPPS) 100 mM (pH 7.5–8.7). All pHs are adjusted at room temperature, and the $\Delta pK_a/\Delta t$ values of 0.000, -0.0085, and -0.011 for acetate, PIPES, and EPPS, respectively, are taken into account for the results.

Metal Studies

Metal ions are removed from TMXI and TNXI by EDTA treatment performed at 4°. TMXI is first dialyzed for 15 hr against buffer A containing 5 mM EDTA and then twice against buffer A. TNXI is first incubated overnight in buffer C containing 10 mM EDTA. It is then dialyzed against

---

[8] Z. Dische and E. Borenfreund, *J. Biol. Chem.* **192**, 583 (1951).

buffer C containing 2 mM EDTA and three times against buffer C. Metal ions are added to the EDTA-treated enzymes, and the mixtures are incubated for 1 hr at room temperature before being tested for activity or stability. All metal ions used are in the chloride form.

Microcalorimetry Studies

Before calorimetry, TNXI is either EDTA treated (see earlier discussion) and dialyzed against 20 mM PIPES (pH 7.0) to remove all metal from the enzyme or it is dialyzed against 20 mM MOPS (pH 7.0) containing 2 mM MgSO$_4$ and 0.2 mM CoCl$_2$. Unfolding is followed by differential scanning calorimetry (DSC) in a Hart Scientific differential scanning calorimeter (apo-TNXI) and in a Nano-DSC differential scanning calorimeter (Calorimetry Sciences Corp., Provo, UT) (TNXI with metals).

TNXI Immobilization

TNXI is covalently immobilized to glass beads (120 to 200 mesh, 200-Å pore size, Sigma Chemical) as follows: The glass beads are silanized with 3-aminopropyltriethoxysilane (pH 4.0) and succinylated with succinic anhydride before being activated by 1-ethyl-3-(3-dimethyaminopropyl)carbodiimide. Solutions of xylose and TNXI in 100 mM MOPS (pH 7.0) are mixed to reach final concentrations of 600 mM xylose and 5 mg/ml TNXI. This solution is circulated through the carbodiimide-activated glass beads for 20 hr at 4°, and the nonadsorbed enzyme is removed by washing the glass beads with 1 M NaCl in 100 mM MOPS (pH 7.0).

Purification and Physical Properties of TMXI and TNXI

XI activity is expressed only when *T. maritima* is grown in the presence of xylose and in the absence of glucose. SDS–PAGE analysis and gel filtration (i.e., the last purification step) indicate that TMXI is a homotetramer composed of 45,000 molecular weight subunits. Its N-terminal sequence (AEFFPEIPKIQFEGKESTNPLAFRFYDPN) is highly similar to that of type II XIs.[9]

TNXI expression in *E. coli* is optimized by cloning the *xylA* gene into a pET vector.[10] With this construct, we are able to routinely obtain 50 mg/liter recombinant protein. Interestingly, TNXI is expressed as a mixture of

[9] S. H. Brown, C. Sjøholm, and R. M. Kelly, *Biotechnol. Bioeng.* **41,** 878 (1993).
[10] J. M. Hess, V. Tchernajenko, C. Vieille, J. G. Zeikus, and R. M. Kelly, *Appl. Environ. Microbiol.* **64,** 2357 (1998).

dimers and tetramers (ratio 20 : 1) in *E. coli*.[10] The two TNXI recombinant forms have biochemical properties similar to those of the native tetrameric enzyme (i.e., thermostability and pH and temperature optima).

Cloning and Characterization of *T. neapolitana xylA* Gene

The PCR-generated DNA probe used to screen the *T. neapolitana* genomic library is a 180-bp DNA fragment encoding a putative peptide, 80% identical to part of *T. thermosulfurigenes* XI. Among 15,000 clones screened, 6 hybridize strongly with the probe. Restriction analysis of these clones reveals that they all overlapped. The 2.5-kb *Sal*I-*Xba*I *T. neapolitana* genomic DNA insert in plasmid pTNE2[2] is sequenced. In this sequence, the 1332 nucleotide *T. neapolitana xylA* gene (GenBank accession No. L38994) encodes a 444 residue polypeptide (calculated molecular weight: 50,892) 70% identical to *T. thermosulfurigenes* XI. The *T. neapolitana xylA* gene is preceded by a Shine–Dalgarno sequence, GGAGGT, that is exactly complementary to the *T. maritima* 16S rRNA sequence 3′-CCUCCA-5′.[11] A potential promoter, TTGAA (−35) TATAAT (−10), corresponding to the consensus defined for *T. maritima*[12] is present 63 bp upstream of the ATG start codon. An inverted repeat located 213 bp downstream of *xylA* might be involved in transcription termination. The amino acid compositions of type II XIs have been compared: with the exception of a significant decrease in the Asn + Gln content, no obvious differences in amino acid composition could be detected between enzymes originating from mesophilic, moderately thermophilic, and hyperthermophilic organisms.

Effects of Temperature, pH, and Metals on TMXI and TNXI Activities and Stabilities

TMXI and TNXI activities increase exponentially with temperature in almost identical fashion (Fig. 1A) (E_a values of 19 kcal/mol [TMXI] and of 17.5 kcal/mol [TNXI]). The two enzymes differ, though, in their temperature ranges: whereas TNXI activity is optimal at 97° and starts decreasing at higher temperatures, TMXI activity increases until 105°, where it is optimally active. The two enzymes are inactive at pH 4.0 and below. Their activities increase with pH up to pH 7.0–8.0, a range in which they are optimally active (Fig. 1B). The effects of Mg^{2+}, Co^{2+}, and Mn^{2+} were tested on TMXI and TNXI activities (on fructose and glucose, respectively) using

[11] L. Achenbach-Richter, R. Gupta, K.-O. Stetter, and C. R. Woese, *Syst. Appl. Microbiol.* **9**, 34 (1987).

[12] D. Lao and P. P. Dennis, *J. Biol. Chem.* **267**, 22787 (1992).

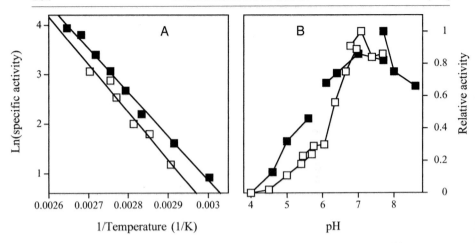

FIG. 1. Effects of temperature (A) and pH (B) on TMXI and TNXI activities.[2,9] ■, TMXI; □, TNXI.

the EDTA-treated enzymes. The three metals activate both enzymes. At any concentration, Co^{2+} is the best activating metal. Activity in the presence of Mg^{2+} is approximately 60% (TMXI) and 40% (TNXI) of the activity observed with the Co^{2+} enzyme. Poorly active, Mn^{2+} enzymes show only 20% (TMXI) and 16% (TNXI) of the activities observed with Co^{2+} enzymes.[9,13]

As shown in Table II, TMXI is a more stable enzyme than TNXI. TNXI inactivation at 120° is too fast to determine the enzyme half-life accurately. The stabilization provided by metals to TNXI is metal specific: the Mn^{2+} enzyme is significantly more stable than the Co^{2+} and Mg^{2+} enzymes at 101° (Table II).[13] The melting point of apo-TNXI (88°)[2]—determined by DSC—is in good agreement with its half-life at 90° (Table II). In the presence of Mg^{2+} and Co^{2+}, TNXI shows two melting transitions (at 99 and 109°). These transitions could correspond either to the existence of two phases in TNXI inactivation[14] or to the presence in the solution of a heterogeneous population of enzymes containing Mg^{2+}, Co^{2+}, or Mn^{2+} as ligand(s). Indeed, the enzyme is not EDTA treated prior to that experiment and, as seen in Table II, metals affect TNXI stability and most certainly its melting temperature to different degrees.

[13] C. Vieille and J. G. Zeikus, unpublished results (1999).
[14] J. M. Hess and R. M. Kelly, *Biotechnol. Bioeng.* **62,** 509 (1999).

TABLE II
EFFECTS OF TEMPERATURE AND METALS ON TMXI
AND TNXI STABILITIES[a]

Enzyme[b]	Metal	Half-life
TMXI	5 mM MgCl$_2$	10 min at 120°
TNXI	None	10 min at 90°
TNXI	5 mM MgCl$_2$	16 min at 101°
TNXI	2 mM CoCl$_2$	23 min at 101°
TNXI	5 mM MnCl$_2$	115 min at 101°

[a] From Brown et al.[9]
[b] Both enzymes were EDTA treated before the thermostability studies.

TMXI and TNXI Kinetic Properties

TMXI and TNXI kinetic parameters were determined on xylose, glucose, and fructose (Table III). As other XIs, TMXI and TNXI have better affinity for xylose than for glucose (K_m values of 15.9 mM versus 88.5 mM at 90°). Kinetic parameters defined for these enzymes are comparable to those defined for other XIs at their respective temperatures.[2] Interestingly, TNXI is a better catalyst than TMXI: its catalytic efficiency is between 1.8 and 3 times higher than that of TMXI, depending on the substrate (Table III).

TABLE III
KINETIC CONSTANTS OF TMXI AND TNXI[a]

Enzyme	Substrate	Temperature (°)	V_{max} (U/mg)	k_{cat}[b] (min^{-1})	K_m (mM)	k_{cat}[b]/K_m (mM^{-1} min^{-1})
TMXI	Xylose	90	68.4	3478	74	47.0
	Glucose	90	16.2	824	118	7.0
	Fructose	90	20.5	1042	225	4.6
TNXI	Xylose	90	52.2	2654	15.9	166.8
		98	75.0	3820	52.1	73.4
	Glucose	90	22.4	1139	88.5	12.9
		98	21.2	1079	159.5	6.8
	Fructose	90	21.0	1070	106.5	10.0
		98	28.5	1449	260.7	5.6
Immobilized TNXI	Xylose	90	56.6	2847	27	105
	Glucose	90	22	1119	115	9.7

[a] From Vieille et al.,[2] Brown et al.,[9] and Hess and Kelly.[14]
[b] The k_{cat} shown is the number of substrate molecules reacted per active site per min.

Properties of Immobilized TNXI

TNXI is immobilized at approximately 22 mg per gram of support.[14] Optimally active at 95° and pH 7.0, the immobilized TNXI has kinetic parameters very similar to those of the soluble enzyme (Table II).

Conclusion

Thermotoga XIs are very thermostable enzymes that may serve as a molecular platform to develop improved, ultrathermostable industrial enzymes by genetic engineering.

[13] β-Mannanases from *Thermotoga* Species

By Swapnil Chhabra, Kimberley N. Parker, David Lam, Walter Callen, Marjory A. Snead, Eric J. Mathur, Jay M. Short, and Robert M. Kelly

Introduction

Mannans are linear polymers of β-1,4-linked mannopyranose units. They may be present as reserve polysaccharides in the endosperm of palm nuts such as ivory nut (*Phytelephas macrocarpa*), in the seeds of leguminous plants such as Lucerne (*Medicago sativa*), in coffee beans, and in roots and tubers of many plant species.[1] In some hetero-1,4-β-D-mannans, i.e., galactomannans, the primary hydroxyl group of the backbone residues (containing β-1,4-linked mannopyranose units) is substituted by α-linked galactose. Galactoglucomannan (present in soft woods) contains β-1,4-linked D-mannopyranose and D-glucopyranose units as backbone residues. The main-chain residues are partially substituted by α-1,6-linked D-galactosyl side groups.[1]

Endo-1,4-β-D-mannanase (EC 3.2.1.78) catalyzes the random cleavage of β-D-1,4-mannopyranosyl linkages within the main chain of galactomannan, glucomannan, galactoglucomannan, and mannan.[2] The complete conversion of galactomannan into D-galactose and D-mannose requires the presence of at least two additional enzymes: α-galactosidase (EC 3.2.1.22)

[1] R. F. H. Dekker and G. N. Richards, *Adv. Carbohydr. Chem. Biochem.* **32,** 277 (1976).
[2] H. Meier and J. S. G. Reid, *in* "Plant Carbohydrates I: Intracellular Carbohydrates" (F. A. Loewus and W. Tanner, eds.), Vol. 13A, p. 418. Springer-Verlag, New York, 1982.

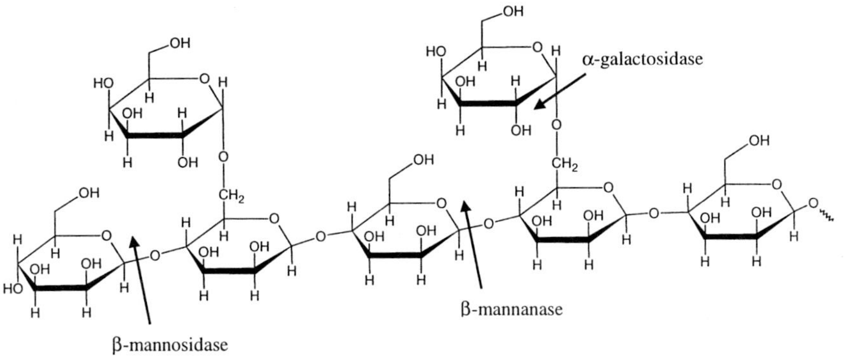

FIG. 1. Schematic representation of galactomannan structure. The nonreducing end of the polysaccharide is shown. Arrows represent glycosidic links recognized by β-mannanase, β-mannosidase, and α-galactosidase. Adapted from G. D. Duffaud, C. M. McCutchen, P. Leduc, K. N. Parker, and R. M. Kelly, *Appl. Environ. Microbiol.* **63**, 169 (1997) (with permission from the publisher).

and β-mannosidase (EC 3.2.1.25).[3] These enzymes catalyze the cleavage of terminal α-1, 6-linked D-galactosyl and β-1,4-linked D-mannopyranosyl residues, respectively (Fig. 1).

Amino acid sequences of mannanases reported to date fall into two distinct families of glycosylhydrolases: family 5 and family 26. Both families belong to the clan GH-A and share a common $(\alpha/\beta)_8$ barrel fold, three conserved residues, and a retaining cleavage mechanism.[4] Examples of mannanases from both families are listed in Tables I and II. Mannanases from family 5 share eight strictly conserved amino acid residues with several endoglucanases from the same family.[5–7] Mannanases from family 26 share five strictly conserved residues with each other.[8] Both families contain multidomain enzymes; i.e., in addition to the catalytic domain these enzymes contain other functional domains.

Thermostable mannanases have been identified from *Thermotoga neapolitana*,[3] *Rhodothermus marinus*,[9] *Bacillus stearothermophilus*,[7] *Thermo-*

[3] G. D. Duffaud, C. M. McCutchen, P. Leduc, K. N. Parker, and R. M. Kelly, *Appl. Environ. Microbiol.* **63**, 169 (1997).

[4] B. Henrissat and G. Davies, *Curr. Opin. Struct. Biol.* **34**, 637 (1997).

[5] Q. Wang, D. Tull, A. Meinke, N. R. Gilkes, R. A. J. Warren, R. Aebersold, and S. G. Withers, *J. Biol. Chem.* **268**, 14096 (1993).

[6] M. Hilge, S. M. Gloor, W. Rypniewski, O. Sauer, T. D. Heightman, W. Zimmermann, and K. Winterhalter, *Structure* **6**, 1433 (1998).

[7] N. Ethier, G. Talbot, and J. Sygusch, *Appl. Environ. Microbiol.* **11**, 4428 (1998).

[8] D. N. Bolam, N. Hughes, R. Virden, J. H. Lakey, G. P. Hazlewood, B. Henrissat, K. L. Braithwaite, and H. J. Gilbert. *Biochemistry* **35**, 16195 (1996).

[9] J. Gomes, and W. Steiner, *Biotechnol. Lett.* **20**, 729 (1998).

TABLE I
β-MANNANASES FROM GLYCOSYLHYDROLASE FAMILY 5

Organism	T_{opt} (°)	pH$_{opt}$	Molecular mass (kDa)	GenBank accession number	Ref.
Agaricus bisporus	—	—	44	Z50095	a
Aspergillus aculeatus	—	—	41	L35487	b
Lycopersicon esculentum	—	—	39	AF017144	c
Bacillus circulans K-1	40	6.0–7.0	62	AB007123	d
Vibrio sp. strain MA-138	45	7.5	43	D86329	e
Streptomyces lividans 66	58	6.7	36	M92297	f
Trichoderma reesei	60	5.0	44	L25310	g
Bacillus stearothermophilus	70	6.5	76	AF038547	h
Thermoanaerobacterium polysaccharolyticum	75	5.8	120	U82255	i
Caldicellulosiruptor saccharolyticus (formerly Caldocellum saccharolyticum)	80	6.0	39	M36063	j
Thermobifida fusca	80	—	38	AJ006227	k
Thermotoga neapolitana 5068	85–90	7.1	77	Ref. 16	l

[a] E. Yague, M. Mehak-Zunic, L. Morgan, D. A. Wood, and C. F. Thurston, *Microbiology* **143,** 239 (1997).

[b] S. Christgau, S. Kauppinen, J. Vind, L. V. Kofod, and H. Dalboge, *Biochem. Mol. Biol. Int.* **33,** 917 (1994).

[c] J. D. Bewley, R. A. Burton, Y. Morohashi, and G. B. Fincher, *Planta* **203,** 454 (1997).

[d] S. Yosida, Y. Sako, and A. Uchida, *Bosci. Biotech. Biochem.* **61,** 251 (1997).

[e] Y. Tamaru, T. Araki, T. Morishita, T. Kimura, K. Sakka, and K. Ohmiya, *J. Ferment. Bioeng.* **83,** 201 (1997).

[f] N. Arcand, D. Kluepfel, F. W. Paradis, R. Morosoli, and F. Shareck, *Biochem. J.* **290,** 857 (1993).

[g] H. Stålbrand, A. Saloheimo, J. Vehmaanpera, B. Henrissat, and M. Penttila, *Appl. Environ. Microbiol.* **61,** 1090 (1995).

[h] N. Ethier, G. Talbot, and J. Sygusch, *Appl. Environ. Microbiol.* **11,** 4428 (1998).

[i] I. Cann, S. Kocherginskaya, M. King, B. White, and R. Mackie, *J. Bacteriol.* **181,** 1643 (1999).

[j] E. Luthi, N. B. Jasmat, R. A. Grayling, D. R. Love, and P. L. Bergquist, *Appl. Environ. Microbiol.* **57,** 694 (1991).

[k] M. Hilge, S. M. Gloor, W. Rypniewski, O. Sauer, T. D. Heightman, W. Zimmermann, and K. Winterhalter, *Structure* **6,** 1433 (1998).

[l] C. M. McCutchen, G. D. Duffaud, P. Leduc, A. R. H. Peterson, A. Tayal, S. A. Khan, and R. M. Kelly, *Biotechnol. Bioeng.* **52,** 332 (1996).

TABLE II
β-MANNANASES FROM GLYCOSYLHYDROLASE FAMILY 26

Organism	T_{opt} (°)	pH_{opt}	Molecular mass (kDa)	GenBank accession number	Ref.
Bacillus subtilis	—	—	37	D37964	a
Thermotoga neapolitana	—	—	40	U58632	b
Pseudomonas fluorescens	37	6.5	47	X82179	c
Piromyces sp.	37	6.5	68	X91857	d
Cellulomonas fimi	42	5.5	100	AF126471	e
Bacillus sp.	60	9.0	55	M31797	f
Caldicellulosiruptor saccharolyticus Rt8B.4	60–65	6.0–6.5	103	U39812	g
Dictyoglomus thermophilum	80	5.0	40	AF013989	h
Rhodothermus marinus	85	5.4	113	X90947	i

[a] N. S. Mendoza, M. Arai, K. Sugimto, M. Ueda, T. Kawaguchi, and L. M. Joson, *Biochim. Biophys. Acta* **1243**, 552 (1995).

[b] D. Yernool, R. Sullivan, and D. Eveleigh, GenBank Accession No. U58632. Direct submission (1997).

[c] K. L. Braithwaite, G. W. Black, G. P. Hazlewood, B. R. S. Ali, and H. J. Gilbert, *Biochem. J.* **305**, 1005 (1995).

[d] C. Fanutti, T. Ponyi, G. W. Black, G. P. Hazlewood, and H. J. Gilbert, *J. Biol. Chem.* **270**, 29314 (1995).

[e] D. Stoll, H. Stålbrand, and R. A. J. Warren, *Appl. Environ. Microbiol.* **65**, 2598 (1999).

[f] T. Akino, C. Kato, and K. Horikoshi, *Appl. Environ. Microbiol.* **55**, 3178 (1989).

[g] M. D. Gibbs, A. U. Elinder, R. A. Reeves, and P. L. Bergquist, *FEMS Microbiol. Lett.* **141**, 37 (1996).

[h] M. D. Gibbs, R. A. Reeves, A. Sunna, and P. L. Bergquist, *Curr. Microbiol.* **39**, 351 (1999).

[i] O. Politz, M. Krah, K. K. Thomsen, and R. Borriss, *Appl. Microbiol. Biotechnol.* **53**, 715 (2000).

anaerobacterium polysaccharolyticum,[10] Caldocellosiruptor saccharolyticus,[11] and Dictyoglomus thermophilum.[12] The β-mannanases from T. neapolitana (family 5) and R. marinus (family 26) are the most thermostable of these. The mannanase from T. neapolitana is less stable at 85°, but more stable at 90°, than the corresponding enzyme from R. marinus. The former has a half-life of 34 hr at 85° and 13 hr at 90°,[3] whereas the latter has a half-life of 45.3 and 4.5 hr at the respective temperatures.[9] Hyperthermophilic mannanases have been found to be useful in several industrial applications

[10] I. Cann, S. Kocherginskaya, M. King, B. White, and R. Mackie, *J. Bacteriol.* **181**, 1643 (1999).

[11] T. Frangos, D. Bullen, P. Bergquist, and R. Daniel, *Int. J. Biochem. Cell Biol.* **31**, 853 (1999).

[12] M. D. Gibbs, R. A. Reeves, A. Sunna, and P. L. Bergquist, *Curr. Microbiol.* **39**, 351 (1999).

where thermostability and thermoactivity are desirable. These include coffee extraction,[13] oil/gas well stimulation,[14] and wood pulp treatment.[15]

This article describes the purification and characterization of a β-mannanase from T. neapolitana,[3] as well as cloning and sequencing of this enzyme from T. neapolitana and T. maritima.[16]

β-D-Mannanase from Thermotoga neapolitana

Bacterial Strains and Culture Conditions

Thermotoga neapolitana strain DSM 5068[17] is obtained from Deutsche Sammlung von Mikroorganismen, Braunschweig, Germany. Artificial seawater (ASW),[18] supplemented with 0.1% (w/v) yeast extract, 0.5% (w/v) tryptone, 0.5% (w/v) lactose, and 0.03% (w/v) guar gum, is used as the culture medium. The medium composition is as follows (per liter): NaCl, 15.0 g; Na_2SO_4, 2.0 g; $MgCl_2 \cdot 6H_2O$, 2.0 g; $CaCl_2 \cdot 2H_2O$, 0.50 g; $NaHCO_3$, 0.25 g; K_2HPO_4, 0.10 g; KBr, 50 mg; H_3BO_3, 20 mg; KI, 20 mg; $Fe(NH_4)_2(SO_4)_2$, 15 mg; $Na_2WO_4 \cdot 2H_2O$, 3 mg; and $NiCl_2 \cdot 6H_2O$, 2 mg. The following compounds are added after sterilization: lactose, guar gum, K_2HPO_4, and $Fe(NH_4)_2(SO_4)_2$. Lactose, K_2HPO_4, and $Fe(NH_4)_2(SO_4)_2$ are filter-sterilized by passage through a 0.2-μm filter prior to addition. Cultures are prepared by heating the media-containing bottles to 98° for 30 min, followed by sparging with N_2 and the addition of $Na_2S \cdot 9H_2O$ (0.5 g/liter) from a 50-g/liter stock solution. These are cooled to 80° and grown in closed bottles (125 ml) under anaerobic conditions.

Biomass Production

Biomass corresponding to 150 liters of culture is generated in the ASW-based media. Lactose (0.5%, w/v) and guar gum (0.03%, w/v) are used as supplements. Cells are grown using an 8-liter fermentor (Bioengineering Lab Fermentor type 1 1523, Basel, Switzerland) in semibatch mode. The

[13] K. K. Y. Wong and J. N. Saddler, in "Hemicellulose and Hemicellulases" (M. P. Coughlan, and G. P. Hazlewood, eds.), p. 127. Portland Press Research Monograph IV, London, 1993.

[14] C. M. McCutchen, G. D. Duffaud, P. Leduc, A. R. H. Peterson, A. Tayal, S. A. Khan, and R. M. Kelly, Biotechnol. Bioeng. 52, 332 (1996).

[15] P. L. Bergquist, M. D. Gibbs, D. J. Saul, R. A. Reeves, D. D. Morris, and V. S. J. Te'o, in "Enzyme Applications in Fiber Processing" (A. Eriksson, ed.), Vol. 687, p. 155. ACS Symposium Series, Washington, DC, 1998.

[16] K. N. Parker, S. R. Chhabra, M. D. Burke, D. Lam, W. Callen, G. D. Duffaud, M. A. Snead, E. J. Mathur, S. A. Khan, and R. M. Kelly, in preparation.

[17] H. W. Jannasch, R. Huber, S. Belkin, and K. O. Stetter, Arch. Microbiol. 150, 103 (1988).

[18] S. H. Brown, C. Sjoholm, and R. M. Kelly, Biotechnol. Bioeng. 41, 878 (1993).

temperature of the mixture and the agitation rate are controlled at 85 ± 2° and 150 rpm, respectively. Anaerobic conditions are maintained by a continuous flow of N_2 at a rate of approximately 5 liter/min. The cells are harvested in late exponential phase ($1.5–2.0 \times 10^8$ cells/ml). Growth is determined by epifluorescent microscopy using acridine orange stain[19] and monitored by cell density enumeration. The culture (150 liters) is chilled and concentrated to approximately 20 liters using a 0.45-μm Pellicon cross-flow filter (Millipore Corp., Bedford, MA). The retentate is centrifuged at 10,000g for 30 min and the pelleted cells are frozen at $-20°$ for later use. Spent media (extracellular material, approximately 130 liters) are further concentrated to 1 liter using a 0.22-μm Pellicon cross-flow filter (Millipore Corp., Bedford, MA) at room temperature.

Purification of Native T. neapolitana β-Mannanase

The native enzyme is purified from extracellular material by fast protein liquid chromatography (FPLC) (Pharmacia, Uppsala, Sweden). All procedures are performed under aerobic conditions and at room temperature. Samples are stored at 4° after processing. All buffers are degassed and filtered through a 0.22-μm filter prior to use. All columns and column media are obtained from Pharmacia.

Step 1. Hydrophobic Interaction Chromatography. Ammonium sulfate (final concentration 1.0 M) is added to the concentrated extracellular material of the *T. neapolitana* cell culture (150 ml) and this material is applied to a column (XK26/70, bed volume 200 ml) of butyl Sepharose previously equilibrated with 1.0 M $(NH_4)_2SO_4$ in 0.1 M sodium phosphate buffer, pH 7.4, and 0.005% (w/v) *N*-lauroylsarcosine (NLS) (buffer A). The column is washed with the same buffer until no protein is detected in the eluent (monitored by measuring absorbance at 280 nm). A two-step gradient is used to elute the bound protein. The concentration of $(NH_4)_2SO_4$ is first decreased to 0.35 M and kept constant at this value for two-column volumes before resuming the gradient.

Step 2. Anion-Exchange Chromatography (Mono Q). Fractions from step 1 containing β-mannanase activity are pooled (20 ml) and loaded on a Mono Q column equilibrated in buffer A. The column is washed with the same buffer until no protein is detected in the eluent (monitored by measuring absorbance at 280 nm). The column is eluted with a multistep NaCl salt gradient in the same buffer A. The multistep salt gradient used is 0–0.08 M (1 ml), 0.08–0.17 M (5 ml), 0.17–0.20 M (1 ml), and 0.20–1.0 M (1 ml). The salt gradients are separated by plateaus at the following salt

[19] J. E. Hobbie, R. J. Daley, and S. Jasper, *Appl. Environ. Microbiol.* **33,** 1225 (1977).

concentrations: 0.08 M (11 ml), 0.17 M (9 ml), and 0.2 M (15 ml). Fractions containing β-mannanase activity are pooled and stored at 4°.

Gel Electrophoresis

SDS–PAGE. Cell extracts and purified protein are boiled for 10 min with sample buffer [Tris, pH 8, 0.1 M; SDS 1% (w/v); 2-mercaptoethanol 10 mM; 10% glycerol (v/v); and bromphenol blue 0.001% (w/v)] and then separated by SDS–PAGE.[20] Proteins are visualized by staining with Coomassie Brilliant Blue. Prestained markers (BRL, Bethesda, MD; Sigma, St. Louis, MO) are used to estimate the molecular mass. A Pharmacia LKB PhastSystem, with 8–25% gradient Phastgels and SDS buffer strips, is used. Trichloroacetic acid (TCA) (10%, w/v) is added to precipitate the purified protein (mixture incubated on ice for 30 min) followed by centrifugation and removal of excess TCA by two acetone washes (0.5 ml). The precipitated protein is dried and resuspended in sample buffer (without glycerol) followed by separation on SDS–PAGE.

Biochemical Assay Method

Locust bean gum and azo-locust bean gum are used as substrates to assay β-mannanase activity. Protein concentrations are estimated by the method of Bradford[21] on samples with high protein concentration and by the bicinchonic acid method (BCA)[22] for samples with low protein concentration using standard assay kits from Bio-Rad (Richmond, CA) and Pierce (Rockford, IL), respectively. Bovine serum albumin is used as a standard in both cases.

Locust Bean Gum (LBG). Assays based on an increase in the concentration of reducing sugar equivalents give a direct measure of the number of glycosidic bonds hydrolyzed. One unit of activity is defined as the amount of enzyme that releases 1 μmol of reducing sugar equivalent per minute at a defined pH and incubation temperature. An adaptation of the Somogyi–Nelson method[23] is used for activity assays of the native β-mannanase. LBG (0.05 to 0.2%, w/v) is used as the substrate. The substrate solution (0.5 ml) is heated at the assay temperature, and a sample of β-mannanase (50 μl) is added to initiate the reaction. The reaction is stopped by immersing the tube in ice water. The standard curve for reducing equivalents is prepared using mannose as the substrate.

[20] U. K. Laemmli, *Nature* **227,** 680 (1970).
[21] M. Bradford, *Anal. Biochem.* **72,** 248 (1976).
[22] P. K. Smith, R. I. Krohn, G. T. Hermanson, A. K. Mallia, and F. H. Gartner, *Anal. Biochem.* **150,** 76 (1985).
[23] B. V. McCleary, *Methods Enzymol.* **160,** 596 (1988).

Azo-LBG. Azo-LBG (Megazyme, Wicklow, Ireland) is used as the chromogenic substrate to measure the activity of native β-mannanase. This substrate is partially depolymerized carob galactomannan dyed with Remazolbrilliant Blue R as the chromogenic substrate (Megazyme). The polysaccharide is dyed to an extent of about one dye molecule per 30 sugar residues. Digestion of the substrate by mannanase breaks the mannose chain, releasing smaller and more soluble galactomannan fragments, which can be monitored spectrophotometrically because of the presence of the chromogenic dye. The assay procedure supplied by manufacturer (http://www.megazyme.com/) is summarized here. The temperature of reaction is changed from the specified 40° to a higher value (60–90°). Preheated enzyme solution (0.25 ml) is added to preheated substrate solution (0.25 ml, 2%, w/v) and the mixture is stirred on a vortex mixer for 5 sec and incubated at the appropriate temperature. The reaction is terminated after 30 min by placing the tubes in ice-cold water, and the high molecular weight substrate is precipitated by the addition of 1.25 ml of ethanol (95%, v/v) with vigorous stirring for 10 sec on a vortex mixer. The samples are then centrifuged at 10,000g for 10 min. The absorbance of the supernatant of the blank and reaction solutions is measured at 590 nm using a UV-Vis spectrophotometer (Lambda2, Perkin-Elmer, Norwalk, CT). One unit of activity is defined as the amount of enzyme that increases the absorbance by 1 unit in 1 min (ΔOD_{590} min^{-1} mg^{-1}). This assay is used for screening column fractions during purification of the β-mannanase as well as determining the pH and temperature optima of the enzyme. Conversion to reducing-sugar equivalents (μmol of reducing-sugar equivalent released per minute and per milligram of protein) is done using a standard curve provided by the manufacturer.

Determination of Biochemical and Biophysical Properties

Temperature and pH Optima. The pH dependence of the enzyme is investigated by measuring the specific activities of the enzymes using azo-LBG as the substrate, with substitution of the appropriate buffer. This substrate is prepared as a 2% (w/w) stock in water and diluted to 1% with the appropriate buffer (100 mM). The buffers used at the indicated pH values are pH 5.5 and 6.7, [2-(N-morpholino)ethanesulfonic acid] (MES) buffer; pH 6.5 and 7.9, [3-(N-morpholino)propanesulfonic acid] (MOPS) buffer; pH 7.3 and 8.7, [N-(2-hydroxyethyl)piperazine-N'-(3-propanesulfonic acid)] (EPPS) buffer; pH 8.6 and 10, [2-(N-cyclohexylamino)ethanesulfonic acid] (CHES); and between pH 9.7 and 11.1, [3-(cyclohexylamino)propanesulfonic acid] (CAPS). NaCl (50 mM) is also added to the buffer. pH values are determined by a pH meter (Fisher Accumet 15,

Fisher Scientific, Pittsburgh, PA) at the appropriate temperature. The specific activities are measured at various temperatures from 40 to 90° and at the optimal pH of the enzyme using the chromogenic substrate azo-LBG. NaCl (50 mM) is also added to the buffer used for the assay. Thermostability is determined by incubating the enzyme for various lengths of time at different temperatures (at the optimal pH), cooling it on ice, and measuring the residual activity.

Isoelectric Point. The isoelectric point is evaluated by isoelectrofocusing on a Phastgel IEF 3-9 (Pharmacia) against IEF markers (Sigma).

Kinetic Parameters. The Michaelis–Menten kinetic parameters, V_{max} and K_m, are calculated for the native mannanase. K_m is estimated from the substrate concentration at which the reaction rate is half of the maximum value. Activity is determined using standard assays. LBG or azo-LBG are used as substrates.

Recombinant β-D-Mannanases from *T. neapolitana* and *T. maritima*

Bacterial Strains and Culture Conditions

Thermotoga neapolitana (DSM 5068) and *T. maritima* (DSM 3109) are obtained from Deutsche Sammlung von Mikroorganismen and Zellkulturen, Braunschweig, Germany. Cell cultures are grown as described earlier. Genomic DNA from *T. neapolitana* and *T. maritima* is purified from 0.5 g each of cell mass using the cetyltrimethylammonium bromide method (Sigma).[24]

Genomic DNA (50 μg) is mechanically sheared by vigorous passage through a 25-gauge double-hub needle (Popper, New Hyde Park, NY) attached to 1-cm^3 syringes. A small aliquot (0.5 μg) is electrophoresed through a 0.8% (w/v) agarose gel to confirm that the majority of the sheared DNA is within the desired range (3–8 kb). The DNA fragments are made blunt ended by treatment with mungbean nuclease (Stratagene Cloning Systems, La Jolla, CA). Naturally occurring *Eco*RI sites are protected by methylation with *Eco*RI methylase (New England Biolabs, Beverly, MA). Double-stranded *Eco*RI linkers (GGAATTCC) are ligated to the blunt DNA fragments with T4 DNA ligase (New England Biolabs) and then cut back with *Eco*RI restriction endonuclease (New England Biolabs). The DNA is size fractionated on a sucrose gradient following a standard protocol.[24] Fractions (500 μl) are collected and 20 μl of each are analyzed on a 0.8% (w/v) agarose gel. Fractions containing DNA fragments of (3–8 kb)

---

[24] P. Gerhardt, R. Murray, W. Wood, and N. Krieg, "Methods for General and Molecular Bacteriology." American Society for Microbiology, Washington, DC, 1994.

are pooled and ligated to prepare λ insertion vectors using T4 DNA ligase (New England Biolabs). These vectors are Lambda ZAP II (Stratagene Cloning Systems) and lambda (λ)gt11 (Stratagene Cloning Systems), which are digested with *Eco*RI and dephosphorylated. The ligation reactions are packaged using Gigapack III λ packaging extract (Stratagene Cloning Systems), and 2.5×10^5 plaque-forming units (pfu) are amplified on NZY plates using the appropriate *Escherichia coli* hosts; Lambda ZAP II is plated on XL1-Blue MRF' *E. coli* strain and λgt11 is plated on Y1088 *E. coli* strain.

The amplified lambda libraries are diluted in SM buffer [0.1 *M* NaCl, 50 m*M* Tris (pH 7.5), 0.02% gelatin, 10 m*M* MgSO₄] to achieve approximately 2000 pfu per plate. The Lambda ZAP II library is plated on the XL1-Blue MRF' host and λgt11 is plated on the Y1090r host in Luria broth (LB) top agarose containing 1 m*M* isopropylthiogalactoside (IPTG) onto LB agar plates and incubated at 39° overnight. Plates are removed from incubation and are replicated on nitrocellulose membranes (Stratagene Cloning Systems) and screened for activity. β-Mannanase expressing clones are identified by applying a substrate overlay to the plates. The substrate overlay is prepared by melting agarose (1%, w/v) in 100 m*M* potassium phosphate buffer (pH 7.0, 100 m*M* NaCl). Azo-Carob galactomannan (Megazyme) is added to 1% (w/v) when the agarose cools to about 70° and is stirred until dissolved. The overlay mixture is cooled to approximately 65° and about 6 ml is applied directly over plaques. Overlays are cooled to room temperature and then incubated at 70° for approximately 2 hr. Clones expressing the enzymes are identified as plaques surrounded by a white halo on a blue background. These are isolated from the nitrocellulose replicas and eluted in SM buffer.

Gene Sequencing and Computer Analysis of Corresponding Amino Acid Sequences

Amino acid sequences of recombinant β-mannanases from *T. neapolitana* and *T. maritima* are deduced from the genes encoding the enzymes. All sequencing reactions are performed by using either the Applied Biosystems dye primer or dye terminator cycle sequencing kit and a Model 373A automated DNA sequencer (Perkin-Elmer Corp., Norwalk, CT). N-terminal amino acid sequences are determined at the University of Georgia's sequencing facility (by Edman degradation). These sequences are compared with other proteins in the GenBank database at http://www.ncbi.nlm.nih.gov/ using the BLAST[25] program available at the same

---

[25] S. F. Altschul, W. Gish, W. Miller, E. W. Myers, and D. J. Lipman, *J. Mol. Biol.* **215**, 403 (1990).

website. Sequence alignments are carried out using CLUSTALW available at http://www2.ebi.ac.uk/clustalw/.

Biochemical and Biophysical Properties

Native β-mannanase is purified from the spent medium of a *T. neapolitana* culture. Table III shows the results for the isolation of the extracellular β-mannanase. The enzyme is purified 4.8-fold in two steps with a 13.6% yield. SDS–PAGE of TCA-precipitated β-mannanase reveals a band corresponding to 65 kDa (data not shown). Note that the molecular masses calculated from the β-mannanase sequences are 76.6 and 76.9 kDa for *T. neapolitana* and *T. maritima*, respectively. Although these values are very close to one another, the molecular mass is slightly higher than that found for the native *T. neapolitana* enzyme. Evidence that the β-mannanase from *T. neapolitana* is extracellular is found in the upstream leader sequence to the *manA* gene in *T. neapolitana*. The signal peptide MRKLVFSFLIVTL-PIVLFAN in the upstream leader sequence is found to compare to typical signal peptides from microorganisms consisting of a positively charged N terminus, a central hydrophobic core, and a C-terminal cleavage region. The cleavage site of the signal peptide and the mature protein is believed to be between the asparagine at amino acid 20 and the serine at amino acid 21.

The native β-mannanase from *T. neapolitana* has a pH optimum of 7.1 (Fig. 2) and an isoelectric point of 5.1 ± 0.1. This enzyme has an optimum temperature of 90–92° (Fig. 2) and exhibits significant activity over a wide range of temperatures; e.g., it exhibits 30% of its maximum activity at 60°. It has a half-life of 34 hr at 85°, 13 hr at 90°, and 35 min at 100°. The Michaelis–Menten parameters, K_m and V_{max}, are estimated to be 0.55 mg/ml and 3.8 μmol reducing-sugar equivalent/min/mg, respectively, using LBG as the substrate. Using azo-LBG, the K_m is estimated to be 0.23 mg/

TABLE III

PURIFICATION OF β-1,4-MANNANASE FROM *Thermotoga neapolitana* 5068

Purification	Activity (units)	Specific activity (units/mg)[a]	Purification (-fold)	Yield (%)
Concentrated culture supernatant	38.5	0.8	1	100
Butyl-Sepharose	12.4	2.6	3.2	32.2
Mono Q	5.2	3.8	4.8	13.6

[a] Units/mg = (μmol reducing sugar-equivalent/min)/mg protein.

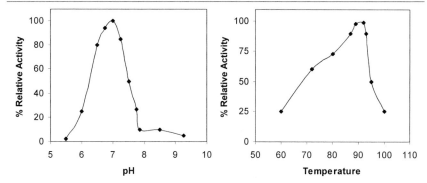

FIG. 2. Effect of pH and temperature on the activity of β-mannanase from *Thermotoga neapolitana* 5068. Adapted from G. D. Duffaud, C. M. McCutchen, P. Leduc, K. N. Parker, and R. M. Kelly, *Appl. Environ. Microbiol.* **63,** 169 (1997) (with permission from the publisher).

ml. Although the value of K_m for the β-mannanase from *T. neapolitana* is within the range of previously reported β-mannanases, the value of V_{max} seems to be significantly lower than that reported for other β-mannanases (Table IV). Kinetic comparisons on galactomannans are difficult as polymeric chains of mannose substituted with galactose, which are used as

TABLE IV
KINETIC CONSTANTS OF β-MANNANASES

Source organism	T_{opt} (°)	K_m (mg/ml)	V_{max} (μmol mannose equivalents per min-mg protein)	Ref.
Aspergillus tamarii	37	0.02	43	a
Streptomyces lividans	58	0.77	210	b
Polyporus versicolor	65	33.3	2500	c
Bacillus stearothermophilus	70	1.5	460	d
Sclerotium rolfsii	74	0.52	475	e
Caldocellum saccharolyticum	80	0.127	63	f
Thermotoga neapolitana	90	0.23–0.55	4	g

[a] A. Civas, R. Eberhard, P. Le. Dizet, and F. Petek, *Biochem. J.* **219,** 857 (1984).
[b] N. Arcand, D. Klucpfel, F. W. Paradis, R. Morosoli, and F. Shareck, *Biochem. J.* **290,** 857 (1993).
[c] K. G. Johnson and N. W. Ross, *Enzyme Microbiol. Technol.* **12,** 960 (1990).
[d] G. Talbot and J. Sygusch, *Appl. Environ. Microbiol.* **56,** 3505 (1990).
[e] G. M. Gubitz, M. Hayn, M. Sommerauer, and W. Steiner, *Biores. Tech.* **58,** 127 (1996).
[f] P. A. Bicho, T. A. Clark, K. Mackie, H. W. Morgan, and R. M. Daniel, *Appl. Microbiol. Biotechnol.* **36,** 337 (1991).
[g] G. D. Duffaud, C. M. McCutchen, P. Leduc, K. N. Parker, and R. M. Kelly, *Appl. Environ. Microbiol.* **63,** 169 (1997).

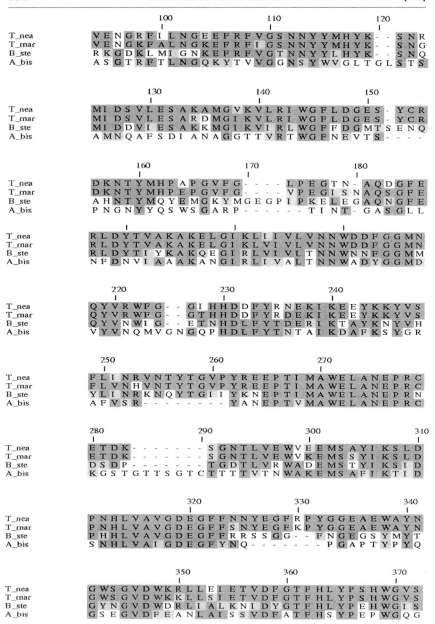

FIG. 3. Alignment of sequences of β-mannanases from *Thermotoga neapolitana* (T_nea) *Thermotoga maritima* (T_mar), *Bacillus stearothermophilus* (B_ste), and *Agaricus bisporus* (A_bis). The alignment is shown only over the conserved regions in each of the sequences.

```
                        380              390              400
                         |               |                |
T_nea   P E N Y A Q W G A K W I  E D H I  K I  A K E V G K P V V L  E E Y
T_mar   P E N Y A Q W G A K W I  E D H I  K I  A K E I G K P V V L  E E Y
B_ste   P E N V E K W G E Q Y I  L D H L  A A G K K A K K P V V L  E E Y
A_bis   - A D A K A W G T Q W I  T D H A  A S M K R V N K P V I  L E E F

                        410              420              430
                         |               |                |
T_nea   G I  P K S A P V N R V A I  Y K L  W N D L V Y N L G G N  G A M F
T_mar   G I  P K S A P V N R T A I  Y R L  W N D L V Y D L G G D  G A M F
B_ste   G I  S A T G V Q N R E M I  Y D T  W N R T M F E H G G T  G A M F
A_bis   G V T T N Q P D - - - - T  Y A E  W F N E I  E S S G L T  G D L I

                        440              450              460
                         |               |                |
T_nea   W M L  A G I  G E G W D - -  R D E K G Y Y P D Y D G F  R I  V N D
T_mar   W M L  A G I  G E G S D - -  R D E R G Y Y P D Y D G F  R I  V N D
B_ste   W L L  T G I  D D N P E S -  A D E N G Y Y P D Y D G F  R I  V N D
A_bis   W Q A G S H L  S T G D T P  N D G Y A V Y P D G P V Y P L V K S
```

FIG. 3. (*continued*)

substrates, vary with length and degree of substitution and are a function of the preparation method and the source of galactomannan.

The amino acid sequences of the mannanases from *T. neapolitana* and *T. maritima* are highly similar, with 85% identity.[16] Another β-mannanase from *T. neapolitana* (family 26) has been previously identified and sequenced,[26] although that sequence has little similarity to the two mannanases discussed here and is also much smaller (346 amino acids as compared to 666 and 669 amino acids). Furthermore, the *T. neapolitana* and *T. maritima* mannanases described here are only 20 and 18% identical, respectively, over a very short amino acid overlap to the mannanase sequence reported by Yernool *et al.*[26] The β-mannanases from *T. neapolitana* and *T. maritima* described here show highest similarity (46% identity) to the β-mannanase (ManF) from the thermophilic bacteria *Bacillus stearothermophilus,* which also has a molecular mass of 76 kDa. ManF, a multidomain enzyme from family 5, has been cloned and expressed in *E. coli.*[7] The *B. stearothermophilus* mannanase has an optimum pH of 7.5 and an optimum temperature of 70°. The enzymes from *T. maritima* and *T. neapolitana* show 35% amino acid sequence identity to the mannanase from the fungi *Agaricus bisporus* and approximately 33% amino acid sequence identities to the mannanases from the fungi *Aspergillus aculeatus* and *Trichoderma reesei,* all of which belong to family 5. A sequence alignment of these mannanases (Fig. 3) shows the eight conserved amino acid residues

[26] D. Yernool, R. Sullivan, and D. Eveleigh, Genbank Accession No. U58632. Direct submission (1997).

in the family 5 group. Following the primary sequence numbering scheme of the β-mannanase from *T. neapolitana* (parentheses indicate the corresponding amino acid from *T. maritima*), these eight residues include the following: Arg-72 (Arg-72), His-152 (His-153), Asn-197 (Asn-198), Glu-198 (Glu-199), His-278 (His-279), Tyr-280 (Tyr-281), Glu-317 (Glu-318), and Trp-350 (Trp-351) (positions 142, 230, 275, 276, 363, 365, 402, and 435 in Fig. 3, respectively).

Concluding Remarks

It is unclear how *Thermotoga* species utilize their mannan-degrading enzymes for processing mannans and galactomannans for nutritional purposes or why these polysaccharides are present in hyperthermophilic environments. These enzymes may be considered, at least *in vitro,* as catalysts that might be useful for the synthesis of β-mannopyranosides,[27,28] which are potential therapeutic agents.[29] *In vivo,* these mannanases may play a role in the degradation of extracellular polysaccharides produced by hyperthermophiles, which, in at least one case, was shown to be composed of mannan.[30]

Acknowledgments

RMK acknowledges the National Science Foundation and the U.S. Department of Agriculture for financial support of this research.

[27] N. Taubken and J. Thiem, *Synthesis* 517 (1991).
[28] H. Itoh and Y. Kamiyama, *J. Ferment. Bioeng.* **80,** 510 (1995).
[29] D. Zopf and S. Roth, *Lancet* **347,** 1017 (1996).
[30] K. D. Rinker and R. M. Kelly, *Appl. Environ. Microbiol.* **62,** 4478 (1996).

[14] β-Mannosidase from *Thermotoga* Species

By Kimberley N. Parker, Swapnil Chhabra, David Lam,
Marjorie A. Snead, Eric J. Mathur, and Robert M. Kelly

Introduction

β-Mannosidase (1,4-β-D-mannoside mannohydrolase, EC 3.2.1.25) is an exo-acting glycosylhydrolase whose function is to cleave mannose residues from the nonreducing termini of mannan oligosaccharides.[1] In microorganisms, β-mannosidases often act in concert with endo-acting β-mannanases to completely hydrolyze mannan-based carbohydrates to be subsequently used for nutritional purposes.[2,3] In mammals, the deficiency of mannosidase can lead to β-mannosidosis, a genetic disorder resulting from the storage and excretion of undegraded substrates.[4,5]

Hyperthermophilic β-mannosidases have been identified in archaea such as *Pyrococcus furiosus*[6] and in the bacteria *Thermotoga maritima*[7] and *T. neapolitana*.[8] The properties of the *P. furiosus* enzyme in relation to other glycosidases in the organism are discussed elsewhere in this volume.[9] Here, the protocols used to purify the β-mannosidase from *T. neapolitana* are presented together with the results of cloning and expression of genes encoding the analogous enzyme of *T. maritima* in *Escherichia coli*.

Methods

Bacterial Strains and Culture Conditions

Thermotoga maritima (DSM 3109) and *T. neapolitana* (DSM 5068) are obtained from Deutsche Sammlung von Mikroorganismen und Zellkulturen, Braunschweig, Germany. Cell cultures are grown on an artificial seawater (ASW)-based medium, supplemented with 0.1% (w/v) yeast extract, 0.5% (w/v) tryptone, and 0.2% (w/v) lactose.[10] The medium composition per liter is NaCl, 15.0 g; Na_2SO_4, 2.0 g; $MgCl_2 \cdot 6H_2O$, 2.0 g; $CaCl_2 \cdot 2H_2O$, 0.5 g; $NaHCO_3$, 0.25 g; K_2HPO_4, 0.1 g; KBr, 50 mg; H_3BO_3, 20 mg; KI, 20 mg; $Na_2WO_4 \cdot 2H_2O$, 3 mg; and $NiCl_2 \cdot 6H_2O$, 2 mg. Lactose and K_2HPO_4 should be added to the autoclaved media after being filtered through a 0.2-μm pore-size filter. Inocula are grown in closed bottles (125 ml) under anaerobic conditions at 80°. Cultures are prepared by heating media containing bottles to 98° for 30 min, sparged with N_2, and reduced

[1] B. McCleary, *Carb. Res.* **111**, 297 (1983).
[2] Y. Oda, T. Komaki, and K. Tonomura, *J. Ferm. Bioeng.* **10**, 353 (1993).
[3] G. M. Gubitz, M. Hayn, M. Sommerauer, and W. Steiner, *Biosource Tech.* **58**, 127 (1996).
[4] D. Wenger, E. Sujansky, P. Fennessey, and J. Thompson, *N. Engl. J. Med.* **315**, 1201 (1986).
[5] F. Percheron, M. Foglietti, M. Bernard, and B. Ricard, *Biochim.* **74**, 5 (1992).
[6] M. W. Bauer, E. Bylina, R. Swanson, and R. M. Kelly, *J. Biol. Chem.* **271**, 23749 (1996).
[7] K. E. Nelson *et al.*, *Nature* **399**, 323 (1999).
[8] G. D. Duffaud, C. M. McCutchen, P. Leduc, K. N. Parker, and R. M. Kelly, *Appl. Environ. Microbiol.* **63**, 169 (1997).
[9] T. Kaper, C. H. Verhees, J. H. G. Lebbink, J. F. T. van Lieshout, L. D. Kluskens, D. E. Ward, S. W. M. Kengen, M. M. Beerthuyzen, W. M. deVos, and J. van der Oost, *Methods Enzymol.* **330** [21] (2001) (this volume).
[10] S. H. Brown, C. Sjoholm, and R. M. Kelly, *Biotechnol. Bioeng.* **41**, 878 (1993).

with $Na_2S \cdot 9H_2O$ (0.5 g/liter) from a 50-g/liter stock solution. Cultures are cooled to 80° and then inoculated.

Purification of Native T. neapolitana β-Mannosidase

Biomass Production. Cells harvested from 150 liters of culture are used to purify the enzyme. The culture is grown in ASW-based media with 0.5% (w/v) lactose and 0.03% (w/v) guar gum as supplements in a semibatch fashion using an 8-liter fermentor (Bioengineering Lab Fermentor Type l 1523, Basel, Switzerland). Anaerobic conditions are established and maintained by a continuous flow of N_2 at approximately 5 liter/min. Temperature and agitation are controlled at $85 \pm 2°$ and 150 rpm, respectively. Growth is monitored by cell density and by epifluorescent microscopy using acridine orange stain.[11] Cells are harvested in late exponential phase ($1.5–2.0 \times 10^8$ cells/ml) and pelleted by centrifugation of the retentate at 10,000g for 30 min. Cells are frozen at $-20°$ for later use. After removing the cells, spent media (extracellular material) are concentrated down to approximately 1 liter using a 0.22-μm Pellicon cross-flow filter (Millipore Corp., Bedford, MA).

Preparation of Cell-Free Extract. Concentrated cell pellets of *T. neapolitana* 5068 are resuspended in 430 ml of 0.1 *M* sodium phosphate buffer, pH 7.4, and disrupted by one passage through a French pressure cell at 18,000 psig. NaN_3 (0.01%, w/v) is added to minimize bacterial contamination. Cell debris is removed by centrifugation at 10,000g for 30 min, and the supernatant is used as the crude enzyme preparation.

Enzyme Assay. β-Mannosidase activity is determined by measuring the release of *p*-nitrophenol (*p*-NP) from the substrate *p*-nitrophenyl-β-D-mannopyranoside (Sigma Chemical, St. Louis, MO) by visible absorption. Spectrophotometric measurements are made with a Lambda 3 spectrophotometer (Perkin-Elmer Corp., Norwalk, CT) equipped with a thermostatted and automated six-cell transport system. A liquid-circulating temperature bath (VWR Scientific Model 1130, Philadelphia, PA), containing a 1:1 mixture of ethylene glycol and water, is used to maintain the desired temperature in the cell holder, monitored by a thermocouple mounted in a cuvette that is placed in the cell transporter. The six-cell transporter is controlled and data are collected and analyzed using Perkin-Elmer computerized spectroscopy software (PECSS, Perkin-Elmer Corp.). Enzymatic assays for β-mannosidase are conducted as follows. Aliquots (1.1 ml) of 10 m*M* substrate in 0.1 *M* sodium phosphate buffer (pH 7.4) are pipetted into capped quartz cuvettes (Uvonic Instruments, Inc., Plainview, NY). The

---

[11] J. E. Hobbie, R. J. Daley, and S. Jasper, *Appl. Environ. Microbiol.* **33,** 1225 (1977).

cuvettes are incubated for at least 10 min to allow the substrate to reach assay temperature. After incubation, 0.1 ml of the enzyme sample is added to the cuvettes and mixed to start the reaction. The release of p-NP is measured by the change in absorbance at 405 nm. A blank containing the same amount of sample in 0.1 M sodium phosphate buffer (pH 7.4) is used as a control. At temperatures below 100° and pH below 9.5, the nonenzymatic release of p-NP is negligible. One unit of β-mannosidase activity is defined as the amount of enzyme releasing 1 μmol of p-NP/min under the specified assay conditions (μmol/min).

Protein determination is performed using two different methods. The method of Bradford[12] using the standard assay kit from Bio-Rad (Richmond, CA), with bovine serum albumin (BSA) as a standard, is used for assays on samples with high protein concentrations such as the cell extract. The bicinchonic acid method[13] using the standard BCA kit from Pierce (Rockford, IL) with bovine serum albumin (BSA) as a standard is used for samples with a low protein concentration, as well as for samples containing detergent.

Purification Protocol. β-Mannosidase is purified from cell-free extracts of *T. neapolitana* using a fast protein liquid chromatography unit (FPLC, Pharmacia, Uppsala, Sweden) to control column flow rates. All procedures are carried out at room temperature under aerobic conditions and samples are stored at 4° after processing. Prior to use, all buffers are filtered through a 0.22-μm filter. All columns and column media have been obtained from Pharmacia unless otherwise stated.

Purification of β-Mannosidase

Ion-Exchange Chromatography. Portions (200 ml) of the crude cell-free extract (0.2 U/mg) are applied to a column (XK50/30, bed volume 500 ml) packed with DEAE-Sepharose CL-6B (Sigma). The column is washed with buffer A [0.1 M sodium phosphate, pH 7.4, 0.005% (w/v) N-lauroylsarcosine (NLS)] until protein is no longer detected in the eluent. The remaining absorbed proteins are eluted with a two-step linear NaCl gradient [0–0.2 M (110 ml), 0.2–1.0 M (70 ml)] in the same buffer A. β-Mannosidase activity elutes at the 0.2 M NaCl (120 ml) plateau between the two gradients.

Hydrophobic Interaction Chromatography. Pooled fractions from the previous step containing β-mannosidase activity (0.9 U/mg) are applied to a column (XK26/70, bed volume 200 ml) packed with Phenyl-650M HIC Sepharose (TosoHaas, Montgomeryville, PA) and previously equilibrated

[12] M. Bradford, *Anal. Biochem.* **72,** 248 (1976).
[13] P. K. Smith, R. I. Krohn, G. T. Hermanson, A. K. Mallia, and F. H. Gartner, *Anal. Biochem.* **150,** 76 (1985).

with 4 M NaCl in buffer A in 100-ml aliquots. The column is washed with the same buffer until unbound protein is no longer detected in the eluent (monitored by measuring absorbance at 280 nm). The column is eluted with a four-step decreasing NaCl gradient [4.0-3.1 M (30 ml), 3.1-2.2 M (40 ml), 2.2-1.6 M (40 ml), 1.6-1.0 M (20 ml)] in buffer A and separated by three plateaus at 3.1 M (40 ml), 2.2 M (30 ml), and 1.6 M (20 ml) NaCl, respectively, in order to optimize peak resolution.

Sephacryl S-300 Gel Filtration Chromatography. Fractions from the previous chromatographic step containing β-mannosidase activity (1.9 U/mg) are pooled and concentrated approximately 10-fold using 10,000 molecular weight cutoff (MWCO) membranes on an Amicon 180 ml (Amicon, Beverly, MA) stirring cell. Portions (3 ml) of the concentrated material are applied to a gel filtration column (XK16/100, bed volume 180 ml, packed with Sephacryl S-300-HR) preequilibrated with 0.2 M NaCl in buffer A.

Anion-Exchange Chromatography (Mono Q). In order to remove NaCl and detergents, pooled fractions with β-mannosidase activity (47.9 U/mg) are dialyzed overnight against 0.1 M sodium phosphate buffer, pH 7.4. Fractions (3 ml) are applied to a 1 ml Mono Q column, preequilibrated with buffer A (without detergent). The column is washed with the same buffer until protein can no longer be detected in the eluent (monitored by measuring the absorbance at 280 nm). The bound proteins are then eluted with a linear NaCl salt gradient (0–0.4 M) in the same buffer A. Fractions containing β-mannosidase activity are analyzed by SDS–PAGE. Fractions showing a single band on the gel are pooled and stored at 4°.

After the anion-exchange step, a decrease in specific activity (34.0 U/mg) is observed, probably due to the removal of detergent from the enzyme preparation. In subsequent experiments where detergent is included prior to anion exchange (Mono Q–Buffer A), up to a 2-fold increase in specific activity is observed. However, because it was difficult to purify the protein to a homogeneous state (data not shown), this approach was abandoned. *T. neapolitana* β-mannosidase is purified 162-fold over the cell extract with a final yield of 35.4% (Table I).

Cloning of Gene Encoding T. maritima β-Mannosidase

Genomic DNA from *T. maritima* is purified from 0.5 g of cell mass using the cetyltrimethylammonium bromide (Sigma) method.[14] Genomic DNA (50 μg) is mechanically sheared by vigorous passage through a 25-gauge double-hub needle (Popper, New Hyde Park, NY) attached to 1 cm³ syringes. A small aliquot (0.5 μg) is electrophoresed through a 0.8% (w/v)

---

[14] F. M. Ausubel *et al.* (eds.), "Current Protocols in Molecular Biology." Wiley, New York, 1995.

TABLE I
PURIFICATION OF NATIVE β-MANNOSIDASE FROM *Thermotoga neapolitana* 5068

Step	Activity (units)	Specific activity (units/mg)	Purification (-fold)	Yield (%)
Cell-free extract	630	0.2	1	100
DEAE Sepharose	590	0.9	4.1	94
Phenyl Sepharose	700	1.9	9.2	111
Sephacryl S-300	460	47.9	228	73
Mono Q	220	34.0	162	35

agarose gel to confirm that the majority of the sheared DNA is within the desired size range (3–8 kb). The DNA fragments are made blunt ended by treatment with mungbean nuclease (Stratagene Cloning Systems, La Jolla, CA). Naturally occurring *Eco*RI sites are protected by methylation with *Eco*RI methylase (New England Biolabs, Beverly, MA) to ensure that the ends are blunt. Double-stranded *Eco*RI linkers (GGAATTCC) are ligated to the blunt DNA fragments with T4 DNA ligase (New England Biolabs). The DNA is size-fractionated on a sucrose gradient following a standard protocol.[14] Fractions (500 μl) are collected and 20 μl of each are analyzed on a 0.8% (w/v) agarose gel. Those fractions containing DNA fragments of 3–8 kb are pooled and ligated to prepared λ insertion vectors using T4 DNA ligase (New England Biolabs). These vectors are Lambda ZAP II (Stratagene Cloning Systems) and lambda (λ)gt11 (Stratagene Cloning Systems), which are digested with *Eco*RI and dephosphorylated. The ligation reactions are packaged using Gigapack III Lambda-packaging extract (Stratagene Cloning Systems), and 2.5×10^5 plaque-forming units are amplified on NZY plates using the appropriate *E. coli* hosts; Lambda ZAP II is plated on XL1-Blue MRF' *E. coli* strain and λgt11 is plated on Y1088 *E. coli* strain. Substrate assay plates are prepared by melting 1% (w/v) agarose in 100 mM potassium phosphate buffer (pH 7.0, 100 mM NaCl) and dissolving 7 mM *p*-nitrophenyl-β-D-mannopyranoside (Sigma Chemical) after cooling to approximately 60°. Substrate is poured into petri dishes and allowed to cool. β-Mannosidase screening is conducted by overlaying nitrocellulose replicas onto substrate assay plates. Nitrocellulose membranes are overlaid onto substrate plates and incubated at 72° for 5 to 20 min. Expressing clones are identified as yellow spots on a white background.

All sequencing reactions are performed by using either the Applied Biosystems dye primer or dye terminator cycle sequencing kit and a Model 373A automated DNA sequencer (Perkin-Elmer Corp.). N-terminal amino

TABLE II
AMINO ACID SEQUENCE HOMOLOGY OF *Thermotoga maritima* β-MANNOSIDASE TO
OTHER REPORTED SEQUENCES

Source	GenBank accession number	Number of amino acid residues in sequence	Identity to T. maritima (%) (sequence length for comparison)
Thermotoga maritima[a]	AE001806.1	785	99 (785)
Aspergillus aculeatus[b]	AB015509.1	937	26 (566)
Bos taurus[c]	Q29444	879	31 (713)
Caenorhabditis elegans[d]	Z78540.1	900	26 (566)
Capra hircus[e]	Q95327	879	32 (712)
Cellulomonas fimi[f]	AF126472.1	842	35 (769)
Streptomyces coelicolor[g] (putative)	AL031514.1	820	35 (705)
Pyrococcus furiosus[h]	U60214.1	510	14 (44)

[a] K. E. Nelson *et al., Nature* **399,** 323 (1999).
[b] G. Takada, T. Kawaguchi, T. Kaga, J. Sumitani, and M. Arai, *Biosci. Biotechnol. Biochem.* **63,** 206 (1999).
[c] H. Chen, J. R. Leipprandt, C. E. Traviss, B. L. Sopher, M. Z. Jones, K. T. Cavanagh, and K. H. Friderici, *J. Biol. Chem.* **270,** 3841 (1995).
[d] L. Matthews, GenBank direct submission accession No. Q93324 (1998).
[e] J. R. Leipprandt, S. A. Kraemer, B. E. Hitchcock, H. Chen, J. L. Dyme, K. T. Cavanagh, K. H. Friderici, and M. Z. Jones, *Genomics* **37,** 51 (1996).
[f] D. Stoll, H. Stalbrand, and R. A. Warren, *Appl. Environ. Microbiol.* **65,** 2598 (1999).
[g] M. Redenbach, H. M. Kieser, D. Denapaite, A. Eichner, J. Cullum, H. Kinashi, and D. A. Hopwood, *Mol. Microbiol.* **21,** 77 (1996).
[h] M. W. Bauer, E. Bylina, R. Swanson, and R. M. Kelly, *J. Biol. Chem.* **271,** 23749 (1996).

acid sequences are determined at the University of Georgia's sequencing facility (by Edman degradation). Sequences are aligned and edited using the Sequencher program (Genecodes, Ann Arbor, MI). Database searches are performed using the BLAST program[15] at the NCBI server (URL address: http://www.ncbi.nlm.nih.gov). Amino acid sequences are analyzed using the University of Wisconsin Genetics Computer Group (GCG) software package[16] and DNA Strider.[17]

Properties of *Thermotoga* β-Mannosidases

The translated amino acid sequence of the gene encoding the *T. maritima* β-mannosidase (TM1624[7]) contains 785 amino acids, with a calculated

[15] S. Altschul, W. Gish, W. Miller, E. Myers, and D. Lipman, *J. Mol. Biol.* **215,** 403 (1990).
[16] J. Devereux, P. Haerberli, and O. Smithies, *Nucleic Acids Res.* **12,** 387 (1984).
[17] C. Marck, *Nucleic Acids Res.* **16,** 1829 (1988).

TABLE III
PROPERTIES OF NATIVE β-MANNOSIDASE FROM
Thermotoga neapolitana 5068

Property	Value
Molecular mass	
SDS–PAGE	95 kDa
Native mobility	~200 kDa (α_2)
pH optimum	pH 7.7 (6.8–8.8)[a]
Isoelectric point	5.6
Temperature optimum	87° (72–96°)[b]
V_{max}	36.9 ± 2.5 (units/mg)
K_m	3.1 ± 0.6 mM
k_{cat}	4070 (min^{-1})
k_{cat}/K_m	1454 (mM^{-1} · min^{-1})

[a] pH range for 50% maximal activity.
[b] Temperature range for 50% maximal activity.

molecular mass of 92.3 kDa. Although the native enzyme from *T. maritima* is not available for direct comparison, the native β-mannosidase from *T. neapolitana* is slightly larger with subunits of 95–100 kDa (estimated by SDS–PAGE).[8] This apparent difference in the size of the two enzymes may be due to the inherent difficulty in determining the size of highly thermostable proteins with SDS–PAGE. Cloning and expression of the gene encoding β-mannosidase from *T. neapolitana* are underway in order to resolve this issue.[18]

Based on sequence analysis, *T. maritima* β-mannosidase has been classified as a glycosylhydrolase family 2 enzyme.[19] Several β-mannosidase sequences have been reported including representatives from each domain of life: archaeal, prokaryotic, and eukaryotic. The sequence of *T. maritima* β-mannosidase shows the highest similarity to those from the bacterial species *Cellulomonas fimi*[20] and *Streptomyces coelicolor* (putative)[21] (see Table II). However, the *T. maritima* enzyme shows low similarity to the hyperthermophilic β-mannosidase from *P. furiosus*.[6] Table II summarizes the amino acid sequence similarities of *T. maritima* β-mannosidases with other mannosidases.

[18] K. N. Parker, S. R. Chhabra, D. Lam, W. Callen, G. D. Duffaud, M. A. Snead, E. J. Mathur, and R. M. Kelly, submitted for publication.

[19] B. Henrissat and A. Bairoch, *Biochem. J.* **316**, 695 (1996).

[20] D. Stoll, H. Stalbrand, and R. A. Warren, *Appl. Environ. Microbiol.* **65**, 2598 (1999).

[21] M. Redenbach, H. M. Kieser, D. Denapaite, A. Eichner, J. Cullum, H. Kinashi, and D. A. Hopwood, *Mol. Microbiol.* **21**, 77 (1996).

The native β-mannosidase purified from *T. neapolitana* is extremely thermostable and thermoactive as might be expected of enzymes from this organism. It has a half-life (based on residual activity) of approximately 18 hr at 85°, 42 min at 90°, and 2 min at 98°. The intracellular localization of this enzyme in *T. neapolitana*[8] is supported by the lack of any obvious signal peptide encoded by the gene of *T. maritima*.[18] The biochemical and biophysical properties of the enzyme are summarized in Table III. The k_{cat}/K_m for the *T. neapolitana* β-mannosidase (1454 mM^{-1} min^{-1}) is comparable to the values reported for corresponding enzymes isolated from *Aspergillus* species: β-mannosidases from *Aspergillus niger*[22] at 870 mM^{-1} min^{-1} and from *Aspergillus awamori*[23] at 360 mM^{-1} min^{-1}. Preliminary efforts with recombinant β-mannosidases from *T. maritima* and *T. neapolitana* indicate that they have biochemical properties similar to those for the native *T. neapolitana* enzyme.[18] Although the role of these hemicellulases in hyperthermophiles is unclear, they may play a role in the degradation of extracellular polysaccharides produced by these organisms.[24]

Acknowledgments

We acknowledge financial support from the National Science Foundation and the Department of Agriculture for support of this research.

[22] S. Bouquelet, G. Spik, and J. Monteuil, *Biochim. Biophys. Acta* **522**, 521 (1978).

[23] K. N. Neustroev, A. S. Krylov, L. M. Firsov, O. N. Abroskina, and A. Y. Knorlin, *Biokhimiya* **56**, 1406 (1991).

[24] K. D. Rinker and R. M. Kelly, *Appl. Environ. Microbiol.* **62**, 4478 (1996).

[15] α-D-Galactosidases from *Thermotoga* Species

By E. S. MILLER, JR., KIMBERLEY, N. PARKER, WOLFGANG LIEBL, DAVID LAM, WALTER CALLEN, MABJORY A. SNEAD, ERIC J. MATHUR, J. M. SHORT, and ROBERT M. KELLY

Introduction

α-Galactosidase (α-D-galactoside galactohydrolase, EC 3.2.1.22) (αGal) is an exo-acting glycosidase that catalyzes the hydrolysis of α-1→6 linked α-D-galactosyl residues from the nonreducing end of simple galactose-containing oligosaccharides, such as raffinose, stachyose, and melibiose, as well as more complex polysaccharides, including galacto(gluco)mannans. Interest in this class of enzymes stems from their potential technological applications. αGal has been used in the sugar beet industry to increase sucrose yields by eliminating raffinose, which prevents normal crystallization of beet sugar syrups.[1] Improved nutritional quality of legume-based foods, such as soybean milk and cowpea meal, can be achieved by the hydrolysis of galactooligosaccharides that are otherwise fermented in the large intestine causing flatulence and gastric distress.[2-4] Modification of guar galactomannans with αGals has been used to improve the gelling and viscosity properties of the polysaccharide, thereby creating modified galactomannan polysaccharides with properties similar to those of the more desirable and expensive locust bean gum.[5] Furthermore, some eukaryotic αGals remove the terminal α-1,3-linked galactose residues from glycans and have potential medical use in transfusion therapy for the conversion of erythrocyte blood group B specificity to the more universally transferable type O.[6]

Intracellular and extracellular αGals are widely distributed in microorganisms, plants, and animals.[7-9] Dey and colleagues showed that αGals can be classified into two groups based on their substrate specificity. One group is specific for low molecular weight α-galactosides, such as melibiose and the raffinose family of oligosaccharides, as well as the artificial substrate, *p*-nitrophenyl-α-D-galactopyranoside. The other group of αGals acts on galactomannans and also hydrolyzes low molecular weight substrates to various extents.[5] Genes encoding αGals have been cloned from various sources, including humans,[10] plants,[6,11] yeasts,[12] filamentous

[1] T. Yamane, *Sucr. Belge/Sugar Ind. Abstr.* **90**, 345 (1971).

[2] E. Cristofaro, F. Mottu, and J. J. Wuhrmann, *in* "Sugars in Nutrition" (H. L. Sipple and K. McNutt, eds.), p. 313. Academic Press, New York, 1974.

[3] R. Cruz and Y. K. Park, *J. Food Sci.* **47**, 1973 (1982).

[4] R. I. Somiari and E. Balogh, *Enzyme Microb. Technol.* **17**, 311 (1995).

[5] P. M. Dey, S. Patel, and M. D. Brownleader, *Biotechnol. Appl. Biochem.* **17**, 361 (1993).

[6] A. Zhu and J. Goldstein, *Gene (Amst.)* **140**, 227 (1994).

[7] P. M. Dey and E. D. Campillo, *Adv. Enzymol. Relat. Areas Mol. Biol.* **56**, 141 (1984).

[8] P. M. Dey and J. B. Pridham, *in* "Advances in Enzymology" (A. Meister, ed.), p. 91. Wiley, New York, 1972.

[9] R. F. H Decker and G. N. Richards, *in* "Advances in Carbohydrate Chemistry and Biochemistry," p. 277. Academic Press, New York, 1976.

[10] D. F. Bishop, D. H. Calhoun, H. S. Bernstein, P. Hantzopoulos, M. Quinn, and R. J. Desnick, *Proc. Natl. Acad. Sci. U.S.A.* **83**, 4859 (1986).

[11] N. Overbeeke, A. J. Fellinger, M. Y. Toonen, D. van Wassenaar, and C. T. Verrips, *Plant Mol. Biol.* **13**, 541 (1989).

[12] M. Sumner-Smith, R. P. Bozzato, N. Skipper, R. W. Davies, and J. E. Hopper, *Gene* **36**, 333 (1985).

fungi,[13–16] and bacteria.[17–19] Based on similarities in primary structure and hydrophobic cluster analyses, αGals have been grouped into three well-conserved families in the general classification of glycosylhydrolases.[20] Those from bacteria have been grouped into families 4 and 36 and those of eukaryotic origin into family 27.

To date, only αGals of the hyperthermophilic bacteria *Thermotoga maritima* (*Tm*GalA) and *T. neapolitana* (*Tn*GalA) have demonstrated activity and prolonged stability above 75°.[21,22] These two enzymes are therefore of considerable interest from a biotechnological standpoint. Potential applications include the high temperature hydrolysis of galactomannans used for well stimulation in the oil and gas industry[23] and oligosaccharide synthesis via glycosyltransferase reactions.[24]

*Tm*GalA and *Tn*GalA show 83% amino acid sequence identity and 90% similarity with each other, as would be expected, but show less than 20 to 25% sequence identity with other family 36 αGals.[21,22] Moreover, this similarity is restricted to a short central part of the *Tm*GalA and *Tn*GalA amino acid sequence. *Thermotoga* enzymes have different subunit compositions and are significantly smaller than their mesophilic counterparts.[22] The degree of unrelatedness of the *Tm*GalA and *Tn*GalA with other αGals of family 36 may be associated with the ancestral origin of the *Thermotoga* species. However, despite the lack of strong homology, both *Thermotoga* enzymes have been assigned to family 36 (http://afmb.cnrs-mrs.fr/~pedro/CAZY/ghf.html).

Genes encoding both *Tm*GalA and *Tn*GalA have been cloned and expressed in *Escherichia coli*.[22,25] Although these enzymes are structurally

[13] I. F. den Herder, A. M. Rosell, C. M. van Zuilen, P. J. Punt, and C. A. vanden Hondel, *Mol. Gen. Genet.* **233**, 404 (1992).

[14] E. M. Margolles-Clark, E. M. Tenkanen, E. Luonteri, and M. Penttila, *Eur. J. Biochem.* **240**, 104 (1996).

[15] H. Shibuya, H. Kobayashi, K. Kasamo, and I. Kusakabe, *Biosci. Biotechnol. Biochem.* **59**, 1345 (1995).

[16] H. Shibuya *et al.*, *Biosci. Biotechnol. Biochem.* **61**, 592 (1997).

[17] J. Aduse-Opoku, L. Tao, J. J. Ferretti, and R. R. B. Russell, *J. Gen. Microbiol.* **137**, 757 (1991).

[18] C. Aslanidis, K. Schmid, and R. Schmitt, *J. Bacteriol.* **171**, 6753 (1989).

[19] P. L. Liljiestrom and P. Liljestrom, *Nucleic Acids Res.* **15**, 2213 (1987).

[20] B. Henrissat and A. Bairoch, *Biochem. J.* **293**, 781 (1993).

[21] G. D. Duffaud, C. M. McCutcheon, P. Leduc, K. N. Parker, and R. M. Kelly, *Appl. Environ. Microbiol.* **63**, 169 (1997).

[22] W. Liebl, B. Wagner, and J. Schellhase, *System. Appl. Microbiol.* **21**, 1 (1998).

[23] C. M. McCutchen *et al.*, *Biotechnol. Bioeng.* **52**, 332 (1996).

[24] K. Hara *et al.*, *Biosci. Biotech. Biochem.* **58**, 652 (1994).

[25] M. R. King, D. A. Yernool, D. E. Eveleigh, and B. M. Chassy, *FEMS Microbiol. Lett.* **163**, 37 (1998).

related, they exhibit different biochemical properties in terms of pH optima, activity, and thermostability. This article discusses the purification, cloning, and expression of recombinant αGal from *T. neapolitana* and *T. maritima*, in addition to some of their biochemical properties.

α-D-Galactosidase of *Thermotoga neapolitana* DSMZ5068

Bacterial Strains and Culture Conditions

Thermotoga neapolitana DSMZ5068 is from Deutsche Sammlung von Mikroorganismen und Zellkulturen, Braunschweig, Germany. Cell cultures are grown on an artificial seawater-based medium supplemented with 0.1% yeast extract, 0.5% tryptone, and 0.2% lactose (w/w).[26] The medium composition per liter is NaCl, 15.0 g; Na_2SO_4, 2.0 g; $MgCl_2 \cdot 6H_2O$, 2.0 g; $CaCl_2 \cdot 2H_2O$, 0.5 g; $NaHCO_3$, 0.25 g; K_2HPO_4, 0.1 g; KBr, 50 mg; H_3BO_3, 20 mg; KI, 20 mg; $Na_2WO_4 \cdot 2H_2O$, 3 mg; and $NiCl_2 \cdot 6H_2O$, 2 mg. Additionally, 5 ml of a 0.2-mg ml^{-1} resazurin stock solution is added as an indicator of O_2 contamination. Lactose and K_2HPO_4 are added to the autoclaved media following filter sterilization through a 0.2-μm filter.

Cultures of *T. neapolitana* DSMZ5068 are grown in 1-liter Pyrex bottles sealed with butyl rubber stoppers at 85° with shaking at 150 rpm. Prior to inoculation, medium (without lactose) is heated at 98° in order to reduce the dissolved O_2 solubility. After 30 min, lactose is added and the medium is reduced by the addition of NaS to a final concentration of 0.5 g/liter with continuous N_2 sparging until the resazurin indicator is reduced (i.e., medium turned clear to yellow). Once reduced, cultures are sealed and inoculated with 25 ml of an overnight culture (16 hr) using a 30-ml syringe fitted with a 25-gauge needle. *T. neapolitana* cultures are stored long term in the same growth medium at 4° in 100-ml serum bottles sealed with butyl rubber stoppers and crimp top caps.

Escherichia coli XL1-Blue MRF′ [Δ(*mcrA*)183, Δ(*mcrCB-hsdSMR-mrr*)173, *endA1*, *supE44*, *thi-1*, *recA1*, *gyrA96*, *relA1*, *lac*, {F′ *proAB*, *lacI*qZΔM15, Tn*10* (tet^r)}] and *E. coli* Y1090 [Δ(*lac*)*U169*, Δ(*lon*), *araD139*, *strA*, *supF*, *mcrA*, *trpC22*::Tn*10*(tet^r), {pMC9amp^{r}tet^r}] (Stratagene Cloning Systems, La Jolla, CA) are used as hosts of Lambda-ZAPII and lambda (λ)gt11 libraries, respectively. Luria broth (1% tryptone, 0.5% yeast extract, and 1% NaCl) (LB) is used for growth and maintenance of all *E. coli* recombinant clones. Cultures are incubated at 37° with shaking at 250 rpm. Cell growth is monitored at 600 nm using a Shimadzu UV160U UV/visible recording spectrophotometer. For solid medium, 18 g/liter agar is added

[26] S. H. Brown, C. Sjoholm, and R. M. Kelly, *Biotechnol. Bioeng.* **41**, 878 (1993).

to LB medium. Plate cultures are stored at 4° for up to 2 weeks and are used as sources of inoculum for starter cultures in recombinant *Tn*GalA fermentations. Liquid cultures are stored long term in LB with 15% (v/v) glycerol at −80°. When appropriate, antibiotics are added to both solid and liquid LB media at the following concentrations: ampicillin, 100 μg ml^{-1}; and tetracycline, 10 μg ml^{-1}.

Cloning of TnGalA

Nucleic Acids Isolation. Genomic DNA from *T. neapolitana* DSMZ5068 is purified from 0.5 g of cell mass using cetyltrimethylammonium bromide (CTAB) (Sigma, St. Louis, MO) by the method described by Gerhardt *et al.*[27]

T. neapolitana Genomic Library Construction. Genomic DNA (50 mg) is mechanically sheared by vigorous passage through a 25-gauge double-hub needle (Popper, New Hyde Park, NY) attached to 1-ml syringes. A small aliquot (0.5 mg) is electrophoresed through a 0.8% agarose gel to confirm that the majority of the sheared DNA was within the desired size range (3–6 kb). The DNA fragments are made blunt ended by treatment with mungbean nuclease (Stratagene Cloning Systems). Naturally occurring *Eco*RI sites are protected by methylation with *Eco*RI methylase (New England Biolabs, Beverly, MA). Double-stranded *Eco*RI linkers (GGAATTCC) are ligated to the blunt DNA fragments with T4 DNA ligase (New England Biolabs). The DNA is size fractionated on a sucrose gradient following a standard protocol.[27] Fractions (500 μl) are collected, and 20 μl of each are analyzed on a 0.8% agarose gel. Fractions containing DNA fragments of 3–8 kb are pooled and ligated into Lambda ZAPII (Stratagene Cloning Systems) and Lambda gt11 (Stratagene Cloning Systems) insertion vectors that have been digested previously with *Eco*RI and dephosphorylated. The ligation reactions are packaged using Gigapack III Lambda-packaging extract (Stratagene Cloning Systems); 2.5 × 10^5 plaque-forming units (pfu) are amplified on NZY plates using the appropriate *E. coli* hosts.

Expression Screening. Unamplified Lambda libraries are diluted in SM buffer [0.1 *M* NaCl, 50 m*M* Tris–HCl (pH 7.5), 0.02% gelatin, 10 m*M* MgSO$_4$] to achieve approximately 2000 pfu per plate. Libraries are plated in LB top agarose containing 1 m*M* isopropyl-β-thiogalactopyranoside (IPTG) onto LB agar plates and incubated at 39° overnight. Plates are

[27] P. Gerhardt, R. Murray, W. Wood, and N. Krieg, *in* "Methods for General and Molecular Biology." American Society for Microbiology, Washington, DC, 1994.

removed from incubation and are replicated with nitrocellulose membranes (Stratagene) and screened for αGal activity.

Substrate assay plates are prepared by melting 1% (w/v) agarose in 100 mM potassium phosphate buffer, pH 7.0, containing 100 mM NaCl. After cooling to ~60°, 7 mM p-nitrophenyl-α-D-galactopyranoside (Sigma) is dissolved in the melted agarose solution, poured into petri dishes, and allowed to solidify. Screening for *Tn*GalA activity is conducted by overlaying nitrocellulose replicas onto substrate assay plates and incubating at 72° for 5 to 20 min. *E. coli* clones expressing functional *Tn*GalA activity are identified as yellow spots on a white background. Positive clones are isolated from original plates and eluted in SM buffer.

Assay Method

Principle. TnGalA activity is assayed spectrophotometrically by monitoring the release of p-nitrophenol (PNP) from p-nitrophenyl-α-D-galacto-pyranoside continuously at 405 nm.

Procedure. Enzymatic assays for *Tn*GalA activity are conducted as follows. The standard reaction mixture contains 1.1 ml of 10 mM p-nitrophenyl-α-D-galactopyranoside in 0.1 M sodium phosphate buffer, pH 7. Substrate and buffer are preincubated for at least 10 min prior to the addition of enzyme to allow the substrate to reach the assay temperature. After incubation, 0.1 ml of appropriately diluted enzyme solution is added to initiate the reaction. A blank consisting of 0.1 ml of 0.1 M sodium phosphate buffer (pH 7) is used as a control.

All assays are conducted using capped quartz cuvettes (Uvonic Instruments, Inc., Plainview, NY). Spectrophotometric readings are taken with a Lambda 3 spectrophotometer (Perkin-Elmer Corp., Norwalk, CT) equipped with a thermostatted six-cell transport system. A liquid-circulating temperature bath (Model 1130; VWR Scientific, Philadelphia, PA) containing a 1:1 mixture of ethylene glycol and water is used to maintain the desired temperature in the cell holder. The temperature is monitored by a thermocouple mounted in a cuvette that is placed in the cell transporter. Computerized spectroscopy software (Perkin-Elmer Corp.) is used to control the six-cell transporter and for data acquisition and analysis.

Definition of Enzyme Unit. One unit of recombinant *Tn*GalA activity (U) is defined as the amount of enzyme releasing 1 μmol of PNP min^{-1} from p-nitrophenyl-α-D-galactopyranoside under the specified assay conditions. Specific activity is defined as units of activity per milligram of protein as measured by the method of Bradford,[28] with bovine serum albumin (BSA) as the standard.

Purification

*Tn*GalA is purified from recombinant *E. coli* biomass by fast protein liquid chromatography (FPLC) (Pharmacia, Uppsala, Sweden). Columns and chromatography media are obtained from Pharmacia unless otherwise stated. Prior to use, buffers are filtered through a 0.22-μm filter. All chromatographic procedures are performed at room temperature. Protein samples are stored at 4°.

Step 1. Preparation of Cell Extract. Seven grams of recombinant *E. coli* cells is resuspended in 3 ml g^{-1} of 20 mM NaCl in 100 mM sodium phosphate buffer, pH 7.0, by gentle shaking at 4°. The resuspended cell pellets are sonicated on ice for 5 min. (30 s pulse, 30 s rest). Following sonication, cell debris is removed by centrifugation at 10,000g for 15 min. Heat-labile host proteins are denatured by heating the clarified mixtures for 20 min at 80° followed by rapid cooling on ice for 5 min. The soluble fraction obtained after a second centrifugation (10,000g, 15 min, 4°) is used as the crude enzyme preparation.

Step 2. Anion-Exchange Chromatography. Portions of the heat-treated crude enzyme preparation (10 ml; 1.6 mg ml^{-1}) are applied to a DEAE-Sepharose CL-6B (Sigma) column (XK 26/20 column; bed volume, 80 ml) equilibrated with 50 mM sodium phosphate (pH 7.0) buffer. The column is washed with the same buffer until protein cannot be detected in the eluent. Bound protein is eluted with a linear NaCl gradient (0 to 1.0 M) in 160 ml of 50 mM sodium phosphate (pH 7.0). *Tn*GalA activity is recovered in the fractions eluting at 0.4–0.6 M NaCl. These fractions are pooled and concentrated 20- to 50-fold using an Amicon 8050 ultrafiltration cell (Amicon, Beverly, MA) fitted with a 10,000 molecular weight cutoff membrane (Amicon).

Step 3. Hydrophobic Interaction Chromatography. The pooled fractions (4.6 ml) are added to 4.6 ml of 1.0 M (NH$_4$)$_2$SO$_4$ in 50 mM sodium phosphate (pH 7) buffer in order to bring the sample to the initial pH and ionic strength of the gradient used for elution. The sample is then filtered through a 0.22-μm filter before being applied to the top of a butyl-Sepharose column (butyl-Sepharose 4 Fast Flow; bed volume, 1.0 ml) equilibrated with 1 M (NH$_4$)$_2$SO$_4$ in 50 mM sodium phosphate buffer, pH 7.0. The column is washed with the same buffer until protein cannot be detected in the eluent. Protein bound to the column is eluted with a linear gradient of decreasing

[28] M. Bradford, *Anal. Biochem.* **72**, 248 (1976).

$(NH_4)_2SO_4$ concentration from 1.0 to 0.0 M in 15 ml of 50 mM sodium phosphate (pH 7). *Tn*GalA activity is present in fractions eluting at 0.2–0.0 M $(NH_4)_2SO_4$.

Biochemical Properties

Molecular Properties. Recombinant *Tn*GalA is a monomeric protein of 554 amino acids. The enzyme is significantly smaller than other bacterial αGals belonging to glycosylhydrolase family 36. It is so far the only monomeric αGal of prokaryotic origin. The enzyme migrates at approximately 61 kDa after electrophoresis on an SDS–polyacrylamide gel and has a calculated molecular mass of 64.1 kDa. The calculated isoelectric point is 5.3.

pH and Temperature Optima. *Tn*GalA is most active in the pH range from 6.4 to 8. At 90°, optimal activity is obtained at pH 7 in 50 mM sodium phosphate buffer. The enzyme has an optimal temperature of 99° in 50 mM sodium phosphate buffer, pH 7.

Stability. At a concentration of 0.06 mg ml^{-1} in 50 mM sodium phosphate buffer (pH 7.0), the enzyme loses less than 50% of its activity within the pH range of 6.4–8 at 98°. *Tn*GalA has a half-life of 2.5 min at 100°.

Specificity. Purified *Tn*GalA has a specific activity of 224 U mg^{-1} with *p*-nitrophenyl-α-D-galactopyranoside as substrate. Activity assays with non-galactose-linked chromogenic substrates reveal that the enzyme is highly specific for galactose as the glycone in the nonreducing end of *p*-nitrophenyl-linked monosaccharides. *Tn*GalA shows barely detectable activity with *p*-nitrophenyl-α-D-glucopyranoside (6.8 U mg^{-1}) and *p*-nitrophenyl-β-D-mannopyranoside (2.5 U mg^{-1}) and no activity with *p*-nitrophenyl-α-D-mannopyranoside and *p*-nitrophenyl-β-D-glucopyranoside. In addition, the enzyme is highly specific for the α anomer of galactose, showing a 43-fold increase in activity with *p*-nitrophenyl-α-D-galactopyranoside over *p*-nitrophenyl-β-D-galactopyranoside (5.2 U mg^{-1}).

*Tn*GalA shows activity with melibiose, [αgal-(1→6)-αglc], (160 U mg^{-1}), representatives of the raffinose family of oligosaccharides, and with branched chain mannans, thus demonstrating its potential in oil well stimulation applications. With the raffinose family of oligosaccharides, *Tn*GalA showed increasing specific activity with decreasing galactose chain length. Of the most widely occurring raffinose oligosaccharides, *Tn*GalA shows the highest specific activity with raffinose, [αgal-(1→6)-αglc-(1→2)-βfru], (147 U mg^{-1}), which is twofold greater than with stachyose, [αgal-(1→6)-αgal-(1→6)-αglc-(1→2)-βfru], (71 U mg^{-1}), and sixfold greater than with verbascose, [αgal-(1→6)-αgal-(1→6)-αgal-(1→6)-αglc-(1→2)-βfru], (25 U mg^{-1}). Comparison of *Tn*GalA specific activity with the just-described substrates and that obtained with the branched chained mannans, Gal^1Man_3 (168 U mg^{-1}) and Gal^2Man_5 (163 U mg^{-1}), indicates that, in addition to galactose containing disaccharides, TnGalA has a slight preference for branched galactosyl residues over straight chain residues. The kinetic constants of TnGalA with p-nitrophenyl-α-D-galactopyranoside, melibiose, and raffinose at 98° in 50 mM sodium phosphate buffer (pH 7.0) are shown in Table I. An Arrhenius plot of the TnGalA catalytic constant as a function of temperature gives an activation energy of 71.3 kJ mol^{-1} for TnGalA hydrolysis of p-nitrophenyl-α-D-galactopyranoside.

α-D-Galactosidase of Thermotoga maritima DSMZ3109

Bacterial Strains and Culture Conditions

Thermotoga maritima DSMZ3109 is from Deutsche Sammlung von Mikroorganismen und Zellkulturen, Braunschweig, Germany. Cell cultures are grown and maintained on an artificial seawater-based medium following the protocols for media preparation, inoculation, and storage described earlier for the growth and maintenance of T. neapolitana DSMZ5608.

Escherichia coli NovaBlue [recAl, endAl, gyrA96, thi-1, hsdR17 (r_{K12}^- m_{K12}^+), supE44, relA1, lac, {F' proAB, lac^{q}ZΔM15: Tn10 (tet^r)}] (Novagen, Madison, WI) is used for routine maintenance and isolation of pET15b and pET24d$^+$ plasmids (Novagen). E. coli BL21(DE3) [F, ompT, hsdS$_B$ (r_B^- m_B^-), gal, dcm (DE3)] (Novagen) is used for the expression of recombinant TmGalA. Protocols for the growth and maintenance of all E. coli plasmid-bearing strains are as described previously.

Cloning of TmGalA

Nucleic Acid Isolation. Genomic DNA from T. maritima DSMZ3109 is purified from 10^{10} cells using a Qiagen genomic DNA isolation kit (Qiagen

TABLE I
KINETIC CONSTANTS OF RECOMBINANT TnGalA[a]

Substrate	K_m (mM)	V_{max} (U mg^{-1})	k_{cat} (min^{-1})	k_{cat}/K_m (mM^{-1} min^{-1})
PNP-α-D-Gal[b]	0.74	298.0	18,175	24,547
Melibiose	15.11	230.4	14,054	930
Raffinose	5.32	155.1	9,464	1,779

[a] At 98° and pH 7.
[b] p-Nitrophenyl-α-D-galactopyranoside.

Inc., Valencia, CA). In order to efficiently lyse *T. maritima* cells, the enzymatic lysis step in the manufacturers protocol is modified essentially as described by Gabelsberger *et al.*[29] Briefly, this modification is as follows: (1) incubate the resuspended cell pellet with 1% lysozyme for 15 min at 37°, followed by incubation at 0° for 10 min and (2) add proteinase K (0.6 mg ml^{-1}, final concentration), and incubate for 30 min at 45°.

Construction of Plasmids for TmGalA Expression in E. coli. Cloning of *Tm galA* is accomplished by polymerase chain reaction (PCR) amplification of the *Tm*GalA gene present on recombinant plasmid pAGT1.[22] This plasmid harbors a 4.6-kb *Sau*3A partial insert of *T. maritima* DSMZ3109 genomic DNA that contains two complete open reading frames encoding α-galactosidase, and galactose-1-phosphate uridylyltransferase.[22] Two synthetic oligonucleotide primers with nucleotide sequences 5'-CCATGCCATGGAAATATTCGGAAAGACC-3' (forward primer) and 5'-CGCGGATCCTATTCTCTCTCACCCTCTTCG-3' (reverse primer) are constructed based on the published *T. maritima* DSMZ3109 α-galactosidase gene sequence (GenBank, accession number 2660640).[22] Restriction sites (underlined sequences) for *Nco*I and *Bam*HI are used for the in-frame fusion of the *galA* open reading frame (ORF) to the first start codon downstream of the P$_{T7}$ promoter and ribosome-binding site of the pET15b expression vector. Using this cloning strategy, the GTG start codon in the *galA* ORF is changed to ATG. The PCR reaction mixture consists of (in 100 μl) linearized pAGT1 DNA template, 100 ng; deoxynucleotide triphosphates, 200 μM each; 50 pmol of each primer; 2.5 U *Pyrococcus furiosus* (*Pfu*) DNA polymerase (Promega Corp.); and 10 μl of 10× *Pfu* reaction buffer (Promega Corp.). The template strands are separated at 94° for 3 min before starting 40 reaction cycles with the following conditions: melt for 30 sec at 94°, anneal for 2 min at 59°, and extension for 4 min at 72°. Following the last cycle, a final 5-min extension at 72° is employed.

A similar cloning strategy based on PCR amplification of *Tm galA* from total *T. maritima* DSMZ3109 genomic DNA can also be used. In this case, the synthetic oligonucleotide primers are 5'-TGAGCTGATCACCATGGAAATATTCGGAAAGACC-3' (forward primer) and 5'-ACCTGACAAGCTTCATTCTCTCTCACCCTCTTC-3' (reverse primer). Choice of restriction sites (underlined sequences) places an *Nco*I site 5' and an *Hind*III site 3' to the *galA* ORF. Use of these sites allows for the in-frame fusion of the *galA* ORF to the first start codon downstream of the P$_{T7}$ promoter and ribosome-binding site of the pET24d$^+$ expression vector. The GTG start codon in the *galA* ORF is also changed to ATG. The PCR reaction mixture consists of (in 100 μl) *T. maritima* DSMZ3109

[29] J. Gabelsberger, W. Liebl, and K. Schleifer, *FEMS Microbiol. Lett.* **109,** 131 (1993).

genomic DNA template, 50 ng; deoxynucleotide triphosphates, 100 μM each; 100 pmol of each primer; 2.5 U *Pfu* DNA polymerase; and 10 μl of 10× *Pfu* reaction buffer. The template strands are separated at 95° for 90 sec before starting 30 reaction cycles with the following conditions: melt for 45 sec at 95°, anneal for 30 sec at 55°, and extension for 3 min and 20 sec at 73°. A final 5-min extension at 73° is employed following the last cycle.

Assay Method

An alternative end point assay method for the determination of *Tm*GalA activity is described. Briefly, substrate and enzyme are incubated at the desired conditions for a specified time interval. The reaction is terminated by the addition of a basic solution, and the absorbance of the released PNP from *p*-nitrophenyl-α-D-galactopyranoside is measured at room temperature at 405 nm. The following method has the advantage in that it can be performed without the need for expensive thermostatted spectrophotometers.

Principle. *Tm*GalA activity is assayed spectrophotometrically by monitoring the release of PNP from *p*-nitrophenyl-α-D-galactopyranoside. The absorbance of PNP is measured at 405 nm in a basic solution.

Procedure. The standard assay mixture contains (in 700 μl) 500 μl of citrate–phosphate buffer, pH 5 (prepared by titration of 50 mM citric acid with 50 mM Na_2HPO_4), 3.8 mM *p*-nitrophenyl-α-D-galactopyranoside, and 100 μl of suitably diluted enzyme solution. After preincubation at 75° for 3 min, the assay is started by the addition of enzyme. The reaction (5–10 min at 75°) is stopped by mixing with 500 μl 1 M Na_2CO_3 and cooling on ice.

Definition of Enzyme Unit. One unit of *Tm*GalA activity is defined as the amount of enzyme releasing 1 μmol of PNP min^{-1} from *p*-nitrophenyl-α-D-galactopyranoside under the specified assay conditions. Specific activity is defined as units of activity per milligram of protein.

Purification[22]

*Tm*GalA is purified from an 18-liter culture of *E. coli* BL21(DE3)/ pET-AGT1 supplemented with 100 μg ml^{-1} ampicillin. All columns and chromatography media are obtained from Pharmacia unless otherwise stated. For induction of the P_{T7} promoter, 0.5 mM IPTG is added during the exponential growth phase at an OD_{600nm} of 0.8 and the culture is incubated for an additional 16 hr at 37°. Under these conditions, *Tm*GalA expression accounts for 20% of the total soluble protein. The purification strategy is summarized in Table II.

Step 1. Preparation of Cell Extract. Fifty grams of *E. coli* cells is resuspended in 1 ml g^{-1} of 20 mM Tris–HCl, pH 8. The resuspended cell pellets

TABLE II
PURIFICATION OF TmGalA FROM RECOMBINANT E. coli BL21(DE3)/pET-AGT1[a]

Purification step	Protein (mg)	Total activity (U)	Specific activity (U mg^{-1})	Purification factor (-fold)	Yield (%)
Crude extract	4020	194,000	48.2	1	100
Heat precipitation and Source 15Q chromatography	1032	149,000	144	3	78
Phenyl Sepharose chromatography	252	56,800	225	4.7	29

[a] From W. Liebl, B. Wagner, and J. Schellhase, *System Appl. Microbiol.* **21**, 1 (1998).

are lysed by threefold passage through a French pressure cell (American Instrument Company, Silver Spring, MD) at 6.9 MPa. Following centrifugation, the clarified supernatant is incubated for 20 min at 75° in order to denature heat-labile host *E. coli* proteins. The soluble fraction obtained after an additional centrifugation is used as the crude enzyme preparation.

Step 2. Anion-Exchange Chromatography. Thirty- to 60-ml portions of the heat-treated crude enzyme preparation ($\sim$16 mg ml^{-1}) are applied to a Source 15Q column (Pharmacia XK26; bed volume, 50 ml) equilibrated with 20 mM Tris–HCl, pH 8. Proteins bound to the column are eluted with a linear NaCl gradient (0 to 1.0 M) in 0.5 liters of 20 mM Tris–HCl, pH 8. TmGalA activity is found in the fractions eluting at 0.3 M NaCl.

Step 3. Hydrophobic Interaction Chromatography. Pooled fractions ($\sim$40 ml) containing TmGalA activity are brought to 5 M NaCl by slowly adding solid NaCl. The sample is then loaded onto a phenyl-Sepharose HP column (Pharmacia XK16/10; bed volume, 20 ml) and eluted with a linear $5 \to 0$ M NaCl gradient in 240 ml of 20 mM Tris–HCl, pH 8, buffer. Most of the TmGalA activity is found in the fractions eluting between 2.5 and 1.5 M NaCl. Active fractions are desalted by dialysis against 20 mM Tris–HCl, pH 8, and stored at $\sim$20°. Two hundred fifty-two milligrams of enzyme with a specific activity of 225 U mg^{-1}, as determined in the standard end point assay described earlier, is obtained following this final purification step (Table II).

Properties[22]

Molecular Properties. Recombinant TmGalA is significantly smaller than and possesses different subunit composition from that of other family 36 α-Gal representatives. It is a homodimeric protein of 552 amino acids that does not require Mn^{2+} or NAD for activity. The enzyme migrates at approximately 62 kDa after electrophoresis on an SDS–polyacrylamide gel, which is close to the calculated molecular mass of 63.7 kDa. Analytical gel permeation chromatography reveals a symmetrical peak with an elution volume corresponding to about 125 kDa, indicating that the enzyme is a homodimer.

pH and Temperature Optima. Maximal activity of purified *Tm*GalA is obtained between pH 5.0–5.5 and 90–95°. The pH optimum of *Tm*GalA is lower than that reported for other bacterial family 36 αGals.[17,21,30–37] Greater than 50% of maximal activity is maintained in the pH range from 4.5 to 7.0 at 75° and between 75 and 95° at pH 5.

Stability. At a protein concentration of 1 mg ml^{-1} in 20 mM Tris–HCl buffer, pH 7, the time required for the enzyme to lose 50% of its activity is 7 days, 48 hr, 6 hr, and 70 min at 75, 80, 85, and 90°, respectively. The enzyme retains greater than 60% of its residual activity when incubated for 5 hr at 75° in 50 mM citrate–phosphate buffer in the pH range of 5–8.

Effect of Cations on Activity. Various divalent metal cations (Ca^{2+}, Mg^{2+}, Mn^{2+}, Ba^{2+}, Co^{2+}, Ni^{2+}, and Zn^{2+}) showed no effect on enzyme activity when added as their chloride salts at a concentration of 1 mM in the standard assay. However, the addition of 1 mM of the chloride salts Cu^{2+} and Fe^{2+} results in 50% inhibition of *Tm*GalA activity in the standard assay; 10 mM EDTA had no significant effect on activity.

Specificity. *Tm*GalA cleaves *p*-nitrophenyl-α-D-galactopyranoside and *o*-nitrophenyl-α-D-galactopyranoside with similar activities at 75°, 225 and 169 U mg^{-1}, respectively, in addition to 5-bromo-4-chloro-3-indolyl-α-D-galactopyranoside. The following chromogenic substrates are not hydrolyzed: *p*-nitrophenyl-β-D-galactopyranoside, *o*-nitrophenyl-β-D-galactopyranoside, *p*-nitrophenyl-α-D-glucopyranoside, *p*-nitrophenyl-β-D-glucopyranoside, *p*-nitrophenyl-β-D-fucopyranoside, *p*-nitrophenyl-α-D-arabinofuranoside, and *p*-nitrophenyl-β-D-xylopyranoside.

*Tm*GalA shows activity with the naturally occurring di- and trisaccharides melibiose and raffinose. Kinetic constants with *p*-nitrophenyl-α-D-galactopyranoside and raffinose at 75° and pH 5 are shown in Table III.

[30] T. Durance and B. Skura, *J. Food Sci.* **50,** 518 (1985).
[31] C. Ganter, A. Bock, P. Buckel, and R. Mattes, *J. Biotechnol.* **8,** 301 (1988).
[32] M. S. Garro, G. S. DeGiori, G. F. DeValdez, and G. Oliver, *Lett. Appl. Microbiol.* **19,** 16 (1994).
[33] M. S. Garro, G. F. DeValdez, G. Oliver, and G. S. DeGiori, *J. Biotechnol.* **45,** 103 (1996).
[34] H. Hashimoto, M. Goto, C. Katayama, and S. Kitahata, *Biosci. Biotech. Biochem.* **55,** 2831 (1991).
[35] M. A. Nadkarni, C. K. K. Nair, V. N. Pandey, and D. S. Pradhan, *J. Gen Appl. Microbiol.* **38,** 23 (1992).
[36] K. Schmid and R. Schmitt, *Eur. J. Biochem.* **67,** 95 (1976).
[37] G. Talbot and J. Sygusch, *Appl. Environ. Microbiol.* **56,** 3505 (1990).

TABLE III
KINETIC CONSTANTS OF RECOMBINANT *Tm*GalA[a]

Substrate	K_m (mM)	V_{max} (U mg^{-1})	k_{cat} (min^{-1})	k_{cat}/K_m (mM^{-1} min^{-1})
PNP-α-D-Gal[b]	0.075	166	10,560	140,800
Raffinose	2.1	103	6,540	3,114

[a] At 75° and pH 5. From W. Liebl, B. Wagner, and J. Schellhase, *System Appl. Microbiol.* **21,** 1 (1998).
[b] *p*-Nitrophenyl-α-D-galactopyranoside.

The enzyme also uses stachyose and verbascose as substrates, but is 20-fold less active than with raffinose.[38] *Tm*GalA does not display activity toward maltose [αglc-(1→4)-glc] sucrose [αglc-(1→2)-βfru], trehalose [αglc-(1→1)-αglc], turanose [αglc-(1→3)-βfru], maltotriose [αglc-(1→4)-αglc-(1→4)-glc], panose [αglc-(1→6)-αglc-(1→4)-glc], melizitose [αglc-(1→3)-βfru-(2→1)-αglc], and the hemicellulose polymers arabinogalactan and β-mannan.

Concluding Remarks

αGals from *Thermus* spp. are the only αGals identified to date of hyperthermophilic origin. The gene encoding αGal from *T. maritima* DSMZ3109 is part of a galactoside utilization gene cluster in this strain.[22] Data presented by Duffaud *et al.*[21] suggest that the enzyme in *T. neapolitana* DSMZ5068 may have similar function. In addition to the αGals discussed here, the genes encoding αGal from another hyperthermophile, *T. neapolitana* NS-E, and from the moderate thermophile, *Thermus* sp. T2, have been cloned and expressed in *E. coli.*[25,39] The properties of the enzyme from *T. neapolitana* NS-E are similar to those of the αGal from *T. neapolitana* DSMZ5068, although its temperature optimum is 5° lower than that reported here.[25]

The biochemical and physical properties of *Tn*GalA and *Tm*GalA are somewhat different than the αGals of glycosylhydrolase family 36. *Tm*GalA is a homodimer,[22] whereas *Tn*GalA is a monomeric protein and the smallest αGal belonging to this family. *Tn*GalA has a slightly higher temperature optima for *p*-nitrophenyl-α-D-galactopyranoside hydrolysis (99°) over that of *Tm*GalA (90–95°). In addition, *Tn*GalA displays optimal *p*-nitrophenyl-α-D-galactopyranoside hydrolysis at pH 7, which is in the range of pH optima (5.8–8.1) reported for other bacterial αGals.[17,21,30–37] In contrast,

[38] E. S. Miller, unpublished observations (1999).
[39] Y. Koyama, S. Okamoto, and K. Furukawa, *Appl. Environ. Microbiol.* **56,** 2251 (1990).

*Tm*GalA displays optimal *p*-nitrophenyl-α-D-galactopyranoside hydrolysis at pH 5.0–5.5, which is similar to that of αGals from eukaryotic origin, particularly fungi, which display optimal *p*-nitrophenyl-α-D-galactopyranoside hydrolysis in the pH range of 3.5–5.[14,34,40,41] Although both *Tm*GalA and *Tn*GalA are cloned from the same genus, the sharing of biochemical properties with enzymes from both prokaryotic and eukaryotic origin may be associated with the ancestral nature of the *Thermotoga* species.

The number of technological applications for αGals in the medical, oil and gas, and food processing industries has generated a large volume of interest in this class of enzymes. The extreme thermostability and high temperature optima exhibited by *Tn*GalA and *Tm*GalA can be exploited to expand the use of these enzymes to applications that would benefit from improved thermostability.[1,23]

Acknowledgments

RMK acknowledges the National Science Foundation, the U.S. Department of Agriculture, and the Novartis Agricultural Biotechnology Research Institute (RTP, NC) for financial support of this research.

[40] S. Zeilinger, D. Kristufek, I. Aristan-Atac, R. Hodits, and C. P. Kubicek, *Appl. Environ. Microbiol.* **59,** 1347 (1993).
[41] S. Rios, A. M. Pedregosa, I. F. Monistrol, and F. Laborda, *FEMS Microbiol. Lett.* **112,** 34 (1993).

[16] α-Glucosidase from *Pyrococcus furiosus*

By Stephen T. Chang, Kimberley N. Parker, Michael W. Bauer, and Robert M. Kelly

Introduction

α-Glucosidase (EC 3.2.1.20; α-D-glucoside glucohydrolase) is a member of the glycosylhydrolases that specifically catalyze the hydrolysis of terminal, nonreducing α-D-glucose residues from the end of various oligosaccharides (exo-acting) with the release of α-D-glucose. The α-1,4 bond is preferentially cleaved, although cleavage of other α bonds is possible depending on the microbial source of the α-glucosidase[1] and typically at a much slower rate.[2]

[1] G. Antranikian, *in* "Microbial Degradation of Natural Products" (G. Winkelmann, ed.), p. 27. VCH Publishers, New York, 1992.
[2] V. Worthington, "Worthington Enzyme Manual: Enzymes and Related Biochemicals." Worthington Biochemical Corporation, Freehold, NJ, 1993.

In general, the hydrolysis rate decreases substantially with increasing substrate size.[2]

Many organisms produce α-glucosidases intracellularly in conjunction with other extracellular amylolytic enzymes, such as α-amylases capable of hydrolyzing the internal α-1,4 bonds of linear polysaccharides (endo-acting) and pullulanases capable of hydrolyzing the α-1,6 bonds of branched polysaccharides (debranching). α-Glucosidase is believed to be the final enzyme involved in the metabolism of these carbohydrates, such as starch, to glucose for cell utilization. It appears that carbohydrates are broken down into oligosaccharides such as maltose before being transported into the cell where they are subsequently hydrolyzed to glucose by α-glucosidase. The oligosaccharide transport system of *Thermococcus litoralis* has been studied and it exhibits a high affinity for maltose and trehalose.[3] The same mechanism of carbohydrate metabolism and transport has been proposed for other hyperthermophilic organisms,[4] including *Pyrococcus furiosus*.[5] Further support that glucose is formed exclusively within the cell for immediate utilization is provided by observations that many hyperthermophiles can only grow efficiently on polymers of glucose but not on glucose itself[4] and mono- and oligosaccharides can be unstable at the high temperatures required for growth.[6] Maltose is an excellent growth substrate for both *P. furiosus* and *T. litoralis*,[5,7] but there is debate, however, on whether both use the same mechanism. Xavier *et al.*[8] have proposed that glucose is produced from maltose in *T. litoralis* through a 4-α-glucanotransferase rather than an α-glucosidase, although an α-glucosidase was found in *T. litoralis*.[6] To further complicate the issue, an extracellular α-glucosidase has been characterized from *Thermococcus* strain AN1,[9] the function of which is not clear.

Hyperthermophilic α-glucosidases could also provide valuable insights into protein function, structure, and stability at high temperatures. Indeed, it is the intrinsic high temperature activity and stability of these proteins that have fueled considerable effort into the development of these and

[3] K. B. Xavier, L. O. Martins, R. Peist, M. Kossmann, W. Boos, and H. Santos, *J. Bacteriol.* **178,** 4773 (1996).

[4] W. M. de Vos, S. W. M. Kengen, W. G. B. Voorhorst, and J. van der Oost, *Extremophiles* **2,** 201 (1998).

[5] H. R. Costantino, S. H. Brown, and R. M. Kelly, *J. Bacteriol.* **172,** 3654 (1990).

[6] L. E. Driskill, K. Kusy, M. W. Bauer, and R. M. Kelly, *Appl. Environ. Microbiol.* **65,** 893 (1999).

[7] S. H. Brown and R. M. Kelly, *Appl. Environ. Microbiol.* **59,** 2614 (1993).

[8] K. B. Xavier, R. Peist, M. Kossmann, W. Boos, and H. Santos, *J. Bacteriol.* **181,** 3358 (1999).

[9] K. Piller, R. M. Daniel, and H. H. Petach, *Biochim. Biophys. Acta* **1292,** 197 (1996).

other glycosylhydrolases for use in starch conversion technology.[1] Currently employed mesophilic enzymes exhibit limited tolerance to the high temperatures and pH variations encountered during starch solubilization and degradation.[1] Furthermore, these mesophilic enzymes often have metal ion requirements for activity, whereas their counterpart hyperthermophilic versions often do not.[5,10] Although pullulanases and glucoamylases (also known as amyloglucosidases) are typically used for saccharification of intermediate starch degradation products to glucose,[1] heat-stable α-glucosidases, together with pullulanases, could theoretically fill that role more efficiently. However, despite the potential impact of hyperthermophilic enzymes on industrial processes, including starch conversion, their application is still largely unrealized. One readily apparent obstacle is developing a cost-efficient method for producing sufficient quantities of enzyme either directly from the source organism or through recombinant means.

The methods that follow detail the experimental protocols for the production and purification of the α-glucosidase from *P. furiosus,* along with the methods used to characterize the enzyme.[5] An alternate purification procedure has been provided by Linke *et al.*[10] for the α-glucosidase from *Pyrococcus woesei* but is not described here. Although a recombinant version of the enzyme has not been reported to date, expression of the gene encoding *P. furiosus* α-glucosidase in *Escherichia coli* has been obtained[11] and will also be discussed briefly.

Methods

Growth of P. furiosus for α-Glucosidase Recovery

Pyrococcus furiosus DSM 3638 is grown in continuous culture with S^0 according to the procedure described by Brown and Kelly[12] with the exception that the culture medium is modified to contain 1 g yeast extract, 2 g tryptone, and 3 g soluble starch per liter. Cells are grown at 98° in batch mode until midlog phase and then switched to continuous mode with a dilution rate of 0.5 hr^{-1}. Further details for growth of *P. furiosus* are contained elsewhere in this volume.[12a] Cell densities are monitored by epifluorescence microscopy with acridine orange stain.[13] Cells are harvested by

[10] B. Linke, A. Rudiger, G. Wittenberg, P. L. Jorgensen, and G. Antranikian, *DECHEMA Biotech. Conf.* **5,** 161 (1992).

[11] K. N. Parker and R. M. Kelly, unpublished results.

[12] S. H. Brown and R. M. Kelly, *Appl. Environ. Microbiol.* **55,** 2086 (1989).

[12a] M. F. J. M. Verhagen, A. L. Menon, G. J. Schut, and M. W. W. Adams, *Methods Enzymol.* **330** [3] (2001) (this volume).

[13] J. E. Hobbie, R. J. Daley, and S. Jasper, *Appl. Environ. Microbiol.* **33,** 1225 (1977).

centrifugation at 12,000g for 20 min and frozen as pellets at $-20°$ until ready for use.

Cell pellets harvested from 80 liter of culture are suspended in 0.1 M sodium phosphate (pH 7.3) and sonicated in 10-ml portions for a total of 2 min in a Tekmar sonic disruptor (Model TM300, Tekmar-Dohrmann, Cincinnati, OH) working at 50% duty cycle. After sonication, sulfur and cell debris are removed by centrifugation at 18,000g for 30 min and are found to be essentially void of α-glucosidase activity.

Purification of α-Glucosidase from Pyrococcus furiosus

Step 1. Ammonium Sulfate Precipitation. Solid ammonium sulfate is added to the cell extract supernatant just described to give 30% saturation at room temperature (176 g/liter) and stirred for 2 hr. The suspension is centrifuged at 15,000g for 30 min, and additional ammonium sulfate is added to the supernatant to yield 70% saturation (449 g/liter). The suspension is stirred for another 2 hr and centrifuged at 15,000g for 30 min. This pellet is resuspended in 20 mM Tris–HCl (pH 8.5) and dialyzed against the same buffer for 4 hr.

Step 2. Anion-Exchange Chromatography. The sample is applied to a DEAE-cellulose column (DE52, 2 × 30 cm, Whatman, Clifton, NJ) equilibrated with 20 mM Tris–HCl (pH 8.5). Proteins are eluted with a linear gradient of 240 ml total volume from 0 to 0.4 M NaCl in the same buffer. Fractions containing α-glucosidase activity are pooled and subsequently concentrated about 10-fold (to approximately 10 ml) in a Filtron Technology Novacell stirred cell (Northborough, MA) with a 10,000 kDa weight cutoff. The retentate is dialyzed overnight against 0.1 M sodium phosphate (pH 7.0) containing 0.1 g/liter sodium azide.

Step 3. Hydroxyapatite Chromatography. The concentrated sample is applied to a hydroxyapatite column (2 × 20 cm, Calbiochem-Novabiochem, San Diego, CA) equilibrated with 0.1 M sodium phosphate (pH 7.0) and eluted with a linear gradient of 240 ml total volume from 0.1 to 0.5 M sodium phosphate. Active fractions totaling 36 ml are pooled and concentrated about fourfold as described earlier and then further concentrated another threefold by centrifugation at 4000g for 30 min in Amicon microconcentrators (Millipore, Bedford, MA) with a molecular weight cutoff of 10,000.

Step 4. Gel Filtration Chromatography. The concentrated sample is applied to a Sephadex G-200 column (3 × 90 cm, Amersham Pharmacia Biotech, Piscataway, NJ) equilibrated with 0.1 M sodium phosphate (pH 7.3) containing 0.1 g/liter sodium azide. Active fractions totaling 36 ml are pooled and concentrated about fourfold using the Novacell stirred cell followed by successive centrifugation in an Amicon microconcentrator to about 0.5 ml.

Step 5. Electrophoresis and Electroelution. Electrophoresis and electroelution are performed to obtain α-glucosidase that is pure enough for subsequent N-terminal amino acid sequencing. The final concentrated sample is prepared for native (nondissociating, discontinuous) polyacrylamide gel electrophoresis (PAGE) by boiling in 20 mM dithiothreitol (DTT) for 10 min. α-Glucosidase activity is measured by incubating the gel in 10 mM p-nitrophenol-α-D-glucopyranoside (pNPG) for 30 min at 70°. Electroelution is then accomplished by cutting the yellow activity band from the gel and eluting with a horizontal electrophoresis system for submerged gel electrophoresis (Bethesda Research Laboratories Model H5, Life Technologies, Rockville, MD). The running buffer is identical to that used in preparing the native gel. The horizontal unit is run at 25 mA for about 12 hr. Following electroelution, the sample is dialyzed against 0.1 M sodium phosphate (pH 6.8). Based on the cell-free extract, the α-glucosidase is purified 310-fold at a yield of 8.5% with a specific activity of 245 U/mg according to the assay described later. From 40 g of cells (wet weight), approximately 0.25 mg of purified enzyme is obtained. Higher yields can be achieved by replacing electroelution with a hydrophobic interaction chromatography step.[14]

Alternate Purification Protocol. Purification of the α-glucosidase from *P. furiosus* can also be accomplished by a series of chromatography steps as reported by Bauer *et al.*[15] According to this procedure, α-glucosidase is separated from β-mannosidase and β-glucosidase after the final chromatography step. The four steps consist of (1) DEAE-Sepharose fast flow chromatography (Amersham Pharmacia Biotech), (2) phenyl-Sepharose 650M chromatography (TosoHaas, Montgomeryville, PA), (3) hydroxyapatite chromatography (Calbiochem-Novabiochem), and (4) gel filtration chromatography (Pharmacia HiLoad 16/60 Superdex 200). The protein solution is concentrated between each chromatography step using an Amicon ultrafiltration cell 202 with a YM10 membrane (Millipore). The specific activity of the purified enzyme against pNPG is 67 U/mg, essentially identical to the value reported from the previous protocol for the step prior to electroelution.

[14] R. M. Kelly, S. H. Brown, I. I. Blumentals, and M. W. W. Adams, *in* "Biocatalysis at Extreme Temperatures: Enzyme Systems Near and Above 100°C" (M. W. W. Adams and R. M. Kelly, eds.), ACS Symposium Series 498, p. 23. American Chem. Society, Washington, DC, 1992.

[15] M. W. Bauer, E. J. Bylina, R. V. Swanson, and R. M. Kelly, *J. Biol. Chem.* **271,** 23749 (1996).

Assays for α-Glucosidase Activity[5]

Standard p-Nitrophenol-α-D-glucopyranoside (pNPG) Assay. This assay employs *p*NPG (Sigma-Aldrich, St. Louis, MO) as the substrate. The reaction is followed spectrophotometrically with a Perkin-Elmer Lambda 3 spectrophotometer (Norwalk, CT) fitted with a thermostatted cell holder. A 1:1 mixture (v/v) of ethylene glycol and water is pumped through the cell holder by a circulating water bath. The temperature inside the cuvette is monitored by a thermocouple. Cuvettes are sealed with rubber septa to keep samples from boiling for temperatures over 100°. Data are collected and analyzed by Perkin-Elmer PECSS software (Norwalk, CT). Each assay is performed by first pipetting 1 ml of assay buffer consisting of 20 mM *p*NPG and 0.1 M sodium phosphate (pH 5.5) into a quartz cuvette. The cuvette is then placed into the cell holder and incubated for 5 to 10 min to allow the assay buffer to reach the desired temperature. The enzyme sample is added to the cuvette to start the reaction, and the production of *p*-nitrophenol (*p*NP) is monitored at 405 nm. Specific activity is expressed as micromole of pNP released per minute per milligram of enzyme. The dependence of the molar absorbance coefficient (A_m) of *p*NP on pH and temperature must be taken into account to determine accurate activity measurements.

Alternate Quench Substrate Assay. This assay can be used to determine enzyme specificity for any number of substrates. An appropriate amount of substrate is incubated in 1 ml of 0.1 M sodium phosphate (pH 5.5) for 10 min at the desired temperature. The enzyme sample is then injected to start the reaction, and after 5 min, the reaction mixture is rapidly cooled on ice for 5 min. The release of glucose is measured with a glucose diagnostic kit (kit 510-A, colorimetric from Sigma-Aldrich, St. Louis, MO).

Biochemical Characterization of Purified α-Glucosidase[5]

Molecular Weight. Sodium dodecyl sulfate–polyacrylamide gel electrophoresis (SDS–PAGE; 10%) is performed on the purified α-glucosidase to determine its molecular mass. Silver staining reveals a protein band at about 125 kDa, which is shown to be α-glucosidase by gel activity staining. This value may be somewhat inaccurate given that this α-glucosidase is particularly difficult to denature. For example, residual activities of 5–10% are measured for samples boiled for 2 min in 1.0% SDS and 100 mM DTT and then assayed in the presence of 0.1% SDS. Without completely denaturing the protein, molecular weights determined by comparison to normal standards could yield at least slight errors. Further, this band could actually represent an intact dimer of the enzyme, giving a monomer molecular mass of about 60 kDa.[16] This value would be similar to the 55-kDa molecular mass calculated from the known amino acid sequence. (However, please note that all specific activities reported here assume one active site per 125-kDa subunit.)

Temperature and pH Optima. The effect of temperature and pH on α-glucosidase activity is determined using the standard pNPG assay with the exception that temperature and pH are varied independently. At pH 5.5, the enzyme has a rather broad temperature optimum between 105 and 115°. A temperature optimum of 105° was measured for the same enzyme from *P. woesei.*[10] Significant activity may be present above 115°, but it is difficult to maintain assay conditions at such high temperatures. At a temperature of 98°, the *P. furiosus* enzyme exhibits maximum activity between pH 5.0 and 6.0.

Substrate Specificity. At 108° (pH 5.5), the specific activities with pNPG and methyl-α-D-glucopyranoside are 287 and 185 U/mg, respectively. The enzyme is more active with maltose and isomaltose, giving values of 1760 and 951 U/mg, respectively. No activity is detectable with starch. Costantino *et al.*[5] reported that the enzyme does not utilize sucrose, but a specific activity of 20 U/mg has since been measured.[17] These results are generally similar to those reported by Linke *et al.*[10] for the same enzyme from *P. woesei.* While this group detected no activity with starch, glycogen, amylopectin, maltodextrin, pullulan, and amylose, significant activity was measured with sucrose and trehalose (155 and 45% of that found with maltose).[10] Interestingly, compared to maltose, activities were higher with substrates having a degree of polymerization (*DP*) of 3 to 5 (maltotriose to maltopentaose) but lower for *DP* 6 and 7 (maltohexaose and maltoheptaose).[10] The α-glucosidase from *P. furiosus* exhibits no calcium ion requirements.

Stability. When incubated at 98° (pH 6.5) for various times, the enzyme has a half-life (loss of 50% activity) of about 48 hr. The enzyme retains more than 80% of its activity when incubated for 30 min at 98° with either 100 mM DTT or 1.0 M urea. Residual activities of 22 and 50% are measured in 1.0 M SDS and 20 mM EDTA, respectively, under the same conditions. Inhibition by EDTA suggests that even though calcium ions are not required for activity, other metal ions may nonetheless be important for stability. The α-glucosidase loses 100% of its activity after 30 min of incubation at 98° with 1.0 M guanidine hydrochloride.

---

[16] R. M. Kelly, unpublished results.

[17] M. W. Bauer and R. M. Kelly, unpublished results.

Preliminary Expression of Recombinant α-Glucosidase[11]

Based on the determined N-terminal amino acid sequence, the α-glucosidase gene was identified[18] within the known *P. furiosus* genome,[19] and primers are designed to amplify the gene using the polymerase chain reaction (PCR). The amplified gene insert is ligated into a previously digested pET-based vector and transformed into a cloning strain of *E. coli*. Positive screens are double-checked by colony PCR and restriction mapping of recovered plasmid DNA. The plasmid encoding the α-glucosidase gene is retransformed into a high-expression *E. coli* host, and the resulting colonies are screened for activity against *p*NPG. Positive colonies are grown in small cultures of 10 ml at 37° and induced with 0.4 m*M* isopropyl-β-D-thiogalactopyranoside at an optical density of 0.5–0.8. Growth is continued for a few more hours before the cells are collected by centrifugation. The pelleted cells are then resuspended in a 50 m*M* sodium phosphate buffer (pH 7), and α-glucosidase activity is confirmed on addition of 10 mg/ml *p*NPG (in 50 m*M* sodium phosphate, pH 6) and heating for 30 min at 95–100°. The monomeric molecular mass of the recombinant enzyme is approximately 55 kDa as determined by SDS–PAGE, essentially identical to what is found for the nonrecombinant α-glucosidase recovered directly from *P. furiosus*.

Comparison of Thermophilic and Hyperthermophilic α-Glucosidases

The α-glucosidases from *P. furiosus* and *T. litoralis* have similar N-terminal amino acid sequences but neither show homology to other families of glycosyl hydrolases.[20] This suggests that these two α-glucosidases may belong to a new family, distinct from the α-glucosidases of the hyperthermophilic bacteria *Thermotoga maritima* and *T. neapolitana*. The latter are currently grouped in family 4 of glycosylhydrolases[21] according to the classification system of Henrisatt.[22] An intracellular α-glucosidase was purified from *Sulfolobus solfataricus* strain 98/2, and it and the *P. furiosus* enzyme exhibit similar substrate specificities and temperature optima, although the *S. solfataricus* enzyme is less thermostable (only 3 hr at 95°).[23] The extracellular α-glucosidase purified from *Thermococcus* strain AN1 is monomeric and has a smaller subunit[9] than the often oligomeric, intracellular

[18] S. H. Brown, Ph.D. dissertation, Johns Hopkins Univ., Baltimore, MD, 1993.

[19] Accessed via Utah Genome Center, Univ. of Utah, http://www.genome.utah.edu.

[20] M. W. Bauer, S. B. Halio, and R. M. Kelly, *Adv. Protein Chem.* **48,** 217 (1996).

[21] M. W. Bauer, L. E. Driskill, and R. M. Kelly, *Curr. Opin. Biotechnol.* **9,** 141 (1998).

[22] B. Henrissat, *Biochem. J.* **280,** 309 (1991).

[23] M. Rolfsmeier and P. Blum, *J. Bacteriol.* **177,** 482 (1995).

TABLE I
COMPARISON OF SOME HEAT-STABLE α-GLUCOSIDASES

Organism	Localization	Number of subunits	Subunit size (kDa)	Temperature optimum (°)	pH optimum	Half-life (hr)[a]	Substrate preference[b]
Pyrococcus furiosus[c]	Intracellular	α_2	55[d]	105–115	5.5	48	Maltose
P. woesei[e]	Intracellular	NR	NR	105	5.0–5.5[f]	NR	Maltopentaose
Sulfolobus solfataricus 98/2[g]	Intracellular	α_4 or α_5	80	105	4.5	2.8–3.0	Maltose
Thermococcus strain AN1[h]	Extracellular	α	60	NR	7.5	0.5	pNPG

[a] At 98°.
[b] Of those substrates tested.
[c] H. R. Costantino, S. H. Brown, and R. M. Kelly, J. Bacteriol. 172, 3654 (1990).
[d] Calculated from amino acid sequence. M. W. Bauer and R. M. Kelly, unpublished results.
[e] B. Linke, A. Rudiger, G. Wittenberg, P. L. Jorgensen, and G. Antranikian, DECHEMA Biotech. Conf. 5, 161 (1992).
[f] C. Leuschner and G. Antranikian, World J. Microbiol. Biotechnol. 11, 95 (1995).
[g] M. Rolfsmeier and P. Blum, J. Bacteriol. 177, 482 (1995).
[h] K. Piller, R. M. Daniel, and H. H. Petach, Biochim. Biophys. Acta 1292, 197 (1996).

versions from P. furiosus,[5] P. woesei,[10] and S. solfataricus.[23,24] A summary of the biochemical properties of these α-glucosidases is provided in Table I.

With regard to the kinetic parameters of these enzymes, the maximum reaction velocity, V_{max}, of the native α-glucosidase from P. furiosus is similar to the values reported for the α-glucosidases from Bacillus caldovelox[25] and Bacillus thermoglucosidus[26] when determined at the optimal growth temperature of each organism, namely 98, 60, and 60°, respectively.[5,27] However, the catalytic efficiency (k_{cat}/K_m) of the P. furiosus enzyme is almost 100-fold higher than that of the B. thermoglucosidus enzyme.[27] The V_{max} of the P. furiosus α-glucosidase increases from 140 to 310 U/mg between 98 and 110° but at some expense to catalytic efficiency (e.g., k_{cat}/K_m), which drops from 4.2×10^5 to 3.2×10^5 per mM-min, respectively.[27]

Differences in substrate specificity can lead to misidentification of α-glucosidase for classification purposes. This is most likely to be the case for the sucrose α-glucohydrolase isolated from P. furiosus. It was named for its high specificity for sucrose,[28] but it is probably the same as P. furiosus α-glucosidase.[29] Such confusion is understandable considering that the

[24] A. Sunna, M. Moracci, M. Rossi, and G. Antranikian, Extremophiles 1, 2 (1997).
[25] M. Giblin, C. T. Kelly, and W. M. Fogerty, Can. J. Microbiol. 33, 614 (1987).
[26] Y. Suzuki, T. Yuki, T. Kishigami, and S. Abe, Biochim. Biophys. Acta 445, 386 (1976).
[27] M. W. W. Adams and R. M. Kelly, Bioorg. Med. Chem. 2, 659 (1994).
[28] H. R. Badr, K. A. Sims, and M. W. W. Adams, Syst. Appl. Microbiol. 17, 1 (1994).
[29] R. M. Kelly, unpublished results.

closely related α-*glucosidase* from *P. woesei* possesses an even higher activity toward sucrose than to maltose.[10] Comparisons between the putative α-glucosidases from different organisms will be facilitated greatly by the availability of the amino acid (gene) sequences of these enzymes.

Acknowledgments

RMK acknowledges the Novartis Agricultural Biotechnology Research Institute (NABRI) (Research Triangle Park, NC), the U.S. Department of Agriculture, and the National Science Foundation for financial support of this work.

[17] Amylolytic Enzymes from Hyperthermophiles

By Costanzo Bertoldo and Garabed Antranikian

Introduction

In recent years, many hyperthermophilic organisms that are able to grow at temperatures up to the normal boiling point of water have been shown to utilize natural polymeric substrates as carbon and energy sources. These extreme thermophilic microorganisms, which belong to Archaea and Bacteria, can facilitate the enzymatic degradation of carbohydrates by producing heat-stable enzymes that are catalytically active above 100°. Such enzymes share the ability to hydrolyze the glycosidic bonds between two or more sugar molecules or between a carbohydrate and a noncarbohydrate moiety. The enzymatic hydrolysis and the modification of polysaccharides are of great interest in the field of food technology. The potential exploitation of this natural source of sugars is not only useful for glucose syrup production, but also for the synthesis of nonfermentable carbohydrates, anticariogenic agents, and antistaling agents in baking.[1] In most hyperthermophilic microorganisms, the activity levels of starch-hydrolyzing enzymes appear to be too low for biotechnological applications. The molecular cloning of the corresponding genes and their expression in heterologous hosts, however, circumvent the problem of insufficient expression in the natural host. The number of genes from hyperthermophiles encoding amylolytic enzymes that have been cloned and expressed in mesophiles has been increasing significantly. In most cases, the thermostable proteins expressed in mesophilic hosts maintain their thermostability, are correctly

[1] D. Cowan, *Tibtech* **14**, 177 (1996).

0076-6879/00 $35.00

folded at low temperature, are resistant to host proteolysis, and can be purified easily using thermal denaturation of the mesophilic host proteins. The degree of enzyme purity obtained is generally suitable for most industrial applications.

This article briefly discusses the enzymatic action and properties of starch-hydrolyzing enzymes and focuses only on those enzymes that have been isolated and characterized from extreme thermophilic and hyperthermophilic Archaea and Bacteria. It also briefly describes their potential biotechnological exploitation. Some of these aspects have been already presented.[2-8] Furthermore, this article describes the general methodology for gene identification and expression and the procedure for the purification of some amylolytic enzymes that have been successfully isolated and extensively studied in the authors' laboratory.

Amylolytic Enzymes

Starch from cultivated plants represents a ubiquitous and easily accessible source of energy. In plant cells or seeds, starch is usually deposited in the form of large granules in the cytoplasm. Starch is composed exclusively of α-glucose units that are linked by α-1,4- or α-1,6-glycosidic bonds. The two high molecular weight components of starch are amylose (15–25%), a linear polymer consisting of α-1,4-linked glucopyranose residues, and amylopectin (75–85%), a branched polymer containing, in addition to α-1,4-glycosidic linkages, α-1,6-linked branch points occurring every 17–26 glucose units. α-Amylose chains, which are not soluble in water but form hydrated micelles, are polydisperse and their molecular weights vary from several hundred to thousands. The molecular weight of amylopectin may be as high as 100 million and in solution such a polymer has colloidal or micellar forms.

Because of the complex structure of starch, cells require an appropriate combination of hydrolyzing enzymes for the depolymerization of starch to

[2] G. Antranikian, in "Microbial Degradation of Natural Products" (G. Winkelmann, ed.), Vol. 2, p. 28. VCH, Weilheim, 1992.

[3] R. Ladenstein and G. Antranikian, Adv. Biochem. Eng./Biotechnol. **61,** 37 (1998).

[4] C. Leuschner and G Antranikian, World J. Microbiol. Biotechnol. **11,** 95 (1995).

[5] A. Rüdiger, A. Sunna, and G. Antranikian, in (D. Waldmann, ed.), "Carbohydrases: Handbook of Enzyme Catalysis in Organic Synthesis," p. 946. VCH, Weilheim, 1994.

[6] A. Sunna, M. Moracci, M. Rossi, and G. Antranikian, Extremophiles **1,** 2 (1996).

[7] R. Müller, G. Antranikian, S. Maloney, and R. Sharp, Adv. Biochem. Eng. Biotechnol. **61,** 155 (1998).

[8] F. Niehaus, C. Bertoldo, M. Kähler, and G. Antranikian, Appl. Microbiol. Biotechnol. **51,** 711 (1999).

oligosaccharides and smaller sugars, such as glucose and maltose. They can be simply classified into two groups: endo-acting enzymes or endohydrolases and exo-acting enzymes or exohydrolases (Fig. 1). Endo-acting enzymes, such as α-amylase (α-1,4-glucan-4-glucanohydrolase; EC 3.2.1.1), hydrolyze linkages in the interior of the starch polymer in a random fashion, which

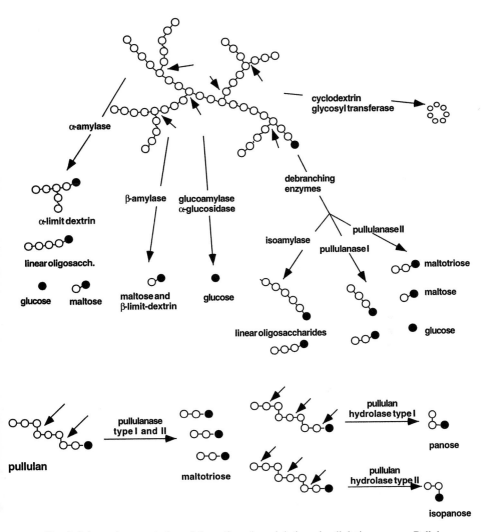

FIG. 1. Schematic presentation of the action of amylolytic and pullulytic enzymes. Pullulanase type I also attacks α-1,6-glycosidic linkages in oligo- and polysaccharides. Pullulanase type II attacks, in addition, α-1,4-linkages in various oligo- and polysaccharides.

leads to the formation of linear and branched oligosaccharides. The sugar-reducing groups are liberated in the α-anomeric configuration. Most starch-hydrolyzing enzymes belong to the α-amylase family, which contains a characteristic catalytic $(\beta/\alpha)_8$-barrel domain. α-Amylases belong to two families, number 13 and 57 of the glycosylhydrolase families.[9] Family 13 has approximately 150 members from Eukarya and Bacteria. Family 57, however, has only three members from Bacteria and Archaea. Throughout the α-amylase family, only eight amino acid residues are invariant, seven at the active site and a glycine in a short turn.[10] On the structural level, there are to date no X-ray structures of amylolytic enzymes derived from hyperthermophiles.

Exo-acting starch hydrolases include β-amylase, glucoamylase, α-glucosidase, and isoamylase. These enzymes attack the substrate from the nonreducing end, producing small and well-defined oligosaccharides. β-Amylase (EC 3.2.1.2), also referred to as α-1,4-D-glucan maltohydrolase or saccharogen amylase, hydrolyzes α-1,4-glucosidic linkages to remove successive maltose units from the nonreducing ends of the starch chains, producing β-maltose by an inversion of the anomeric configuration of the maltose. β-Amylase belongs to family 14 of the glycosylhydrolases, having 11 members from Eukarya and Bacteria.

Glucoamylase (EC 3.2.1.3) hydrolyzes terminal α-1,4-linked-D-glucose residues successively from nonreducing ends of the chains, releasing β-D-glucose. Glucoamylase, which is typically a fungal enzyme, has several names: α-1,4-D-glucan hydrolase, amyloglucosidase, and γ-amylase. Most forms of the enzyme can hydrolyze α-1,6-D-glucosidic bonds when the next bond in sequence is α-1,4. However, in vitro this enzyme hydrolyzes α-1,6- and α-1,3-D-glucosidic bonds in other polysaccharides with high molecular weights.

α-Glucosidase (EC 3.2.1.20), or α-D-glucoside glucohydrolase, attacks the α-1,4 linkages of oligosaccharides that are produced by the action of other amylolytic enzymes. Unlike glucoamylase, α-glucosidase liberates glucose with an α-anomeric configuration. α-Glucosidases are members of family 15 and the very diverse family 31 of the glycosylhydrolases.[9] Isoamylase (EC 3.2.1.68), or glycogen 6-glucanohydrolase, is a debranching enzyme specific for α-1,6 linkages in polysaccharides, such as amylopectin, glycogen, and β-limit dextrin, but it is unable to hydrolyze the α-1,6 linkages in pullulan or branched oligosaccharide; therefore, it has limited action on α-limit dextrin.

[9] B. Henrissat, *Biochem. J.* **280,** 309 (1991).
[10] T. Kuriki and T. Imanaka, *J. Biosci. Bioeng.* **87,** 557 (1999).

Pullulytic Enzymes

Pullulan is a linear α-glucan consisting of maltotriose units joined by α-1,6-glycosidic linkages and it is produced by *Aureobasidium pullulans* with a chain length of 480 maltotriose units. Enzymes capable of hydrolyzing α-1,6 glycosidic bonds in pullulan are defined as pullulanases (Fig. 1). On the basis of substrate specificity and product formation, pullulanases have been classified into two groups: pullulanase type I and pullulanase type II. Pullulanase type I (EC 3.2.1.41) specifically hydrolyzes the α-1,6 linkages in pullulan as well as in branched oligosaccharides, and its degradation products are maltotriose and linear oligosaccharides, respectively. Pullulanase type I is unable to attack α-1,4 linkages in α-glucans and belongs to family 13 of the glycosylhydrolases. Pullulanase type II, or amylopullulanase, attacks α-1,6-glycosidic linkages in pullulan and α-1,4 linkages in branched and linear oligosaccharides (Fig. 1). The enzyme has a multiple specificity and is able to fully convert polysaccharides (e.g., amylopectin) to small sugars ($DP1$, $DP2$, $DP3$; DP is the degree of polymerization) in the absence of other enzymes, such as α-amylase or β-amylase.

In contrast to the previously described pullulanases, pullulan hydrolases type I and type II are unable to hydrolyze α-1,6-glycosidic linkages in pullulan or in branched substrates. They can attack α-1,4-glycosidic linkages in pullulan-forming panose or isopanose. Pullulan hydrolase type I or neopullulanase (EC 3.2.1.135) hydrolyze pullulan to panose (α-6-D-glucosylmaltose). Pullulan hydrolase type II or isopullulanase (EC 3.2.1.57) hydrolyzes pullulan to isopanose (α-6-maltosylglucose) (Fig. 1).

Finally, cyclodextrin glycosyltransferase (EC 2.4.1.19), or α-1,4-D-glucan α-4-D-(α-1,4-D-glucano)transferase, is an enzyme that is generally found in Bacteria and has been discovered in Archaea. This enzyme produces a series of nonreducing cyclic dextrins from starch, amylose, and other polysaccharides. α-, β-, and γ-cyclodextrins are rings formed by 6, 7, and 8 glucose units that are linked by α-1,4 bonds, respectively (Fig. 1).

Enzyme Production by Hyperthermophilic Microorganisms

Hyperthermophilic Bacteria and Archaea that are able to grow on starch at temperatures over 70° have been isolated and the corresponding starch-degrading enzymes have been purified and characterized. In several cases, genes encoding these enzymes have been isolated, cloned, and overexpressed in heterologous hosts (Table I).

TABLE I

STARCH-HYDROLYZING ENZYMES FROM EXTREMELY THERMOPHILIC AND HYPERTHERMOPHILIC
ARCHAEA AND BACTERIA

Enzyme	Organism[a]	Enzyme properties[a]			
		Optimal temperature	Optimal pH	Molecular mass (kDa)	Remarks
α-Amylase	*Desulfurococcus mucosus* (85)	85	5.5	74	Purified/cloned
	Pyrococcus furiosus (100)	100	6.5–7.5	129	Purified/cloned/intracellular
		100	7.0	68	Purified/cloned/extracellular
	Pyrococcus sp. KOD1	90	6.5	49.5	Purified/cloned/extracellular
	P. woesei (100)	100	5.5	68	Purified/extracellular
	Pyrodictium abyssi (98)	100	5.0	—[b]	Crude extract
	Staphylothermus marinus (90)	100	5.0	—	Crude extract
	Sulfolobus solfataricus (88)	—	—	240	Extracellular
	Thermococcus celer (85)	90	5.5	—	Crude extract
	T. profundus DT5432 (80)	80	5.5	42	Purified/cloned/"Amy S"
	T. profundus (80)	80	4.0–5.0	42	Purified/"Amy L"
	T. aggregans (85)	95	6.5	—	Cloned
	Dyctyoglomus thermophilum Rt46B.1 (73)	90	5.5	75	Purified/cloned/cytoplasmic fraction
	Thermotoga maritima MSB8 (90)	85–90	7.0	61	Purified/cloned/lipoprotein
Pullulanase type I	*Fervidobacterium pennivorans* Ven5 (75)	80	6	190 (93)	Purified/cloned
	T. maritima MSB8 (90)	90	6.0	93 (subunit)	Cloned/type I[c]
	Thermus caldophilus GK24 (75)	75	5.5	65	Purified/cell associated
Pullulanase type II	*Desulfurococcus mucosus* (88)	85	5.0	74	Purified/cloned
	P. woesei (100)	100	6.0	90	Purified/cloned/cell associated
	P. abyssi (98)	100	9.0	—	Crude extract
	T. celer (85)	90	5.5	—	Crude extract
	Thermococcus litoralis (90)	98	5.5	119	Purified/extracellular/glycoprotein
	T. hydrothermalis (80)	95	5.5	128	Purified/extracellular/glycoprotein
	T. aggregans (85)	100	6.5	83	Purified/cloned
Glucoamylase	*Thermoplasma acidophilum* (60)	90	6.5	141	Purified
	Picrophilus oshimae (60)	90	2.0	140	Purified
	P. torridus (60)	90	2.0	133	Purified
CGTase	*Thermococcus* sp. (75)	100	2.0	83	Purified
α-Glucosidase	*Thermococcus* strain AN1 (80)	130	—	63	Purified/extracellular/glycoprotein
	Thermococcus hydrothermalis (80)	—	—	—	Cloned

[a] Values in parentheses give the optimal growth temperature for each organism in degrees Celsius.
[b] Not determined.
[c] Unpublished results.

Heat-Stable Amylases and Glucoamylases

Extremely thermostable α-amylases have been characterized from *Pyrococcus furiosus*,[11] *Pyrococcus woesei*,[12] and *Thermococcus profundus*.[13] Optimal temperatures for the activity of these enzymes are 100°, 100°, and 80°, respectively. In addition, either amylase and/or pullulanase activities have been observed in hyperthermophilic Archaea of the genera *Sulfolobus, Thermophilum, Desulfurococcus, Thermococcus,* and *Staphylothermus*.[14,15] Molecular cloning of the corresponding genes and their expression in heterologous hosts circumvent the problem of insufficient expression in the natural host. The gene encoding an extracellular α-amylase from *P. furiosus* has been cloned and the recombinant enzyme expressed in *B. subtilis* and *Escherichia coli*.[16,17] This is the first report on the expression of an archaeal gene derived from an extremophile in a *Bacillus* strain. The high thermostability of the pyrococcal extracellular α-amylase (thermal activity even at 130°) in the absence of metal ions, together with its unique product pattern and substrate specificity, makes this enzyme an interesting candidate for industrial application. In addition, an intracellular α-amylase gene from *P. furiosus* has been cloned and sequenced.[18] It is interesting that the four highly conserved regions usually identified in α-amylases are not found in this enzyme. α-Amylases with lower thermostability and thermoactivity have been isolates from the archaea *T. profundus*,[19] *Pyrococcus* sp. KOD1,[20] and the bacterium *Thermotoga maritima*.[21] Genes encoding these enzymes were successfully expressed in *E. coli*. Similar to the amylase from *B. licheniformis,* which is commonly used in liquefaction, the enzyme from *T. maritima* requires Ca^{2+} for activity. Further investigations have shown that

[11] R. Koch, K. Spreinat, K. Lemke, and G. Antranikian, *Arch. Microbiol.* **155,** 572 (1991).

[12] Y. C. Chung, T. Kobayashi, H. Kanai, T. Akiba, and T. Kudo, *Appl. Environ. Microbiol.* **61,** 1502 (1995).

[13] J. T. Lee, H. Kanai, T. Kobayashi, T. Akiba, and T. Kudo, *J. Ferment. Bioeng.* **82,** 432 (1996).

[14] J. M. Bragger, R. M. Daniel, T. Coolbear, and H. W. Morgan, *Appl. Microbiol. Biotechnol.* **31,** 556 (1989).

[15] F. Canganella, C. Andrade, and G. Antranikian, *Appl. Microbiol. Biotechnol.* **42,** 239 (1994).

[16] G. Dong, C. Vieille, A. Savchenko, and J. G. Zeikus, *Appl. Environ. Microbiol.* **63,** 3569 (1997).

[17] S. Jørgensen, C. E. Vorgias, and G. Antranikian, *J. Biol. Chem.* **272,** 16335 (1997).

[18] K. A. Laderman, K. Asada, T. Uemori, H. Mukai, Y. Taguchi, I. Kato, and C. B. Anfinsen, *J. Biol. Chem.* **268,** 24402 (1993).

[19] Y. S. Kwak, T. Akeba, and T. Kudo, *J. Ferment. Bioeng.* **86,** 363 (1998).

[20] Y. Tachibana, L. M. Mendez, S. Fujiwara, M. Takagi, and T. Imanaka, *J. Ferment. Bioeng.* **82,** 224 (1996).

[21] W. Liebl, I. Stemplinger, and P. Ruile, *J. Bacteriol.* **179,** 941 (1997).

the extreme hyperthermophilic archaeon *Pyrodictium abyssi* can grow on various polysaccharides and also secretes a heat-stable amylase that is optimally active above 100° and has a wide functional pH range (unpublished results).

Unlike α-amylase, the production of glucoamylase seems to be very rare in extremely thermophilic and hyperthermophilic Bacteria and Archaea. Among thermophilic anaerobic bacteria, glucoamylases have been purified and characterized from *Clostridium thermohydrosulfuricum* 39E,[22] *Thermoanaerobacterium thermosaccharolyticum* DSM 571,[23] and *Clostridium thermosaccharolyticum*.[24] The latter enzyme is optimally active at 70° and pH 5.0. It has been shown that the thermoacidophilic archaea *Thermoplasma acidophilum, Picrophilus torridus,* and *Picrophilus oshimae* produce heat- and acid-stable glucoamylases. The purified archaeal glucoamylases are optimally active at pH 2 and 90°. Catalytic activity is still detectable at pH 0.5 and 100°. This represents the first report on the presence of glucoamylases in thermophilic Archaea (unpublished results).

α-Glucosidases have been found in thermophilic Archaea and Bacteria. An extracellular α-glucosidase from the thermophilic archaeon *Thermococcus* strain AN1[25] was purified and its molecular characteristics determined. The monomeric enzyme (60 kDa) is optimally active at 98°. The purified enzyme has a half-life around 35 min, which is increased to around 215 min in the presence of 1% (w/v) dithiothreitol (DTT) and 1% (w/v) bovine serum albumin (BSA). The substrate preference of the enzyme is *p*NP-α-D-glucoside > nigerose > panose > palatinose > isomaltose > maltose > turanose. No activity was found with starch, pullulan, amylose, maltotriose, maltotetraose, isomaltotriose, cellobiose, and β-gentiobiose. The enzyme was active at 130°. The gene encoding an α-glucosidase from *Thermococcus hydrothermalis*[26] was cloned by complementation of a *Saccharomyces cerevisiae* deficiency maltase-deficient mutant strain. The cDNA clone isolated encodes an open reading frame (ORF) corresponding to a protein of 242 amino acids. The protein shows 42% identity to a *Pyrococcus horikoshii* unknown ORF, but no similarities were obtained with other polysaccharidase sequences.

[22] H. H. Hyun and J. G. Zeikus, *J. Bacteriol.* **164**, 1146 (1985).
[23] D. Ganghofner, J. Kellermann, W. L. Staudenbauer, and K. Bronnenmeier, *Biosci. Biotechnol. Biochem.* **62**, 302 (1998).
[24] U. Specka, F. Mayer, and G. Antranikian, *Appl. Environ. Microbiol.* **57**, 2317 (1991).
[25] K. Piller, R. M. Daniel, and H. H. Petach, *Protein Struct. Mol. Enzymol.* **1**, 197 (1996).
[26] A. Galichet and A. Belarbi, *FEBS Lett.* **2**, 188 (1999).

Thermoactive Pullulanases and CGTases

Thermostable and thermoactive pullulanases from extreme thermophilic microorganisms have been detected in *Thermococcus celer, Desulfurococcus mucosus, Staphylothermus marinus,* and *Thermococcus aggregans.*[15] Temperature optima between 90 and 105°, as well as remarkable thermostability, even in the absence of substrate and calcium ions, have been observed. Most thermoactive pullulanases identified to date belong to the type II group, which attack α-1,4- and α-1,6-glycosidic linkages. They have been purified from *P. furiosus* and *T. litoralis,*[27] *T. hydrothermalis,*[28] and *Pyrococcus* strain ES4.[29] Pullulanases type II from *P. furiosus* and *P. woesei* have been expressed in *E. coli.*[30,31] The unfolding and refolding of pullulanase from *P. woesei* have been investigated using guanidinium chloride as the denaturant. The monomeric enzyme (90 kDa) was found to be very resistant to chemical denaturation, and the transition midpoint for guanidinium chloride-induced unfolding was determined to be 4.86 ± 0.29 M for intrinsic fluorescence and 4.90 ± 0.31 M for far-UV circular dichroism (CD) changes. The unfolding process was reversible. Reactivation of the completely denatured enzyme (in 7.8 M guanidinium chloride) was obtained on removal of the denaturant by stepwise dilution; 100% reactivation was observed when refolding was carried out via a guanidinium chloride concentration of 4 M in the first dilution step. Particular attention has been paid to the role of Ca^{2+}, which activates and stabilizes this archaeal pullulanase against thermal inactivation. The enzyme binds two Ca^{2+} ions with a K_d of 0.080 ± 0.010 μM and a Hill coefficient H of 1.00 ± 0.10. This cation significantly enhances the stability of the pullulanase against guanidinium chloride-induced unfolding. The refolding of pullulanase, however, was not affected by Ca^{2+}.[32] Genes encoding pullulanases from *T. hydrothermalis*[33] and *T. aggregans* (unpublished results) have been isolated and expressed in mesophilic hosts. The aerobic thermophilic bacterium *Thermus caldophilus* GK-24 produces a thermostable pullulanase of type I when grown on

[27] S. H. Brown and R. M. Kelly, *Appl. Environ. Microbiol.* **59,** 2614 (1993).

[28] H. Gantelet and F. Duchiron, *Appl. Microbiol. Biotechnol.* **49,** 770 (1998).

[29] J. W. Schuliger, S. H. Brown, J. A. Baross, and R. M. Kelly, *Mol. Mar. Biol. Biotechnol.* **2,** 76 (1993).

[30] A. Rüdiger, P. L. Jørgensen, and G. Antranikian, *Appl. Environ. Microbiol.* **61,** 567 (1995).

[31] G. Dong, C. Vieille, and J. G. Zeikus, *Appl. Environ. Microbiol.* **63,** 3577 (1997).

[32] R. M. Schwerdtfeger, R. Chiaraluce, V. Consalvi, R. Scandurra, and G. Antranikian, *Eur. J. Biochem.* **264,** 479 (1999).

[33] M. Erra-Pujada, P. Debeire, F. Duchiron, and M. J. O. Donohue, *J. Bacteriol.* **181,** 3282 (1999).

starch.[34] This enzyme debranches amylopectin by attacking specifically α-1,6-glycosidic linkages. The pullulanase is optimally active at 75° and pH 5.5, is thermostable up to 90°, and does not require Ca^{2+} for either activity or stability. The first debranching enzyme from an anaerobic thermophile was identified in the bacterium *Fervidobacterium pennivorans* Ven5 and finally cloned and expressed in *E. coli*.[35] In contrast to the pullulanase from *P. woesei* (specific to both α-1,6 and α-1,4 glycosidic linkages), pullulanase type II or amylopullulanase, the enzyme from *F. pennivorans* ven5 attacks exclusively the α-1,6-glycosidic linkages in polysaccharides (pullulanase type I). This thermostable debranching enzyme leads to the formation of long chain linear polysaccharides from amylopectin.

Thermostable CGTases are produced by the members of the genus *Thermoanaerobacter* and *Thermoanaerobacterium thermosulfurigenes*[36,37] and *Anaerobranca bogoriae*.[38] A CGTase has been purified from a newly isolated archaeon, *Thermococcus* sp. The optimum temperatures for starch-degrading activity and cyclodextrin synthesis are 110 and 90–100°, respectively. This is the first report on the presence of a thermostable CGTase in a hyperthermophilic archaeon.[39] The ability of extreme thermophiles and hyperthermophiles to produce heat-stable glycosylhydrolases is summarized in Table I.

Methods

Cloning and Expression of Hyperthermophilic Amylolytic Enzymes in Mesophilic Hosts

Recombinant starch-hydrolyzing enzymes from several extreme thermophilic and hyperthermophilic microorganisms have been expressed in *E. coli* as soluble cytosolic proteins. In general, the recombinant enzyme produced in mesophilic hosts maintains the properties of the wild-type enzyme. The gene encoding amylolytic enzyme can be identified and ampli-

[34] C. H. Kim, O. Nashiru, and J. H. Ko, *FEMS Microbiol. Lett.* **138,** 147 (1996).

[35] C. Bertoldo, F. Duffner, P. L. Jørgensen, and G. Antranikian, *Appl. Environ. Microbiol.* **65,** 2084(1999).

[36] S. Pedersen, B. F. Jensen, L. Dijkhuizen, S. T. Jørgensen, and B. W. Dijkstra, *Chemtech* **12,** 19 (1995).

[37] R. Wind, W. Liebl, R. Buitlaar, D. Penninga, A. Spreinat, L. Dijkhuizen, and H. Bahl, *Appl. Environ. Microbiol.* **61,** 1257 (1995).

[38] S. Prowe, J. van de Vossenberg, A. Driessen, G. Antranikian, and W. Konnings, *J. Bacteriol.* **178,** 4099 (1996).

[39] Y. Tachibana, A. Kuramura, N. Shirasaka, Y. Suzuki, T. Yamamoto, S. Fujiwara, M. Takagi, and T. Imanaka, *Appl. Environ. Microbiol.* **65,** 1991 (1999).

fied by polymerase chain reaction (PCR) using conserved sequences that are known for amylolytic enzymes.[18,31] In many cases, it is possible to detect a starch-hydrolyzing activity after a phenotypical screening of a genomic library, created by shotgun cloning. By using sensitive screening methods, it is possible to detect positive clones producing very low levels of thermostable enzymes. By employing this technique, it is possible to identify new genes expressing novel hydrolytic enzymes. Such procedures have been used to isolate heat-stable α-amylases,[16,17,20,21] as well as pullulanase type II from the archaeon *P. woesei*[30] and pullulanase type I from the anaerobic bacterium *F. pennivorans* Ven5.[35] Figure 2 outlines the clone selection procedure used to isolate these genes, which is based on the secretion of amylolytic/pullulytic activity from recombinant cells and the detection of enzymatic activity by producing a clear halo around the colonies. The next section focuses on the general methodology used to clone amylolytic and pullulytic enzymes from extreme thermophilic Archaea and Bacteria.

Screening of Genomic Library. Chromosomal DNA is isolated from thermophilic Archaea and Bacteria according to Pitcher *et al.,*[40] and 100 μg of DANN is partially digested with 20 U of *Sau*3A for 10 min. The digestion is terminated by a phenol : chloroform extraction, and the DNA is ethanol precipitated, size fractionated on a sucrose gradient and after fragments between 3 and 7 kb are pooled. A plasmid library is constructed in the vector pSJ933[41] and the host strain *E. coli* MC100 using the following method. The cloning vector is digested with *Bam*HI, and a 3.8-kb fragment is purified from an agarose gel. Approximately 0.75 μg of vector fragment is ligated to 4 μg of size-fractionated chromosomal DNA and used to transform *E. coli* MC100 by electroporation. Red dyed substrate (amylopectin or pullulan) is made by suspending 50 g substrate (Hayashibara Biochemical Laboratories) and 5 g Cibachron Rot B (Ciba Geigy) in 500 ml 0.5 M NaOH and incubating at room temperature under constant stirring for 16 hr. The pH is adjusted to 7.0 with 4 N H_2SO_4. The dyed substrates are precipitated under constant stirring on the addition of 600 ml ethanol, harvested by centrifugation, and then resuspended in 500 ml distilled water. This precipitation procedure is repeated three times and the final dyed substrate is resuspended in 500 ml of distilled water and autoclaved. Red-dyed substrate is added to solid medium at a concentration of 1% (v/v). Positive clones expressing active α-amylase or pullulanase can be identified by forming a clear halo around the colonies. Halos are formed when the

[40] D. G. Pitcher, N. A. Saunders, and R. J. Owen, *Lett. Appl. Microbiol.* **8,** 151 (1989).
[41] G. Antranikian, P. L. Jørgensen, M. Wümpelmann, and S. T. Jørgensen, Patent WO 92/02614.

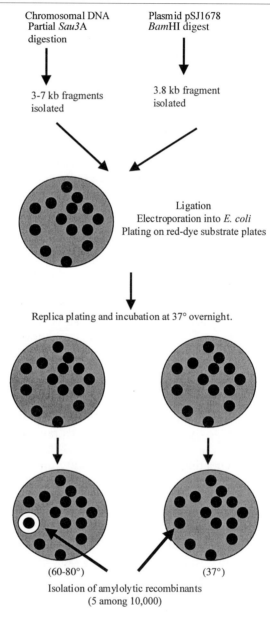

Chromosomal DNA
Partial *Sau*3A
digestion

Plasmid pSJ1678
*Bam*HI digest

3-7 kb fragments
isolated

3.8 kb fragment
isolated

Ligation
Electroporation into *E. coli*
Plating on red-dye substrate plates

Replica plating and incubation at 37° overnight.

(60-80°)   (37°)

Isolation of amylolytic recombinants
(5 among 10,000)

FIG. 2. Outline of the methodology applied to isolate, clone, and detect DNA fragments of chromosomal DNA from hyperthermophilic Archaea and Bacteria exhibiting amylolytic or pullulytic activity.

dye amylopectin or pullulan complex is attacked by the enzyme, causing the release of the dye.

Escherichia coli transformants are plated on LB agar containing 10 μg/ml chloramphenicol and, after 16 hr of incubation at 37°, approximately 14,000 colonies are observed. These are replica plated onto a new set of LB plates containing 2% agar, 6 μg chloramphenicol/ml, and 1% dyed amylopectin or pullulan and grown overnight at 37°. The plates are then incubated at 60° for 4 hr, and the positive clones are identified by halo formation. *E. coli* clones expressing the recombinant extracellular α-amylases and pullulanase are grown aerobically at 37° in LB medium[42] containing 10 μg/ml chloramphenicol. The plasmid containing the gene of interest is isolated using a kit from Qiagen (Hilden, Germany).

Gene Expression and Characterization

Using the just mentioned method, we were able to isolate *E. coli* clones containing the α-amylase gene from *P. furiosus*,[16] the pullulanase type II (amylopullulanase) from *P. woesei*,[30] and the debranching enzymes (pullulanase type I) from *F. pennivorans*.[35] This section reports on the expression and characterization of the gene encoding pullulanase type I from *F. pennivorans* Ven5 and the purification of the recombinant enzyme after expression in *E. coli*.

After shogun cloning, *E. coli* clone PL2125 producing thermostable pullulanase type I from *F. pennivorans* Ven5 is further characterized. The entire 8.1-kb insert is sequenced in both directions and three large open reading frames (ORFs) are identified. ORF1 is confirmed as encoding a pullulanase by subcloning into pUC18 and observing activity of the *E. coli* transformants on red-dyed pullulan plates. The subcloning is performed by carrying out PCR amplification using the Expand Long Template PCR System (Boehringer Mannheim) using the following temperature profile: 94°, 2 min and 30 cycles of 94°; 10 sec, 45°; 45 sec, 68°, 4 min. Cloning of PCR-amplified fragments is carried out using the TA cloning kit (Invitrogen). Pullulanase activity can be seen after overnight growth of the positive clones on LB medium containing red-dyed pullulan at 37° and after heat treatment at 70° for 16 hr. The plasmid pSE420 containing the IPTG-inducible *trc* promoter (Invitrogen) is used for expression.

The pullulanase gene of *F. pennivorans* Ven5 (*pulA*) is 2550 bp in length and encodes a protein of 849 amino acids with a predicted molecular mass of 96.6 kDa before processing. A Shine–Dalgarno-like sequence of AGGAGG is present at positions -10 to -15 from the ATG site. The

[42] J. Sambrook, E. F. Fritsch, and T. Maniatis, "Molecular Cloning: A Laboratory Manual," 2nd Ed. Cold Spring Harbor Laboratory Press, Cold Spring Harbor, NY, 1989.

G + C content of *pul*A is 41.9%. A signal sequence of 28 amino acids is present with a cleavage site between the amino acids Ala and Glu. This was confirmed by N-terminal sequencing of the mature pullulanase isolated from *F. pennivorans* Ven5 (ETELIIHYHRW). The *pul*A gene is subcloned without its signal sequence and overexpressed in *E. coli* (under the control of the *trc* promoter). It is interesting to note that this clone, *E. coli* FD748, produces two proteins (93 and 83 kDa) with pullulanase activity. This is not due to proteolytic hydrolysis of the recombinant mature protein but due to the presence of a second start site (TTG) identified 128 amino acids downstream from the ATG start site, with a Shine–Dalgarno-like sequence (GGAGG). The TTG translation initiation codon was then mutated to produce only the mature 93-kDa protein. Mutation of the leucine codon, TTG, was carried out by PCR according to the method described by Nelson and Long.[43] The primers are as follows: mutation primer, GTG GCT CTT ACA AGG AAT AG; nonsense, CGA TCG ATC GAG GAT CCT TA; reverse + nonsense, CGA TCG ATC GAG GAT CCT TAT TAA TTA CCT TTG TAC ATT ACC; and forward, ATA AAC ATG TCG AAA ACA GAG CTG ATT ATC.

The PCR product is cloned into pCR2.1 and the mutation is confirmed by sequencing. The fragment containing the mutation is digested with *Afl*III and *Bam*HI and cloned into pSE420/*Nco*I and *Bam*HI. The mutation is again confirmed by sequencing. Pullulanase-containing clones are detected on pullulan–red agar plates. *E. coli* FD748 containing the pullulanase cloned without the signal peptide is cultivated in LB or TB medium[42] containing 100 μg/ml ampicillin and induced with 1 m*M* IPTG, when the absorbance of the culture at 600 nm is approximately 0.8. The pullulanase expression level of this clone is 40 times higher than that from *E. coli* PL2125.

Purification of Recombinant Pullulanase from E. coli FD748

In general, heat treatment of the cell extract is ideal for the purification of heat-stable recombinant proteins. All purification steps are performed at room temperature. *E. coli* cells (10 g) expressing pullulanase type I from *F. pennivorans* Ven5 are washed with 25 m*M* Tris–HCl, pH 7.4 (buffer A), and then resuspended with 50 ml of the same buffer. Cells are disrupted by sonication, and cell debris is removed by centrifugation for 20 min at 30,000*g*. The supernatant is heat treated at 75° for 60 min and the denatured host proteins are pelleted by centrifugation (15 min at 30,000*g*). The pullulanase remains in the clear supernatant. The specific activity of the purified

[43] R. M. Nelson and G. L. Long, *Anal. Biochem.* **180,** 147 (1989).

recombinant pullulanase of *F. pennivorans* Ven5 expressed in *E. coli* FD748m (r*pul*A) is 3 U/mg (Table II).

The pullulanase preparation after heat treatment is applied to a β-cyclodextrin–epoxy-activated Sepharose column (1 × 10 cm) equilibrated in 50 m*M* sodium acetate buffer, pH 6.0 (buffer C). The column is washed stepwise with 50 ml of buffer C and then with the same buffer containing 1 *M* NaCl until no absorbance at 280 nm is detectable. Pullulanase activity is eluted with 1% pullulan in buffer C, containing 1 *M* NaCl. Active fractions are pooled, concentrated by ultrafiltration (cutoff 10 kDa), and dialyzed against 1000 volumes of buffer A. The protein solution is then applied to a Mono Q HR 5/5 column (Pharmacia LKB, Freiburg, Germany), which is equilibrated with buffer B, and elution is carried out with the same buffer at a flow rate of 0.5 ml/min. After anion-exchange chromatography on Mono Q, a single protein band is observed. The recombinant full-length pullulanase is purified 25-fold with a specific activity of 75 U/mg and a final yield of about 11.7%. (Table II). The pure enzyme on SDS–PAGE has a molecular mass of 93 kDa. The estimated molecular weight of the native recombinant *pul*A, calculated in a native gradient gel electrophoresis, is 190 kDa. Accordingly, the enzyme, which is optimally active at 80° and pH 6.0, is a homodimer.

β-Cyclodextrin-Sepharose affinity chromatography is a suitable step to purify pullulanases because cyclodextrins are competitive inhibitors of these enzymes. The β-cyclodextrin-Sepharose affinity matrix is prepared by coupling β-cyclodextrin of epoxy-activated Sepharose 6B according to the following protocol following the instructions of the manufacturer (Affinity Chromatography, Pharmacia Fine Chemicals, Uppsala, Sweden). The ep-

TABLE II
PURIFICATION OF RECOMBINANT PULLULANASE FROM *E. coli* FD748m CLONE[a]

Step	Total protein (mg)	Total activity (U)[b]	Specific activity (U/mg)	Yield (%)	Purification (-fold)
Crude extract	747	2268	3	100	
Heat treatment (1 hr at 75°)	172	2184	12.7	96	4.2
β-Cyclodextrin-Sepharose	16.2	873.6	53.9	23	17.9
Mono Q	3.56	267	75	11.7	25

[a] After the aerobic growth of *E. coli* at 37°, a 5-liter culture was centrifuged (6 g of cells wet weight) and the cells were disrupted by sonication.

[b] One unit of pullulanase catalyzes the formation of 1 μmol of reducing sugars/min under the defined conditions; maltose was used as a standard.

oxy-activated Sepahrose 6B (5 g) is washed with distilled water (1 liter) and reswelled in 0.1 M carbonate buffer, pH 11. To this gel suspension, 40 ml of β-cyclodextrin (40 μmol $\times$ g resin) is added and mixed. The mixture is left under agitation on a shaker (60 rpm) at 40° for 16 hr. Excess ligand is removed by washing with carbonate buffer (pH 11) and then with 0.1 M phosphate buffer, pH 8. The remaining epoxy groups are blocked with 1 M ethanolamine dissolved in 0.1 M phosphate buffer, pH 8. After incubation for 16 hr, the resulting β-cyclodextrin-Sepharose is washed with water and then equilibrated with 50 mM sodium acetate buffer, pH 5.5.

Analytical Methods

Assay for Amylolytic/Pullulytic Enzymes

Amylolytic/pullulytic activity is determined by measuring the amount of reducing sugars released during incubation with a substrate. To 50 μl of 1% (w/v) substrate (starch for amylolytic and pullulan for pullulytic activity) dissolved in 50 mM sodium acetate buffer (pH 6.0), 25 or 50 μl of enzyme solution is added, and the samples are incubated at different temperatures for 10 to 60 min. The reaction is stopped by cooling on ice, and the amount of reducing sugars released is determined by the dinitrosalicylic acid (DNS) method. During starch degradation, monomeric or oligomeric sugars are released. The reducing ends of the released sugars bind covalently to (DNS) at 100°, forming a red complex. The absorption of this complex is determined photometrically at 546 nm and is linearly proportional to the amount of reducing sugars.

One unit (U) is defined as the amount of enzyme required to liberate 1 μmol of reducing sugar (with maltose as standard) per minute at a temperature, which is optimal for enzymatic activity. α-Amylase or pullulanase is measured in 50 mM sodium acetate buffer at pH 6.0 and 0.5% (w/v) substrate (starch or pullulan). In order to calculate the enzyme concentration (U/ml), the following equation is used:

$$U/ml = \frac{\Delta EFV}{t}$$

where ΔE is extinction at 546 nm against a blank value, F is the factor from the maltose standard curve (1.02), V is the factor considering the sample volume employed in the assay,* and t is the incubation time at optimal temperature in minutes.

---

* Units should be calculated for 1-ml samples. Sample blanks are extremely important and are used to correct for the nonenzymatic release of reducing sugars that can take place under assay conditions.

Gel Electrophoresis

In order to determine the size of the native enzyme, native polyacryl-amide gels containing a gradient of 5 to 27% polyacrylamide are prepared as described by Koch et al.[44] Gels are run at 300 V for 24 hr at 4. High molecular weight marker proteins (Pharmacia Biotech) are used as standards. In order to examine the subunit composition of the pullulanase, protein samples are also analyzed by sodium dodecyl sulphate–polyacrylamide gel electrophoresis (12% SDS–PAGE) as described by Laemmli[45] after heating the samples at 100° for 5 min. Low molecular weight marker proteins (Pharmacia Biotech) are used as standards. Following native and SDS–PAGE, the proteins are stained with Coomassie blue. Zymogram staining for pullulytic activity is performed according to Furegon et al.[46] The red dye–pullulan, prepared as described previously, is incorporated in the resolving native or SDS gel at a final concentration of 1% (v/v). For electrophoresis, samples are diluted in buffer according to Laemmli.[45] After electrophoresis performed at room temperature under 20 mA, the gels are immersed in 100 ml of buffer at the optimum pH for the amylolytic activity preheated at 70° and incubated with gentle agitation at the same temperature in a water bath. When SDS gels are used, the buffer is changed after the first 10 min of incubation. After a period of time that varies according to the activity of the enzyme assayed, pullulanase is detectable as a yellow band on a red background (Fig. 3). When the desired intensity of the yellow band is reached, the reaction is stopped by immersion of the gel in 7.5% (v/v) acetic acid. Amylolytic activity is detected as follows: after rinsing the native or SDS gel with water, it is soaked in 50 mM sodium acetate (buffer C) containing 1% starch (w/v) at 4° for 60 min. The gel is further incubated at the optimum temperature required for enzymatic activity in buffer C. After 2 min of incubation, the gel is soaked in a solution containing 0.15% (w/v) iodine and 1.5% (w/v) potassium iodide solution until clear bands on a dark brown-blue background become visible (Fig. 4). The stained gels are stored in 10% (v/v) of ethanol at room temperature.

Characterization of Hydrolysis Product

Hydrolysis products arising after the action of amylolytic enzymes on various linear and branched polysaccharides are analyzed by high-perfor-

[44] R. Koch, F. Canganella, H. Hippe, K. D. Jahnke, and G. Antranikian, *Appl. Environ. Microbiol.* **63**, 1088 (1997).
[45] U. K. Laemli, *Nature* **227**, 680 (1970).
[46] L. Furegon, A. Curioni, and D. B. A. Peruffo, *Anal. Biochem.* **221**, 200 (1994).

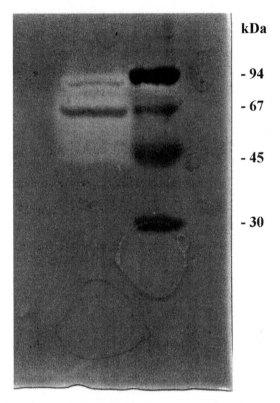

Fig. 3. SDS–PAGE of a partially purified pullulanase from *Desulfurococcus mucosus*. The halo indicating pullulytic activity (lane 1) was obtained by loading 20 μg of protein onto SDS–PAGE, followed by electrophoresis, removal of SDS, and staining for heat-stable pullulytic activity. Lane 2, molecular markers. Protein bands were then detected with Coomassie blue (0.1%).

mance liquid chromatography (HPLC) with an Aminex HPX-42A column (300 by 78 mm) (Bio-Rad, Hercules CA). A refractometer is used to detect the peaks. Double distilled water is used as the mobile phase at a flow rate of 0.3 ml/min. Various oligosaccharide peaks (*DP*7 to *DP*1) are eluted between 9 and 18 min. The purified pullulanase is incubated at 65° with 0.5% (w/v) pullulan, starch, glycogen, amylopectin, maltodextrin, panose, and 0.2% (w/v) amylose.

Samples are withdrawn at different time intervals, and the reaction is stopped by incubation on ice. Figure 5 shows the mode of action of pullula-

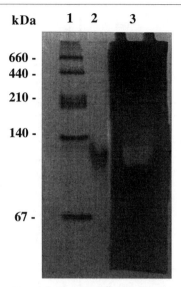

Fig. 4. Native polyacrylamide gel (left) with corresponding zymogram (right) of the purified recombinant α-amylase from *Pyrococcus woesei*. For the zymogram, 20 mU of α-amylase activity was loaded onto a native gel followed by electrophoresis and staining for heat-stable α-amylase activity as described in the text. Lane 1, molecular marker; lane 2, purified α-amylase from *Pyrococcus woesei;* lane 3, zymogram.

nase type I from *F. pennivorans* Ven5. The thermostable pullulanase hydrolyzes more than 98% of pullulan after 1 hr of incubation at 80° (Fig. 5a). The hydrolysis pattern, after its action on pullulan, reveals the complete conversion of pullulan to maltotriose in an endo-acting fashion by attacking the α-1,6-glycosidic linkages. In order to confirm that the hydrolysis product from pullulan is maltotriose (possessing α-1,4-glycosidic linkages) and not panose or isopanose (possessing two α-1,4- and α-1,6-glycosidic linkages), incubation of the products of pullulan hydrolysis is performed in the presence of α-glycosidase from yeast. The formation of glucose as the main product confirms the formation of maltotriose (and not panose) from pullulan (Fig. 5b). No degradation of amylose is observed after 16 hr of incubation at 65° with the recombinant pullulanase demonstrating the low affinity of the purified enzyme to α-1,4-glycosidic linkages. In contrast to this, incubation of amylose with purified recombinant pullulanase type II at 100° from *P. woesei*[30] leads to the formation of oligosaccharides of different degrees of polymerization and glucose, thus indicating activity toward α-1,6- and α-1,4-glycosidic linkages (Fig. 5c). After 72 hr of incubation with the purified rpulA, very low levels of maltose and maltotriose are detectable in the

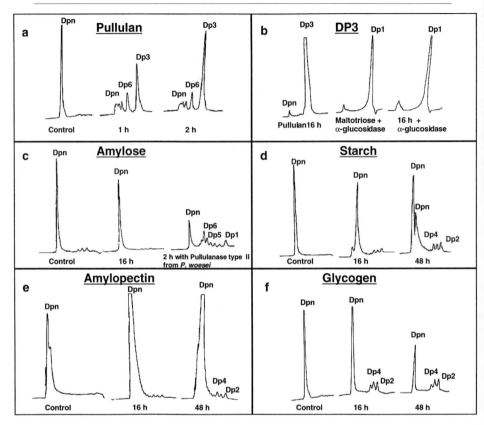

Fig. 5. HPLC analysis of hydrolysis products formed after incubation of the purified recombinant pullulanase from *Fervidobacterium pennivorans* Ven5 in the presence of different substrates: (a) 0.5% pullulan, (d) 0.5% starch, (e) 0.5% amylopectin, and (f) 0.5% glycogen. Samples were incubated at 65° and aliquots were withdrawn and analyzed on an Aminex HPX 42-A column for oligosaccharides at different time intervals. (b) The hydrolysis product formed after incubation of 0.5% pullulan with pullulanase from *F. pennivorans* Ven5 (16 hr pullulan) and 0.5% maltotriose was incubated with commercial α-glucosidase and then analyzed by HPLC. (c) HPLC analysis of hydrolysis products formed after incubation of 0.2% amylose in the presence of recombinant pullulanases from *F. pennivorans* Ven5 (65° for 16 hr) and from *P. woesei* (75° for 2 hr). DP, degree of polymerization; DP1, glucose; DP2, maltose; and so on.

hydrolysis product of soluble starch (Fig. 5d), amylopectin (Fig. 5e), and glycogen (Fig. 5f). According to these results, r*pul*A attacks specifically α-1,6 linkages of pullulan and branched oligosaccharides and is classified as pullulanase type I.

Biotechnological Relevance

The finding of extremely thermostable starch-hydrolyzing enzymes such as amylases and pullulanases that are active under similar conditions will significantly improve the industrial starch bioconversion process, i.e., liquefaction, saccharification, and isomerization. Because of the lack of novel thermostable enzymes that are active and stable above 100° and at acidic pH values, the bioconversion of starch to glucose and fructose has to be performed under various conditions. This multistage process (step 1: pH 6.0–6.5, 95–105°; step 2: pH 4.5, 60–62°; step 3: pH 7.0–8.5, 55–60°) is accompanied by the formation of undesirably high concentrations of salts. In the final step, where high fructose syrup is produced, salts have to be removed by expensive ion exchangers. In addition, by using robust starch-modifying enzymes from hyperthermophilic microorganisms, innovative and environmentally friendly processes can be developed aimed at the formation of products of high added value from native starch for the food industry. New and enhanced functionality can be obtained by changing the structural properties of starch. In order to prevent retrogradation, starch-modifying enzymes can be used at a higher temperature. The use of extremely thermostable amylolytic enzymes can lead to valuable products, which include innovative starch-based materials with gelatine-like characteristics and defined, linear dextrins that can be used as fat substitutes, texturizers, aroma stabilizers, and prebiotics.[1,47]

CGTases are used for the production of cyclodextrins that can be used as a gelling, thickening, or stabilizing agent in jelly desserts, dressing, confectionery, and dairy and meat products. Because of the ability of cyclodextrins to form inclusion complexes with a variety of organic molecules, cyclodextrins improve the solubility of hydrophobic compounds in aqueous solution. This is of interest for pharmaceutical and cosmetic industries. Cyclodextrin production is a multistage process in which starch is first liquefied by a heat-stable amylase, and in the second step a less-thermostable CGTase from *Bacillus* sp. is used. Due to the low stability of the latter enzyme, the second step must run at lower temperatures. The application of heat-stable CGTase in jet cooking, where temperatures up to 105° are achieved, will allow liquefaction and cyclization to take place in one step.[48]

[47] W. D. Crabb and C. Mitchinson, *Tibtech* **15,** 349 (1997).
[48] B. E. Norman and S. T. Jørgensen, *Denpun Kagaku* **39,** 101 (1992).

[18] Cellulolytic Enzymes from *Thermotoga* Species

By WOLFGANG LIEBL

Introduction

β-Glucans are important natural polymers consisting of β-glycosidically linked glucose residues. These polysaccharides mostly fulfill structural roles. The β-1,4-glucan cellulose and mixed-linkage β-1,4/1,3-glucans, such as barley β-glucan and lichenan, occur as cell wall components of higher plants and lichens, and β-1,3-glucans are found in the cell walls of yeast and filamentous fungi and as structural and storage polysaccharide of the marine macroalga *Laminaria saccharina* (laminarin). Thermostable enzymes for the degradation of these types of β-glucans and/or their corresponding genes have been isolated from different strains of the heterotrophic strictly anaerobic bacterial genus *Thermotoga*.[1-10] The following is an overview of β-glucan-cleaving enzymes found in *Thermotoga* species and describes the preparation and characteristics of an endoglucanase and a β-glucosidase of *Thermotoga maritima*.

β-Glucan-Degrading Enzymes of *Thermotoga* Species

In *T. maritima,* a hyperthermophile with an optimum growth temperature of 80°, enzymes capable of the hydrolysis of β-1,4-glucans and mixed-

[1] W. Liebl, P. Ruile, K. Bronnenmeier, K. Riedel, F. Lottspeich, and I. Greif, *Microbiology* **142,** 2533 (1996).

[2] K. Bronnenmeier, A. Kern, W. Liebl, and W. Staudenbauer, *Appl. Environ. Microbiol.* **61,** 1399 (1995).

[3] O. N. Dakhova, N. E. Kurepina, V. V. Zverlov, V. A. Svetlichnyi, and G. A. Velikodvorskaya, *Biochem. Biophys. Res. Commun.* **194,** 1359 (1993).

[4] L. D. Ruttersmith and R. M. Daniel, *Biochem. J.* **277,** 887 (1991).

[5] L. D. Ruttersmith and R. M. Daniel, *Biochim. Biophys. Acta* **1156,** 167 (1993).

[6] J.-D. Bok, D. A. Yernool, and D. E. Eveleigh, *Appl. Environ. Microbiol.* **64,** 4774 (1998).

[7] J.-D. Bok, S. K. Goers, and D. E. Eveleigh, *in* "Enzymatic Conversion of Biomass for Fuels Production" (M. E. Himmel, J. O. Baker, and R. P. Overend, eds.), p. 54. American Chem. Society, Washington, DC, 1994.

[8] D. E. Eveleigh, J. D. Bok, H. El-Dorry, S. El-Gogary, K. Elliston, A. Goyal, C. Waldron, R. Wright, and Y.-M. Wu, *Appl. Biochem. Biotech.* **51/52,** 169 (1995).

[9] V. V. Zverlov, I. Y. Volkov, T. V. Velikodvorskaya, and W. Schwarz, *Microbiology* **143,** 1701 (1997).

[10] V. V. Zverlov, I. Y. Volkov, T. V. Velikodvorskaya, and W. Schwarz, *Microbiology* **143,** 3537 (1997).

linkage β-1,4/1,3-glucans and β-glucosidases, which release glucose from β-linked oligosaccharides, have been identified.[1,2,11,12] Also, the *T. maritima* genome sequencing project[13] has revealed a variety of genes encoding different (putative) β-glucan-degrading enzymes [open reading frame (ORF) designations, approximate size, and enzyme families according to the classification of Henrissat[14] in parentheses]: the endoglucanases CelA and CelB (TM1524 and TM1525; about 30 kDa; glycosylhydrolase family 12), five additional small endoglucanases (TM1751, TM1752, TM1048, TM1049, and TM1050; about 36 to 39 kDa; glycosylhydrolase families 5 and 60), one large endoglucanase (TM0305; about 78 kDa), a laminarinase (TM0024; about 73 kDa; glycosylhydrolase family 16), a β-glucosidase (TM0025; 81 kDa; glycosylhydrolase family 3), and a cellobiose phosphorylase (TM1848; about 93 kDa). The amino acid sequences of the latter three enzymes are highly similar to the *T. neapolitana* laminarinase LamA,[9] β-glucosidase/laminaribiase BglB,[10] and cellobiose phosphorylase CepA (unpublished sequence).

Two cellulases, designated cellulase I and cellulase II, have been purified from cells of *T. maritima* grown on cellobiose and xylose.[2] Also, two endoglucanases, called CelA and CelB, which share about 50% amino acid sequence identity and are encoded by a tandem pair of genes isolated from a *T. maritima* gene library, were purified and characterized from a recombinant source.[1] In contrast to CelA, CelB contains an N-terminal signal peptide. By N-terminal sequencing of cellulase I, it has been shown that this enzyme resembles CelA.[1] It is not clear if *T. maritima* 29-kDa cellulase II, which displayed avicelase activity, could be CelB, but the fact that crude recombinant CelB did not have avicel-degrading ability[1] argues against this possibility. Unfortunately, an N-terminal sequence for cellulase II is not available. Apart from CelA, CelB is the only other cellulase found in the genome with a size similar to that of cellulase II. All other putative cellulases deduced from the genome with similarity to known cellulase sequences have a larger predicted size, i.e., between 36 and 78 kDa. Cellulase II could represent a processing product of one of these enzymes or could be encoded by a further cellulase gene not yet recognized.

[11] J. Gabelsberger, W. Liebl, and K. H. Schleifer, *FEMS Microbiol. Lett.* **109,** 131 (1993).
[12] J. Gabelsberger, W. Liebl, and K. H. Schleifer, *Appl. Microbiol. Biotechnol.* **40,** 44 (1993).
[13] K. E. Nelson, R. E. Clayton, S. R. Gill, M. L. Gwinn, R. J. Dodson, D. H. Haft, E. K. Hickey, J. D. Peterson, W. C. Nelson, K. A. Ketchum, L. McDonald, T. R. Utterback, J. A. Malek, K. D. Linher, M. M. Garrett, A. M. Stewart, M. D. Cotton, M. S. Pratt, C. A. Phillips, D. Richardson, J. Heidelberg, G. G. Sutton, R. D. Fleischmann, J. A. Eisen, O. White, S. L. Salzberg, H. O. Smith, J. C. Venter, and C. M. Fraser, *Nature* **399,** 323 (1999).
[14] B. Henrissat, *Biochem. J.* **280,** 309 (1991).

β-Glucosidase activity, which releases glucose from cello-oligosaccharides or other low molecular mass β-glucosides, has also been found in *T. maritima*. The broad specificity β-glycosidase BglA, which is active against aryl-β-glucoside, -galactoside, and -fucoside substrates, as well as the disaccharides cellobiose and lactose, was purified from an *Escherichia coli* clone of a *T. maritima* gene library.[12] Another β-glucosidase called BglB, which cleaved β-glucosidic but not β-galactosidic bonds, was identified in *T. maritima* cell lysates.[11] It remains to be shown if this activity is encoded by the putative β-glucosidase gene (ORF TM0025) identified in the *T. maritima* genome sequence.[13] Surprisingly, the *bglA* gene, which was cloned and studied by Liebl *et al.*,[15] was not detected in the *T. maritima* genome. However, an approximately 300-bp sequence immediately upstream of the published *bglA* gene is present (100% nucleotide sequence identity) in the genome sequence where it represents the 3' end of the putative cellobiose phosphorylase gene (ORF TM1848; *cepA*) of *T. maritima*. It appears as if *bglA* was originally part of a cellulolytic gene cluster containing *cepA* and *bglA*, but was lost from the *T. maritima* genome by deletion.

Strains of *Thermotoga neapolitana*, a close relative of *T. maritima*, have been reported to produce several cellulolytic enzymes, including enzymes with endoglucanase, β-glucosidase, and β-glucan glucohydrolase activities.[3,6-8,16] The endoglucanases CelA (29 kDa, pH_{opt} 6.0) and CelB (30 kDa, pH_{opt} 6.0-6.6) of *T. neapolitana* NS-E are 86 and 94% similar to CelA and CelB, respectively, of *T. maritima* MSB8. Despite the close structural association of the *celA* and *celB* genes (the start codon of *celB* overlaps the stop codon of *celA* in *T. maritima* as well as in *T. neapolitana*), Northern blot experiments surprisingly indicate the separate, monocistronic transcription of these genes in cellobiose-grown *T. neapolitana* NS-E cells. Expression of *celA* and *celB* appears to be completely repressed in the presence of glucose.[6] Not only *celA* and *celB* expression, but also β-glucosidase synthesis in *T. neapolitana* NS-E, has been reported to be induced during growth on cellobiose.[16] A β-glucosidase gene designated *bglB* of *T. neapolitana* Z2706-MC24, which is very similar to the open reading frame TM0025 of the *T. maritima* genome, has been studied. Characterization of the purified recombinant 81-kDa enzyme, which belongs to the glycosylhydrolase family 3, revealed that it had a 14-fold higher specific activity and a 5-fold lower K_m with laminaribiose (β-1,3-linked disaccharide) than with cellobiose. Therefore, it was classified as a laminaribiase, which is in agreement with the genetic context of the gene: *bglB* is located immediately upstream of the laminarinase gene *lamA*.[10]

[15] W. Liebl, J. Gabelsberger, and K. H. Schleifer, *Mol. Gen. Genet.* **242**, 111 (1994).
[16] M. Vargas and K. M. Noll, *Microbiology* **142**, 139 (1996).

Thermotoga sp. strain FjSS3-B1 also produces cellulolytic enzymes during growth on cellobiose. A 36-kDa cellulase, which was classified as a cellobiohydrolase, was isolated from a cell-free culture supernatant of this strain.[5] This enzyme was reported to have a pH optimum of around 7.3 and to be stabilized by 0.8 M NaCl. It is important to note that this cellobiohydrolase, just like the avicelase activity (cellulase II) of *T. maritima* MSB8, displays relatively high activity with carboxymethyl cellulose,[2] a typical endoglucanase substrate. A 75-kDa β-glucosidase characterized from strain FjSS3-B1 cells[5] seems to have a similar substrate profile as BglB of *T. neapolitana* Z2706-MC24,[10] but laminaribiase activity was not checked for the FjSS3-B1 enzyme.

Purification of Recombinant and Authentic *T. maritima* CelA

The *E. coli* strain BL21(pT7-7-*celA*) can be used for recombinant CelA production. pT7-7-*celA* contains the *celA* open reading frame fused to the ATG start codon of expression vector pT7-7.[1] Isopropylthiogalactoside (IPTG)-mediated induction of the chromosomally integrated T7 RNA polymerase gene leads to severe growth impairment of *E. coli* BL21(pT7-7-*celA*). Therefore, the strain is grown without IPTG addition. For large-scale enzyme preparation, 2.5-liter portions of 2× LB medium [2% (w/v), peptone; 1% (w/v), yeast extract; 1% (w/v), NaCl; pH 7.2] are prepared in 5-liter baffled flasks, sterilized, supplemented with 100 μg/ml ampicillin, and inoculated at 1% with *E. coli* BL21(pT7-7-*celA*). Growth of the strain is conducted on a rotary shaker at 35° for about 14 hr, after which the cells are harvested, washed once with 20 mM Tris–HCl buffer (pH 7), and suspended in a small volume (0.5–1 ml/g of wet cell mass) of the washing buffer. Cell disruption is carried out by threefold passage through a French pressure cell (American Instrument Company, Silver Spring, MD) at 6.9 MPa and one freeze/thaw cycle. Cellular debris is removed by centrifugation (27,000g, 30 min, 4°). Starting from a culture volume of 10 liter, about 63 g wet cell paste can be obtained, yielding about 4 g total soluble protein. A large proportion (90–95%) of the host cell proteins can be removed by heat treatment (75° for 20 min) followed by centrifugation. For chromatographic purification, the cleared supernatant is adjusted to pH 8.0 with 1 M Tris–HCl, pH 9.5, and loaded onto a Q-Sepharose Fast Flow XK 26/10 column (Pharmacia, Freiburg, Germany). Bound protein is eluted with a 10 bed volume linear gradient of 0 to 0.5 M NaCl in 20 mM Tris–HCl, pH 8.0. Fractions of 5 ml are collected and tested for thermostable carboxymethyl cellulase activity. The active fractions are combined, adjusted to 1.5 M NH$_4$Cl–50 mM Tris–HCl, pH 8.0, and loaded onto a Phenyl-Sepharose HP HiLoad 16/10 column (Pharmacia). Elution of bound enzyme is achieved with a 10 bed volume gradient of 1.5 to 0 M ammonium chloride in 50 mM Tris–HCl, pH 8.0. Using the procedure outlined earlier, the yield of highly purified enzyme typically amounts to more than 40% of the total activity initially present in the crude lysate (Table I).

The next section describes the purification of native CelA (cellulase I[1,2]) from *T. maritima* MSB8. Expression of cellulolytic enzymes in *T. neapolitana* NS-E has been reported to be induced during growth on cellobiose.[6,16] Also, growth media used for the cultivation of *Thermotoga* sp. strain FjSS3-B.1 for isolation of a cellobiohydrolase and a β-glucosidase employed cellobiose as a substrate.[4,5] Finally, purification of CelA from *T. maritima* was achieved with cells propagated in cellobiose-containing medium.[2] Thus, growth in the presence of cellobiose seems to be appropriate if the isolation of cellulolytic enzymes from cells of *Thermotoga* species is intended.

Thermotoga maritima MSB8 is grown anaerobically at 80° in a medium containing (per liter) 0.5 g KH_2PO_4, 2 mg $NiCl_2 \cdot 6H_2O$, 20 g NaCl, 0.5 g yeast extract, 1 mg resazurin, 0.5 g $Na_2S \cdot 9H_2O$, 15 ml trace element solution,[2] 250 ml artificial seawater, 5 g cellobiose, and 0.5 g xylose. Typical cell yields in batch cultures amount to about 0.1 to 0.2 g/liter. The cells are harvested and washed twice with 20 mM Tris–HCl, pH 8.0–1 M NaCl, suspended in the same buffer without salt, and disintegrated by French pressure cell treatment. After centrifugation (45,000g, 40 min, 4°), the lysate supernatant is dialyzed against 20 mM Tris–HCl, pH 8.0, and loaded onto a Q-Sepharose Fast Flow XK 26/10 column. Bound protein is eluted with a 10 bed volume linear gradient of 0 to 0.5 M NaCl in 20 mM Tris–HCl, pH 8.0, while collecting 5-ml fractions. Under these conditions, β-glucanase, carboxymethyl cellulase, avicelase, laminarinase, and β-glucosidase activi-

TABLE I
PURIFICATION OF RECOMBINANT CelA[a]

Step	Protein (mg)	Total activity[b] (U^c)	Specific activity (U/mg)	Yield (%)
Cleared lysate	4002	11,090	2.76	100
Heat treatment (20 min at 75°)	228	8,740	40.4	79
Q-Sepharose	56.4	6,470	71.9	58
Phenyl Sepharose	16.8	4,900	291	44

[a] Modified, with permission, from W. Liebl, P. Ruile, K. Bronnenmeier, K. Riedel, F. Lottspeich, and I. Greif, *Microbiology* **142**, 2533 (1996). Copyright Society for General Microbiology, Reading, UK.

[b] Enzyme activities are expressed as carboxymethyl cellulose-hydrolyzing activities measured in 30-min assays at pH 6.0 at 85°.

[c] International units (amount of enzyme needed for the formation of 1 μmol product/min).

ties elute between 0.2 and 0.4 M NaCl, whereas xylanase activity, which is also present in the crude lysate, is eluted at 0.2 to 0.25 M NaCl. The combined β-glucanase fractions are adjusted to 1.2 M NH$_4$Cl, and loaded onto a phenyl-Sepharose HP HiLoad 16/10 column equilibrated with 1.2 M NH$_4$Cl–50 mM Tris–HCl, pH 8.0. Elution of bound enzyme is achieved with a 10 bed volume gradient of 1.2 to 0 M NH$_4$Cl in 50 mM Tris–HCl, pH 8.0. CelA (cellulase I) elutes at around 0.6 M NH$_4$Cl and is separated from the other β-glycanases as well as from p-nitrophenyl (pNP)-β-glucoside-, pNP-β-xyloside, and pNP-α-arabinofuranoside-cleaving activities, which are released from the column at lower NH$_4$Cl concentrations or after completion of the gradient (0 M NH$_4$Cl). As the final purification step, the CelA (cellulase I)-containing fractions of the hydrophobic interaction chromatography step are dialyzed against 20 mM Tris–HCl, pH 6.0, and applied to a Mono Q HR 5/5 column equilibrated with the same buffer. Elution is achieved with a 20 bed volume linear gradient of 0 to 0.4 M NaCl in 20 mM Tris–HCl, pH 6.0, at a flow rate of 1 ml/min.

Fractions from the various purification steps using SDS–PAGE are analyzed followed by staining for carboxymethyl cellulase/β-glucanase activity. For this purpose, enzyme samples are separated in a 12.5% SDS–polyacrylamide gel containing 0.1% carboxymethyl cellulose. The detergent is removed by washing the gel at ambient temperature for 30 min in a mixture (4:1, v/v) of 100 mM succinate buffer, pH 6.0, and 2-propanol and then for another 30 min in 100 mM succinate buffer, pH 6.0. The wash solution should be renewed every 15 min. Then, the gel is incubated for 30–120 min at 60° in 100 mM succinate buffer, pH 6.0. Subsequent staining with 0.1% Congo red for 20 min and removal of excess dye by washing with 1 M NaCl give rise to bands of endoglucanase activity against a red background.

The CelA-mediated liberation of reducing groups can be determined with the dinitrosalicylic acid (DNSA) method.[17] Enzyme assays are done at 85° in 0.5-ml reaction mixtures containing 50 mM MES buffer, pH 6.0, and 0.5% (w/v) carboxymethyl cellulose (low viscosity, Sigma, Deisenhofen, Germany). One unit of enzyme activity is defined as the amount of enzyme needed to release 1 μmol of glucose-equivalent reducing groups per minute. Nonenzymatic hydrolysis of the substrates at elevated temperatures is corrected for with appropriate controls in every experiment.

Properties of CelA

The mobility of CelA during SDS–PAGE analysis (apparent molecular mass about 30 kDa) is in accordance with the size deduced from its primary

[17] P. Bernfeld, *Methods Enzymol.* **1,** 149 (1955).

structure (29,732 Da). With the substrate carboxymethyl cellulose, the optimum pH for recombinantly produced CelA is around pH 5, with near-maximum (>75%) activity between pH 4.5 and 6.5. However, with the substrates β-glucan, pNP-β-cellobioside, or xylan, CelA is most active near neutrality with high activity (>75%) between pH 6.0 and pH 8.0.[1] Substrate-dependent shifts in the pH activity profiles of enzymes, which may be caused by pH-dependent variations of the substrate-binding affinities for different substrates, have also been reported for other β- and α-glucanases.[18–20] Using a 30-min assay in 50 mM MES buffer, pH 6.0, with 1% carboxymethyl cellulose, the "temperature optimum" of CelA is about 90°. An endoglucanase (EglA) of the hyperthermophilic archaeon, *Pyrococcus furiosus*, which like CelA belongs to glycosylhydrolase family 12 and which displays about 40% amino acid sequence identity with CelA and CelB of *T. maritima* and *T. neapolitana*, has been produced recombinantly and characterized.[21] This enzyme is even more thermoactive than the *Thermotoga* endoglucanases in accordance with the higher optimum growth temperature of the producer *P. furiosus*. The three-dimensional structure of mesophilic *Streptomyces lividans* endoglucanase of family 12 has been elucidated,[22] which may help unravel features involved in stabilization of the related hyperthermostable enzymes of *Thermotoga* and *Pyrococcus*.

Purified recombinant CelA is remarkably resistant against thermoinactivation. Dilute enzyme (1.6 μg/ml in 50 mM MES, pH 6.0) incubated at 70° does not suffer significant inactivation over a period of at least 56 hr. At 85°, CelA still displayed more than 40% of its initial activity after 48 hr of incubation, whereas at 95° the enzyme is inactivated rapidly within about 30 min (Fig. 1). Interestingly, thermoinactivation of CelA at high temperatures can be reduced dramatically by the addition of salt. In the presence of 5 M NaCl, the activity of CelA only drops slightly during an incubation period of 6 hr at 95° (Fig. 1). Protein-stabilizing effects brought about by the addition of certain cosolvents such as $MgSO_4$ or NaCl can be explained on the basis of preferential exclusion of these salts from the protein.[23]

Thermotoga maritima CelA is most active against soluble β-glucans

[18] J. Hitomi, J.-S. Park, M. Nishiyama, S. Horinouchi, and T. Beppu, *J. Biochem.* **116,** 554 (1994).

[19] K. Ishikawa, I. Matsui, and K. Honda, *Biochemistry* **29,** 7119 (1990).

[20] K. Ishikawa, I. Matsui, S. Kobayashi, H. Nakatani, and K. Honda, *Biochemistry* **32,** 6259 (1993).

[21] M. W. Bauer, L. E. Driskill, W. Callen, M. A. Snead, E. J. Mathur, and R. M. Kelly, *J. Bacteriol.* **181,** 284 (1999).

[22] G. F. Sulzenbacher, R. Shareck, R. Morosoli, C. Dupont, and G. J. Davies, *Biochemistry* **36,** 16032 (1997).

[23] S. N. Timasheff, *Annu. Rev. Biophys. Biomol. Struct.* **22,** 67 (1993).

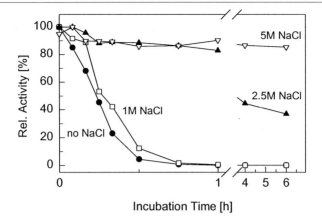

FIG. 1. Thermoinactivation of purified CelA at 95° at various salt concentrations. Taken and modified, with permission, from W. Liebl, P. Ruile, K. Bronnenmeier, K. Riedel, F. Lottspeich, and I. Greif, *Microbiology,* **142,** 2533 (1996). Copyright Society for General Microbiology, Reading, UK.

with 1,4-glycosidic linkages (Table II). α-Glucans such as soluble starch or pullulan are not attacked. Like many other β-glucanases, CelA displays higher activity with barley β-glucan, a β-1,4-/β-1,3-mixed linkage substrate, than with the modified cellulose model substrate carboxymethyl cellulose. The initial products of barley β-glucan hydrolysis, as analyzed via thin-layer chromatography (not shown), are a mixture of oligosaccharides with a degree of polymerization (DP) of 3 and larger. After extended incubation, tri- and tetrasaccharides prevail, but a substantial amount of DP 2 and some glucose are also formed. Hydrolysis of pNP-cellobioside, which is often used as a model substrate for exo-β-glucanases, was also observed. However, the differentiation of exo- and endocellulases on the basis of their specificities using small soluble substrates does not appear to be a very useful method[24] and, thus, does not contradict the classification of CelA as an endo-β-glucanase [1,4-β-D-glucan glucanohydrolase (EC 3.2.1.4)]. The enzyme apparently is not strictly specific for β-1,4-D-glucosidic bonds because oat spelts xylan, a nonglucose polysaccharide with a backbone of β-1,4-linked xylose units, was also attacked, although with a very low catalytic efficiency (Table II; however, it cannot be definitely ruled out that impurities in the xylan could have been the actual substrate). Xylan hydrolysis has also been observed with *Erwinia carotovora* CelS,[25] a distantly related endoglucanase belonging to the same enzyme family as CelA.

[24] M. Claeyssens and B. Henrissat, *Protein Sci.* **1,** 1293 (1992).
[25] H. T. Saarilahti, B. Henrissat, and E. T. Palva, *Gene* **90,** 9 (1990).

TABLE II
Substrate Specificity of CelA[a]

Substance	Specific activity[b] (U/mg)	Relative activity (%)
Barley β-glucan	1785	100
Carboxymethyl cellulose	302	17
pNP-cellobioside	7.4	0.42
Acid-swollen Avicel	8.0	0.48
Avicel	0	0
Cellulose MN300	0	0
Oat spelts xylan	2.5	0.14

[a] Modified, with permission, from W. Liebl, P. Ruile, K. Bronnenmeier, K. Riedel, F. Lottspeich, and I. Greif, *Microbiology* **142**, 2533 (1996). Copyright Society for General Microbiology, Reading, UK.

[b] Assays were performed at substrate concentrations of 4 mM (pNP-cellobioside), 1% (nonsoluble polymeric substrates) or 0.5% (soluble polymers) in 50 mM MES buffer, pH 6.0, at 85° using 81 ng/ml of purified CelA. Incubation was caried out for 10–60 min for β-glucan, carboxymethyl cellulose, and pNP-cellobioside; 24 hr for xylan and acid-swollen Avicel; and up to 48 hr for Avicel and cellulose MN300. Activity on barley β-glucan was defined as 100%.

Purification of Recombinant *T. maritima* BglA

The *E. coli* strain MC1061/pJTG9ex, which bears the *T. maritima* β-glucosidase gene on a recombinant plasmid under the transcriptional control of the P_{tac} promoter, is a convenient source for isolation of BglA.[12] The strain is grown at 37° in LB medium [1% (w/v), peptone; 0.5% (w/v), yeast extract; 0.5% (w/v), NaCl; pH 7.2] with ampicillin selection (100 μg/ml). At an $OD_{600\,nm}$ of about 0.5, IPTG is added at 0.1 mM. After 16 hr, the cells are harvested, washed with 50 mM Tris–HCl, pH 7.5, suspended in 20 ml of the washing buffer, and lysed mechanically by twofold passage through a French pressure cell at 6.9 MPa. The lysate is cleared by centrifugation (40,000g, 30 min, 4°), and the pellet is suspended in 50 mM Tris–HCl containing 0.5% Triton X-100 and then recentrifuged. Streptomycin sulfate is added to the combined supernatants at a final concentration of 4%, followed by stirring on ice for 45 min and centrifugation (20,000g, 20 min, 4°). The supernatant is dialyzed, subjected to a heat treatment (75°, 15 min), and then cleared of precipitated host proteins by centrifugation. About 20-mg portions of the supernatant proteins are loaded onto a Q-Sepharose Fast Flow HR 16/10 anion-exchange chromatography column (Pharmacia) equilibrated with 10 mM Bis–Tris–HCl, pH 6.2. Elution is achieved with a linear 100–230 mM NaCl gradient in the same buffer. Fractions containing thermostable β-glucosidase activity are pooled and dialyzed against 25 mM Bis–Tris–HCl, pH 6.2. Chromatofocusing on a Mono P HR 5/5 column (Pharmacia) equilibrated with 25 mM Bis–Tris–HCl, pH 6.2, serves as the final purification step. Elution from this column is carried out at a flow rate of 0.5 ml/min with 10-fold diluted Polybuffer 74 (Pharmacia) titrated to pH 3.0 with HCl. The purification steps summarized in Table III give yield to a highly pure recombinant enzyme preparation.

The following assay system can be used to measure the hydrolysis of chromogenic aryl glycosides by β-glucosidase: an enzyme sample (0.2–0.7 μg/ml) is preequilibrated at 75° in 1 ml 50 mM sodium phosphate buffer, pH 6.2, before starting the reaction by adding 200 μl of 13 mM pNP- or oNP-β-D-glucopyranoside solution. The reaction is stopped with 0.5 ml 1 M Na$_2$CO$_3$ and cooled on ice. The amounts of nitrophenol derivatives released can be calculated from the absorbancies measured at 420 nm on the basis of the molar absorption coefficients for o-nitrophenol and p-nitrophenol of 4500 and 13,500 M^{-1} cm^{-1} (determined under the reaction conditions used here), respectively.

TABLE III
PURIFICATION OF RECOMBINANT BglA FROM *E. coli* MC1061/PJTG9EX[a]

Step	Protein (mg)	Total activity[b] ($\times 10^3$ U[c])	Specific activity (U/mg)	Yield (%)
Crude extract	1619	77.7	48.0	100
Streptomycin sulfate precipitation	1024	52.7	51.5	68
Heat treatment (15 min at 75°)	146	36.5	250	47
Q-Sepharose	49	15.1	308	19
Mono P	31	10.1	324	13

[a] Modified, with permission, from J. Gabelsberger, W. Liebl, and K. H. Schleifer, *Appl. Microbiol. Biotechnol.* **40,** 44 (1993). Copyright Springer-Verlag, Heidelberg.

[b] Total and specific activities are expressed as oNP-β-D-glucopyranoside-hydrolyzing activity and were determined in 50 mM sodium phosphate buffer, pH 6.2, at 75°.

[c] International units (amount of enzyme needed for the formation of 1 μmol product/min).

Properties of BglA

In SDS–PAGE, purified BglA has a mobility (47 kDa) that is slightly different from the size expected from the primary structure (51.5 kDa). After SDS–PAGE, the enzyme can be visualized with a zymogram staining method, but only if the sample was not boiled prior to application to the gel (this results in an altered mobility corresponding to about 54 kDa[12]). On a Superdex 200 column, most of the enzyme elutes as a dimeric from.[12] At its optimum pH of 6.0–6.2, BglA displayed maximal hydrolytic activity at 85–95° (2-min assay). The enzyme is extremely insensitive toward thermal inactivation. In 50 mM sodium phosphate buffer, pH 6.2, at an enzyme concentration of 50 μg/ml and an incubation period of 2 hr, BglA suffered an activity loss of about 5, 20, and 40% at 90, 95, and 98°, respectively. BglA is a broad specificity β-glycosidase, which is active against various β-glycosidically linked saccharides, such as (relative activities in parentheses; determined at pH 6.2 at 75° at substrate concentrations of 2.2 mM for nitrophenol derivatives, 18 mM for salicin and arbutin, 40 mM for disaccharides, and 0.5% for polysaccharides; cellobiose-cleaving activity was defined as 100%) oNP-β-D-glucopyranoside (379%), oNP-β-D-galactopyranoside (353%), pNP-β-D-glucopyranoside (237%), pNP-β-D-galactopyranoside (156%), pNP-β-D-fucopyranoside (216%), salicin (47%), and arbutin (31%), as well as the disaccharides cellobiose (100%) and lactose (65%). Low activity was also observed with oNP-β-D-xylopyranoside (13%), pNP-β-D-xylopyranoside (6%), and the polymeric substrates laminarin (a 1,3-β-glucan; 8%) and lichenan (a 1,4/1,3-mixed linkage β-glucan; 4%). The transferase activity of BglA has been observed during incubation of the enzyme with 20% solutions of cellobiose or lactose (not shown).

Acknowledgment

This work was supported by the Deutsche Forschungsgemeinschaft.

Note Added in Proof

After completion of this manuscript Yernool et al.[26] described a glucooligosaccharide catabolic gene cluster consisting of genes for cellobiose phosphorylase (cbpA) and β-glucosidase (gghA) in Thermotoga neopolitana. The β-glucosidase has >90% amino acid sequence identity to T. maritima BglA.[12,15] Both enzymes are highly similar in terms of size, pH optimum, substrate specificity, and K_m for cellobiose.

[26] D. A. Yernool, J. K. McCarthy, D. E. Eveleigh, and J.-D. Bok, J. Bacteriol. 182, 5172 (2000).

[19] Hyperthermophilic Xylanases

By Peter L. Bergquist, Moreland D. Gibbs, Daniel D. Morris,
Dion R. Thompson, Andreas M. Uhl, and Roy M. Daniel

Introduction

Xylan is a major component of plant hemicellulose and is the second most abundant polysaccharide in nature. It is a complex polymer consisting of a β-D-1,4-linked xylanopyranoside backbone, which can be acetylated or substituted with arabinosyl and glucuronic acid side groups. The complete enzymatic hydrolysis of xylan into assimilable sugars requires a number of enzymes, including endo-β-1,4-xylanase (EC 3.2.1.8). More than 80 sequences of the genes coding for xylanases have been reported and many more xylanases from bacteria and fungi have been purified and characterized. A substantial amount of the interest engendered in this class of enzyme stems from some well-publicized applications of xylanases, particularly as an aid to the bleaching of paper pulp,[1] for the modification of the nutritional content of grains used as livestock feedstuffs, and in the baking industry.[2]

Enzymes involved in the metabolism of plant carbohydrate polymers have been grouped into more than 60 different families on the basis of primary and tertiary sequence homologies.[3] Endo-β-1,4-D-xylanases comprise families 10 and 11. The only similarity between members of these two families is their ability to hydrolyze acetylmethylglucuronoxylans of hardwoods and arabinomethylxylans of softwoods, as they are unrelated biochemically and structurally. Like most other cellulolytic and hemicellulolytic enzymes, xylanases are highly modular in structure and may be composed of either a single domain or a number of distinct domains broadly classified as catalytic or noncatalytic. Linker peptides typically delineate the individual domains of multidomain enzymes into discrete and functionally independent units. The catalytic domain of a xylanase determines the hydrolytic activity and hence governs the classification of the enzyme as belonging to family 10 or 11.

A striking feature of many microbial xylanolytic systems is the presence of multiple family 10 and/or 11 xylanases that are produced as discrete

---

[1] L. Viikari, A. Kantelinen, J. Sundquist, and M. Linko, *FEMS Microbiol. Rev.* **13,** 335 (1994).

[2] K. K. Y. Wong and J. N. Saddler, *in* "Hemicellulose and Hemicellulases" (M. P. Coughlan and G. P. Hazelwood, eds.), p. 127. Portland Press, London (1993).

[3] http//expasy.hcuge.ch/cgi-bin/lists?glycosid.txt

gene products.[4] Detailed comparisons of the catalytic properties of xylanases isolated from multixylanase systems have revealed distinctions in the hydrolytic activities of the individual xylanases present, including differences in the yields, rates of hydrolysis, and hydrolytic products from different xylan substrates of varying complexity. These observations suggest that xylanase multiplicity is a mechanism employed by both bacteria and fungi to enhance their xylanolytic capabilities on complex xylan substrates. This article focuses on xylanases from bacteria that grow above 80° and is essentially confined to the enzymes that have been cloned from *Dictyoglomus* and *Thermotoga*. The *Thermotoga* xylanase described in 1991[5] remains the most stable xylanase known. Only one Archeal xylanase is known, from *Thermococcus zilligii* isolate AN1, and so far, the gene for this enzyme has been refractory to cloning so the information is for the nonrecombinant form.

Genetics and Gene Cloning

The endoxylanase component of *Dictyoglomus thermophilum* Rt46B.1[6] comprises a family 10 xylanase (XynA) and a family 11 xylanase (XynB). XynA is a single-domain enzyme, whereas XynB is a multidomain enzyme, consisting of an N-terminal family 11 xylanase catalytic domain and a C-terminal domain of unknown function (see Fig. 1). XynA and XynB appear to be the only xylanases produced by this bacterium. *Thermotoga maritima* strain FjSS3B.1 is isolated from a marine environment in Fiji[7] and has an enzyme complement similar to the type strain and *T. neapolitana*.[8] The xylanase complement of *T. maritima* FjSS3B.1 consists of a single domain family 10 enzyme XynA[9] and two multidomain family 10 enzymes, XynB and XynC. The latter enzyme appears to be largely nonfunctional and is assumed to be coded for by a pseudogene.[10] There are no sequences for family 11 enzymes on the genome of this bacterium. Others have cloned

[4] K. K. Y. Wong, L. U. L. Tan, and J. N. Saddler, *Microbiol. Rev.* **52,** 305 (1988).

[5] H. D. Simpson, U. R. Haufler, and R. M. Daniel, *Biochem. J.* **277,** 413 (1991).

[6] B. K. Patel, H. W. Morgan, J. Weigel, and R. M. Daniel, *Arch. Microbiol.* **147,** 21 (1987).

[7] B. A. Huser, B. K. C. Patel, R. M. Daniel, and H. W. Morgan, *FEMS Microbiol. Lett.* **37,** 121 (1986).

[8] V. Zverlov, K. Piotukh, O. Dakhova, G. Velikodvorskaya, and R. Borriss, *Appl. Microbiol. Biotechnol.* **45,** 245 (1996).

[9] D. J. Saul, L. C. Williams, R. A. Reeves, M. D. Gibbs, and P. L. Bergquist, *Appl. Environ. Microbiol.* **61,** 4110 (1995).

[10] R. A. Reeves, D. D. Morris, M. D. Gibbs, D. J. Saul, and P. L. Bergquist, *Appl. Environ. Microbiol.* **66,** 1532 (2000).

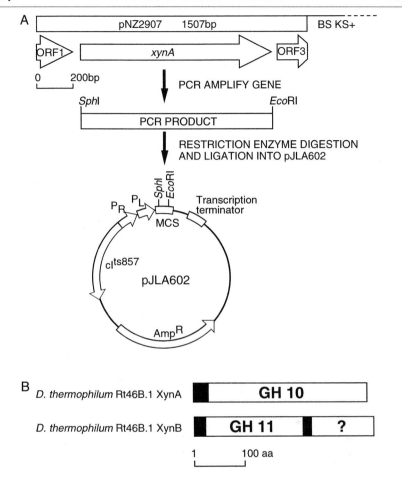

FIG. 1. PCR strategy used to isolate and clone *Dictyoglomus thermophilum* Rt46B.1 *xynA* and *xynB* genes from genomic DNA. (A) Oligonucleotide primers incorporating restriction sites were used to PCR amplify each gene. PCR products were restriction enzyme digested and then ligated directionally into the controlled expression plasmid pJLA602. (B) Domain structures of *D. thermophilum* Rt46B.1 XynA and XynB. GH10, glycosylhydrolase family 10 domain; GH11, glycosylhydrolase family 11 domain; N-terminal black box, signal peptide sequence; internal black box, linker peptide; ?, domain of unknown function.

and characterised family 10 xylanases from *T. maritima*[10,11] and *T. neapolitana*[7] that are the counterparts of our XynB. The pH optimum and half-lives of XynB differ from those reported for XynA from *Thermotoga* FjSSB.1, which has both a broader pH range for 50% activity and is significantly more thermostable.[5,9]

[11] C. Winterhalter, P. Heinrich, A. Candussio, G. Wich, and W. Liebl, *Mol. Microbiol.* **15,** 43 (1995).

Dictyoglomus thermophilum xynA

Chromosomal DNA from Rt46B.1 is partially digested with *Sau*3AI and 5- to 8-kb fragments are purified. Partially backfilled λZAPII vector (Stratagene, San Diego, CA) and genomic DNA are ligated and packaged using Gigapack XL packaging extracts (Stratagene). One plaque, which exhibits the strongest xylanase activity based on the size of the zone of clearing, is converted into a pBS(SK-) plasmid using the ExAssist excision procedure (Stratagene). The plasmid pNZ2900 contains the complete *xynA* gene isolated by the polymerase chain reaction (PCR) and ligated into pJLA602[12] digested with *Sph*I and *Eco*RI. Two DNA oligonucleotide primers (DictxynF 5′ ATGGTAGCATGCTTAACCAAAGGTTTTCTATC-3′ and DictxynR 5′ CCTATAGAATTCAAAACTTTACAATCTCCC-3′) have been designed to allow the PCR amplification of the entire *xynA* gene from Rt46B.1 genomic DNA. The primers DictxynF and DictxynR have been designed to incorporate restriction enzyme sites *Sph*I and *Eco*RI, respectively, at their 5′ ends, allowing the directional cloning of the PCR product into the expression vector.

Dictyoglomus thermophilum xynB

The two-step PCR approach used for the identification and cloning of the Rt46B.1 *xynB* gene involves a consensus PCR step (step one) to identify any family 11 xylanase genes within the Rt46B.1 genome, followed by a subsequent genomic walking PCR strategy (step two) to sequence the desired family 11 xylanases. The rationale behind the design and implementation of the genomic-walking PCR (GWPCR) technique has been described.[13,14]

Linker Assembly. The general-purpose *Nco*I-blunt synthetic DNA linker is assembled by annealing 75 pmol of the *Nco*I-blunt upper strand oligonucleotide (5′ CAT GGC GCA GGA AAC AGC TAT GAC CGG T 3′) with 75 pmol of the universal lower strand oligonucleotide (5′ CGC GTC CTT TGT CGA TAC TGG CCA 3′) in 50 μl TE buffer (10 mM Tris–HCl, 0.1 mM EDTA, pH 7.5) at 50° for 30 min following a 4-min denaturation step at 94°.

Linker Library Construction. Rt46B.1 restriction fragments are generated by incubating 1 μg of RNase A-treated Rt46B.1 genomic DNA over-

[12] B. Schauder, H. Blöcker, R. Frank, and J. E. G. McCarthy, *Gene* **52,** 27 (1987).

[13] D. D. Morris, R. A. Reeves, M. D. Gibbs, D. J. Saul, and P. L. Bergquist, *Appl. Environ. Microbiol.* **61,** 226 (1995).

[14] D. D. Morris, C. W. J. Chin, M.-H. Koh, K. K. Y. Wong, R. W. Allison, P. J. Nelson, and P. L. Bergquist, *Appl. Environ. Microbiol.* **64,** 175 (1998).

night at 37° with 20 U of the *Nco*I, *Dra*I, *Eco*RV, *Hinc*II, *Hpa*I, *Pvu*II, or *Ssp*I restriction endonucleases. Rt46B.1 restriction fragment/linker libraries are then prepared by ligating 5 μl of digested Rt46B.1 DNA with 1 μl of the *Nco*I-blunt DNA linker in a 10-μl overnight reaction using T4 DNA ligase in 1/10th volume of the buffer supplied by the manufacturer (Boehringer Mannheim). Rt46B.1 *Nco*I restriction fragments are extracted with phenol due to the regeneration of the *Nco*I restriction site on linker ligation. Following ligation, the library volume is made up to 50 μl with TE buffer.

Genomic Walking PCRs (GWPCRs). GWPCRs are performed in standard 50-μl PCR mixtures using 12 pmol of linker primer, 12 pmol of the forward (dictGF, 5' GGT ACT ATT GAT CAA ATT ACT CTT TGT GTT G 3') or reverse (dictGR, 5' GTA ATT GGC GTC CAC CAG GTG CAA CC 3') walking primer, and 1 μl of the appropriate Rt46B.1 restriction fragment/linker library. The *xynB* genomic walking primers are positioned to provide maximum novel sequence information from the *xynB* GWPCR products and minimum overlap between the *xynB* consensus PCR fragment and GWPCR products for accurate sequence alignment. In addition, to help ensure high-fidelity PCRs, the lengths of the walking primers are optimized with respect to their GC composition for a targeted T_m of 72°, the theoretical melting temperature of the linker primer. In this manner, the walking and linker primers exhibit similar annealing profiles, and consequently, the likelihood of background amplification from mismatched primers is reduced.

xynA from Thermotoga maritima FjSS3B.1 (see Fig. 2)

A gene library is prepared as described for *D. thermophilum* Rt46B.1. It is usual for xylanolytic organisms to produce more than one active xylanase and at least 3000 plaques are screened. Sequence data show the fragment to possess a single, complete open reading frame, designated *xynA*.[9]

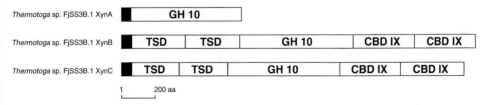

FIG. 2. Domain structures of *Thermotoga* sp. FjSS3B.1 xylanases. GH10, glycosylhydrolase family 10 domain; TSD, thermostabilizing domain; CBD IX, family IX cellulose-binding domain; black box, signal peptide sequence.

xynB and xynC from Thermotoga maritima FjSS3B.1 (see Fig. 2)

Chromosomal DNA is isolated and digested with several restriction enzymes to generate linker libraries for genomic walking as described by Morris *et al.*[13,14] The two-step PCR method as just described is used for the identification and sequencing of the FjSS3B.1 *xynB* and *xynC* genes. The full-length genes and the DNA coding for the catalytic domains of XynB and XynC are cloned into the heat-inducible vector pJLA602 as described previously. For each xylanase gene, *E. coli* strain DH5α or C600 is transformed by the recombinant pJLA602/*xynA, B,* and *C* plasmids with selection for ampicillin resistance.

Assays

Plate Assays for Xylanase Activity

λZAPII plaques containing genomic inserts are screened for the expression of xylanase activity using the Congo Red assay of Teather and Wood.[15] Plaques are overlayed with 0.4% oat spelts xylan dissolved in 0.4% agarose and then incubated at 70° for at least 5 hr. Following incubation, plates are stained with 1% Congo Red and destained with 1 *M* NaCl. Positive plaques are identified by a zone of clearing around those expressing xylanase. A similar procedure is used to identify recombinant bacteria expressing the cloned xylanases

Enzyme Assays in Liquid Media

Reducing Sugar Assay. The generation of reducing sugars from 0.25% oat spelts xylan solutions by the thermophilic xylanases is determined quantitatively using the *p*-hydroxybenzoic acid hydrazide (PHBAH) colorimetric assay,[16] which allows the measurement of the release of reducing sugars from polymers. The reagent should be prepared fresh daily by combining the following stock solutions: 10 ml 0.5 *M* sodium citrate, 10 ml 1 *M* sodium sulfite, 10 ml 0.2 *M* calcium chloride, and 10 ml 5 *M* sodium hydroxide. Make to 100 ml with water and add 1.52 g *p*-hydroxybenzoic acid hydrazide. It is important to follow the order given and to stir continuously to avoid precipitation. To prepare the substrate, 1 g oat spelts xylan (Sigma, St. Louis, MO) is suspended in 100 ml of the appropriate buffer and insoluble xylan is removed by centrifugation. The amount of xylan in

[15] R. M. Teather and P. J. Wood, *Appl. Environ. Microbiol.* **43,** 77 (1987).
[16] M. Lever, *Biochem. Med.* **7,** 27 (1973).

solution can be increased by boiling the preparation, but this will also increase the presence of background reducing sugars. A 10-μl sample of cell extract (approximately 0.005 XU, where one international unit of xylanase activity corresponds to the release of 1.0 μmol of xylose reducing sugar per minute[17]) is added to 190 μl of buffered 0.25% oat spelts xylan solution and incubated in 1.5-ml microcentrifuge tubes at the desired temperature for 10 min. Following incubation, the samples are made up to 600 μl with PHBAH, boiled for 5 min, and the A_{405} of 200-μl portions from each sample are measured in flat-bottom microtiter trays. A net A_{405} of 1.000 is equivalent to 0.067 μmol of xylose. Sodium acetate is used to buffer solutions between pH 4.0 and 6.0, Bis–Tris propane (BTP, 1,3-bis[tris(hydroxymethyl)methylamino]propane) for pH 6.0–9.0 and CAPS (3-[cyclohexylamino]-1-propanesulfonic acid) for pH 9.0–11.0.

Colorimetric Assay. The most convenient assay for *Thermotoga* and *Dictyoglomus* XynA utilizes the ability of family 10 xylanases to tolerate a wider range of substrates than their family 11 counterparts, including artificial substrates.[18] Five microliters of enzyme is added to 250 μl of 10 mM *o*-nitrophenol-β-D-xylopyranoside (Sigma) dissolved in an appropriate buffer (phosphate/citrate buffer gives a wide pH range and good temperature stability). It is necessary to warm the buffer gently to get the *o*NPXyl to dissolve fully. The reaction is run at 80° for 10–20 min, stopped by the addition of 1 ml of cold 1 M Na$_2$CO$_3$, and the absorbance is read at 420 nm.

This method only gives a reasonably accurate estimation of XynA activity if the conditions are kept constant. The estimated K_m of the *Thermotoga* enzyme (15 mM) is above the solubility limits of *o*NPXyl so that any change affecting the K_m (e.g., temperature or pH) will produce different results. Additionally, it is important that the reaction should not be allowed to proceed to an extent that the substrate concentration in the reaction mixture is lowered significantly (i.e., above an absorbance of 1, equivalent to a decrease in substrate concentration of about 2%). Thus, although convenient, this method is not suitable for kinetic or accurate characterization studies. *p*NPXyl is considerably more soluble than its ortho counterpart, allowing the reaction to proceed closer to its V_{max}, but the V_{max} is about 10-fold less and it is subject to transferase activity.[18] The apparent activity increases markedly in the presence of xylan; *o*NPXyl assays show no deviation in the presence of xylan. For accurate determination of activity, the PHBAH method should be used.

[17] M. J. Bailey, P. Biely, and K. Poutanen, *J. Biotechnol.* **23,** 257 (1992).
[18] P. B. Beily, M. Vrsanska, M. Tenkanen, and D. Kluepfel, *J. Biotechnol.* **57,** 151 (1997).

Temperature Stability Assays

Temperature stability assays are all performed in 12.5 mM BTP buffer, pH 6.5, in the absence of substrate. Enzyme samples are incubated in tightly capped 1.5-ml microcentrifuge tubes. At the appropriate time points, residual xylanase activity is measured immediately from 10-μl triplicate samples using the PHBAH assay described earlier. The percentage remaining activity is expressed relative to the residual xylanase activity of a control xylanase sample retained either on ice or at room temperature.

Activity Gels Following SDS–Polyacrylamide Electrophoresis

Cellular proteins (approximately 100 μg) are separated on Laemmli gels.[19] The method of Beguin[20] is used to renature protein in SDS–acrylamide gels for the identification of enzyme activity. Gels are then soaked in a 0.2% solution of oat spelts xylan for 2 hr at 70°, removed, wrapped in cling film, and incubated a further 12 hr at 70°. Following incubation, gels are stained in 1% Congo Red for 1 hr and destained 4–5 hr in 1 M NaCl. Clear regions in the gel indicate the presence of xylanase activity.

Substrate Specificity

The composition of xylose oligomers solubilized from 0.25% oat spelts xylan are determined by paper chromatography methods using butanol/acetic acid/H$_2$O (12:3:5) as the running solvent. Purified xylobiose and other low molecular weight standards are from Megazyme Ltd., Ireland.

Cell Growth

Growth of Escherichia coli Carrying Xylanases Inserted into pJLA602 Expression Plasmid

The example given here is for XynA of *D. thermophilum* Rt46B.1: crude extracts of other xylanases are prepared in a similar manner. The XynA enzyme is produced as follows: 100–500 μl of an overnight culture of C600 carrying pNZ2900 (grown at 30° in L broth, containing 60 mg/ml ampicillin) is used to inoculate a fresh 600-ml culture, which is grown to an A_{600} of 1.0 and then transferred to 42° for 2 hr to induce XynA production. Cells are pelleted by centrifugation at 3000g for 5 min at 4° and then resuspended in 50 ml ice-cold TES buffer (0.05 M Tris, pH 8.0, 0.05 M NaCl, and

[19] U. K. Laemmli, *Nature* **227**, 680 (1970).
[20] P. Beguin, *Anal. Biochem.* **134**, 333 (1983).

0.005 M EDTA), centrifuged again at 3000g for 5 min, and the cell pellet resuspended in 20 ml TES. Cells are then ruptured in a French pressure cell, and the resulting lysate is heated to 80° for 30 min. After heating, the cell lysate is centrifuged at 20,000g for 30 min to pellet denatured mesophilic proteins and cell debris to give a supernatant containing a relatively pure XynA enzyme.

Large-Scale Growth of E. coli Expressing Thermotoga maritima FjSS3B.1 xynA

Thermotoga maritima FjSS3B.1 XynA is produced in larger quantities by a 10-liter batch fermentation of *E. coli* strain DH5α. Cultures are started from freeze-dried ampoules by suspension in 10 ml Luria broth (5 g/liter tryptone, 5 g/liter yeast extract, 5 g/liter NaCl, 50 mg/liter ampicillin) and incubation with shaking at 28° overnight. Seven milliliters of starter culture is used to inoculate 8.5 liters of BGM2 media (5.475 g/liter NH_4Cl, 1.785 g/liter KH_2PO_4, 0.23 g/liter $MgSO_4 \cdot 7H_2O$, 73 mg/liter K_2SO_4, 5 mg/liter $FeSO_4$, 17 g/liter tryptone, 8 g/liter yeast extract, 17 g/liter glycerol, 1.7 mg/liter thiamin, and 10 ml antifoam). The fermenter (LH 500 series III) is run at 28°, 4.8 liter/min air flow, and 60% max rpm stirring. At the end of log-phase growth, approximately 24–30 hr and at an OD_{650} of 15–20, 1 liter of a 10 times concentrated BGM2 medium is added. Cells are harvested 6–9 hr later when xylanase production has reached a plateau. Typically, about 200 g wet weight of cells are produced.

Production and Purification of Fully Deuterated Thermotoga Xylanase

Fully deuterated xylanase is produced by a modification of the method described earlier. The *E. coli* starting culture is transferred several times through media made up in increasing percentages of D_2O before being used as an inoculum. The fermenter culture medium consists of 5 g/liter KH_2PO_4, 1 g/liter NH_4Cl, 1 g/liter $(NH_4)_2SO_4$, 50 ml trace elements solution, and 30 ml vitamins solution (Deutsche Sammelung von Mikroorganismen medium 141) made up in D_2O, 0.2 g/liter ampicillin, and 0.25 g/liter of a deuterated peptone, Celltone-d (Martek Biosciences, Columbia, MD). The salts are all exchanged with 10 times their weight of D_2O and freeze dried, twice, before use. The carbon source is 10 g/liter deuterated sodium succinate. The sodium succinate is synthesized by the method of Lemaster and Richards[21] and yields 83% enriched deuterated product. The media constituents are made up in 99% deuterated D_2O (Cambridge Isotope Laboratories, Andover, MA). The 9-liter fermenter culture yielded about

---

[21] D. M. Lemaster and F. M. Richards, *Anal. Biochem.* **122,** 238 (1982).

100 g wet weight of cells, containing about 35 mg deuterated xylanase. This material is purified by the same method as the nondeuterated enzyme.

Expression of Dictyoglomus and Thermotoga Xylanase Genes in Kluyveromyces lactis

The rationale behind the construction of the secretion vectors is described for the *xynA* gene of *Dictyoglomus*.[22] Similar considerations hold for the *xynA* gene from *Thermotoga* FjSS3.B1, although the final expression plasmid is constructed in a slightly different way (described in Walsh *et al.*[23]). Plasmid pSPGK-xyn contains an in-frame fusion between the *Dictyoglomus xynA* structural gene without the DNA coding for the leader sequence and the secretion signal of the *K. lactis* killer toxin. The fusion is created at codon Met-30 in the full-length XynA amino acid sequence in order to exclude the predicted bacterial signal sequence encoded by residues 1–29.[22] The *K. lactis* secretion signal in pSPGK-xyn is followed by the dipeptide Lys-Arg, which is a potential cleavage site for the Kex1 endopeptidase of *K. lactis*,[24] and an Ile-Arg pair partly encoded by the *Eco*RI cloning site. In this construction, *xynA* expression is under the control of the *S. cerevisiae* PGK promoter, which also functions in *K. lactis*.[25–27] DNA encoding XynA joined to the secretion signal for the *K. lactis* killer toxin is amplified by PCR and inserted into a plasmid carrying the pLAC4 promoter. The DNA fragment coding for the promoter, signal sequence, bacterial structural gene, and LAC4 terminator is then cut out using *Hin*dIII and *Sal*I and ligated into plasmid pCJK1[22] to give pCXJK-xyn.

Plasmid pCXJK-xyn is constructed to allow regulated expression of *Dictyoglomus xynA* under the control of the promoter from the *K. lactis* LAC4 gene, which is induced up to 100-fold by galactose and lactose.[27] In order to ensure inclusion of all essential regulatory elements from the 5'-flanking region of *LAC4,* a 1.1-kb fragment containing all three UAS elements, which potentially interact with the *trans*-acting LAC9 protein,[28] is amplified by PCR. The expression cartridge is designed as a *Sal*I/*Hin*dIII fragment to facilitate direct introduction into the corresponding sites in

[22] D. J. Walsh and P. L. Bergquist, *Appl. Environ. Microbiol.* **63,** 329 (1997).

[23] D. J. Walsh, M. D. Gibbs, and P. L. Bergquist, *Extremophiles* **2,** 9 (1998).

[24] C. Tanguy-Rougeau, M. Wésolowski-Louvel, and H. Fukuhara, *FEBS Lett.* **234,** 464 (1988).

[25] R. Fleer, X. J. Chen, N. Amellal, P. Yeh, A. Fournier, F. Guinet, N. Gault, D. Faucher, F. Folliard, H. Fukuhara, and J.-F. Mayaux, *Gene* **107,** 285 (1991).

[26] I. G. Macreadie, L. A. Castelli, A. C. Ward, M. J. R. Stark, and A. A. Azad, *Biotechnol. Lett.* **15,** 213 (1993).

[27] R. C. Dickson and J. S. Markin, *J. Bacteriol.* **142,** 777 (1980).

[28] J. A. Leonardo, S. M. Bhairi, and R. C. Dickson, *Mol. Cell. Biol.* **7,** 4369 (1987).

pCXJ1. The kanamycin resistance gene is added to allow plasmid selection in wild-type strains, and the final recombinant plasmid is introduced into several mutant and wild-type *K. lactis* strains, including CBS1065 and CBS2359.

Kluyveromyces lactis CBS1065 carrying the gene for *Thermotoga xynA* (pCWK-xyn) is inoculated into 10 ml of YPD medium (1% yeast extract, 2% Bacto-peptone, 2% glucose) plus 200 μgm/ml G418 and grown overnight at 30°. The overnight culture is subcultured into 200 ml of YPD medium plus G418 to 5×10^5 cells/ml and again grown overnight at 30°. On the third day, 2 liter of YPD plus G418 is inoculated to 5×10^6 cells/ml in a fermenter and grown with pH control to a cell density of between 5×10^8 and 10^9 cells/ml. Expression of the xylanase is induced by the addition of 200 ml of 20% lactose (it took from 11 to 24 hr to reach this cell density). The concentration of glucose is tested at the point of induction and, ideally, should be about 1%; glucose is then added as required. After 120 hr, the cells are removed by centrifugation and the supernatant is concentrated by diafiltration. XynA secreted into the medium is more than 90% pure as isolated from the supernatant, but may be purified further using the same procedures as for the enzymes produced in *E. coli*. Examples of untreated supernatants from several strains of *K. lactis* expressing *Dictyoglomus* XynA and *Thermotoga* XynA grown in shake flask culture are shown in Fig. 3.

At this stage, the enzyme is heavily contaminated with a brown pigment apparently derived from the medium. This pigment can be removed as follows: NaCl is added to the sample until the concentration is 4 *M*. A phenyl-Sepharose column is preequilibrated with unbuffered 4 *M* NaCl (200 ml of phenyl-Sepharose may be used for amounts of protein from 200 mg to 2 g). The sample is then loaded onto the column, and the unretained fraction is collected (140 ml) and discarded. The column is washed with 4 *M* NaCl, and the wash is collected as two fractions (40 and 110 ml) and discarded. The enzyme is then eluted with water and collected as a single fraction (160 ml). Eighty percent of the activity absorbed to the column is found in the final fraction, and the pigment is removed in the washes.

Purification of Hyperthermophilic Xylanases Expressed in *E. coli*

Purification of D. thermophilum XynA

A heat-treated cellular extract containing xylanase (10 ml) is passed through a PD-10 G-25 Sephadex column (Pharmacia, Uppsala, Sweden) to remove salt and smaller molecules. A 26/10 HiLoad Q-Sepharose column (Pharmacia) equilibrated with 10 m*M* HEPES (pH 7.5) is used to purify

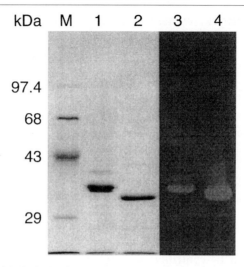

FIG. 3. Untreated shake flask culture supernatant from recombinant *K. lactis* strain CBS1065 expressing *D. thermophilum* Rt46B.1 XynA and *Thermotoga* sp. FjSS3B.1 XynA. Lane M, molecular mass ladder. Lane 1, *D. thermophilum* Rt46B.1 XynA. Lane 2, *Thermotoga* sp. FjSS3B.1 XynA. Lanes 3 and 4, xylanase activity assayed by zymogram analysis, respectively, for *D. thermophilum* Rt46B.1 XynA and *Thermotoga* sp. FjSS3B.1 XynA.

XynA. Enzyme activity is eluted at 70 mM NaCl with a linear 0–150 mM salt gradient. The enzyme is estimated to be over 95% pure after purification based on SDS–PAGE analysis (data not shown).

Purification of D. thermophilum XynB

The wild-type gene gives low yields in *E. coli* and is toxic for cell growth. Removal of the first 24 amino acids (including most of the leader sequence) and the introduction of an N-terminal methionine substantially improve expression and thermostability of the enzyme.[14] XynB is purified to electrophoretic homogeneity using cation-exchange chromatography. A 2.5-ml sample of heat-treated *E. coli* cell extract prepared in 10 mM sodium acetate (pH 5.5) is passed through a PD-10 G-25 Sephadex column (Pharmacia, Auckland, New Zealand) to remove the low molecular weight compounds from the cell extracts. The PD-10 fraction (3.5 ml) is then applied to an S15 "SartoBind" cation-exchange disk (Sartorius Australia, Victoria, Australia) attached to a Pharmacia fast protein liquid chromatography (FPLC) workstation. The S15 disk is washed with 15 ml of equilibration buffer (10 mM sodium acetate, pH 5.5), and the bound proteins are then eluted over a 0–0.3 M NaCl gradient. Rt46B.1 XynB proteins elute at approximately 0.05 M NaCl.

Large-Scale Purification of Thermotoga XynA

Ten liters of cells and media are concentrated to 1 liter and exchanged into extraction buffer (10 mM Bis–Tris, pH$_{20}$7) using a 0.1-μm hollow fiber cartridge and then frozen. After thawing for 15 min, sonication is used to lyse the *E. coli* cells. A further 1 liter of extraction buffer is added, which also contains 20 mM Mg^{2+} ions and 0.05% DNase I. The temperature is increased to 35° and is held for 1 hr before being increased to 75° over 2 hr. Heat treatment causes the majority of the protein contaminants to precipitate and the DNase I removes the *E. coli* DNA. The precipitated proteins are removed by centrifugation (10,000g, 4°, 60 min) after standing at 0–5°. Best results are obtained with a 2- to 4-day standing period; because protein precipitation is so slow, samples that have been centrifuged too soon continue to precipitate during subsequent column runs.

An equal volume of 2 M ammonium sulfate is added to approximately 2 liters of the heat-treated supernatant (containing about 7.6 mg protein/ml) to bring the salt concentration to 1 M. Alternatively, a final concentration of 4 M NaCl can be used, but at high protein concentrations, although there may be problems with aggregation and consequent loss of activity. A second centrifugation may be necessary to remove new precipitates caused by the high salt concentration depending on the completeness of the heat treatment. The clear supernatant is loaded onto a 700-ml phenyl-Sepharose column, washed with 1.5 column volumes of 1 M ammonium sulfate, and eluted using 10 mM extraction buffer. Active fractions are identified and desalted to <2 mS and concentrated by ultrafiltration to 1 liter. A 3-kDa cutoff membrane is required because XynA is not completely retained by a 10-kDa membrane. The pH of the sample is adjusted to 6.7 and loaded onto a 150-ml Fast Flow Q-Sepharose column (Pharmacia) and eluted with a 2-liter, 0 to 150 mM, NaCl gradient in 10 mM Bis–Tris, pH$_{20}$6.7. The final purification step is performed on a 30-ml Mono S column (Pharmacia) in small quantities. Approximately 100 ml of the Fast Flow Q sample is dialyzed (6–8 kDa molecular weight cutoff) first in 5 liters of H$_2$O and then in 1 liter 10 mM MES, pH$_{20}$5.3, and is then loaded onto the Mono S column and eluted with a 360-ml, 0 to 100 mM, NaCl gradient. This procedure gives sufficient purification so that only one band is visible on silver-stained SDS–PAGE. The results of the purification are given in Table I.

Purification of Thermotoga XynB and XynC

Thermotoga sp. FjSS XynB1, XynB4, XynC1, and XynC4[10] xylanases are purified using anion-exchange chromatography. Curde cell extracts containing XynB1, XynB4, XynC1, and XynC4 are prepared from *E. coli* DH5α strains harboring recombinant pJLA602 plasmids as described pre-

TABLE I
PURIFICATION OF *Thermotoga maritima* FjSS3B.1 XynA

Step	Total activity (kU)[a]	Total protein (g)[b]	Specific activity (U/mg)	Recovery	Purification (-fold)
Harvest	18	28	0.64	100	1
Sonicated supernatant	16.3	24	0.68	90.5	1.1
Heat treatment	11	2.7	4.07	61	6.4
Phenyl-Sepharose	11.8	0.95	12.4	66	19.4
Fast Flow Q Sepharose	10.1	0.11	96.8	59	151
Mono S Sepharose	9.7	0.07	139	54	217

[a] One unit is defined as the amount of xylanase required to release 1 μmol of nitrophenol per minute.
[b] Protein was estimated using absorbance at 280 nm with purified xynA used as a standard.

viously.[29] These extracts are concentrated 10-fold by centrifugation across a 10,000 molecular weight cutoff size-exclusion membrane (Vivascience, Binbrook, Lincoln, UK) to a final volume of 2.5 ml. Low molecular weight materials, which appear to interfere with liquid chromatography, are removed by passing the concentrated extracts down a Pharmacia PD10 superfine G-25 column equilibrated with buffer A (12.5 mM Bis–Tris propane buffer, pH 7.3). The 3.5-ml PD10 eluate is applied to a Pharmacia 5-ml HiTrap-Q column (equilibrated with buffer A), which is then washed with 50 ml of buffer A. Xylanases are eluted from the column using a stepwise NaCl gradient involving 0, 25, 50, 75, 100, 125, and 150 mM NaCl in buffer A. Three 5-ml fractions are collected at each step. Reducing sugar assays are utilized to monitor the elution of XynB1 and XynB4 from the HiTrap-Q column, whereas SDS–PAGE is used for XynC1 and XynC4 due to the absence of detectable xylanase activity in these samples. XynB1 and XynC1 eluted from the column in 125 mM NaCl, whereas XynB4 and XynC4 eluted in 100 mM NaCl.

Purification of Archaeal Xylanase

Xylanase from *Thermococcus zilligii* strain AN1[30] is the first archaeal xylanase (or hemicellulase) described to date. The enzyme from approximately 60 liters of extracellular supernatant of *T. zilligii* strain AN1, grown anaerobically in AN1 medium[31] at 75° in 2-liter Schott bottles, is purified

[29] M. D. Gibbs, R. A. Reeves, and P. L. Bergquist, *Appl. Environ. Microbiol.* **61,** 4403 (1995).
[30] A. Uhl and R. M. Daniel, *Extremophiles* **4,** 263 (1999).
[31] K. U. Klages and H. W. Morgan, *Arch. Microbiol.* **162,** 261 (1994).

initially on a phenyl-Sepharose (Pharmacia) column. The column (44 mm diameter, 200 mm long) is equilibrated with 1 M $(NH_4)_2SO_4$ before the extracellular supernatant [taken to 1 M $(NH_4)_2SO_4$, pH 7.0] is loaded. After washing the column with 1 M NaCl, the activity is eluted sequentially with 20 mM Bis–Tris propane, pH 7.0, and 50% ethylene glycol/20 mM Bis–Tris propane, pH 7.0. The combined fractions containing xylanase are concentrated, and the buffer is exchanged to 20 mM Tris–HCl, pH 7.5, using an Amicon YM30 membrane. The concentrate is applied to a Pharmacia Mono Q 10/10 column equilibrated with the same buffer and eluted with a 0–1.0 M NaCl gradient. Active fractions are combined, equilibrated in 10 mM sodium phosphate buffer, pH 6.8, containing 10 mM $CaCl_2$, and applied to a Bio-Rad HPHT (hydroxylapatite) column equilibrated in the same buffer. Protein is eluted using a gradient of 10–350 mM sodium phosphate, pH 6.8. The enzyme elutes between 50 and 140 mM sodium phosphate. The final step utilizes a Phenomenex Biosep-SEC 3000 HPLC gel filtration column run isocratically with 20 mM Bis–Tris propane, pH 7.0. Purity and molecular size are determined on silver-stained (Bio-Rad method) SDS PhastGels (8–25%; Pharmacia).

Reasons for the large losses in activity during purification (Table II) are not obvious. It may be that synergistically acting enzyme(s) (e.g., a xylosidase) are being separated away from the xylanase. Enzyme instability seems an unlikely cause. The enzyme is very stable at room temperature, and AN1 proteases are not active at the relatively low temperatures (25°) at which the purification and concentration steps occur. The addition of agents, such as 10 mM EDTA, 12% $(NH_4)_2SO_4$, 30 mM $MgCl_2$, 30 mM $MgCl_2$, and 0.1% (w/v) dithiothreitol (DTT), did not prevent the losses. However, the addition of 2% bovine serum albumin or 0.1% DTT stabilized the activity during storage, handling, and activity assays, and 0.1% DTT is used routinely in storage and handling. The purified xylanase gives a single band on a silver-stained SDS electrophoresis gel. The R_f of the enzyme on both the HPLC column and the electrophoresis gel corresponds to a

TABLE II
PURIFICATION OF *Thermococcus zilligii* AN1 XYLANASE

Purification step	Total activity (U)	Total protein (mg)	Specific activity (U mg^{-1})	Yield (%)	Purification (-fold)
Crude extract	718	6730	0.107	100	1.0
Phenyl-Sepharose	157	943	0.166	22	1.6
Mono Q	23	181	0.128	3	1.2
HPHT	5.6	1.1	5.07	0.8	47.6

TABLE III
BIOCHEMICAL CHARACTERIZATION OF *Dictyoglomus* Rt46B.1 XYLANASES

Enzyme	XynA	XynB
Thermostability: half-life at 85° (no substrate)	>24 hr	>24 hr
pH optimum	6.5	6.5
pH range for 50% activity	5.5–9.5	5.5–7.5
Release of reducing sugars from *Pinus radiata* kraft pulp (μmol/g dry weight/6 hr/75°)	0.0875	0.0625
Products of hydrolysis of *P. radiata* kraft pulp	Xylose; xylobiose	Xylobiose; xylotriose; high molecular weight xylo-oligo-saccharides

molecular mass of 95 kDa, so there was no evidence for multiple subunits. The specific activity of the purified enzyme (Table II) is low, but comparable with those of some bacteria.[32,33]

Biochemical Characteristics of Hyperthermophilic Xylanases

XynA and XynB from Dictyoglomus thermophilum Rt46B.1

Both enzymes exhibit optimal xylanase activity at pH 6.5, 85°; XynA is active over a broad pH range with over 50% activity between pH 5.5 and 9.5, but XynB is more limited, with 50% activity observed between pH 5.5 and 7.5. In addition, the two xylanases show more or less equivalent activity on both isolated and fiber-bound xylan substrates, as measured by the release of reducing sugars (Table II). Both enzymes are very thermostable and show no loss of activity on incubation for 24 hr at 85°. Each enzyme produces different products on digestion of *Pinus radiata* kraft pulp. XynA produces predominantly xylose and xylobiose (suggesting that it also has aryl β-xylosidase activity) whereas XynB digestion gives xylobiose, xylotriose, and high molecular weight xylo-oligosaccharides as products (Table III).

XynA and XynB of Thermotoga maritima FjSS3B.1

Under the assay conditions used, the pH for optimal activity of XynB1 at 85° is 6.5, with 50% of maximum activity being retained between pH 5.5 and 7.4 (data not shown). The pH optimum and half-lives of XynB1 differ

[32] S. F. Lee, C. W. Forsberg, and J. B. Rattray, *Appl. Environ. Microbiol.* **53**, 644 (1987).
[33] F. Uchimo and T. Nakane, *Agric. Biol. Chem.* **45**, 1121 (1981).

from those reported previously for XynA from *Thermotoga* FjSSB.1,[9] which has both a broader pH range for 50% activity and is significantly more thermostable (Table IV). The full-length XynC has low activity, but it is possible to measure residual enzyme activity. It shows a similar pH profile to XynB1. The catalytic domain of XynB has a much narrower pH range with optimal activity at pH 7.0 and the truncated enzyme is much less thermostable, with a temperature optimum of 70° and a significantly reduced half-life.

Properties of *Thermococcus* AN1 Xylanase

In 50 mM sodium citrate buffer (pH 2–6) and 50 mM Bis–Tris propane buffer (pH 6–9.5), the optimum pH for activity is pH 6, with 50% activity

TABLE IV
TEMPERATURE STABILITIES AND pH CHARACTERISTICS OF *Thermotoga*
FjSS3B.1 XYLANASES XynA AND XynB[a]

Enzyme	Temperature	Half-life
XynA	95°	12 hr
	90°	22 hr
	85°	No detectable loss after 16 hr
XynB1	95°	<10 min
	90°	8 hr
	85°	No detectable loss after 16 hr
XynB4	94°	1.5 min
	90°	4 min
	85°	3 h

	pH optimum	50% activity
XynA	6.3	5.1–8.1
XynB1	6.5	5.5–7.4
XynB4	7.0	6.1–7.4
XynC1	6.5	5.8–7.4

[a] Enzymes were incubated in 25 mM Bis–Tris-propane in the absence of substrate. These are minimal values: substrate protection may increase the apparent thermostability figures. pH optima were determined in 25 mM Bis–Tris propane adjusted to pH at 85°. Enzyme activity was measured as the release of reducing sugars from oats spelt xylan as described previously.[28] XynB1 and XynC1 refer to the full-length enzymes expressed in *E. coli*. XynB4 refers to the catalytic domain expressed without other domain structures. Assays used to measure thermostability were performed at 85° for XynA and XynB1 and at 67° for XynB4 (the temperature for maximal activity under the assay conditions employed).

at pH 3.9 and 8.3. The enzyme is active against all xylans tested, and HPLC of the reaction products indicates endo action. The enzyme also yields smaller xylo-oligosaccharides from xylohexaose and xylopentaose.[30] The K_m (and V_{max}) values for xylans from different sources are larchwood, 0.028% (40.5 U/mg); oat spelts, 0.12% (30.4 U/mg); wheat arabinoxylan, 0.36% (44.4 U/mg); and birchwood, 1.58% (69.0 U/mg). These K_m values are typical of those for microbial xylanases. The enzyme is also active against beechwood xylan. The enzyme shows no detectable activity in PAHBAH assays against carboxymethyl cellulose, Avicel (Sigma-Aldrich, Sydney, Australia) starch, amylopectin, Konjac gum (galactoglucomannan), locust bean gum (galactomannan), and guar gum. It is not active against xylobiose, cellobiose, or nitrophenylsaccharides.

At 85°, 80% activity remains after an 8-hr incubation period. At 95°, after an incubation time of 3 hr, more than 60% of the activity could still be recovered, corresponding to a $t_{1/2}$ of 4 hr. At 100° the $t_{1/2}$ is 8 min.

Applications

Apart from their uses in bleaching applications in the pulp and paper industry[1,34] and in the preparation of dissolving pulps for the manufacture of viscose rayon,[35] xylanases have potential use in the modification of baked grain products to provide goods of potential commercial value. Among the properties that can be modified by xylanase treatment is the specific volume of bread.[2] Xylanases may also be of use for improving the nutritional value of feeds for nonruminant animals such as poultry. The pentosan components (β-glucans and arabinoxylans) of grains such as wheat and rye are known to be responsible for poor nutrient uptake and sticky guano. Both problems are thought to result from the high viscosity of undigested pentosans, which hamper the diffusion of nutrients and bind water to make excreta watery. These polysaccharides from viscous gel-like structures in the small intestine and, in so doing, trap nutrients such as starch, proteins, and fats, which would normally be accessible to the digestive enzymes and transport systems of the animal. Several studies have shown that these problems can be alleviated using crude xylanase preparations. Xylanase action was shown to improve both the weight gain of chicks and their feed conversion efficiency (summarized in Wong and Saddler[2]).

---

[34] T. W. Jeffries and L. Viikari (eds.), "Enzymes for Pulp and Paper Processing." American Chem. Society, Washington, DC, 1996.
[35] G. M. Gübitz, T. Lischnig, D. Stebbing, and J. N. Saddler, *Biotechnol. Lett.* **19,** 491 (1997).

Acknowledgments

The work reported from P. L. Bergquist's laboratory was supported in part by grants from the Public Good Science Fund (New Zealand), an Australian Research Council Small Grant, and the Macquarie University Research Grants Fund. Research from R. M. Daniel's laboratory was supported in part by grants from the Foundation for Research, Science, and Technology and by the award of a James Cook Fellowship to RMD from the Royal Society of New Zealand.

[20] Chitinase from Thermococcus kodakaraensis KOD1

By TADAYUKI IMANAKA, TOSHIAKI FUKUI, and SHINSUKE FUJIWARA

Introduction

Chitins are a large family of glycans that are β-1,4-linked, insoluble linear polymers of N-acetylglucosamine (GlcNAc). They are present in the walls of higher fungi and also exist in the exoskeletons of insects, arachnids, and many other groups of invertebrates and as an extracellular polymer of some bacteria. Chitin is the second most abundant organic compound (after cellulose). It is estimated that an annual formation rate and steady-state amount of chitin is in the order of 10^{10} to 10^{11} tons.[1] Therefore, the application of thermostable chitin-hydrolyzing enzymes (chitinases) can be expected for the effective utilization of this abundant biomass.

Various chitinases and their genes from Eucarya and Bacteria have been investigated. However, studies on archaeal chitinase have been limited to those from hyperthermophilic archaeal strains *Thermococcus chitonophagus*[2] and *Thermococcus kodakaraensis* KOD1[3,4] (previously reported as *Pyrococcus kodakaraensis* KOD1[5,6]). *T. chitonophagus* produces a chitinase and utilizes chitin as a carbon source. The chitinoclastic enzyme system of the strain is oxygen stable, cell associated, and inducible by chitin. *T.*

[1] G. W. Gooday, *in* "Biochemistry of Microbial Degradation" (C. Ratledge, ed.), p. 279. Kluwer Academic, Dordrecht, 1994.

[2] R. Huber, J. Stöhr, S. Hohenhaus, R. Rachel, S. Burggraf, H. W. Jannasch, and K. O. Stetter, *Arch. Microbiol.* **164**, 255 (1995).

[3] M. Takagi, T. Tanaka, S. Fujiwara, and T. Imanaka, *in* "Genetics, Biochemistry and Ecology of Cellulose Degradation" (K. Ohmiya *et al.*, eds.), p. 683. UNI Publishers, Tokyo, 1999.

[4] T. Tanaka, S. Fujiwara, S. Nishikori, T. Fukui, M. Takagi, and T. Imanaka, *Appl. Environ. Microbiol.* **65**, 5338 (1999).

[5] M. Morikawa, Y. Izawa, N. Rashid, T. Hoaki, and T. Imanaka, *Appl. Environ. Microbiol.* **60**, 4559 (1994).

[6] S. Fujiwara, M. Takagi, and T. Imanaka, *Biotechnol. Annu. Rev.* **4**, 259 (1998).

kodakaraensis KOD1 also produces a cell-associated thermostable chitinase. The *Tk-chiA* gene encoding the chitinase of KOD1 strain has been cloned and sequenced.[3,4] The *Tk-chiA* gene is composed of 3645 nucleotides encoding a large protein (1215 amino acids) with a molecular mass of 134,259 Da. Sequence analysis indicates that *Tk*-ChiA is divided into two distinct regions with respective active sites, as shown in Fig. 1. N-terminal (region A) and C-terminal half (region B) regions show sequence similarity with chitinase A1 from *Bacillus circulans* WL-12 and chitinase from *Streptomyces erythraeus* ATCC 11635, respectively; both are classified into family 18 of glycosylhydrolases. Furthermore, *Tk*-ChiA possesses unique chitin-binding domains (CBDs), CBD1 and CBD2, 3, which show sequence similarity with families V and II of cellulose-binding domains, respectively. Recombinant *Tk*-ChiA and various deletion mutants are pro-

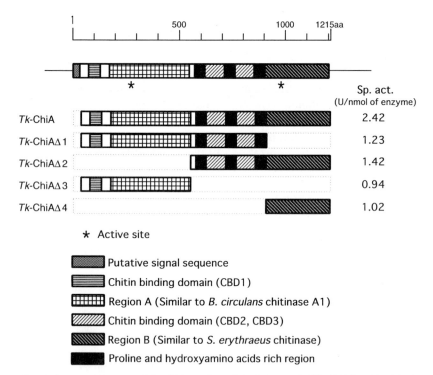

FIG. 1. Structural features of the chitinase from *T. kodakaraensis* (*Tk*-ChiA) and its deletion mutants, and specific activities toward colloidal chitin. The putative signal sequence, regions A and B, three chitin-binding domains, and three proline and hydroxyamino acid residue-rich regions are indicated. The specific activities of *Tk*-ChiA and its deletion mutants toward colloidal chitin were determined at 80°.

duced in *Escherichia coli,* and their biochemical properties are examined. All constructs show extreme thermostability. Both N-terminal and C-terminal halves are functional as a chitinase, and CBDs play an important role for insoluble chitin binding and hydrolysis.

This section describes basic procedures for the purification of recombinant chitinase, substrate preparation, enzyme assay, and determination of reaction product. In addition, a simple method to examine chitin-binding activity is introduced.

Purification of Recombinant Chitinase

Vector and Strain for Chitinase Expression

Tk-ChiA contains a hydrophobic core region in the N-terminal portion, which may be functional as a signal sequence. Efficient expression of *Tk*-ChiA in *E. coli* has not been achieved from the recombinant plasmid harboring *Tk-chiA* with its entire original signal sequence (data not shown), probably due to detrimental effects of the archaeal secretion signal on efficient expression in bacteria. Hence, the putative signal region of *Tk*-ChiA (29 amino acid residues) has been replaced with a bacterial signal sequence by using an expression plasmid pET-25b(+) (Novagen, Madison, WI). The *Tk-chiA* gene has been fused with the *pelB* leader of pET-25b(+) by inserting the fragment into *Nco*I and *Bam*HI sites, which we refer to as pET-ChiA. Significant chitinase activity has been detected from cells harboring this recombinant plasmid. As mentioned earlier, *Tk*-ChiA possesses unique structural features with an unusually large molecular mass. For functional analysis of each domain in *Tk*-ChiA, several recombinant plasmids for the production of deletion mutants (*Tk*-ChiAΔ1, ChiAΔ2, ChiAΔ3, and ChiAΔ4, as shown in Fig. 1) have also been constructed.

Overexpression and Purification

Escherichia coli strain BL21(DE3), which possesses a stable chromosomal copy of the T7 RNA polymerase gene under control of the *lacUV5* promoter, is used as the host strain for expression of the *Tk-chiA* gene. Gene expression in *E. coli* cells harboring pET-ChiA is induced by 1 mM isopropyl-β-D-thiogalactopyranoside (IPTG) at midexponential growth phase and incubated for 4 hr at 37°. Cells are harvested by centrifugation (7000g for 10 min at 4°), washed with buffer A (50 mM Tris–HCl, pH 7.5, 1 mM EDTA), and then resuspended in buffer A. The cells are disrupted by sonication, and a clear supernatant is obtained by centrifugation (24,000g for 20 min at 4°). The supernatant is incubated at 70° for 15 min and centrifuged (15,000g for 15 min at 4°) to obtain a heat-stable crude extract. The crude supernatant is brought to 40% saturation with ammonium sulfate, followed by stirring overnight at 4°. The suspension is centrifuged (12,000g for 30 min at 4°), and the resulting precipitate is dissolved in buffer A. The solution is then dialyzed overnight against the same buffer and centrifuged (12,000g for 30 min at 4°) again to remove insoluble proteins formed during the dialysis. A Resource Q column (Amersham Pharmacia Biotech, Uppsala, Sweden) is equilibrated with buffer A, and the resulting supernatant is applied to the column. Tk-ChiA is eluted by a linear gradient of NaCl (0–0.5 M) using FPLC (fast protein liquid chromatography, ÄKTA explorer 10S, Amersham Pharmacia Biotech). Peak fractions are concentrated with Centricon-100 (Millipore, Bedford, MA), and the sample is applied to a Superdex-200 HR 10/30 column (Amersham Pharmacia Biotech) equilibrated with buffer B (50 mM Tris–HCl, pH 7.5, 150 mM NaCl). Active fractions are collected, and the purified Tk-ChiA is stored at 4°. Purification procedures for the deletion mutants of Tk-ChiA are conducted with slight modifications. The heat-stable crude extract of Tk-ChiAΔ4 is brought to 50% saturation with ammonium sulfate. For concentrating samples prior to gel filtration, Centricon-50 (Tk-ChiAΔ1, ChiAΔ2, and ChiAΔ3) or Centricon-30 (Tk-ChiAΔ4) is used. A Superdex-75 HR 10/30 column (Amersham Pharmacia Biotech) is used for Tk-ChiAΔ4 instead of Superdex-200 HR 10/30. The protein concentration is determined by the Bio-Rad protein assay system (Bio-Rad, Hercules, CA) with bovine serum albumin as a standard.

Preparation of Colloidal Chitin

Powdered chitin (10 g) (Wako Pure Chemical Industries, Osaka, Japan) is mixed with 500 ml of 85% phosphoric acid and stirred for 24 hrs at 4°. The suspension is poured into 5 liters of deionized water and centrifuged (12,000g for 10 min). The resulting precipitate is washed with deionized water until the pH is maintained at 5.0 and is then neutralized by the addition of 6 N NaOH. The suspension is centrifuged (12,000g for 10 min) and washed with 3 liters of deionized water for desalting. The resulting precipitate is suspended with deionized water. The chitin content in the suspension is determined by measuring the weight of a dried sample, and the final concentration is adjusted to 1%.

Methods for Chitinase Assay

Chitinase activity is examined by two conventional methods. Method 1 detects the reducing groups produced by the chitinase reaction. Method

2 is a fluorometric assay with labeled N-acetylchitooligosaccharides as substrates.

Method 1

Chitinase can be assayed by a modification of the Schales procedure[7,8] with chitin or colloidal chitin as the substrate. One unit of chitinase activity is defined as the amount of enzyme that produces 1 μmol of reducing sugar per minute. The method used is as follows.

1. Preincubate 0.83 ml of reaction mixture [0.5 ml of 100 mM sodium acetate buffer (pH 5.0) and 0.33 ml of 0.5% colloidal chitin] in a 1.5-ml Eppendorf tube at a specified reaction temperature.
2. Add enzyme solution (0.17 ml) and incubate the mixture at the specified temperature for 10 min.
3. Transfer the mixture to an ice bath.
4. Centrifuge the tube (3000g, 10 min at 4°) and obtain the supernatant as a sample solution.
5. Transfer 0.5 ml of the sample solution into a fresh tube and add 0.67 ml of color reagent solution. The color reagent solution can be made by dissolving 0.5 g of potassium ferricyanide in 1 liter of 0.5 M sodium carbonate and storing in a brown bottle.
6. Incubate the tube in boiling water for 15 min.
7. After cooling the tube in an ice bath, measure the absorbance at 420 nm.

In order to prepare standard curves, incubate mixtures of 2.0 ml of color reagent and 3.0 ml of GlcNAc solution in boiling water for 15 min and measure the decrease in A_{420} at various GlcNAc concentrations ranging from 0 to 0.5 μmol/ml.

The concentration of sodium carbonate is sufficiently high so as not to be affected by the buffer of the sample solution. The color development is not affected by the high concentration of sodium carbonate. The color is also not affected when the mixture is allowed to stand for a few hours before or after boiling. A blank value of 0.85 (A_{420}) is obtained within 1% deviation.

Method 2

Chitinase activity can be measured by a fluorometric assay with 4-methylumbelliferyl-N-acetyl-β-D-glucosaminide (GlcNAc-4MU), 4-meth-

[7] O. Schales and S. S. Schales, Arch. Biochem. **8**, 285 (1945).
[8] T. Imoto and K. Yagishita, Agric. Biol. Chem. **35**, 1154 (1971).

ylumbelliferyl-β-D-N,N'-diacetylchitobioside (GlcNAc$_2$-4MU), and 4-methylumbelliferyl-β-D-N',N'',N''-triacetylchitotrioside (GlcNAc$_3$-4MU) (Sigma, St. Louis, MO), based on a reported procedure[9] with slight modifications. The fluorescence of liberated 4-methylumbelliferone (4MU) is measured (350 nm excitation, 440 nm emission) in a fluorescence spectrophotometer F-2000 (Hitachi, Tokyo, Japan). One unit of chitinase activity is defined as the amount of enzyme that liberates 1 μmol of 4MU per minute. The following protocol is used.

1. Prepare 1 mM GlcNAc$_n$-4MU solutions by dissolving each powdered sample in 50% (v/v) methanol.

2. Mix 10 μl of the GlcNAc$_n$-4MU solution with 990 μl of 100 mM sodium acetate buffer (pH 5.0) and preincubate at a specified reaction temperature.

3. Add 20 μl of enzyme solution and incubate at the specified temperature.

4. Take 100 μl from the reaction tube at various reaction periods and mix with 900 μl of ice-cold 100 mM glycine–NaOH buffer (pH 11).

5. Measure the fluorescence directly (350 nm excitation, 440 nm emission) in a fluorescence spectrophotometer and plot against reaction time.

Standard curves can be prepared by measuring the value of various concentrations of 4MU. Mix 100 μl of standard solutions of 4MU (2 to 10 μM) with 900 μl of ice-cold 100 mM glycine–NaOH buffer (pH 11) and measure the fluorescence directly.

The optimal temperature and pH of Tk-ChiA for colloidal chitin are 85° and pH 5.0, respectively. The specific activity of Tk-ChiA and its deletion mutants are shown in Fig. 1. All deletion mutants show hydrolytic activity toward colloidal chitin, indicating that the respective active sites in region A and region B are independently functional. The sum of the specific activities of mutants containing region A (Tk-ChiAΔ3) or region B (Tk-ChiAΔ2) is nearly equivalent to the activity of Tk-ChiA. This result suggests that the hydrolytic effects of region A and B are additive and not synergistic in this case. The thermostability of Tk-ChiA and the four deletion mutants is higher in the following order: ChiAΔ2 > ChiAΔ4 > ChiAΔ1 > ChiA > ChiAΔ3. Tk-ChiAΔ2 and ChiAΔ4, both containing only region B, exhibit high thermostability and retain more than 70% activity even after heat treatment at 100° for 3 hr.

The release of 4MU by Tk-ChiA on monomeric-, dimeric-, and trimeric-labeled substrates is 1, 258, and 961 U/μmol of enzyme, respectively. Typical endo-type chitinases are known to show higher activity for GlcNAc$_3$-4MU than for GlcNAc$_2$-4MU.[9] Experimental results show that Tk-ChiA does

[9] P. W. Robbins, C. Albright, and B. Benfield, *J. Biol. Chem.* **263**, 443 (1988).

not hydrolyze GlcNAc-4MU and that the enzyme reaction velocity is faster for $GlcNAc_3$-4MU than for $GlcNAc_2$-4MU, suggesting that Tk-ChiA is an endo-type enzyme.

Product Analysis by Thin-Layer Chromatography

In order to identify final and intermediate products of the chitinase reaction, products for various N-acetylchitooligosaccharides ($GlcNAc_n$, $n = 1$–6) and colloidal chitin are analyzed by silica gel thin-layer chromatography (TLC). The procedure is as follows.

1. Prepare the samples for TLC. As for $GlcNAc_n$ substrates, incubate the reaction mixture (50 μl) containing 0.7 mg of $GlcNAc_n$ substrates in 70 mM sodium acetate buffer (pH 5.0) with enzyme (0.45 μg for $GlcNAc_{1-3}$ or 0.9 μg for $GlcNAc_{4-6}$) at a specified temperature. In the case of colloidal chitin as a substrate, incubate the reaction mixture (1 ml) containing 0.16 mg of colloidal chitin in 50 mM sodium acetate buffer (pH 5.0) with enzyme (0.6 μg) at a specified temperature. Centrifuge the tube and obtain the supernatant (250 μl each).

2. Spot aliquots (1 μl) of the reaction mixtures onto a silica gel plate (Kieselgel 60, Merck, Berlin, Germany). Develop with n-butanol–methanol–25% ammonia solution–water [5:4:2:1 (v/v/v/v)].

3. Detect the products by spraying the plate with an aniline–diphenylamine reagent (mixture of 4 ml of aniline, 4 g of diphenylamine, 200 ml of acetone, and 30 ml of 85% phosphoric acid). Bake at 180° for 3 min.

Typical patterns of TLC analysis for various N-acetylchitooligosaccharides and colloidal chitin are shown in Figs. 2 and 3. When $GlcNAc_{4-6}$ are used as substrates for Tk-ChiA, the major product is chitobiose ($GlcNAc_2$), along with a little amount of GlcNAc. $GlcNAc_2$ and GlcNAc are formed from $GlcNAc_3$; however, Tk-ChiA does not hydrolyze $GlcNAc_2$ (Fig. 2). When colloidal chitin is used as a substrate, $GlcNAc_2$ is detected on TLC about 1.5 hr after initiation of the reaction (Fig. 3). These results indicate that the major reaction product by the digestion of Tk-ChiA is chitobiose ($GlcNAc_2$). In addition, the TLC pattern for reaction products of $GlcNAc_6$ did not show major $GlcNAc_4$ spots in the course of the enzyme reaction, indicating that Tk-ChiA is not an exo-type enzyme.

Chitin-Binding Experiment

Unique conserved regions, known as substrate-binding domains, are found in various cellulases, xylanases, and chitinases.[10] Tk-ChiA possesses

---

[10] E. Brun, F. Moriaud, P. Gans, M. J. Blackledge, F. Barras, and D. Marion, *Biochemistry* **36**, 16074 (1997).

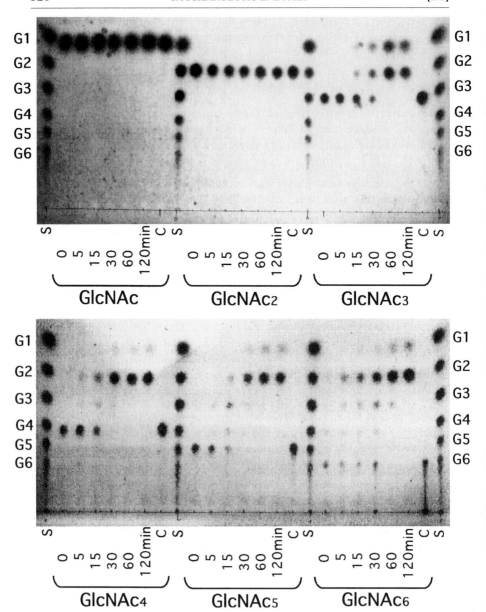

FIG. 2. Thin-layer chromatograms of restriction products by *Tk*-ChiA from various *N*-acetylchitooligosaccharides. The products at 0, 5, 15, 30, 60, and 120 min were sampled and analyzed. Lane C represents a negative control that was incubated without enzyme at 80° for 120 min. Lane S represents standard *N*-acetylchitooligosaccharides ranging from GlcNAc (G1) to GlcNAc$_6$ (G6).

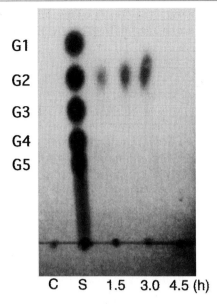

FIG. 3. Thin-layer chromatogram of restriction products by *Tk*-ChiA from colloidal chitin. The reaction was carried out at 80° for 1.5, 3.0, and 4.5 hr. Lane C represents a negative control that was incubated without enzyme at 80° for 4.5 hr. Lane S represents standard *N*-acetylchitooligosaccharides ranging from GlcNAc (G1) to GlcNAc$_5$ (G5).

three conserved regions: one is found in front of region A (CBD1) and the other two are between regions A and B (CBD2 and CBD3 with almost identical amino acid sequence). In order to confirm that these regions are actually functional as CBDs, chitin-binding assays are performed based on the procedure for the cellulose-binding assay[10–13] with some slight modifications.

1. Prepare enzyme extracts. In addition to purified chitinase, crude extract is also applicable for this experiment. [In our experiment, recombinant chitinase is expressed in *E. coli* cells (200-ml culture), and enzyme extracts are obtained by sonication in 2 ml of 50 m*M* sodium phosphate buffer (pH 7.0). The extracts are then heat treated and soluble supernatants are obtained by centrifugation.]

[11] P. Tomme, A. Boraston, B. McLean, J. Kormos, A. L. Creagh, K. Sturch, N. R. Gilkes, C. A. Haynes, R. A. J. Warren, and D. G. Kilburn, *J. Chromatogr. B Biomed. Sci. Appl.* **715,** 283 (1998).

[12] P. Tomme, R. A. J. Warren, and N. R. Gilkes, *Adv. Microbiol. Physiol.* **37,** 1 (1995)

[13] N. R. Gilkes, R. A. J. Warren, R. C. Miller, Jr., and D. G. Kilburn, *J. Biol. Chem.* **263,** 10401 (1988).

2. Mix 35 μl of enzyme extract in 50 mM sodium phosphate buffer (pH 7.0) and 35 μl of 1% colloidal chitin suspension. Incubate for 1 hr at room temperature with occasional stirring.

3. Separate the supernatant containing unadsorbed protein from colloidal chitin by centrifugation (15,000g for 10 min at 4°). Mix 15 μl of supernatant with an equal amount of sample buffer [50 mM Tris–HCl, pH 6.5, 10% (v/v) glycerol, 2% (w/v) sodium dodecyl sulfate (SDS), 2% (v/v) 2-mercaptoethanol].

4. Wash the precipitate twice with 50 mM sodium phosphate buffer (pH 7.0) and resuspend into 35 μl of sample buffer as a chitin-binding fraction.

5. Boil both fractions for 5 min and then apply to SDS–PAGE and detect objective bands under the standard procedure.

Results of typical experiments performed on partially purified Tk-ChiAΔ2 and ChiAΔ4 are shown in Fig. 4. Tk-ChiAΔ2 possesses a C-terminal active site with two CBD regions (CBD2 and CBD3), but Tk-ChiAΔ4 lacks the two CBDs. Both deletion mutants showed chitinase activities using colloidal chitin as a substrate. It was found that Tk-ChiAΔ2 was detected in the chitin-binding fraction, whereas Tk-ChiAΔ4 existed as a soluble protein that was not adsorbed on chitin. In addition, Tk-ChiAΔ1 and ChiAΔ2 showed slightly higher activity toward colloidal chitin than Tk-ChiAΔ3 and ChiAΔ4, respectively (Fig. 1). These results indicate that

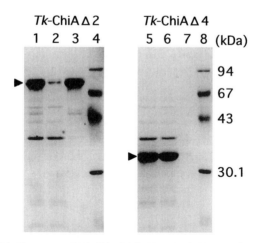

FIG. 4. Chitin-binding assay of Tk-ChiAΔ2 (left) and Tk-ChiAΔ4 (right). Lanes 1 and 5, enzyme extract used for chitin-binding experiment; lanes 2 and 6, unbound fraction to colloidal chitin; lanes 3 and 7, bound fraction to colloidal chitin; and lanes 4 and 8, molecular mass standards, namely rabbit muscle phosphorylase b (94 kDa), bovine serum albumin (67 kDa), egg white ovalbumin (43 kDa), and bovine erythrocyte carbonic anhydrase (30.1 kDa).

CBD2 and/or CBD3 plays an important role in binding to and hydrolysis of the insoluble substrate.

Acknowledgment

This work was supported by a grant from CREST (Core Research for Evolutional Science and Technology, Japan).

[21] Characterization of β-Glycosylhydrolases from *Pyrococcus furiosus*

By Thijs Kaper, Corné H. Verhees, Joyce H. G. Lebbink,
Johan F. T. van Lieshout, Leon D. Kluskens, Don E. Ward,
Servé W. M. Kengen, Marke M. Beerthuyzen, Willem M. de Vos,
and John van der Oost

Introduction

Enzymes from bacterial and archaeal hyperthermophiles have been studied extensively for their catalytic properties and stability at extremely high temperatures, as well as their biotechnological potential as biocatalysts at elevated temperatures.[1] Considerable attention has been given to sugar-converting enzymes from heterotrophic hyperthermophilic archaea, mainly in members of the archaeal orders Sulfolobales (i.e., *Sulfolobus solfataricus*) and Thermococcales (i.e., *Pyrococcus furiosus*). The aim here is to give an overview of methods that have been instrumental in the molecular and biochemical characterization of enzymes from *Pyrococcus* species that catalyze the hydrolysis of β-linked sugars, the β-glycosylhydrolases (βGHs). These studies form an essential basis for a next generation of experiments, the engineering of βGHs for optimal exploitation of their potential.

Gene Cloning

A variety of methods have been used for the isolation of βGH-encoding genes from *P. furiosus* and related species (Table I). In the absence of either orthologs or paralogs, a classical reversed genetics approach may be used. This is realized through biochemical enrichment of the activity,

---

[1] F. Niehaus, C. Bertoldo, M. Kähler, and G. Antraninkian, *Appl. Microbiol. Biotechnol.* **51**, 711 (1999).

TABLE I
PYROCOCCAL β-GLYCOSYLHYDROLASES

Parameter	CelB	BmnA	BglA	BglB	BglC	LamA	EglA	EglB
Method[a]								
Reverse genetics	+							
Expression library	+	+				+	+	
PCR			+	+				
Genomics		+	+	+	+			
Types[b]								
GH family	1	1	1	1	35	16	12	5
Exo/endo	exo	exo	exo	exo	(exo)	endo	endo	(endo)
Specificity (β)	glu*	man*	glu	man*	(gal)	lam	lich	(cell)
	1,3–1,4	1,4(1,3)	1,4(1,3)	1,4(1,3)	(1,4)	1,3	1,4	(1,4)
P. furiosus	+	+	+	+	+	+	+	−
P. horikoshii	−	−	+	+	+	−	−	+
P. abyssi	−	+	−	+	+	+	+	+

[a] Methods of cloning βGH-encoding genes from *P. furiosus*.
[b] Different types of βGHs in *Pyrococcus* spp. Features in parentheses are based on analogy with members from the same GH family; exo/endo, β-exoglucanase or β-endoglucanase activity; substrate specificity is based on V_{max} and, when available (*), catalytic efficiency (k_{cat}/K_m) [glu, glucose; man, mannose; gal, galactose; lam, laminarin (β-1,3-glucan); lich, lichenan (β-1,3–1,4-glucan); cell, cellulose (β-1,4-glucan)], and hydrolysis of β-1,3- and/or β-1,4-glycosyl linkage.

eventually resulting in purification and the N-terminal sequencing of the enzyme of interest. Subsequently, oligonucleotide primers can be designed for screening phage- or plasmid-based genomic libraries, as have been used for the identification of *P. furiosus celB* and *lamA* genes (Table I).[2,3] Alternatively, a genomic library is constructed in a heterologous host (usually an *Escherichia coli* strain) that is plated on solid media and screened for selected activity. This approach has been applied for the isolation of CelB- and BmnA-encoding genes (Table I).[4] When potential homologous sequences are available, degenerate primers can be designed for cloning the counterpart by performing polymerase chain reaction (PCR) on genomic DNA from the organism of interest. The latter method has been used to screen 21 isolates that belong to the order Thermococcales, using primers

[2] W. G. Voorhorst, R. I. Eggen, E. J. Luesink, and W. M. de Vos, *J. Bacteriol.* **177**, 7105 (1995).
[3] Y. Gueguen, W. G. Voorhorst, J. van der Oost, and W. M. de Vos, *J. Biol. Chem.* **272**, 31258 (1997).
[4] M. W. Bauer, E. J. Bylina, R. V. Swanson, and R. M. Kelly, *J. Biol. Chem.* **271**, 23749 (1996).

that were based on conserved regions of thermostable family 1 βGHs. As a result, a number of homologous fragments were obtained that appear to form a subclass of GH family 1 (BglB; Table I).[5]

At present, the genome sequences of 16 archaea have been or are being completed, three of which belong to the genus *Pyrococcus: P. horikoshii, P. abyssi,* and *P. furiosus.*[6] This genomics era has led to a spectacular increase in the release of sequence information in public databases and has subsequently resulted in the identification of many more βGHs in *Pyrococcus* spp. (Table I).[7]

Gene Expression

The analysis of gene expression in prokaryotes is often hampered by the quality of mRNA. In contrast to eukaryotes, the stability of mRNA from bacteria and archaea is relatively low. Partially degraded mRNA samples, however, can still be used in dot-blot experiments and in RT-PCR analysis. The latter approach is attractive because of its sensitivity; however, quantification requires very careful analysis. To enable Northern blot analysis, high-quality RNA preparations are required that may be obtained using the method described in this article.

For the isolation of total RNA of *P. furiosus,* a 250-ml culture of early to mid-exponentially grown cells is harvested and washed once in TE buffer [10 mM Tris, 1 mM EDTA (pH 8.0)]. The cells are resuspended in 0.5 ml of ice-cold TE buffer to which 3.75 ml of guanidine thiocyanate solution is added.[8] After a 5-min incubation at room temperature, 375 ml of 2 M sodium acetate (pH 4.5) and an equal volume of acid phenol–chloroform (5:1, pH 4.5) is added. The mixture is vortexed and incubated on ice for 5 min, and phase separation is obtained by centrifugation (10,000g for 20 min, 4°). The aqueous phase is extracted twice with an equal volume of phenol–chloroform–isoamyl alcohol (25:24:1) (pH 8.0) and once with chloroform–isoamyl alcohol (24:1). The aqueous phase is removed and the RNA is precipitated by the addition of 1/10 volume of 3 M sodium acetate (pH 5.5) and 2.5 volume of ice-cold 96% (v/v) ethanol, followed by a 2-hr incubation at −80°. The pelleted RNA is washed three times with 70% (v/v) ethanol, dried, and resuspended in 10 mM Tris (pH 8.5). This method routinely yields 350–400 μg total RNA from a 250-ml *P. furiosus*

[5] T. Kaper and J. Van der Oost, unpublished results.
[6] N. Kyrpides, http://geta.life.uiuc.edu/~nikos/genomes.html#1 (2000).
[7] L. Driskill, K. Kusy, M. Bauer, and R. Kelly, *Appl. Environ. Microbiol.* **65,** 893 (1999).
[8] J. DiRuggiero, L. A. Achenbach, S. H. Brown, R. M. Kelly, and F. T. Robb, *FEMS Microbiol. Lett.* **111,** 159 (1993).

culture ($A_{260/280}$ 1.8–2.0). For Northern blot analysis, 15 μg of total RNA is separated on a 1.5% formaldehyde agarose gel and, following electrophoresis, is transferred to a Hybond N$^+$ membrane. Probes are routinely generated by PCR, the product of which is purified by Qiaquick columns (Qiagen), and labeled with [α-^{32}P]dATP by nick translation.[9] Hybridization is performed in sodium phosphate buffer (pH 7.4), with 1% bovine serum albumin (BSA), 1 mM EDTA, and 7% sodium dodecyl sulfate (SDS, overnight, 68°), and washed in 6× SSC (15 min, 37°) and 2× SSC (2× 15 min, 68°). Subsequently, the filter is exposed using a phospho screen (Molecular Dynamics) for 6–24 hr.[10]

For several *P. furiosus* GH-encoding genes, the transcriptional start site has been determined by primer extension analysis. For this purpose, up to 10–50 μg of total *P. furiosus* RNA and 5 pmol of labeled primer (infrared label IRD800; MWG-Biotech, Ebersberg, Germany) are dissolved in 10 μl RT buffer: 50 mM Tris (pH 8.3), 75 mM KCl, 3 mM MgCl$_2$, and 10 mM dithiothreitol (DTT). The RNA/primer mix is incubated at 70° for 5 min and slowly cooled to room temperature. After the addition of 3 μl RT buffer containing 0.2 mM dNTPs, 22 U of anti-RNase (Ambion), and 200 U of Superscript II RT (Life Technologies), the sample is incubated at 48° for 30 min. Optionally, cDNA synthesis is followed by RNase A treatments before the sample is ethanol precipitated (as described earlier) and the pellet is washed once with 70% ethanol. Subsequently, the pellet is dried, dissolved in 2 μl formamide-loading buffer, applied on a 6% denaturing sequence gel (0.25-mm spacers), and analyzed on a Li-Cor 4000L DNA sequencer.

An *in vitro* system consisting of the basal elements of the transcription machinery of *P. furiosus* has been established.[11] *In vitro*-generated transcripts of the *celB* gene and the *lamA* operon of *P. furiosus* are in perfect agreement with the "*in vivo*"-transcribed mRNA (primer extension, Northern blot), resulting in a detailed description of the corresponding intergenic promoter region (Fig. 1).[12]

Enzyme Overproduction

Genes that encode the *P. furiosus* βGHs listed in Table I have all been isolated, and attempts have been made to overproduce them in *E. coli*

[9] J. Sambrook, E. F. Fritsch, and T. Maniatis, Cold Spring Harbor Laboratory Press, Cold Spring Harbor, NY, 1989.

[10] D. Ward, S. Kengen, J. Van der Oost, and W. De Vos, *J. Bacteriol.* **182,** 2559 (2000).

[11] C. Hethke, A. C. Geerling, W. Hausner, W. M. de Vos, and M. Thomm, *Nucleic Acids Res.* **24,** 2369 (1996).

[12] W. Voorhorst, W. G. B. Voorhorst, Y. Gueguen, A. C. M. Geerling, G. Schut, I. Dahlke, M. Thomm, J. van der Oost, and W. M. de Vos, *J. Bact.* **181,** 3777 (1999).

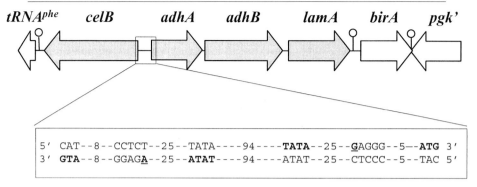

```
5'  CAT--8--CCTCT--25--TATA----94----TATA--25--GAGGG--5--ATG 3'
3'  GTA--8--GGAGA--25--ATAT----94----ATAT--25--CTCCC--5--TAC 5'
```

FIG. 1. Organization of the the *P. furiosus celB-lamA* locus. Gene products: *celB*, β-glucosidase; *adhA* and *adhB*, alcohol dehydrogenases; *lamA*, endoglucanase. The intergenic region between the *celB* gene and the *lamA* operon is shown in more detail, indicated by the TATA box (boldface type), the ATG start codon (boldface), and the transcriptional start site (boldface, underlined), which in both instances is linked to potential ribosome-binding sites. Numbers indicate spacing in base pairs. CelB and LamA are coregulated at the transcriptional level. The 94-bp linking sequence contains a number of repeats, which have been proposed to be potential recognition sites of a transcriptional regulator.[12]

using either pUC-derived vectors[13] or the T7 expression system[14] (Table II). Stable overproduction of the *P. furiosus* β-glucosidase (CelB), β-mannosidase (BmnA), and laminarinase (LamA) is obtained when their genes are amplified by PCR using *Pfu* DNA polymerase and as *Nco*I–*Bam*HI (*celB, lamA*) or *Bsp*HI–*Blp*I (*bmnA*) fragments fused translationally to the T7 promoter of the pET9d vector (Novagen), which carries a kanamycin resistance (Kan^R) gene. The resulting plasmids are transformed to *E. coli* BL21(DE3), in which the gene coding for T7-RNA polymerase has been integrated in its chromosome under control of the *lacUV5* promoter.[14] In contrast to derivatives of plasmids with ampicillin resistance as the selective marker, enzyme production of the three pyrococcal βGHs is stable in culture volumes of 1–8 liter. Optimal expression (50 mg/liter in rotating Erlenmeyer or 90 mg/liter in fermentor) is observed when cultures are grown overnight, without the need of isopropyl-β-D-thiogalactopyranoside (IPTG) induction.[5]

A drawback of an expression system based on the strong T7 promoter may be the formation of inclusion bodies, as was observed in the expression of the *P. furiosus* genes *eglA, bglA,* and *P. horikoshii bglB* (see Table I), of which the majority of the protein (75–95%) is inactive and insoluble. In

[13] C. Yanisch-Perron, J. Vieira, and J. Messing, *Gene* **33**, 103 (1985).
[14] F. W. Studier, A. H. Rosenberg, J. J. Dunn, and J. W. Dubendorff, *Methods Enzymol.* **185**, 60 (1990).

TABLE II
HETEROLOGOUS PRODUCTION SYSTEMS FOR *P. furiosus* GHs[a]

Source	CelB	BmnA	BglA	BglB	BglC	LamA	EglA
E. coli pUC	+	+					
E. coli pET	+ +	+ +	+	+	(+)	+	+
L. lactis	+						

[a] Putative β-galactosidase BglC is placed in parentheses because expression has been obtained, but no activity could be measured with a series of chromogenic substrates.[5]

the case of the *P. furiosus bglA* gene (coding for a β-glycosidase) and *P. horikoshii bglB* gene (coding for a β-mannosidase), a high level of expression is observed as well, but only 5% of the recombinant protein is found in the soluble fraction of the cell-free extract. Optimization might be accomplished by (i) cultivation at suboptimal conditions (e.g., lower temperature; lower aeration rate; minimal medium),[15] (ii) use of a low-copy T7 expression variant,[16] (iii) a different vector, such as high copy (pUC-derivatives) or low copy (pBR322 derivatives), with the gene of interest under control of a variety of inducible promoters (lactose, tryptophan/lactose, arabinose, rhamnose),[17–20] and (iv) a different host/vector system (see later).

Lactococcus lactis

For the application of enzymes in food, it is necessary to produce them in a food-grade manner, i.e., nontoxic and safe. The lactic acid bacterium *Lactococcus lactis* has a long history of safe use as a starter for industrial dairy fermentation and can be employed for the food-grade production of enzymes. A NICE expression system (nisin-controlled expression) for this organism has been developed.[21–23] For nisin-producing strains, the presence of the peptide in the medium is an activator for the transcription of the

[15] C. H. Verhees and J. Van der Oost, personal communication and unpublished results.
[16] D. E. Ward, R. P. Ross, C. C. Van der Weijden, J. L. Snoep, and A. Claiborne, *J. Bact.* **181,** 5433 (1999).
[17] F. Fuller, *Gene* **19,** 43 (1982).
[18] E. Amann, J. Brosius, and M. Ptashne, *Gene* **25,** 167 (1983).
[19] L. Guzman, D. Belin, M. Carson, and J. Beckwith, *J. Bacteriol.* **177,** 1421 (1995).
[20] N. Krebsfänger, F. Zocher, J. Altenbuchner, and U. T. Bornscheuer, *Enzyme Microbiol. Technol.* **22,** 641 (1998).
[21] P. G. de Ruyter, O. P. Kuipers, and W. M. de Vos, *Appl. Environ. Microbiol.* **62,** 3662 (1996).
[22] O. P. Kuipers, P. G. de Ruyter, M. Kleerebezem, and W. M. de Vos, *Trends Biotechnol.* **15,** 135 (1997).
[23] W. De Vos, *Int. Dairy J.* **9,** 3 (1999).

nis operon. In addition to its own structural gene (*nisA*), this includes genes coding for proteins that are involved in nisin maturation and immunity against nisin, as well as genes that code for a sensor histidine kinase and a response regulator of a two-component regulatory system (*nisRK*).[24] Integration of the genes that encode the nisin sensor–regulator system (*nisRK*) in the genome of a nisin negative *L. lactis* strain allows for controlled expression of genes under control of the *nisA* promoter.[21]

In a pilot experiment for the production of CelB in *L. lactis,* a NcoI–BamHI-digested PCR fragment of the *celB* gene is ligated in pNZ8037, a 3.1-kb *E. coli/L. lactis* shuttle vector that contains the nisin promoter adjacent to a multiple cloning site and a chloramphenicol resistance gene.[21] An aliquot of the desalted ligation mix is added to 40 μl electrocompetent *L. lactis* NZ9000 cells (*nisRK*) and mixed before transfer to an electroporation cuvette. The cells are pulsed in a Gene pulser (Bio-Rad, Hercules, CA, set at 2000 V, 25 μF, and 200 Ω), resuspended in 1 ml L-M17B medium (Oxoid) containing 0.5% glucose, 20 mM MgCl$_2$, and 2 mM CaCl$_2$, incubated at 30° for 1.5 hr, and plated on L-M17B agar supplemented with 0.5% glucose and 10 μg/ml chloramphenicol. After 2 days at 30°, colonies are picked and grown overnight in L-M17B medium containing 0.5% glucose and 10 μg/ml chloramphenicol (M17GC). The expression of *celB* is tested by inoculating 25 ml M17GC medium with 0.5 ml of an overnight culture of *L. lactis* NZ9000/pTK1. Cultures are incubated statically at 30° until an OD$_{600}$ of 0.5 is reached. Subsequently, nisin is added to one culture with a final concentration of 5 ng/ml, and cultures are further incubated. After 2 hr, cells are harvested, resuspended in 1 ml 150 mM sodium citrate (pH 5.0), and lysed. After a heat incubation (15 min at 80°) that denatures all the *L. lactis* proteins, β-glucosidase activity is detected in extracts derived from cultures that are induced with nisin; the activity assay is described later. Analysis of these extracts by SDS–PAGE reveals CelB as a major protein band. The level of expression could be modulated by the amount of nisin added to the culture medium (Fig. 2). Under optimal conditions, CelB made up 2% of the total cell protein. For food-grade expression the use of antibiotics should be omitted, and instead a plasmid with a food-grade marker should be used.[22,23,25] However, the described experiment shows that it is possible to produce thermostable, archaeal enzymes in *L. lactis.*

[24] M. Kleerebezem, L. E. Quadri, O. P. Kuipers, and W. M. de Vos, *Mol. Microbiol.* **24,** 895 (1997).
[25] C. Platteeuw, I. van Alen Boerrigter, S. van Schalkwijk, and W. M. de Vos, *Appl. Environ. Microbiol.* **62,** 1008 (1996).

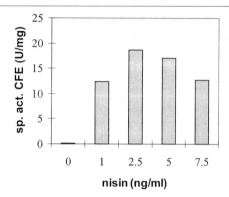

FIG. 2. Functional production of *P. furiosus* CelB in *Lactococcus lactis*. β-Glucosidase activity at 90° in heat-stable, cell-free extracts of *L. lactis* NZ9000 harboring pTK1 (nisin promoter, P_{nisA}; chloramphenicol resistance gene, Cm^R; *P. furiosus* β-glucosidase gene, *celB*) after induction with different amounts of nisin.

Alternative Production Systems

As an alternative to the aforementioned bacterial expression systems, genetic systems have been developed for thermophilic archaea, such as *Sulfolobus* and *Pyrococcus*.[26–29] Although promising, these systems still require optimization. To date, no eukaryal expression systems have been reported for the heterologous production of *P. furiosus* glycosylhydrolases. The β-glycosidase (LacS) of *Sulfolobus solfataricus* has been functionally expressed in yeast (*Saccharomyces cerevisiae*).[30]

Protein Purification

Glycosidases from thermophilic organisms can be isolated from crude cell extracts using conventional isolation techniques. In a *P. furiosus* culture grown on cellobiose or laminarin, CelB has been purfied from the cytoplasmic fraction, whereas the extracellular LamA is partly membrane associated and partly in the medium. In general, the isolated proteins are very stable and can be stored at 4° in the presence of 0.02% azide for long periods of time, without loss of activity.[5,31] When the enzymes are produced

[26] C. Aagaard, I. Leviev, R. N. Aravalli, P. Forterre, D. Prieur, and R. A. Garret, *FEMS Microbiol. Rev.* **18**, 93 (1996).
[27] M. Elferink, C. Schleper, and W. Zillig, *FEMS Microbiol Lett.* **137**, 31 (1996).
[28] R. Cannio, P. Contursi, M. Rossi, and S. Bartolucci, *J. Bacteriol.* **180**, 3237 (1998).
[29] K. Stedman, C. Schleper, E. Rumpf, and W. Zillig, *Genetics* **152**, 1397 (1999).
[30] M. Moracci, A. La Volpe, J. F. Pulitzer, M. Rossi, and M. Ciaramella, *J. Bacteriol.* **174**, 873 (1992).
[31] S. W. Kengen, E. J. Luesink, A. J. Stams, and A. J. Zehnder, *Eur. J. Biochem.* **213**, 305 (1993).

in mesophilic hosts such as *Escherichia coli,* however, they can be purified easily by employing their superior thermostability: heat incubation followed by centrifugation usually results in at least 90% pure protein. In general, subsequent purification is performed by ion-exchange chromatography and by either gel filtration or hydrophobic interaction chromatography.

LamA has been purified from *E. coli* by resuspending the harvested cells from a 1-liter culture of *E. coli* BL21(DE3) cells harboring pLUW532 (*lamA* cloned in pET9d) in 20 ml 150 m*M* sodium citrate buffer (pH 5.0) and a threefold passage through a French Press (100 MPa). The cell-free extract is subjected to a heat incubation (45 min at 80°), after which the precipitated proteins are removed by centrifugation (10,000*g*, 10 min). A heat incubation in citrate buffer is significantly more efficient than one in a Tris buffer due to stabilizing properties of the latter (see later, Thermal Inactivation). The resulting heat-stable, cell-free extract is dialyzed against 20 m*M* Tris–HCl (pH 8.0) and is loaded on an anion-exchange column (Q-Sepharose Fast Flow; Äkta-FPLC; Amersham-Pharmacia) equilibrated with the same buffer. During a linear gradient of NaCl (0.0–1.0 *M*, in the same Tris buffer), the active fractions (as determined by the DNS Method, see later) eluted around 0.5 *M* NaCl. Pooled LamA-containing fractions are concentrated (Centricon concentrator, 10-kDa cut off; Amicon), and 0.5 ml is loaded on a gel filtration column (Superdex 200, Amersham-Pharmacia), which removes traces of contaminating DNA. On this column, LamA migrates as a 31-kDa monomer.

The *P. furiosus* enzymes CelB and BmnA (β-mannosidase, Table I) could be produced in the same T7 expression system and are purified similarly. In a larger scale CelB production (8-liter fermentor), the gel filtration step in the purification method is replaced by phenyl-Sepharose chromatography. The contaminating DNA appears in the flow through, whereas CelB elutes from the column at the end of a (NH$_4$)$_2$SO$_4$ gradient (1.0–0.0 *M*).[32]

Activity Assay

A variety of methods are available for βGH activity measurements, depending on the substrate specificity as well as the level of purification [colony on agar plate, lysate on polyacrylamide gel, or (partially) purified in solution] (Table III). This section describes different techniques based on either artificial chromogenic substrates or "natural" sugar compounds.

---

[32] J. H. G. Lebbink, T. Kaper, S. W. M. Kengen, J. Van der Oost, and W. M. De Vos, (this volume).

TABLE III
ACTIVITY ASSAYS OF EXO- AND ENDO-β-GLYCOSYLHYDROLASES[a]

Medium	Chromogenic substrates	Oligo-/disaccharides	Polysaccharides
Agar plates	X-Gly		Congo red
Solution	Np-Gly (Fig. 3)	HPLC, glucose	DNS, HPLC
Native PAGE	Mb-Gly		Congo red

[a] X-Gly, 5-bromo-4-chloro-3-indole glycoside; Np-Gly, o-/p-nitrophenol glycopyranoside; Mb-Gly, methylumbelliferyl glycoside; DNS, dinitrosalicylic acid.

X-Glycosides

The dye 5-bromo-4-chloro-3-indole oxidizes and precipitates as a blue complex when free in solution. This dye is the coloring agent in all the X-linked substrates commonly used for the detection of exoglycosidase activity in bacterial colonies on agar plates. The well-known blue-white screening system for E. coli, based on cloning vectors with "α-complementation" of the truncated β-galactosidase LacZ, makes use of the hydrolysis of X-β-D-galactopyranoside (X-Gal) to discriminate between transformant colonies containing a plasmid with an insert (often white phenotype) or without an insert (blue phenotype). Several derivatives of the X dye with different colors are available, enabling simultaneous detection of different glycosidase activities (Biosynth, Staad, Switzerland), e.g., in the case of P. furiosus β-glucosidase CelB and E. coli β-glucuronidase GusA.[33]

The P. furiosus β-glucosidase (CelB) has its optimum temperature for activity at 100°. Nevertheless, colonies of the LacZ-deficient host E. coli JM109(DE3) that are transformed with a celB expression vector (see earlier discussion) turn blue when grown at 37° on agar plates containing X-β-D-glucopyranoside [X-Glu, 0.16% (w/v)]. This feature has been used to construct a library of E. coli cells that contains expression vectors with randomly mutagenized celB genes.[32] In contrast to E. coli LacZ (GH family 2), the P. furiosus CelB has a broad substrate specificity, capable of hydrolyzing X-Glu, X-Gal, and, to a lesser extent, X-β-xyloside. BmnA has been characterized as a β-mannosidase, but when produced in E. coli, colonies turn blue on X-Glu- or X-Gal-containing agar plates, which is in agreement[4,5] with the generally broad substrate specificity of family 1 GHs. Because LacZ specifically hydrolyzes X-Gal, even E. coli strains that have an intact lacZ gene, such as BL21(DE3), can be screened for functional β-

[33] A. Sessitsch, K. J. Wilson, A. D. Akkermans, and W. M. de Vos, Appl. Environ. Microbiol. **62,** 4191 (1996).

glucosidase activity by culturing on agar plates containing X-Glu. When selecting for mutant enzymes with altered hydrolytic characteristics, however, the screening of glycosidase activity on X substrate-containing agar plates by analysis of the color intensity should not be regarded as a quantitative measure of mutant enzyme activity. In the case of *celB* variants, blue color formation after overnight growth in general corresponds to initial enzyme activity, but discrepancies have been noted for a significant number of clones.[32]

Nitrophenol Glycosides

Nitrophenol is a commonly used chromogen in kinetic measurements with hydrolytic enzymes such as glycosidases. Many different mono- and oligoglycosides are available as *o*-nitrophenyl (*o*Np) and *p*-nitrophenyl (*p*Np) derivatives (Sigma). The substrates can be used in either discontinuous or continuous assays at the elevated temperatures at which thermostable glycosidases operate. The *P. furiosus* β-glucosidase (CelB) activity has been analyzed at 90° using both assays. Discontinuous and continuous assays are performed in a 150 mM sodium citrate buffer of pH 5.0 (set at room temperature; theoretically, the change in pH of the citrate buffer at 90° is expected to be less than 0.2 units[34]), the pH at which CelB has its optimum for activity. In the discontinuous assay, 495 μl citrate buffer with 10 mM *p*Np-β-D-glucopyranoside (*p*Np-glu) in a 1.5-ml Eppendorf reaction vial is preheated for at least 2 min at 90°, after which the reaction is started by the addition of 5 μl enzyme solution. After a defined period of time, usually 15 min, the reaction is stopped by the addition of 1.0 ml ice-cold 0.5 M Na_2CO_3. This causes the pH to rise to about 9–10, which terminates the reaction and enhances the specific absorption coefficient of the liberated nitrophenol (see later). The absorption of the reaction mixture is measured at 405 nm. The measured activity should always be corrected for nonenzymatic hydrolysis of the substrate. The discontinuous assay has been mainly used in assays with a fixed substrate concentration, such as pH optima determination, qualitative assays in protein purification, and thermostability experiments.

The continuous assay has been used for the determination of kinetic parameters of *P. furiosus* CelB. The liberation of nitrophenol in the reaction mixture is directly followed at 405 nm using a UV/VIS spectrophotometer equipped with a temperature controller. After at least 2 min of preheating 995 μl buffer with *p*Np-Glu (0.05–10 mM) at 90° in a quartz cuvette, the nonenzymatic hydrolysis of the substrate is recorded for about 2 min. Then

[34] R. G. Bates and M. Paabo, *in* "Handbook of Biochemistry: Selected Data for Molecular Biology" (H. A. Sober, ed.). The Chemical Rubber Company, Cleveland, Ohio, 1970.

5 μl of enzyme is added and the increase of absorption is monitored for about 3 min. The activity is determined from the initial increase in absorbance and is expressed as units per milligram, in which 1 unit (U) is defined as 1 μmol nitrophenol released per minute. It should be emphasized that the specific absorption coefficient (ε) of nitrophenol depends on both pH and temperature and that ε is different for oNp and pNp. Therefore, ε should be determined for each nitrophenol derivative under the assay condition used. As an example, the specific absorption coefficients for pNp at 90° have been determined at different pH values (Fig. 3A). When the pH is raised from 4 to 8, the specific absorption coefficient for pNp increases more than 100-fold. This should be taken into account when determining pH optima for activity. As an example, the pH optimum for the *P. furiosus* β-mannosidase BmnA has been determined with and without correction

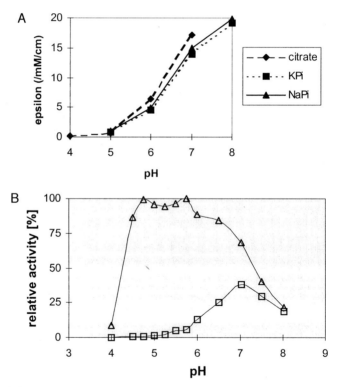

FIG. 3. (A) Specific absorption coefficients (ε) for p-nitrophenol (pNp) at 405 nm in sodium citrate (◆), KP$_i$ (■), and Nap$_i$ (▲) buffer (100 mM each) at different pH values at 90°. (B) pH optimum curve for BmnA for the hydrolysis of pNp-man with correction (△) and without correction (□) for pH.

of the specific absorption coefficient. Without correction the pH optimum appears to be at a pH value that is about 2 units higher than the actual optimum (Fig. 3B).

Methylumbelliferylglycosides

Whereas 5-bromo-4-chloro-3-indole (X) and nitrophenol (Np) develop a color on hydrolysis, free 4-methylumbelliferone emits light with an optimum at 460 nm (fluorescent blue) after excitation with UV light. This makes the dye more sensitive in activity staining in native PAGE gels compared to X substrates and nitrophenol substrates. Again, mono- and oligoglycosides are available as conjugates with 4-methylumbelliferone (Biosynth). To detect β-glucosidase activity in *P. furiosus* cell free extracts, a sample is loaded and run on a native PAGE gel. The gel (1.5 ml) is soaked in 10 ml (1.7 g/liter) 4-methylumbelliferyl-β-D-glucopyranoside, preheated at 90°. After 15 min of incubation at 90°, the gel is analyzed under UV light (standard DNA visualizing equipment), and activity is visible as bright spots on the gel.

Disaccharides

For reasons of convenience, the activity of exo-β-glycosidases is often measured using several artificial substrates, but it is important to keep in mind that the hydrolysis of physiological substrates (di- and oligosaccharides) can make quite a difference (see later). The activities of β-glucosidase CelB and β-mannosidase BmnA for the disaccharides cellobiose and lactose are determined by incubation at 90° in 150 mM sodium citrate buffer (pH 5.0) with different concentrations of lactose or cellobiose. Lactose and cellobiose could be dissolved to at least 250 and 500 mM, respectively. For the detection of free glucose in solution, several glucose oxidase kits are available commercially, which can be used to measure the activity on glucose-containing disaccharides, such as maltose, cellobiose, laminaribiose, lactose, and sucrose, or exoglucosidase activity on glucose polymers, such as starch, pullulan, cellulose, laminarin, or lichenan. In Eppendorf tubes, 495 μl of 0–300 mM lactose in 150 mM sodium citrate buffer (pH 5.0) is preheated for 5 min at 90°. Next, 5 μl of a CelB or BmnA dilution is added and after 15 min the reaction is stopped by cooling on ice. A 10-μl sample of the assay mix is transferred to microtiter plate wells, which each contain 200 μl of the Peridochrom glucose GOD-GAP kit (Boehringer). After 30 min, the developed color is analyzed at 492 nm in an ATTC Elisa reader (SLT lab instruments). A standard series of known glucose concentrations (typically 0–10 mM) is always included and used to calculate activities. The signals of lactose, cellobiose, and galactose are negligible in this assay. The kinetic parameters for the hydrolysis of cellobiose and lactose are compared to those of their chromogenic counterparts pNp-Glu and pNp-Gal. In both cases, the efficiencies of CelB for the hydrolysis of the disaccharides are about 40 to 80 times lower compared to the pNp substrates. This was mainly because the affinity of CelB for cellobiose and lactose is lower (high K_m, Table IV). For BmnA, the difference in efficiency is smaller, as the affinity of BmnA for pNp-Glu and pNp-Gal is just a little higher than for cellobiose and lactose (2- to 15-fold difference). Hence, the activity on chromogenic substrates gives only an approximate indication for the enzyme's hydrolytic efficiency of natural substrates.

Congo Red

Hydrolytic enzyme activity toward β-linked sugar polymers such as laminarin (β-1,3), lichenan (β-1,3–1,4), and carboxymethyl-cellulose (CMC; β-1,4) can be visualized qualitatively using Congo Red. This red dye will attach to the polysaccharide and enzyme activity can be detected by the appearance of clearing zones ("halo"). For the detection of *P. furiosus* endo-β-1,3-glucanase (laminarinase, LamA), the assay is carried out on (granulated) agar plates that, just before solidifying (50°), are mixed with a sugar polymer solution (final concentration 3 g/liter) in 50 mM phosphate buffer (pH 6.0). CMC can be added in concentrations up to 10 g/liter (final concentration) due to a higher solubility. Subsequently, 5 μl of enzyme

TABLE IV
KINETIC PARAMETERS OF CelB AND BmnA AT 90° ON pNp-Glu, pNp-Gal, pNp-Man, CELLOBIOSE, AND LACTOSE[5,38]

Substrate	Kinetic parameter	CelB	BmnA
pNp-Glu	V_{max} (U/mg)	1800	486
	K_m (mM)	0.42	69
	k_{cat}/K_m (mM^{-1} sec^{-1})	3900	6.9
pNp-Gal	V_{max} (U/mg)	2600	496
	K_m (mM)	5.0	29
	k_{cat}/K_m (mM^{-1} sec^{-1})	480	16.8
pNp-Man	V_{max} (U/mg)	72.4	190
	K_m (mM)	1.3	0.99
	k_{cat}/K_m (mM^{-1} sec^{-1})	54.7	198
Cellobiose	V_{max} (U/mg)	720	260
	K_m (mM)	14	100
	k_{cat}/K_m (mM^{-1} sec^{-1})	48	2.6
Lactose	V_{max} (U/mg)	1500	101
	K_m (mM)	120	89
	k_{cat}/K_m (mM^{-1} sec^{-1})	11	1.1

extract is applied onto the plate and incubated at 55° for 2 hr. Enzyme activity is visualized by staining the plates with 1 g/liter Congo Red in deionized water for 30 min and destaining twice with 2 M NaCl for 10 min. For a sharper contrast of the clearing zones, 1 M HCl is added after destaining, which causes a color shift from deep red to dark blue. Because of differences in substrate specificity, it is possible to detect independently endo-β-1,3-glucanase LamA activity and endo-β-1,4-glucanase EglA activity in concentrated culture supernatant or resuspended cell debris of *P. furiosus* cultures. Laminarin is only hydrolyzed by LamA, whereas EglA activity could be detected using CMC.[7]

DNS Method

The action of endo-glucamases on sugar polymers results in an increase of sugar reducing ends. The accumulation of sugars with reducing ends can be determined colorimetrically by the addition of dinitrosalicylic acid (DNS), which results in a change of color from yellow toward brown and can be detected spectrophotometrically.[35] The DNS reagent (per 100 ml: 1.6 g NaOH, 1 g 3,5-dinitrosalicylic acid, 30g potassium sodium tartrate, 0.2 g phenol, 0.05 $NaSO_3$) should be stored at room temperature and protected from light. For *P. furiosus* LamA, a 5-mg/ml substrate solution of laminarin and lichenan, respectively, is made in a 50 mM phosphate buffer of pH 6.0. Enzyme is added and the solution is incubated at 80° for 30 min and subsequently transferred to 0°. A 100-μl aliquot of the enzyme reaction is added to 100 μl DNS reagent. As a standard, 100 μl of a glucose dilution series is taken and assayed similarly. The samples are boiled for 5 min, after which 1 ml of MQ water is added. Color development is determined spectrophotometrically at 575 nm.

HPLC

High-performance liquid chromatography (HPLC) can be used to monitor reactions of glycosylhydrolases both qualitatively and quantitatively. Depending on the column that is used, saccharides with a different degree of polymerization (*DP* 1–20) are resolved. Because glycosylhydrolases change the degree of polymerization of saccharides, HPLC is well suited to observe not only changes in the amount of substrates and products, but also the range/variation of products. The β-glucosidase CelB and β-glucanase LamA were tested for their single and combined ability to hydrolyze the β-1,3-linked glucose polymer laminarin. A 0.5% (w/v) solution of laminarin is incubated without enzyme, with 5 μg/ml CelB, with 5 μg/

[35] T. M. Wood and K. M. Bhat, *Methods Enzymol.* **160**, 87 (1988).

ml LamA, and with both 5 μg/ml CelB and 5 μg/ml LamA for 4 hr at 80°. The reactions are stopped by adding H_2SO_4 up to a final concentration of 0.1 M. This causes the pH to lower to approximately pH 2, which terminates the reactions and makes the mixtures suitable for loading on the HPLC column. Samples are run on an Aminex HPX-87-H column (Bio-Rad) at a temperature of 30° with a flow of 0.6 ml/min in 0.01 M H_2SO_4. Eluates are analyzed by a refractive index detector. By calculating the relative peak areas in the chromatogram, the amount of product formed and substrate converted are determined (Fig. 4).[3]

Thermal Inactivation

Routinely, 200 μl of a 50-μg/ml CelB or LamA solution is transferred to glass vials closed with a screw cap with a Teflon inlay. The vials are then submerged in a silicon oil bath, set at 95 to 105°. At constant time intervals, a vial is transferred from the oil bath and cooled down. Residual activity is subsequently determined, e.g., with pNp-Glu or laminarin as a substrate using either the continuous or the discontinuous activity assay (see earlier discussion). This kind of inactivation usually displays a first-order relationship with time. Half-life values for thermal inactivation are obtained from data fits.

The chemical environment, such as buffer components, pH, and protein concentration, has a significant influence on the thermostability of enzymes. The buffers Tris [tris(hydroxymethyl)aminomethane], PIPES (1,4-piperazine-diethanesulfonic acid), and imidazole have a stabilizing effect on β-

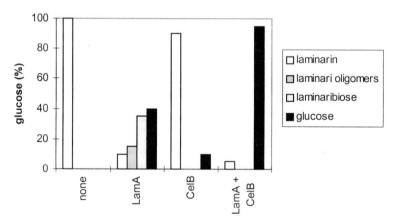

FIG. 4. Action of LamA and CelB on laminarin [0.5% (w/v)]. Laminarin was incubated for 4 hr at 80° with 5 μg/ml LamA and/or CelB. Samples were analyzed by HPLC.[3]

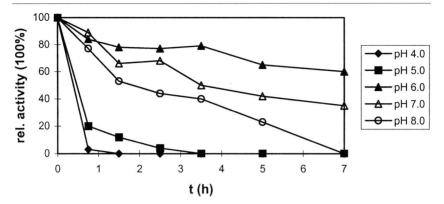

FIG. 5. Thermal stability of BglA at 100° in McIlvain buffer at different pH values. Residual activity was determined with the continuous assay at 90° using *p*Np-Glu.

glucosidase CelB.[36] The effect of pH on the thermostability of enzymes has been investigated with *P. furiosus* β-glucosidase BglA. The enzyme is incubated at 100° in McIlvain's buffer with different pH values (pH 4–8). Optimal stability is observed at pH 6 (Fig. 5), which is close to the optimal pH for activity (pH 5.0).[15]

The irreversible loss of activity due to prolonged incubation at high temperatures is most likely caused by the irreversible unfolding of protein; however, inactivation might also be the result of a limited, local effect. Thermal inactivation experiments are, therefore, only an indication of the conformational stability of a protein. Usually, the unfolding of peptide chains by high temperature or incubation in guanidinium hydrochloride or urea is evaluated by analysis of the fluorescence spectrum or circular dichroism absorption.[37]

Concluding Remarks

At present, many sugar-converting enzymes involved in the hydrolysis of either α- or β-linked sugar substrates have been characterized genetically and biochemically, as well as biophysically. An obvious next step is to gain knowledge concerning the molecular basis that determines the substrate specificity as well as extreme thermal and chemical stablity of this class of enzymes. Two distinct approaches might give rise to a higher level of understanding the different features of GHs: (i) screening for natural variants and (ii) optimization of currently available enzymes. The latter ap-

[36] S. Kengen and A. Stams, *Biocatalysis* **11,** 79 (1994).
[37] J. Pouwels, M. Moracci, B. Cobuzzi-Ponzano, G. Perugino, J. van der Oost, T. Kaper, J. H. G. Lebbink, W. M. de Vos, M. Ciaramella, and M. Rossi, *Extremophiles* **4,** 157 (2000).

proach can be divided into directed and random engineering. Currently, several *P. furiosus* GHs are being engineered with a directed mutagenesis approach, a random method of "molecular evolution,"[32,38] or a combination thereof. The obvious goal is to optimize the GHs in such a way that they efficiently catalyze the conversion of a specific substrate at desired physical conditions.

Acknowledgments

This work has been partially supported by Contracts BIO-4-CT96-0488, FAIR CT96-1048, and BRPR-CT97-0484 of the European Union.

[38] J. Lebbink, T. Kaper, P. Bron, J. Van der Oost, and W. M. Vos, *Biochemistry* **39**, 3656 (2000).

[22] β-Endoglucanase from *Pyrococcus furiosus*

By Susan G. Cady, Michael W. Bauer, Walter Callen,
Marjory A. Snead, Eric J. Mathur, J. M. Short,
and Robert M. Kelly

Introduction

Glycosylhydrolases have been isolated from a variety of heterotrophic hyperthermophiles and include glucanases, hemicellulases, and cellulases.[1,2] *Pyrococcus furiosus,* a hyperthermophilic heterotroph isolated by Fiala and Stetter[3] from geothermal regions of Vulcano Island, Italy, grows on a wide range of α- and β-linked carbohydrates, a property supported by the enzyme inventory revealed in its genome sequence.[4] Driskill *et al.*[4] showed that *P. furiosus* can grow on β-1,4- and β-1,3-linked carbohydrates, including cellobiose, pachyman, laminarin, lichenan, and barley glucan, although little or no growth was noted on cellulose (type 20) and carboxymethylcellulose (CMC). In most cases, growth yields were better if the carbohydrate source was added continuously, presumably because this minimizes deleterious side reactions arising from the thermolability of these growth substrates.[4] In particular, the degradation of β-1,4- and β-1,3-polyglucans by this organ-

[1] M. W. Bauer, L. E. Driskill, and R. M. Kelly, *Curr. Opin. Biotechnol.* **9**, 141 (1998).
[2] A. Sunna, M. Moracci, and G. Antranikian, *Extremophiles* **1**, 2 (1997).
[3] G. Fiala and K. O. Stetter, *Int. J. Syst. Bacteriol.* **36**, 573 (1986).
[4] L. E. Driskill, K. Kusy, M. W. Bauer, and R. M. Kelly, *Appl. Environ. Microbiol.* **65**, 893 (1999).

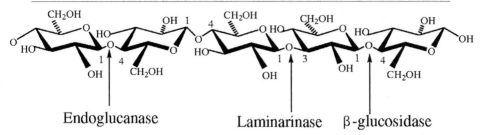

FIG. 1. Mixed β-1,4–β-1,3 linkage polysaccharide showing positions of *P. furiosus* and CelB,[6] LamA,[7] and EglA[8] cleavage.

ism can be accomplished *in vitro* by three enzymes acting in concert[5]: β-glucosidase (CelB),[6] laminarinase (LamA),[7] and β-endoglucanase (EglA)[8] (see Fig. 1). The first two of these enzymes are discussed elsewhere in this volume.[9] The third glucanase, EglA, is an extracellular enzyme capable of hydrolyzing β-1,4 bonds in β-1,4/β-1,3-mixed linkage polysaccharides, such as barley glucan and lichenan.[8]

This article describes the approaches used for the cloning and expression in *Escherichia coli* of the *eglA* gene, which encodes the *P. furiosus* endoglucanase, and protocols used to study the biochemical properties of the recombinant enzyme.

Methods

Expression Screening and Subcloning[8]

A genomic library is constructed in lambda (λ) phage by shearing and size selecting the *P. furiosus* chromosomal DNA, followed by ligation to *Eco*RI linkers. The resulting fragments are cloned into *Eco*RI-digested λgt11 arms and packaged as recommended by the manufacturer (Stratagene, La Jolla, CA). The λgt11 library is then used to transfect Y1090

[5] L. E. Driskill, M. W. Bauer, and R. M. Kelly, *Biotechnol. Bioeng.* **66**, 51 (1999).

[6] S. W. M. Kengen, E. J. Luesink, J. M. Stams, and A. J. B. Zehnder, *Eur. J. Biochem.* **213**, 305 (1993).

[7] Y. Gueguen, W. G. B. Voorhorst, J. van der Oost, and W. M. de Vos, *J. Biol. Chem.* **272**, 31258 (1997).

[8] M. W. Bauer, L. E. Driskill, W. Callen, M. A. Snead, E. J. Mathur, and R. M. Kelly, *J. Bacteriol.* **181**, 284 (1999).

[9] T. Kaper, C. H. Verhees, J. H. G. Lebbink, J. F. T. van Lieshout, L. D. Kluskens, D. E. Ward, S. W. M. Kengen, M. M. Beerthuyzen, W. H. de Vos, and J. van der Oost, *Methods Enzymol.* **330** [21] (2001) (this volume).

cells (Stratagene). These are plated in molten 0.7% NZ agar with 0.2% Remazo Brilliant Blue R (RBB)-barley glucan (Megazyme International Ireland Limited, Bray, Co., Wicklow, Ireland) onto NZ agar plates (1.5%) and screened for endoglucanase activity.[10] Any *P. furiosus* gene inserted in the λ vector will be expressed along with the viral components. When enough endoglucanase is produced, it degrades the dyed barley glucan around the plaque to produce a clearing zone. Such positive plaques are individually cored from the plate using a sterile pipette tip, suspended in sterile medium, and plated as described earlier to produce a collection of single λ vectors with various sizes of DNA containing the *eglA* gene.

The endoglucanase-containing DNA insert is amplified using polymerase chain reaction (PCR) and primers that annealed to the λgt11 sequences flanking either side of the *Eco*RI site (5'-GGTGGCGACTCCTGGAG-CAGCCCG-3' and 5'-TTGACACCAGACCAACTGGTATG-3'). The PCR product is isolated on an agarose gel and cloned into the plasmid pCR2.1 with a commercial TOPO TA cloning kit (Invitrogen, Carlsbad, CA). The nucleotide sequence has been determined with an ABI Prism sequencing kit and an AB1377 sequencer (Perkin-Elmer/Applied Biosystems Division, Foster City, CA). The open reading frame was identified by BLAST analysis. Primers used to PCR amplify the endoglucanase gene are 5'-AATAACAATTGAAGGAGGAATTTAAATGGCTTATCA-TACCTCTGAGGACAA-G3' and 5'-AATAAGTCGACTTAGGAAA-TAAGAGGTCTATC-3'.

Production and Purification of Recombinant Endoglucanase Protein[8]

For expression of archaeal genes containing rare *E. coli* codons, a modified QE30 plasmid system (Qiagen, Chatsworth, CA) is formed by cloning a tRNA gene (for AGG codons) into the *Xba* site. The PCR-amplified endoglucanase gene product is then cloned into the *Eco*RI and *Sal*I sites of the modified QE30 plasmid system and overexpressed in *E. coli* M15 cells. Cells are grown overnight at 37° in 200 ml of LBamk, i.e., Luria broth with 100 μg of ampicillin/ml, 80 μg of methicillin/ml, and 50 μg of kanamycin/ml. This starter culture is used to inoculate 2 liters of LBamk. Cultures are grown at 37° until the optical density at 600 nm reaches 0.8 and then isopropyl-β-D-thiogalactopyranoside (IPTG) is added to a final concentration of 1 mM. Cultures are harvested after 6 hr and centrifuged (30 min, 10,000g, 4°). The cell pellet is resuspended in 20 ml of 50 mM sodium phosphate buffer (pH 8.0), passed twice through a French press,

[10] C. Chen, R. Adolphson, J. F. Dean, K. E. Eriksson, M. W. Adams, and J. Westpheling, *Enzyme Microb. Technol.* **20,** 39 (1997).

and centrifuged for 20 min at 30,000g, 4°. The supernatant is heated at 80° for 20 min to denature *E. coli* proteins and centrifuged for 20 min at 30,000g. This heat-treated supernatant is loaded onto a column of DEAE-Sepharose CL-6B (5 cm × 40 cm) that has been equilibrated previously with 50 mM sodium phosphate buffer, pH 8.0. A 4.0-liter linear gradient, containing zero to 1 M NaCl in the starting buffer, is applied to the column. RBB-barley glucan (2%, w/v) is used as the substrate to measure endoglucanase activity in column fractions. The assay is carried out at 95° (see later).[11] Fractions containing activity are pooled and equilibrated with 50 mM sodium phosphate buffer (pH 7.0) containing 150 mM NaCl before being loaded onto a column (1.6 × 60 cm) of Superdex 200 (Pharmacia, Uppsala, Sweden). Those fractions containing EglA activity are pooled and concentrated. The enzyme is judged to be homogeneous by SDS–PAGE. Protein concentrations are determined by the Bradford method[12] with bovine serum albumin used as a standard.

Determination of Endoglucanase Substrate Specificity and Activity

Various dye-linked cellulose or modified cellulose products available commercially can serve as substrates for endoglucanases. These colorimetric substrates allow one to detect enzyme activity rapidly. Assays, which are modifications of the manufacturer's protocols for *P. furiosus* EglA activity on two azo-dyed substrates and cellulose azure, are given later. These are in addition to the information provided by Bauer *et al.*[8]

Azocarboxymethylcellulose Assay[13]

Principle. CMC is partially deploymerized and dyed with RBB (Megazyme International, Ireland Ltd.) to the extent that there is about one dye molecule per 20 glucose residues. The 1,4-β-D-glucoside links are hydrolyzed by the endoglucanase to form low molecular weight fragments. When the precipitant solution is added, these small fragments will remain in solution whereas larger fragments will precipitate on centrifugation. The visible absorbance of the soluble dyed fragments is measured and compared with a standard curve to determine the extent of azo-CMC hydrolysis.

Reagents

Substrate solution: Prepare a 2% (w/v) slurry of azocarboxymethylcellulose (Megazyme International Ireland Ltd.) in 100 mM sodium acetate buffer, pH 4.5.

[11] C. Malet and A. Planas, *Biochemistry* **36,** 13838 (1997).
[12] M. Bradford, *Anal. Biochem.* **72,** 248 (1976).
[13] Product literature from Megazyme International Ireland Ltd.

Precipitant solution: Dissolve sodium acetate trihydrate (40 g) and zinc acetate (4 g) in 150 ml deionized water. Adjust the pH to 5.0 with 5 M HCl and bring to 200 ml with deionized water. Mix this (200 ml) solution with 800 ml ethanol (95%, v/v) and store at 4° in a Parafilm-sealed bottle. Mix the solution thoroughly and bring to room temperature prior to use.

Method. This is a modified version of the procedure suggested by the manufacturer (Megazyme[13]).

The 2% (w/v) slurry of azo-CM-cellulose is stirred slowly at room temperature prior to use. Screw cap tubes (1.5 ml) are placed on ice, and 200 μl of the viscous azo-CMC substrate solution is placed quickly in each tube using a positive displacement (Eppendorf) multipipettor. A 200-μl sample of the *P. furiosus* enzyme solution (or dilutions of the enzyme stock) is added to each substrate solution. Two reference samples are prepared by adding water, instead of enzyme, to two tubes containing the azo-CMC. The mixture is vortexed for 10 sec and floated in an 80° oil bath for various amounts of time (at least 20 min to overnight). During long incubations, it helps to invert the tubes now and then to resuspend the settled azo-CMC.

After the incubation period, the tubes are removed from the bath and the reaction is stopped by adding 1 ml of the precipitant solution. (For kinetic studies over short periods of time, tubes are incubated without caps. To stop the reaction, the precipitate solution is added to the tubes, which are then removed from the bath.) Vortex the covered tubes for 10 sec and allow the tubes to cool to room temperature for at least 10 min. Centrifuge the tubes for 10 min. Remove 0.8 ml of the solution without disturbing the pellet or precipitate on the tube walls and measure its absorbance at 590 nm. Use the two reference solutions (azo-CMC without enzyme) as blanks.

Using the measured values of absorbance at 590 nm for each sample set, the manufacturer's standard curve can be employed to determine the extent of hydrolysis or the number of milli-Somogyi units per assay.[14] One Somogyi (international) unit of enzyme activity equals 1 μmol of glucose-reducing sugar equivalent released per minute at 30° and pH 4.6.

Azocellulose Assay

The method just described can also be used with less soluble polysaccharides, such as azocellulose (Megazyme International Ireland Ltd.), but longer incubation times are often needed. The manufacturer provides no standard azocellulose curve so it is necessary to determine the concentration of reducing ends[14] to compare this substrate with other substrates.

[14] M. Somogyi, *J. Biol. Chem.* **195,** 19 (1952).

Cellulose Azure Assay

Because cellulose azure is less expensive than azocellulose and azo-CMC, it can be used initially to determine qualitatively if endoglucanase activity is present. This assay is similar in principle to that for other azo-dyed substrates. Upon centrifugation, the digested dyed fragments remain soluble in the postincubation alcohol stop solution and are measured spectrophotometrically.

Reagents

Buffer: 10 mM sodium citrate buffer, pH 5.0. Dissolve anhydrous citric acid in deionized water and adjust the pH at 37° by adding 1 M NaOH.

Stock substrate solution: Cellulose azure solution, 2% (w/v). Dissolve 0.5 g cellulose azure in 25 ml 10 mM sodium citrate buffer, pH 5.0.

Alcohol precipitation solution: Prepare a methanol solution saturated with calcium chloride by adding small amounts of solid anhydrous calcium chloride to 0.5 ml of absolute methanol until no more dissolves. Centrifuge the mixture to obtain a clear solution and add the saturated solution to 100 ml of 95% (v/v) ethanol.

Endoglucanase solution: Immediately before use, dilute the *Pfu* EglA stock to the desired concentration in cold sodium citrate buffer.

Method. The previously described assay[15] has been modified as follows to utilize smaller volumes of the solutions and to replace the filtration step with a more convenient centrifugation step.

Typically, 50 μl of the cellulose azure stock solution is added to 1.5-ml Eppendorf tubes. Then, 500 μl of 10 mM citrate buffer is added to each tube, followed by the enzyme solution (25 μl). At time zero, the tubes are placed in a stationary, 80° oil bath and removed at specified times to measure reaction kinetics. In tubes containing active enzyme the reaction mixture turns a different color (violet, pink) compared to the references (no enzyme added, blue). To stop the reaction, 1 ml of the alcohol precipitant solution is added to each tube and the tubes are mixed by inversion. Larger cellulose fragments will precipitate, whereas smaller dyed fragments will remain soluble in the alcohol solution. After allowing the tubes to sit for a few minutes, the tubes are centrifuged for 4 min at 3000 rpm. The absorbance of the supernatant is measured at 575 nm in disposable cuvettes [polystyrene, 1-cm path length, 1.5 ml semimicro, UV (340–750 nm) Fisherbrand]. If the absorbance is off scale, place it back in the tube with the precipitate, mix it, and recentrifuge it for a longer time. If the absorbance is still

---

[15] Sigma Chemical Company, technical report SSCE02, St. Louis, Missouri, March 1996.

off scale, dilute the supernatant with the alcohol precipitate solution and remeasure the absorbance. (If glass cuvettes are not used, be careful not to use too much alcohol precipitate solution. The polystyrene cuvettes will "fog up" as the alcohol dissolves the plastic.) No extinction coefficient or standard curve is provided by the manufacturer. Reducing ends[14] must be determined to compare this substrate with other endoglucanase substrates.

Characteristics of EglA from *P. furiosus*

Nucleotide and Amino Acid Sequence for EglA

The nucleotide sequence of the *Pfu* EglA gene predicts a 319 amino acid protein with a calculated molecular mass of 35.9 kDa. A 19 amino acid signal peptide is associated with the protein but there is no evidence of a cellulose-binding domain. The *P. furiosus* endoglucanase shares significant amino acid sequence similarities with endoglucanases from glycosylhydrolase family 12,[16] including putative catalytic residues (E197 and E290) (see Fig. 2). The sequence shown in Fig. 2 (GenBank accession number AF181032) includes two corrections made since the original report.[8] The nucleotide sequence for glycine (G), amino acid 143, is GGC, not GCC. Amino acid 280 is leucine (L), not lysine (K).

Biochemical and Biophysical Properties

As mentioned, *P. furiosus* EglA is very active toward β-1,4–β-1,3 mixed-linked glucans, such as barley glucan and lichenan, as well as CMC. Lower activity was noted on microcrystalline cellulose, Whatman paper, and cotton linter. Subsite mapping done with cellooligosaccharides revealed that EglA has three glucose-binding subsites ($-$I, $-$II, and $-$III) for the nonreducing end and two glucose-binding subsites ($+$I and $+$II) for the reducing end relative to the cleaved glycosidic linkage. Highest activity was toward cello-pentaose, cellohexaose, (961 and 865 U/mg, respectively) with significant activity also toward cellotetraose (208 U/mg). Very little activity was noted toward cellobiose and cellotriose, and there was no indication that the enzyme would hydrolyze β-1,3 linkages in laminarin. The enzyme had temperature and pH optima of 100° and 6.0, respectively. Like other *P. furiosus* glycosylhydrolases,[1] this enzyme is very thermostable with a half-life of 40 hr at 95° and a melting temperature of 112°, as determined from differential scanning microcalorimetry.[8]

[16] B. Henrisatt, *Biochem. J.* **280**, 309 (1991).

M S K K K F V I V S I L T I L L V Q A I Y F V E K Y H T S E

D K S T S N T S S T P P Q T T L S T T K V L K I R Y P D D G

E W P G A P I D K D G D G N P E F Y I E I N L W N I L N A T
                                          *   *

G F A E M T Y N L T S G V L H Y V Q Q L D N I V L R D R S N

W V H G Y P E I F Y G N K P W N A N Y A T D A P I P L P S K
                    *

V S N L T D F Y L T I S Y K L E P K N G L P I N F A I E S W

L T R E A W R T T G I N S D E Q E V M I W I Y Y D G L Q P A
                              *   * *   *

G S K V K E I V V P I I V N G T P V N A T F E V W K A N I G
*

W E Y V A F R I K T P I K E G T V T I P Y G A F I S V A A N
                                            *

I S S L P N Y T E L Y L E D V E I G T E F G T P S T T S A H
                                      *

L E W W I T N I T L T P L D R P L I S

FIG. 2. Amino acid sequence of endoglucanase (EglA) from *Pyrococcus furiosus* as deduced from the nucleic acid sequence (GenBank accession number AF181032). The putative signal peptide is underlined. The 10 conserved amino acid residues characteristic of glycosylhydrolase family 12 are denoted by stars underneath the residue. The two putative catalytic amino acid residues (E197 and E290), based on family 12 homology, are shadowed. Note that two corrections have been made since the original report.[8] The codon for glycine (G143) is GGC, not GCC. Amino acid 280 is leucine (L), not lysine (K).

Pyrococcus furiosus EglA appears to act in concert with other glucanases in *P. furiosus* to hydrolyze β-1,4- and β-1,3-linked glucans for nutritional purposes.[4] The significant synergism that exists between these enzymes in the hydrolysis of β-linked glucans has been studied.[5] It is interesting that *P. furiosus* and other heterotrophic hyperthermophilic archaea do not seem to produce enzymes for the hydrolysis of nonglucan polysaccharides, such as mannan or xylan, even though *Thermotoga maritima,* a hyperthermophilic bacterium, does.[1,2] Whether this is a distinguishing feature of archaeal growth physiology remains to be seen. It has been reported that a *Thermococcus* species produces a xylanse, although only preliminary biochemical characterization for this enzyme has been completed.[17] As additional

[17] A. M. Uhl and R. M. Daniel, *Extremophiles* **3,** 263 (1999).

genome sequence data for hyperthermophiles become available, the diversity of glycosidases within this group of organisms will be more evident.

Acknowledgments

RMK acknowledges the National Science Foundation, the U.S.D.A, and FMC Corporation for support of this work.

[23] α-Amylases and Amylopullulanase from *Pyrococcus furiosus*

By ALEXEI SAVCHENKO, CLAIRE VIEILLE, and J. GREGORY ZEIKUS

Introduction

Amylopullulanase and α-amylase belong to the large group of enzymes—O-glycosyl hydrolases (EC 3.2.1–3.2.3)—that hydrolyze the glycosidic bonds in starch and related α-glycans, such as pullulan and glycogen. Amylopullulanase and α-amylase are endo-acting enzymes that hydrolyze α-1,4-glycosidic linkages in starch at random. In contrast to α-amylase, amylopullulanase is also active on α-1,6-glycosidic linkages in starch and pullulan. α-Amylases are widely used in the industrial starch conversion, brewing, baking, and textile industries.[1–4] These broad applications make them a major product in the enzyme market and have led to the characterization of a large variety of these enzymes in the past few years. *Bacillus licheniformis* α-amylase (BLA), one of the most thermostable eubacterial α-amylases, is widely used in industrial starch processing.[4] With α-1,4 and α-1,6 specificities, amylopullulanases have potential as debranching enzymes and as catalysts for conversion of starch into fermentation syrups and as a detergent additive. Because contemporary industrial starch processing requires enzymes that are active and stable at high temperatures (up to

[1] T. Godfrey, *in* "Industrial Enzymology" (T. Godfrey and S. West, eds.), 2nd Ed., p. 359. Stockton Press, New York, 1996.

[2] T. Godfrey, *in* "Industrial Enzymology" (T. Godfrey and S. West, eds.), 2nd Ed., p. 87. Stockton Press, New York, 1996.

[3] T. O'Rourke, *in* "Industrial Enzymology" (T. Godfrey and S. West, eds.), 2nd Ed., p. 103. Stockton Press, New York, 1996.

[4] I. S. Bentley and E. C. Williams, *in* "Industrial Enzymology" (T. Godfrey and S. West, eds.), 2nd Ed., p. 339. Stockton Press, New York, 1996.

110°), considerable commercial interest has been placed on characterizing glycosylhydrolases from hyperthermophilic microorganisms.[5–7]

One of the best characterized hyperthermophiles is *Pyrococcus furiosus,* an anaerobic marine heterotroph that grows optimally at 100°. Starch-degrading activity was reported in *P. furiosus* cell lysate and in its spent growth medium.[8–10] Three amylolytic enzymes—extracellular amylopullulanase, extracellular α-amylase, and intracellular α-amylase—have been characterized from this hyperthermophile.[8,11–14]

This article mainly focuses on *P. furiosus* amylopullulanase (PFAPU) and extracellular α-amylase (PFA). Properties of *P. furiosus* intracellular α-amylase (PFIA) are only described briefly.

Pyrococcus furiosus Growth Conditions

Pyrococcus furiosus DSM 3638 is cultivated in artificial sea water[10] supplemented with 0.25% (w/v) soluble starch, 2.5% (w/v) tryptone, 2% (w/v) yeast extract, and 0.1% (w/v) elemental sulfur. The pH of the medium is adjusted to pH 7.0 with 1 *M* NaOH. The fermentation is performed in a 15-liter vessel with a 5% inoculum in a working volume of 10 liter (B. Braun Biotech, Bethlehem, PA) at 90° for 20 hr under constant gassing with N_2. Cells are harvested in the stationary growth phase and stored at $-20°$ before use.

Cloning of *Pyrococcus furiosus* Amylolytic Enzymes in *Escherichia coli*

Pyrococcus furiosus chromosomal DNA extraction and genomic library construction are performed as reported.[15,16] *Escherichia coli* Sure strain

[5] A. R. S. of Japan (ed.), "Handbook of Amylases and Related Enzymes, Their Sources, Isolation Methods, Properties and Applications." Pergamon Press, Oxford, 1988.

[6] C. Vieille, D. S. Burdette, and J. G. Zeikus, *Biotechnol. Annu. Rev.* **2,** 1 (1996).

[7] J. G. Zeikus and B. C. Saha, *Trends Biotechnol.* **7,** 234 (1989).

[8] K. A. Laderman, B. R. Davis, H. C. Krutzsch, M. S. Lewis, Y. V. Griko, P. L. Privalov, and C. B. Anfinsen, *J. Biol. Chem.* **268,** 24394 (1993).

[9] R. Koch, P. Zablowski, A. Spreinat, and G. Antranikian, *FEMS Microbiol. Lett.* **71,** 21 (1990).

[10] S. H. Brown, H. R. Costantino, and R. M. Kelly, *Appl. Environ. Microbiol.* **56,** 1985 (1990).

[11] K. A. Laderman, K. Asada, T. Uemori, H. Mukai, Y. Taguchi, I. Kato, and C. B. Anfinsen, *J. Biol. Chem.* **268,** 24402 (1993).

[12] S. Jørgensen, C. E. Vorgias, and G. Antranikian, *J. Biol. Chem.* **272,** 16335 (1997).

[13] G. Dong, C. Vieille, and J. G. Zeikus, *Appl. Environ. Microbiol.* **63,** 3577 (1997).

[14] G. Dong, C. Vieille, A. Savchenko, and J. G. Zeikus, *Appl. Environ. Microbiol.* **63,** 3569 (1997).

[15] F. M. Ausubel, R. Brent, R. E. Kingston, D. D. Moore, J. G. Seidman, J. A. Smith, and K. Struhl (eds.), "Current Protocols in Molecular Biology." Greene Publishing & Wiley-Interscience, New York, 1993.

[16] C. Vieille, J. M. Hess, R. M. Kelly, and J. G. Zeikus, *Appl. Environ. Microbiol.* **61,** 1867 (1995).

(Stratagene, La Jolla, CA) transformed with the ligation mixture is plated on 1.5% agar LB ampicillin (100 μg/ml) plates. After 16–20 hr incubation at 37°, colonies are replicated onto a new set of LB ampicillin plates containing 1% phytagel (Sigma, St. Louis, MO) instead of agar and 0.2% soluble starch or 0.2% reactive red dye–pullulan. After overnight growth, the plates are incubated at 80° for 8–10 hr. Starch and pullulan degrading activities are detected by flooding the plates with I_2/KI and looking for clearing zones around individual colonies.

Amylopullulanase Purification

All purification steps are performed at room temperature under aerobic conditions. The native PFAPU is purified from the supernatant of five 10-liter *P. furiosus* cultures by ultrafiltration in a Pellicon cassette cell harvester (Millipore, Bedford, MA) equipped with a 0.45-μm filter. The supernatant is then concentrated with the same system equipped with a 100,000 molecular weight cutoff membrane. The concentrated crude enzyme is loaded onto a Q-Sepharose column (2.5 × 20 cm) equilibrated with 50 mM Tris–HCl, pH 8.0 (buffer A). The column is washed with buffer A, and proteins are eluted with a 0.1–0.6 M linear NaCl gradient in buffer A. Fractions with pullulanase activity are pooled and loaded directly onto a hydroxylapatite column (1.5 × 16 cm) equilibrated with buffer A. After washing with buffer A, the enzyme is eluted with a 0.0–0.4 M linear potassium phosphate gradient (pH 8.0). The active fractions are dialyzed against 50 mM sodium acetate (pH 5.6) (buffer B) and applied to an α-cyclodextrin-Sepharose affinity column (2.5 × 16 cm). Before loading the enzyme, the column is washed with 6 M urea to remove all noncovalently bound α-cyclodextrin and then equilibrated with buffer B.[17] After loading the enzyme, the column is washed with buffer B and then with 1 M NaCl in buffer B. PFAPU is finally eluted with 0.5% α-cyclodextrin in buffer B. The purified enzyme is dialyzed against buffer B and concentrated in an ultrafiltration cell (Amicon, Danvers, MA).

When expressed in *E. coli*, the recombinant PFAPU is not secreted into the medium. *E. coli* Sure strain carrying plasmid pSK211[13] is grown in LB-ampicillin medium. Cell homogenates are prepared by French press treatment. After a 30-min heat treatment at 85°, the cell homogenate is centrifuged at 16,300g for 20 min. The recombinant enzyme is purified from the supernatant by ion-exchange and affinity chromatographies as described earlier.

[17] S. P. Mathupala, S. E. Lowe, S. M. Podkovyrov, and J. G. Zeikus, *J. Biol. Chem.* **268,** 16332 (1993).

α-Amylase Purification

All purification steps are performed at room temperature under aerobic conditions. When expressed in *E. coli*, PFA is not secreted into the medium. Cells carrying plasmid pS4[14] are grown in LB ampicillin (100 μg/ml) medium. After centrifugation (5000g for 10 min), 40-g cells (wet weight) are resuspended in 120 ml Tris–HCl 50 mM (pH 7.5). A cell homogenate is prepared by twice passing the cell suspension through a French press cell at 15,000 lb/in.^{2} After heat treatment at 80° for 15 min, the cell homogenate is centrifuged at 16,300g for 20 min. The enzyme is precipitated by adding 60% $(NH_4)_2SO_4$ to the supernatant, and the pellet is resuspended in 50 mM sodium acetate (pH 6.0).

The concentrated crude enzyme is loaded onto a phenyl-Sepharose (Pharmacia, Uppsala, Sweden) column (1.5 × 18 cm) equilibrated with 50 mM sodium acetate (pH 6.0). The column is washed with the same buffer and then with 50 mM Tris–HCl (pH 8.0). PFA is eluted with 6 M urea in 20 mM Tris–HCl (pH 9.4). After concentration in an ultrafiltration cell equipped with a 30,000 molecular weight cutoff membrane (Amicon, Beverly, MA), the enzyme is dialyzed against 50 mM Tris–HCl (pH 6.0), and hydrophobic interaction chromatography is repeated as described earlier. Fractions with α-amylase activity are pooled and concentrated by ultrafiltration (see earlier discussion).

The concentrated enzyme is loaded onto a Sephacryl S200 (Pharmacia) column (1.5 × 80 cm) equilibrated with 20 mM Tris–HCl (pH 9.4) containing 5% (v/v) glycerol. Active fractions are concentrated by ultrafiltration (see earlier discussion) and dialyzed against buffer B.

Enzyme Assays and EDTA Treatment

Pullulanase and α-amylase activities are determined by measuring the amount of reducing sugar released during enzymatic hydrolysis of 1% pullulan or soluble starch in buffer B containing 0.5 mM Ca^{2+} (PFAPU) and in the absence of Ca^{2+} (PFA). A control without enzyme is used. The amount of reducing sugar produced is measured by the dinitrosalicylic acid method.[18] One unit of pullulanase or α-amylase activity is defined as the amount of enzyme that releases 1 μmol of reducing sugar as glucose per minute under the assay conditions. Protein concentration is determined as described.[19]

EDTA (10 mM) is added to the recombinant PFAPU in buffer B. After

[18] P. Bernfeld, *Methods Enzymol.* **1,** 149 (1955).
[19] M. M. Bradford, *Anal. Biochem.* **72,** 248 (1976).

1 hr incubation, the mixture is dialyzed against buffer B containing 2 mM EDTA and then three times against buffer B. Different metal ions are added to the EDTA-treated enzyme, and the mixtures are incubated for 1 hr at room temperature. All metal ions used are in the chloride form.

Cloning and Characterization of *P. furiosus apu* and *amy A* Genes

Among about 10,000 colonies tested for amylase activity at 80°, two are positive and are shown to contain the *P. furiosus amyA* gene.[14] The *P. furiosus amyA* gene (GenBank accession No. AF001268) encodes a 460 residue polypeptide containing a 26 residue signal peptide. One of two sequences reminiscent of the *E. coli* consensus promoter, and located 80 to 52 and 58 to 29 bp upstream of the *P. furiosus amyA* start codon, is probably responsible for PFA expression in *E. coli*. The amyA stop codon is immediately followed by a stretch of pyrimidines containing the sequence TTTTTTCT, typical of archaeal transcription termination signals.

Among approximately 20,000 *E. coli* colonies tested for pullulanase activity, one colony shows both pullulanase and α-amylase activities.[13] The *P. furiosus* genomic DNA fragment propagated in that recombinant *E. coli* strain is sequenced. Both pullulanase and α-amylase activities are encoded by a single 2559-bp gene (GenBank accession No. AF016588). The corresponding 853 residue polypeptide contains a 27 residue signal peptide. As was the case for PFA, a sequence almost identical to the *E. coli* consensus promoter located 72 bp upstream of PF *apu*'s start codon probably accounts for *P. furiosus apu* gene expression in *E. coli*.

Substrate Specificity and Hydrolysis Product Analysis

Although PFA hydrolyzes a wide variety of substrates (e.g., glycogen and oligosaccharides) beside starch, it does not hydrolyze pullulan, cyclodextrins, sucrose, or maltose. The presence of cyclodextrins in the reaction mixture does not significantly decrease PFA activity on starch.[12] Like BLA, PFA is a liquefying enzyme. The main products of starch hydrolysis (in glucose units: G_1, glucose; G_2, maltose, etc.) are G_2–G_7 (Table I). A low amount of glucose is formed after long hydrolysis periods. PFA hydrolyzes long-chain oligosaccharides faster than shorter chain oligosaccharides, as interpreted from the quantitation of products formed after short versus long hydrolysis periods.[14]

PFAPU is active on pullulan, amylopectin, amylose, glycogen, and oligosaccharides. Pullulan hydrolysis by PFAPU is significantly inhibited by cyclodextrins, with the inhibitory effect increasing from γ- to α-cyclodextrin. The starch hydrolyzing activity of PFAPU is less affected by cyclodextrins

TABLE I

BIOCHEMICAL PROPERTIES OF *P. furiosus* α-AMYLASE (PFA), AMYLOPULLULANASE
(PFAPU), AND *B. licheniformis* α-AMYLASE (BLA)

Properties	PFAPU	PFA	BLA
Mature enzyme[a] (size in amino acids)	827	434	482
Molecular mass (kDa)	89	52	62
Quaternary structure	Monomeric	Monomeric	Monomeric
Specific activity on starch[b] (U/mg)	58.5	3900	2000
Specific activity on pullulan[b] (U/mg)	100.5	No activity	No activity
Optimal pH	5.5–6	5.5–6	7.0–8.0
Optimal temperature (°)	105	100	90
Half-life at 90° in absence of Ca^{2+} [c] (hr)	20.4	20.6	0.36
Half-life at 90° in presence of Ca^{2+} (hr)	44.3	20.6	9.8
End products (with starch as the substrate)	G_2–G_6	G_2–G_7	G_1–G_7

[a] Without leader sequence.
[b] Specific activities were measured at the optimal temperature and pH for each enzyme.
[c] Enzymes were not treated with EDTA prior to the experiment.

(γ-cyclodextrin is not inhibitory at all). Products of pullulan degradation by PFAPU are maltotriose (87%) and maltose (10%), which are typical products for pullulanase-type enzymes. Soluble starch is degraded down to G_1–G_6 (Table I), identifying PFAPU as a liquefying enzyme. The rate of oligosaccharide hydrolysis by PFAPU also depends on their length: the longer the oligosaccharide chain, the higher its hydrolysis rate.

Effect of Metal Ions on Activity and Stability

Most glycosylhydrolases, including α-amylases and amylopullulanases, are Ca^{2+} dependent.[20] Surprisingly, Ca^{2+} has no effect on PFA stability and activity. Its half-lives at 90° (Table I) and 98° (not shown) are the same in the presence or in the absence of Ca^{2+}, and EDTA does not affect its activity (Fig. 1).

Pullulan hydrolysis by recombinant PFAPU decreases almost three times after extensive EDTA treatment (Table II). Adding Ca^{2+} to the EDTA-treated enzyme completely restores its pullulanase activity. The effect of Ca^{2+} on PFAPU starch hydrolysis activity is slightly different. Adding 0.5 m*M* Ca^{2+} to the holoenzyme slightly decreases its starch hydrolysis activity. Still, the EDTA-treated PFAPU loses approximately 50% of

[20] S. Janecek, *Prog. Biophys. Mol. Biol.* **67,** 67 (1997).

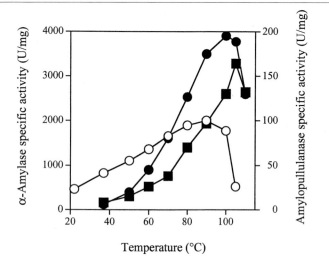

Fᴵɢ. 1. Effect of temperature on the activities of recombinant *P. furiosus* amylopullulanase (■), α-amylase (●), and *B. licheniformis* α-amylase (○). Enzyme activities were assayed in standard reaction buffer with (○, ■) or without (●) 0.5 m*M* Ca²⁺. Enzymes were pretreated with EDTA as described in the text.

its starch hydrolysis activity, and this activity is nearly completely restored on adding 0.5 m*M* Ca²⁺ to the reaction mixture.[13] Neither Ca²⁺ nor EDTA affects the optimal temperature for PFAPU activity. The most dramatic effect of Ca²⁺ on PFAPU catalysis is the 12-fold increase of the enzyme affinity for pullulan (K_m decreasing from 1.6 to 0.13 mg/ml), leading to a 14.5-fold increase in its catalytic efficiency (V_{max}/K_m increasing from 83 to 1205). In comparison, PFAPU affinity for soluble starch increases only 2-fold in the presence of Ca²⁺. Other divalent cations were tested for their effect on PFAPU pullulanase (Table II). Their effect follows nearly exactly the Irving–Williams[21] order $Ba^{2+} < Sr^{2+} < Ca^{2+} < Mg^{2+} < Mn^{2+} < Fe^{2+} < Co^{2+} < Ni^{2+} < Cu^{2+} < Zn^{2+}$, in which the ionic radius decreases and the ionization potential increases from Ba^{2+} to Zn^{2+}. Ca^{2+} is the most activating cation; Ni^{2+}, Cu^{2+}, and Zn^{2+} are inhibitory. These results emphasize the specificity of the PFAPU–Ca^{2+} interaction. PFAPU is also highly stabilized by Ca^{2+} ions: the half-life of EDTA-treated PFAPU at 90° increased 4-fold (from 11 to 44 hr) in the presence of 5 m*M* Ca²⁺ (Fig. 1).

[21] H. Irving and R. J. P. Williams, *Chem. Soc. J.* 3192 (1953).

TABLE II
EFFECT OF DIVALENT CATIONS (2 mM)
ON EDTA-TREATED PFAPU
PULLULANASE ACTIVITY[a]

Metal ion	Specific activity (U/mg)
None	35.4
Ca^{2+}	129.8
Sr^{2+}	119.5
Ba^{2+}	64.5
Mg^{2+}	40.8
Mn^{2+}	40.8
Co^{2+}	27.2
Ni^{2+}	12.0
Zn^{2+}	0.0
Cu^{2+}	0.0

[a] Adapted from Dong et al.[13]

Influence of Denaturing Agents

Both enzymes (PFAPU and PFA) are highly resistant to denaturing reagents. PFA retains 55% activity in the presence of 1.5 M urea and 61% in 0.3 M guanidinium hydrochloride.[12] Incubation of PFAPU with 1 M guanidinium hydrochloride or 5 M urea does not cause any activity loss. PFAPU activity on pullulan doubles after its preincubation with 0.1–5.0% Triton X-100. The same effect is observed after incubation with Tween 80 and polyethylene glycol (PEG) 8000.[13] In the presence of 1 mM sodium dodecyl sulfate (SDS), pullulanase activity increases by 77%. Higher SDS concentrations, however, reduce PFAPU activity.

Sequence Analysis

PFA shows 45–56% similarity and 20–35% identity to other amylolytic enzymes of the α-amylase family (glycoside hydrolase family 13) and contains the four consensus regions characteristic of that enzyme family. PFA most probably adopts the three-domain fold typical of α-amylase: an (α_8/β_8) barrel, the barrel's β_3–α_3 loop forming a separate B domain, and a C-terminal domain. Among the α-amylases of known structure, PFA is most similar to BLA (55.7% similarity and 35.7% identity).[14] Because PFA is significantly more thermostable than BLA (see Table I), the two enzyme sequences were carefully compared to look for potentially stabilizing elements in PFA. Conservation is not uniform along the whole sequence:

sequences corresponding to the (α_8/β_8) barrel, B, and C domains show 40, 22, and 30% identity, respectively. PFA is 10% shorter than BLA. Areas affected by deletions in PFA are mostly regions with little secondary structure or higher flexibility such as loops $\beta_7 \rightarrow \alpha_7$ and $\beta_8 \rightarrow \alpha_8$ in the (α_8/β_8) barrel, or the B domain[22] (29% shorter in PFA than in BLA). Loop shortening has been suggested as a protein-thermostabilizing factor.[23] Here, the role of a shorter B domain in PFA stabilization is questionable as the B domain of the mesophilic barley Amy2 α-amylase is also much shorter than the BLA B domain. The amino acid compositions of the two enzymes (not shown) differs in three ways. (i) PFA is more negatively charged (net charge of -21) than BLA (net charge of -8). This difference is mainly due to a lower number of Lys plus Arg residues in PFA (35 compared to 54 in BLA) and is probably responsible for the two-units difference between the isoelectric points (pI values of 4.78 and 6.83 for PFA and BLA, respectively) of the enzymes and for the two-pH-units difference between their optimum pH values for activity (pH 5.5 for PFA and pH 7.5 for BLA, respectively) (see later). It is not clear, though, if this charge difference affects the stability of PFA. (ii) PFA contains 5% more aromatic residues (18.5% against 13.7%) than BLA. Aromatic residues have been shown to form networks of potentially stabilizing aromatic interactions in some thermostable enzymes.[24] (iii) PFA contains five cysteines while BLA contains none. It is now well known that cysteine residues are among the residues most sensitive to degradation at high temperatures[25] and that they are usually rare in highly thermostable enzymes.[26,27] The presence of five cysteines in PFA is, therefore, surprising. For comparison, PFAPU contains no cysteines. The potential role of cysteines in PFA thermal stabilization is under investigation.

In contrast to PFA, the PFAPU sequence shares very low homology (17–21% identity) with other PFAPUs and enzymes of family 13. In particular, none of the consensus regions present in the α-amylase family could be identified. Instead, PFAPU shows similarity to four proteins: *P. furiosus*

[22] M. Machius, G. Wiegand, and R. Huber, *J. Mol. Biol.* **246,** 545 (1995).

[23] M. K. Chan, S. Mukund, A. Kletzin, M. W. Adams, and D. C. Rees, *Science* **267,** 1463 (1995).

[24] A. V. Teplyakov, I. P. Kuranova, E. H. Harutyunyan, B. K. Vainshtein, C. Frommel, W. E. Hohne, and K. S. Wilson, *J. Mol. Biol.* **214,** 261 (1990).

[25] D. B. Volkin and C. R. Middaugh, *in* "Stability of Protein Pharmaceuticals" (T. J. Ahern and M. C. Manning, eds.), p. 215. Plenum Press, New York, 1992.

[26] J. DiRuggiero, F. T. Robb, R. Jagus, H. H. Klump, K. M. Borges, M. Kessel, X. Mai, and M. W. Adams, *J. Biol. Chem.* **268,** 17767 (1993).

[27] P. Zwickl, S. Fabry, C. Bogedain, A. Haas, and R. Hensel, *J. Bacteriol.* **172,** 4329 (1990).

intracellular α-amylase, *Dictyoglomus thermophilum* α-amylase A, and *Pyrococcus* KOD1 and *Thermococcus litoralis* 4-α-glucanotransferases, all belonging to glycoside hydrolase family 57.[13] Because no structural information is available on this family, it is not known if these enzymes share a similar folding and catalytic mechanism with family 13 enzymes.

P. furiosus Intracellular α-Amylase

The native PFIA is purified from the cytosolic fraction of a *P. furiosus* cell extract.[8] A 157-kDA dimeric enzyme is optimally active at 100° and pH 6.5–7.5. The products of starch hydrolysis by PFIA are a mixture of G_1 to G_6. PFIA is able to hydrolyze smaller polysaccharides, including maltotriose. The production of G_4 and G_6 from maltose and from G_4, G_5, and G_6 from maltotriose indicates that PFIA might have transglycosylation or condensation activity.

The gene encoding PFIA is cloned by DNA/DNA hybridization. The probe is synthesized by polymerase chain reaction using degenerate primers designed according to the N-terminal and internal sequences of PFIA.[11] As is the case for PFAPU, PFIA belongs to glycosylhydrolase family 57. The absence of a typical signal peptide in the 649 residue PFIA confirms that this enzyme is indeed intracellular.

Conclusion

The *P. furiosus* extracellular amylolytic enzymes PFAPU and PFA are most probably involved in starch degradation *in vivo*. A putative integral membrane protein[13] might participate in transporting the starch hydrolysis product inside the cells where α-glucosidase hydrolyzes them to G_1.[28] The function of PFIA is not clear. It is the only intracellular α-amylase characterized in starch-degrading hyperthermophiles. Its function might be close to that of *Pyrococcus* KOD1 and *T. litoralis* 4-α-glucanotransferases.

PFA has attractive features for use in industrial starch degradation, as higher activity and thermostability at more acidic pH ranges than contemporary commercial enzyme (BLA) and it does not require Ca^{2+} for stability or activity at high temperatures.

[28] H. R. Costantino, S. H. Brown, and R. M. Kelly, *J. Bacteriol.* **172,** 3654 (1990).

[24] β-Glucosidase CelB from *Pyrococcus furiosus*: Production by *Escherichia coli,* Purification, and *in Vitro* Evolution

By JOYCE H. G. LEBBINK, THIJS KAPER, SERVÉ W. M. KENGEN, JOHN VAN DER OOST, and WILLEM M. DE VOS

Introduction

One of the key enzymes of the hyperthermophilic archaeon *Pyrococcus furiosus* involved in growth on β-linked sugars is the inducible β-glucosidase (CelB). This enzyme serves as a model system for studying the molecular mechanisms that are employed by hyperthermophilic organisms to optimize enzyme stability and catalysis. Presented here are (i) an overview of the state of the art on this hyperthermostable enzyme, (ii) protocols for heterologous production and enzyme purification, and (iii) development and application of a directed evolution procedure that has resulted in the isolation and characterization of an active site mutant.

Characterization and Application Potential of CelB

The cytoplasmic CelB was purified from *P. furiosus,* showed a molecular mass of 230 kDa, and is composed of four identical subunits.[1] The enzyme is one of the most thermostable glycosylhydrolases described to date and is optimally active at 102–105° and pH 5.0.[1] It shows high activity on aryl glucosides and aryl galactosides, as well as on the β-(1,4)-linked disaccharides cellobiose and lactose and the β-(1,3)-linked disaccharide laminaribiose.[1,2] It furthermore has low β-mannosidase and β-xylosidase activity.[1] High catalytic efficiency (4200 sec^{-1} M^{-1}) is reported for the hydrolysis of *p*-nitrophenyl-β-D-glucopyranoside (*p*Np-Glu), with a V_{max} of 700 U/mg and a K_m for this substrate of 0.15 mM at 90°.[1] The enzyme is not dependent on bivalent cations, and thiol groups are not essential for activity. Finally, the enzyme is competitively inhibited by several ground-state analogs such as glucose (K_i 40 mM) and transition-state analogs such as gluconolactone

[1] S. W. M. Kengen, E. J. Luesink, A. J. Stams, and A. J. Zehnder, *Eur. J. Biochem.* **213,** 305 (1993).

[2] M. W. Bauer, L. E. Driskill, W. Callen, M. A. Snead, E. J. Mathur, and R. M. Kelly, *J. Bacteriol.* **181,** 284 (1999).

(K_i 0.080 mM), as well as tris(hydroxymethyl)aminomethane (Tris) (K_i 1.3 mM).[3]

The *CelB* gene was cloned and sequenced, and the deduced amino acid sequence showed high homology with β-glycosidases that belong to glycosylhydrolase family 1.[4] Overexpression in *Escherichia coli* resulted in high-level production of a thermostable β-glucosidase that was purified and found to have similar kinetic and stability properties as CelB purified from *P. furiosus*. This indicated that the *E. coli*-produced enzyme could replace the native enzyme in studying features of hyperthermostability and -activity. Furthermore, the heterologous expression system allowed for the first protein engineering studies of a hyperthermostable enzyme. It was found that by changing glutamate-372, which is conserved among family 1 enzymes, to an aspartate or a glutamine, a 200- and 1000-fold reduction in specific activity were found, respectively. This confirmed that this conserved glutamate residue is the active site nucleophile involved in catalysis above 100°.[4] A comparison of kinetic properties of CelB with those of the β-glucosidase BglA from *Agrobacterium faecalis* indicated that these homologous enzymes share a common catalytic mechanism.[3] Substrate hydrolysis occurs via a double displacement mechanism, involving a covalent glucosyl–enzyme intermediate and results in retention of the configuration at the anomeric carbon atom.[5,6] A three-dimensional CelB model, based on data derived from crystal diffraction to 3.25 Å, shows the complete conservation of the active site architecture compared to other family 1 enzymes.[7] Evidently, during the evolution of family 1 enzymes from organisms growing optimally at different temperatures, the overall protein structure has been adapted to the changing environmental conditions while the integrity of the active site, and thereby substrate binding and enzyme turnover, has been maintained.

It has been reported that CelB production is highest when *P. furiosus* is grown on β-linked sugars and may reach 5% of total cell protein in this organism.[1] This production is controlled at the transcriptional level, and gene expression is induced within 10 min after addition of cellobiose.[8]

[3] M. W. Bauer and R. M. Kelly, *Biochemistry* **37,** 17170 (1998).

[4] W. G. B. Voorhorst, R. I. L. Eggen, E. J. Luesink, and W. M. de Vos, *J. Bacteriol.* **177,** 7105 (1995).

[5] J. B. Kempton and S. G. Withers, *Biochemistry* **31,** 9961 (1992).

[6] Q. Wang, D. Trimbur, R. Graham, R. A. Warren, and S. G. Withers, *Biochemistry* **34,** 14554 (1995).

[7] T. Kaper, J. H. G. Lebbink, J. Pouwels, J. Kopp, G. E. Schulz, J. Van der Oost, and W. M. de Vos, *Biochemistry* **39,** 4963 (2000).

[8] W. G. B. Voorhorst, Y. Gueguen, A. C. M. Geerling, G. Schut, I. Dahlke, M. Thomm, J. van der Oost, and W. M. de Vos, *J. Bacteriol.* **181,** 3777 (1999).

Although CelB is highly active on the β-(1,4)-linked disaccharide cellobiose, it is not able to liberate glucose from β-(1,4)-linked polysaccharides such as cellulose. It does, however, show low activity on the β-(1,3)-linked polymer laminarin.[1] In fact, a strong synergy between CelB and the secreted endo-β-(1,3)-glucanase (LamA) from *P. furiosus* on laminarin has been described.[9] This indicates that these two enzymes cooperate to enable *P. furiosus* to grow efficiently on β-(1,3–1,4)-glucans, which may be ubiquitously available in its marine environment as constituents of the cell wall of eukaryotic algae (laminarin) or methanogenic archaea (pseudopeptidoglycan).[9,10] A second endoglucanase (EglA) has been characterized, with specificity for β-(1,4) linkages. LamA and EglA may work in concert to hydrolize β-(1,3–1,4)-glucan, delivering short oligosaccharides to the intracellular CelB.[2]

CelB is the most thermostable member of family 1 glycosylhydrolases. Half-life values for thermal inactivation of 85 and 13 hr at 100 and 110° were reported, respectively, and an apparent melting temperature was determined by differential scanning microcalorimetry of 108°.[1,3] Half-life values for thermal inactivation were lowered considerably when the incubation buffer was changed from 140 mM Tris (pH 8.5) to 150 mM sodium citrate (pH 5.0). This stabilizing effect of Tris became apparent during SDS–PAGE analysis, which shows that CelB can only be completely denatured when boiled in nonconventional sample buffer, in which Tris has been replaced with citrate as buffer component. The stabilizing effect of Tris was attributed to its structural analogy with several compatible solutes that are present in hyperthermophiles and have been shown to act as thermostabilizers.[1,11] Alternatively, the observed stabilization may be related to the fact that Tris is a competitive inhibitor of CelB.[3] Interaction between this organic compound and active site residues may stabilize the enzyme against thermal inactivation. Incubation in high concentrations of denaturants, such as 4 M urea, 1 M guanidine hydrochloride, or 8.5 M ethanol, and subsequent assaying for activity in the absence of these compounds revealed no loss of activity, indicating that CelB is also very resistant against chemical treatments.[11]

The nearest family 1 relative of CelB is the β-glycosidase (LacS) from the hyperthermophilic archaeon *Sulfolobus solfataricus*.[12] This enzyme has

[9] Y. Gueguen, W. G. B. Voorhorst, J. van der Oost, and W. M. de Vos, *J. Biol. Chem.* **272,** 31258 (1997).

[10] H. Konig, M. Kandler, M. Jensen, and E. T. Rietschel, *Hoppe-Seyler's Z. Physiol. Chem.* **364,** 627 (1983).

[11] S. W. M. Kengen and A. J. M. Stams, *Biocatalysis* **11,** 79 (1994).

[12] F. M. Pisani, R. Rella, C. A. Raia, C. Rozzo, R. Nucci, A. Gambacorta, M. de Rosa, and M. Rossi, *Eur. J. Biochem.* **187,** 321 (1990).

a half-life of inactivation of 3 hr at 85° and shares with CelB a similar broad substrate specificity.[12,13] The stability of LacS toward high temperature, SDS, and alkaline pH has been studied extensively.[14–16] Elucidation of the three-dimensional structure of LacS revealed that thermostability is achieved by a relatively high number of large ion pair networks and solvent-filled, hydrophilic cavities. These were proposed to confer resilience to the enzyme, thereby protecting against denaturation on large conformational fluctuations at high temperature.[17] Despite their high degree of amino acid homology and three-dimensional structure similarity, the temperature at which CelB and LacS inactivate by 50% within 1 hr is at least 20° higher for CelB. The molecular basis for this difference in thermostability between CelB and LacS is currently under investigation.[18]

CelB is an enzyme with an interesting potential for industrial or diagnostic applications. A series of transposons carrying the celB gene as a genetic marker has been developed for gram-negative bacteria in order to study plant–bacterium interactions.[19] In plants or plant–root ecosystems, endogenous background β-glucosidase and β-galactosidase activity are heat inactivated easily and the combination of celB and the gusA markers allows for simultaneous detection of multiple-strain occupancy of plants and individual nodules by rhizobia. This application has been developed into the CelB gene marking kit that is marketed by FAO/IAEA.[19] CelB is a very suitable enzyme to perform transglycosylation and glucoconjugation with retention of configuration at the anomeric carbon atom. Due to its ability to accept a wide variety of aglycones such as primary and tertiary alcohols, CelB seems a promising biocatalyst for regio- and stereoselective sugar derivative synthesis.[20] In addition, efficient production of β-galacto-oligosaccharides

[13] R. Nucci, M. Moracci, C. Vaccaro, N. Vespa, and M. Rossi, *Biotechnol. Appl. Biochem.* **17**, 239 (1993).

[14] R. Nucci, S. D'Auria, F. Febbraio, C. Vaccaro, A. Morana, M. de Rosa, and M. Rossi, *Biotechnol. Appl. Biochem.* **21**, 265 (1995).

[15] S. D'Auria, M. Moracci, F. Febbraio, F. Tanfani, R. Nucci, and M. Rossi, *Biochimie* **80**, 949 (1998).

[16] S. D'Auria, A. Morana, F. Febbraio, C. Vaccaro, M de Rosa, and R. Nucci, *Protein Expr. Purif.* **7**, 299 (1996).

[17] C. F. Aguilar, I. Sanderson, M. Moracci, M. Ciaramella, R. Nucci, M. Rossi, and L. Pearl, *J. Mol. Biol.* **271**, 789 (1997).

[18] J. Pouwels, M. Moracci, B. Cobucci-Ponzano, G. Perugino, J. van der Oost, T. Kaper, J. H. G. Lebbink, W. M. de Vos, M. Ciaramella, and M. Rossi, *Extremophiles* **4**(3), 157 (2000).

[19] A. Sessitsch, K. J. Wilson, A. D. Akkermans, and W. M. de Vos, *Appl. Environ. Microbiol.* **62**, 4191 (1996).

[20] L. Fischer, R. Bromann, S. W. M. Kengen, W. M. de Vos, and F. Wagner, *BioTechnology* **14**, 88 (1996).

from lactose has been described.[21] At present, its performance in high-temperature lactose hydrolysis, oligosaccharide synthesis, and application in biosensors is being tested (T. Kaper and S. W. M. Kengen, unpublished data, 2000).

Small- and Large-Scale Heterologous Production of CelB in *Escherichia coli*

High-level overexpression of the *celB* in *E. coli* MC1061 was initially achieved by cloning a 1.9-kb *Ssp*I–*Sma*I DNA fragment into the *Sma*I site of vector pTTQ19 under control of the vector-located *tac* promoter.[4] On induction with isopropyl β-D-thiogalactopyranoside (IPTG), CelB levels in *E. coli* MC1061 harboring this plasmid pLUW510 amounted up to 20% of total cell protein.[4] However, high selection pressure had to be applied in order to achieve these levels of overexpression (500 μg/ml ampicillin) and even then in medium-scale fermentations low expression levels, probably due to structural instability of the plasmid, were often observed. For this reason, a new and stable expression system was developed that (i) would preferably confer resistance to an antibiotic that is metabolized *inside* the cells, thereby exerting stronger selection pressure than ampicillin, which is detoxified *outside* the cell, (ii) would reduce the amount of antibiotic to be added greatly, and (iii) would result in high-level overexpression and allow large-scale fermentations without the problems of plasmid instability.

The coding region of the *celB* gene was polymerase chain reaction (PCR)-amplified using primers BG238 and BG239, which overlap the start and stop codons of the gene, respectively, and have the following sequence (with the start and stop codons in italics and restriction sites underlined):

BG238
 5′-GCGCG<u>CC*ATG*G</u>CAAAGTTCCCAAAAAACTTCATGTTTG
BG239 5′-CGCGC<u>GGATCC</u>C*TA*CTTTCTTGTAACAAATTTGAGG

BG238 is homologous to the coding strand and introduces a *Nco*I restriction site (CC*ATG*G), which overlaps the start codon. In order to be able to introduce this restriction site, three extra bases were introduced (GCA) that encode for an alanine residue that is now inserted in between the N-terminal methionine and lysine at position 2 in the enzyme. This alanine will be referred to as residue number 1a, and its introduction slightly increased the rate of thermal inactivation of CelB at 106°.[22] BG239 is homolo-

---

[21] M. A. Boon, J. van der Oost, W. M. de Vos, A. E. M. Janssen, and K. van't Riet, *Appl. Biochem. Biotechnol.* **75,** 269 (1998).

[22] J. H. G. Lebbink, Ph.D. Thesis, Wageningen University, 1999.

gous to the noncoding strand and introduces a *Bam*HI restriction site immediately downstream of the stop codon. The introduction of the *Nco*I and *Bam*HI restriction sites allowed for a translational fusion of the *celB* gene to the ϕ10 translation initiation and termination signals on expression plasmid pET9d.[23] The resulting plasmid pLUW511 carries *celB* under control of the bacteriophage T7 promoter and could be stably maintained in *E. coli* BL21(DE3) [*hsdS gal* (λcIts857 *ind*1Sam7 *nin*5 *lac*UV5-T7 gene 1)], which contains a T7 RNA polymerase gene in its chromosome under control of the *lac*UV5 promoter.[24] pLUW511 confers kanamycin resistance to BL21(DE3), and relatively low amounts of this antibiotic (30–50 μg/ml) are required for stable plasmid maintenance. Induction of a log-phase culture ($OD_{600\,nm}$ 0.5) with IPTG resulted in high-level overexpression, with CelB amounting to at least 20% of total cell protein. The specific activity of CelB produced from this plasmid amounted to 260 U/mg in cell-free extract, compared to 220 U/mg for the enzyme produced from pLUW510. Overnight growth without induction resulted in a lower expression level but a higher total CelB yield because of much higher cell densities. This indicates that the *lac*UV5 promoter is not completely blocked by the repressor and that, under these conditions, T7 DNA polymerase gene expression is sufficient for high-level overproduction of CelB. Because of practical considerations and the high cell densities that can be reached during overnight growth without induction, this procedure was employed as standard protocol. The new construct pLUW511 was found to be completely stable during selective and nonselective growth for at least 50 generations and is, therefore, very well suited for large-scale fermentations. In an 8-liter fermentor, a 20-hr culture of BL21(DE3) harboring pLUW511 yielded 140 g wet weight *E. coli,* from which 720 mg pure CelB was obtained (see later).

Purification of Recombinant CelB

Purification of CelB from *E. coli* has been reported as a simple two-step method, including denaturation of most *E. coli* proteins by heat incubation and further purification using anion-exchange chromatography.[4] A standard procedure included in this method is the incubation of the cell extract with DNase and RNase. However, despite this treatment, oligonucleotides were sometimes found to coelute with the β-glucosidase from the anion-exchange column. The comparison of kinetic and stability parameters

[23] A. H. Rosenberg, B. N. Lade, D. S. Chui, S. W. Lin, J. J. Dunn, and F. W. Studier, *Gene* **56,** 125 (1987).
[24] F. W. Studier and B. A. Moffatt, *J. Mol. Biol.* **189,** 113 (1986).

of mutated variants of CelB with the wild-type enzyme, which is performed routinely in our laboratory, requires identical enzyme preparations that are completely free of oligonucleotides and other contaminants. We, therefore, adapted the following purification procedure.

Purification of CelB from 1-Liter Batch Cultures

One liter of TY medium (1% w/v tryptone, 0.5% w/v yeast extract, 0.5% w/v NaCl), containing 30 μg/ml kanamycin, is inoculated with a 10-ml overnight culture of E. coli BL21(DE3), harboring expression plasmid pLUW511, and cultured overnight in 2-liter baffled Erlenmeyer flasks at 37° in a rotary shaker. Cells (typically 6 g wet weight/liter) are collected at 5400g, resuspended in 20 ml 20 mM sodium citrate, (pH 4.8), and disrupted using a French press at 110 MPa (Aminco, Silver Spring, MD). The resulting cell lysate is incubated for 20 min at 80°, and cell debris and denatured E. coli proteins are removed by centrifugation at 48,000g for 30 min at 4°. The use of 20 mM sodium citrate (pH 4.8), instead of 50 mM sodium phosphate (pH 7.5)[4] or 20 mM Tris–HCl (pH 8.0), results in a more efficient removal of E. coli proteins in this step. The resulting heat-stable, cell-free extract is dialyzed overnight against 20 mM Tris–HCl (pH 8.0). A 15-ml Q-Sepharose anion-exchange column (Pharmacia, Uppsala, Sweden), operated by the HiLoad System with Pump and Gradient Kit P50 (Pharmacia), is equilibrated with the same buffer. The dialyzed heat-stable, cell-free extract is loaded onto this column, and after extensive washing with 20 mM Tris–HCl (pH 8.0), a linear 0.0–1.0 M sodium chloride gradient in the same buffer is applied. CelB elutes at 0.5 M NaCl under these conditions. Active fractions are pooled and concentrated to less than 1 ml total volume using Centricon-30 devices (Amicon, Beverly, MA). The resulting sample is applied onto a 300-ml Superdex 75 gelfiltration column (Pharmacia) equilibrated with 20 mM Tris–HCl (pH 8.0), 50 mM NaCl, in order to remove last traces of E. coli-contaminating proteins and oligonucleotides. Purity of the fractions is analyzed by SDS–PAGE. Pure fractions typically show the presence of a protein band with the expected molecular weight for the monomer of 58 kDa, as well as a higher molecular weight form presumably corresponding to the tetrameric form, as described earlier.[1,11] Absence of oligonucleotides is verified spectroscopically using a Hitachi U-2010 UV-VIS spectrophotometer, by the location of the UV absorption maximum at 280 nm and the absence of a shoulder in the UV spectrum at 260 nm. Pure fractions are pooled and dialyzed against an appropriate buffer. The exchange of Tris–HCl to another buffer is required because Tris is a competitive inhibitor of CelB and other glycosylhydrolases with an inhibition constant around 1 mM.[3,5] Sodium citrate (pH 4.8) buffer should be avoided because, at higher protein concentrations (starting at approximately 1 mg/ml), CelB reversibly precipitates in this buffer. CelB has a pI of 4.40[1] and apparently the charge distribution on the enzyme surface at pH 4.8 is such that aggregation occurs at high protein concentrations. A preferred buffer is sodium phosphate at pH 6 or 7. Protein concentrations are determined at 280 nm using an extinction coefficient for one subunit of $\varepsilon_m^{280\,nm} = 1.28 \times 10^5\,M^{-1}\,cm^{-1}$, calculated according to Gill and von Hippel.[25] Typically 50–80 mg of pure CelB is obtained from a 1-liter batch culture using this protocol.

Purification of CelB from 8-Liter Fermentations

A fermentor containing 8 liter TY medium with 50 μg/ml kanamycin and 55 mM glucose is inoculated with 50 ml of an overnight culture of BL21(DE3) harboring pLUW511. A continuous air flow of 18 liter hr and a stirring speed of 400 rpm are employed, and to prevent excessive foaming, 600 μl antifoam is added (Antifoam 289, Sigma). The pH is kept constant at 7.0. After growth at 37° for approximately 20 hr, an OD$_{660\,nm}$ of 8 is reached, and cells are collected using a continuous centrifuge (Heraeus Biofuge28RS, Germany). Cells (140 g wet weight) are resuspended in 420 ml of a 20 mM sodium citrate buffer (pH 4.8) and broken using a French press at 110 MPa. The resulting cell lysate is incubated for 60 min at 80°, and cell debris and denatured *E. coli* proteins are removed by centrifugation at 48,000g at 4° for 30 min. After dialysis against 20 mM Tris–HCl (pH 8.0), the heat-stable, cell-free extract is loaded onto a 50-ml Q-Sepharose column (Pharmacia) equilibrated with the same buffer. After extensive washing with the equilibration buffer, CelB is eluted using a 0.0–1.0 M NaCl gradient in the same buffer. CelB elutes at 0.45 M NaCl. The large volume of the pooled samples (330 ml containing 810 mg protein) rules out gel filtration as the subsequent step. The pooled fractions are saturated with 1 M ammonium sulfate and loaded onto a phenyl-Sepharose column (Pharmacia), after which a linear decreasing ammonium sulfate gradient (1.0–0.0 M) in 20 mM Tris (pH 8.0) is applied. CelB elutes at the end of the gradient, separated from oligonucleotides that do not bind to the column. The last traces of contaminating *E. coli* proteins are removed by a 30-min heat treatment at 100°. The final yield amounts to 60% and 720 mg pure CelB is obtained.

Construction of Random CelB Library

Directed evolution by random mutagenesis and *in vitro* recombination is a powerful approach for studying many characteristics of a model system

[25] S. C. Gill and P. H. von Hippel, *Anal. Biochem.* **182,** 319 (1989).

and for evolving a desired new property.[26–30] This approach has so far been restricted to enzymes from mesophiles and thermophiles and has not yet been applied to hyperthermostable enzymes. CelB is a very suitable candidate to be used as a model in the development of such a procedure for hyperthermostable enzymes. The enzyme is efficiently overexpressed in *E. coli* and therefore easily accessible for the introduction of random mutations by genetic techniques. The enzyme is extremely thermostable and resistant to inactivation by chemical denaturants.[11] Furthermore, because of its broad substrate specificity, the enzyme is able to hydrolyze a variety of chromogenic substrate analogs. Therefore, high-throughput screening and selection methods can be applied that are based on the production and analysis of large libraries consisting of random CelB variants.

Description of Model System

As starting material for a random CelB library, we use the expression plasmids pET9d and its derivative pLUW511 in *E. coli* strain JM109(DE3). The genotype of JM109(DE3) is *recA1 supE44endA1 hsdR17 gyrA96 relA1 thi Δ(lac-proAB)* F′ [*traD36 proAB⁺ lacI*q *lacZDM15*] (λ*cI*ts857 *ind1Sam7 nin5 lacUV5*-T7 gene *1*) and it is, therefore, deficient in β-galactosidase (LacZ) activity, in contrast to BL21(DE3) (see earlier discussion). In combination with vector pET9d, which does not code for the α peptide of the *E. coli* β-galactosidase, this strain will result in white colony formation on agar plates containing the chromogenic substrates 5-bromo-4-chloro-3-indolyl-β-D-glucopyranoside (X-Glu) or 5-bromo-4-chloro-3-indolyl-β-D-galactopyranoside (X-Gal) and will not hydrolyze the artificial substrates *p*Np-Glu and *p*Np-Gal. When a gene coding for a functional CelB is cloned into pET9d, colonies will develop a blue color due to hydrolysis of either the X-Glu or the X-Gal substrates. Alpha (α) complementation between the truncated LacZ produced by *E. coli* JM109(DE3), and parts of, or the complete CelB enzyme, has not been observed and is thought to be highly unlikely; LacZ and CelB belong to different families of glycosylhydrolases that do not share amino acid or structural homology. In family 1 glycosylhydrolases the active site is buried at the inside of an $(\alpha\beta)_8$ barrel within each individual monomer; however, the active site in LacZ involves residues from different subunits.[17,31,32] Even in the absence of an inducer, it is evident

[26] W. P. C. Stemmer, *Proc. Natl. Acad. Sci. U.S.A.* **91,** 10747 (1994).

[27] W. P. C. Stemmer, *Nature* **370,** 389 (1994).

[28] A. Crameri, S. A. Raillard, E. Bermudez, and W. P. C. Stemmer, *Nature* **391,** 288 (1998).

[29] O. Kuchner and F. H. Arnold, *Trends Biotechnol.* **15,** 523 (1997).

[30] F. H. Arnold, *Acc. Chem. Res.* **31,** 125 (1998).

[31] C. Wiesmann, W. Hengstenberg, and G. E. Schulz, *J. Mol. Biol.* **269,** 851 (1997).

[32] R. H. Jacobson, X. J. Zhang, R. F. DuBose, and B. W. Matthews, *Nature* **369,** 761 (1994).

that the *lacUV5* promoter in JM109(DE3) is not fully repressed and CelB is produced, as described earlier for BL21(DE3). Although the activity of the thermostable CelB at 37° is only 2% of its activity at 90°, this is sufficiently high to hydrolyze the chromogenic substrates during overnight growth. Blue color formation in general correlates with the activity of the CelB variant that is expressed; wild-type CelB-producing colonies develop a darker blue color than, for example, colonies harboring plasmid pLUW513, coding for mutant E417S CelB that has a 10-fold reduced specific activity on *p*Np-glucose.[7] A major advantage of using the pET9d expression plasmid is the possibility of constructing translational fusions, using restriction enzyme recognition sites that are overlapping the start and stop codons of the gene of interest. This means that solely the coding region of the *celB* gene is mutagenized and that the library will not contain random clones with mutations in promoter or terminator sequences or mutations affecting plasmid replication, stability, or copy number. These kinds of mutations may result in fluctuating expression levels of the gene and thereby total enzyme activity, and were shown previously to be responsible for part of the increase in activity found for an *in vitro*-evolved fucosidase.[33]

Random Mutagenesis of celB Gene

Random mutagenesis and DNA shuffling of the *celB* gene are essentially performed according to the protocol of Stemmer[26] with the optimization described by Lorimer and Pastan.[34] Individual steps in the procedure are visualized in Fig. 1. For the preparation of template for DNA shuffling, the *celB* gene is PCR amplified using the *Taq* DNA polymerase, thereby exploiting the lack of proofreading activity of this polymerase to introduce random mutations. Primers are developed that just overlap the start and stop codons of the *celB* gene. This allows mutagenesis of the complete coding region except for the start and stop codons themselves without contaminating mutations in the flanking regions. The *celB* gene is PCR amplified using pLUW511 as template and primers BG238 and BG309 (5'-GTTAGCAGCC<u>GGATCC</u>C*TA*; with the *Bam*HI site underlined and the stop codon in italics). While BG238 is the same primer as used for the construction of expression clone pLUW511, BG309 resembles BG239 but allows amplification (and mutagenesis) of all codons up to, but not including, the stop codon. In later procedures BG238 has been replaced by primer BG417 (5'-CTTTAAGAAGGAGATATA<u>CC*ATG*</u>), which allows mutagenesis of all codons immediately downstream of the start codon. PCR is

[33] J. H. Zhang, G. Dawes, and P. W. C. Stemmer, *Proc. Natl. Acad. Sci. U.S.A.* **94,** 4504 (1997).

[34] I. A. Lorimer and I. Pastan, *Nucleic Acids Res.* **23,** 3067 (1995).

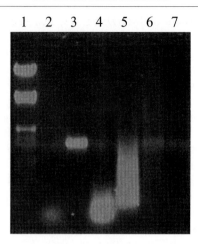

Fig. 1. A 2% agarose gel showing DNA shuffling of the *P. furiosus celB* gene. Lanes 1 and 2, markers, phage λ DNA digested with *Hin*dIII and *Eco*RI and a 36-bp oligonucleotide, respectively; lane 3, amplified *celB* gene; lane 4, DNase I-treated *celB* gene; lane 5, reassembled *celB* gene; lane 6, enrichment of *celB* gene by conventional PCR on assembly product; and lane 7, reassembled *celB* gene after digestion with restriction enzymes *Nco*I and *Bam*HI and gel purification.

performed in the following reaction mixture (PCR-MIX): 10 ng pLUW511, 100 ng of the primers BG238 and BG309, 0.2 mM dNTPs, *Taq* DNA polymerase buffer, and 5 units *Taq* DNA polymerase (Pharmacia). Thirty cycles of 30 sec, 94°; 30 sec, 45°; 90 sec, 72° on a Perkin-Elmer (Norwalk, CT) GeneAmp 2400 are performed, preceded by 5 min denaturation at 94° and followed by 7′ extension at 72°. PCR products obtained in 24 independent amplification reactions are pooled and purified using the QIAquick PCR purification kit (Qiagen, Westburg, Leusden, The Netherlands). PCR products (200 ng/μl) are incubated with 1 ng/μl DNase I (Sigma, Zwijndrecht, The Netherlands) in 50 mM Tris–HCl (pH 7.4) containing 10 mM MnCl$_2$ at 20°. The replacement of magnesium with manganese in this step (i) results in the introduction of double-stranded nicks into the DNA instead of single-stranded nicks, which means that also after denaturation the single-stranded fragments will have the length as visualized before denaturation on the agarose gel and (ii) lowers the affinity of DNase I for fragments smaller than 50 bp, which makes the timing of the DNase I treatment less critical.[34] The reaction is stopped after 10 min by the addition of 0.1 volume of 0.5 M EDTA, pH 8.0. This treatment typically results in the formation of 50- to 200-bp fragments (Fig. 1). These fragments are purified using the QIAquick nucleotide removal kit (Qiagen, Westburg), according to the supplied protocol except for centrifugation at both washing steps, which is performed at 6000 rpm. Alternatively, DNA fragments of 50–150 bp are excised from an agarose gel and purified using the QIAEX II gel extraction kit. The concentration of DNA fragments after purification is increased by the removal of excess water in a vacuum excicator. The resulting gene fragments are assembled in a PCR reaction without primers (Fig. 1). Incubation mixtures containing *Taq* DNA polymerase buffer, 0.2 m*M* dNTPs, 20 ng/μl *celB* fragments, and 5 units *Taq* DNA polymerase (Pharmacia) are subjected to 40 cycles of 30 sec, 94°; 30 sec 55°; 90 sec, 72°, preceded by 5 min, 94° and followed by 7 min, 72°. Correct fragments are enriched in a subsequent PCR reaction, including primers BG238 and BG309 as follows: the mixtures are diluted 10-fold in PCR-MIX lacking pLUW511 and subjected to 30 cycles of amplification as described for the first PCR reaction (see earlier discussion). It is critical in the whole procedure to avoid contamination with intact *celB* during reassembly steps. A total of 100 cycles of PCR amplification using *Taq* DNA polymerase is performed in this way. The resulting products, a smear of DNA fragments equal to and larger than 1.5 kb, are digested with *Nco*I and *Bam*HI (Fig. 1). Fragments of approximately 1.5 kb length are isolated from an agarose gel using the QIAEX II gel extraction kit and ligated into *Nco*I–*Bam*HI-linearized pET9d using a vector:insert molar ratio of 1:3 *E. coli* strain JM109(DE3) is transformed with the ligation mixture and plated onto TY agar plates containing 30 μg/ml kanamycin and 16 μg/ml X-Glu (Biosynth, Switzerland). Transformants containing functional *P. furiosus* CelB are identified by blue color formation. Approximately 5000 transformants/μg of transformed linearized vector DNA are obtained, from which one-fourth appear to be blue and contain a complete *celB* gene, in contrast to the white colonies that contain high-frequency *celB* inserts smaller than 50–100 bp.

Mutation Frequency

Nine functional CelB clones are picked randomly, and plasmid DNA is isolated and sequenced in order to determine the mutation frequency. The number of mutations ranges from 0 to 5 and shows the following distribution: no mutations (2), one mutation (1), two mutations (2), three mutations (2), four mutations (1), and five mutations (1). These mutations are distributed randomly along the entire *celB* gene. The average mutation frequency is 2.3 base changes per *celB* gene, which results in one to two amino acid changes per CelB enzyme. This is an acceptable compromise between having a minimum of wild-type sequences in the library and a maximum of amino acid substitutions per enzyme while preventing the occurrence of compensatory mutations starting to interfere with desired phenotypic changes.

Construction and Screening of Random celB Library

Single blue colonies are picked and transferred to microtiter plates containing 200 μl of TY medium with 30 μg/ml kanamycin and 10% glycerol. Each microtiter plate contains negative and positive controls, namely JM109(DE3) harboring pET9d, pLUW511, or pLUW513 coding for CelB mutant E417S.[7] Seventy microtiter plates are prepared (containing approximately 6200 random clones), incubated overnight at 37°, and stored at −80°. For screening on chromogenic substrates, the library is replica plated using a 96-pin replicator stamp onto TY agar plates containing 30 μg/ml kanamycin and 16 μg/ml X-Glu or X-Gal. Plates are incubated overnight at 37°, and the intensity of blue color formation is assayed by visual inspection. This results in blue color formation of similar intensity on both substrates for the wild-type CelB, whereas clones with empty vector pET9d remain completely white and the E417S mutant develops only a faint blue color on X-Glu. Multiple clones are identified that show unchanged activity on X-Glu, but largely reduced or completely abolished color formation on X-Gal.

Characterization of CelB Mutant N206S

One of the clones displaying a higher ratio in color formation on X-Glu over X-Gal is selected and a single colony is picked. DNA sequencing reveals a single base change of A to G at position 617 of the *celB*-coding region. This mutation results in an amino acid substitution of asparagine-206 to serine in the CelB enzyme. Investigation of this amino acid position in a multiple amino acid sequence alignment of family 1 glycosylhydrolases reveals that this asparagine is a conserved residue and is located next to active site glutamate-207, which is proposed to act as the acid/base catalyst during substrate hydrolysis. The putative role of asparagine-206 has been deduced from an investigation of the three-dimensional structures from a mutant form of the 6-phospho-β-galactosidase LacG from *Lactococcus lactis* and from BglA from *Bacillus polymyxa*.[31,35] These two family 1 glycosylhydrolases are cocrystallized with the substrate 6-phosphogalactose and the inhibitor gluconate in their active site, respectively. Analysis of the three-dimensional structures reveals that the amide side chain of the corresponding asparagine of the CelB Asn-206 is in direct contact with the substrate, with a hydrogen bond presumably formed between the side-chain nitrogen and the hydroxyl group at the C-2 position of the sugar

[35] J. Sanz-Aparicio, J. A. Hermoso, M. Martinez-Ripoll, J. L. Lequerica, and J. Polaina, *J. Mol. Biol.* **275,** 491 (1998).

moiety of the substrate (Fig. 2). This interaction would be formed in the case of glucose and galactose moieties at this position in the substrate because both sugars have the C-2 hydroxyl group in an equatorial position. In the case of mannose, however, this hydroxyl group is axial, and visual inspection of the models suggests that in this case no hydrogen bond can be formed between the C-2 hydroxyl group and the amide side chain of Asn-206. Rather, the mannose C-2 hydroxyl group may be stabilized by an interaction with Glu-207. In mutant N206S the asparagine side chain is shortened to a serine, which is still a polar side chain, but because of the increased distance, a favorable interaction between the C-2 hydroxyl of the substrate and this side chain is considered highly unlikely.

In order to investigate the validity of the analysis described earlier,

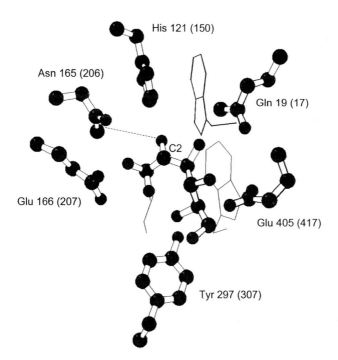

FIG. 2. Model of the active site of *B. polymyxa* β-glucosidase complexed with the inhibitor gluconate. Residue numbering is according to the *B. polymyxa* enzyme with CelB numbering in parentheses. The amide group of asparagine-165 (N206 in CelB) is hydrogen bonding the hydroxyl group at the C-2 position of substrates containing glucose and galactose, but not mannose. In the random CelB mutant N206S, this hydrogen bond cannot be formed. The affinity for glucose and galactose substrates in this mutant is much reduced compared to wild-type CelB, whereas the affinity for mannose is virtually unchanged. Prepared using the program MOLSCRIPT.[41]

TABLE I
KINETIC PARAMETERS FOR WILD-TYPE AND MUTANT N206S CelB[a]

| CelB | pNp-Glu | | pNp-Gal | | pNp-Man | |
	V_{max} (U/mg)	K_m (mM)	V_{max} (U/mg)	K_m (mM)	V_{max} (U/mg)	K_m (mM)
Wild-type	1800	0.42	2600	5.0	78	3.5
N206S	340	12	1200[b]	200[b]	17	4.7

[a] Determined at 90° on aryl glucosides pNp-glucose, pNp-galactose, and pNp-mannose.
[b] Maximum substrate concentration employed for pNp-Gal is 40 mM, and the calculated K_m value solely serves to indicate that N206S no longer has noteworthy affinity for this substrate.

wild-type CelB and mutant N206S are purified and kinetic parameters on the artificial substrates pNp-Glu, pNp-Gal, and pNp-mannose (pNp-Man) are determined. Wild-type and mutant CelB are purified from 1-liter cultures according to the protocol described previously. Kinetic parameters at 90° are determined on a Hitachi U-2010 spectrophotometer equipped with a SPR-10 temperature controller (Hitachi, Tokyo, Japan). A detailed description of the CelB activity assay is described elsewhere.[36] Initial free pNp formation is followed continuously at 405 nm in 150 mM sodium citrate (pH 4.8; set at room temperature) containing increasing concentrations of the aryl glucosides pNp-Glu (0.05–40 mM), pNp-Gal (0.2–40 mM), or pNp-Man (0.4–26 mM). An extinction coefficient of 0.561 M^{-1} cm^{-1} is used for free pNp in the assay buffer. V_{max} and K_m values are derived from fitting the data according to the Michaelis–Menten equation corrected, if appropriate, for substrate inhibition, using the nonlinear regression program Tablecurve (Tablecurve 2D for Windows, version 2.03 Jandel Scientific). Data are listed in Table I. A more detailed description of kinetic data for the wild-type enzyme is described elsewhere.[37] On all three substrates, the maximum activity of the mutant drops considerably compared to wild-type CelB, suggesting that Asn-206 is not only involved in substrate binding, but also plays a role in catalysis. Indeed, it has been described for several β-glycosidases from mesophilic organisms that interactions at the C-2 position of the substrate strongly stabilize the transition states for both

[36] T. Kaper, C. Verhees, J. H. G. Lebbink, J. van Lieshout, L. Kluskens, D. E. Ward, S. W. M. Kengen, W. M. de Vos, and J. van der Oost, Methods Enzymol. 330 [21] 2000 (this volume).
[37] J. H. G. Lebbink, T. Kaper, P. Bron, J. van der Oost, and W. M. de Vos, Biochemistry 39, 3656 (2000).

glycosylation and deglycosylation of the enzyme.[38–40] However, its role in substrate binding appears to be more important, as the increase in K_m for pNp-Glu and pNp-Gal is much more dramatic than its effect on V_{max} values. Because the maximum pNp-Gal concentration used (40 mM) is far below the calculated K_m value for this substrate for mutant N206S, the calculated kinetic parameters for this mutant/substrate combination only serve to illustrate that affinity for this substrate has been lost in the mutant. This is in agreement with the observation that *E. coli* JM109(DE3) producing mutant N206S remains almost white after overnight growth on agar plates containing X-Gal. The effects on K_m values can be directly correlated to the removal of the hydrogen bond acceptor in the mutant N206S. This is supported by the fact that the K_m for pNp-Man in the mutant is virtually unchanged with respect to the wild-type CelB, which indicates that the asparagine is not a ligand for the axial C-2 hydroxyl group of this substrate.

Conclusions

CelB from *P. furiosus* serves as a very suitable model glycosylhydrolase to study substrate specificity as well as adaptations of stability and activity to extreme temperatures. Stable production in *E. coli* and an efficient purification protocol allow for simple and rapid preparation of pure wild-type and mutant CelB enzymes. These routine procedures have been applied in studying substrate recognition, thermostability, and low- and high-temperature-dependent activity.[7,37] The *in vitro* evolution procedure that has been developed allows for easy screening of the existing mutant CelB library for altered characteristics and enables further rounds of directed evolution. The described results on CelB mutant N206S demonstrate (i) the successful construction of the random CelB library, (ii) the feasibility to select mutants with changed catalytic properties and/or substrate specificity, and (iii) the applicability of LacG and BglA crystal structures in developing models for interactions between CelB and its substrates.

Acknowledgments

This work has been partially supported by Contracts BIOT-CT93-0274, BIO 4-CT96-0488, FAIR CT96-1048, and BRPR-CT97-0484 of the European Union. The authors are very grateful to Peter Bron and Melike Balk for their contribution to the work described in this article.

[38] M. N. Namchuk and S. G. Withers, *Biochemistry* **34,** 16194 (1995).
[39] J. D. McCarter, M. J. Adam, and S. G. Withers, *Biochem. J.* **286,** 721 (1992).
[40] T. Mega and Y. Matsushima, *J. Biochem.* **94,** 1637 (1983).
[41] P. J. Kraulis, *J. Appl. Crystallogr.* **24,** 946 (1991).

Section III

Proteolytic Enzymes

[25] Purification, Characterization, and Molecular Modeling of Pyrolysin and Other Extracellular Thermostable Serine Proteases from Hyperthermophilic Microorganisms

By WILLEM M. DE VOS, WILFRIED G. B. VOORHORST, MARCEL DIJKGRAAF, LEON D. KLUSKENS, JOHN VAN DER OOST, and ROLAND J. SIEZEN

A variety of hyperthermophilic Archaea and Bacteria are capable of growth on proteinaceous substrates. While the exact nature of their natural substrates is unknown, growth may be supported by mixtures of proteins or peptides, such as casein, peptone, or tryptone. In general, these heterotrophic microorganisms contain a set of proteolytic enzymes that, in conjunction with transport systems, are involved in generating a pool of intracellular peptides and amino acids. The key enzymes in this degradation are extracellular endoproteases that in many cases are classified as serine proteases according to their sensitivity to various inhibitors or conserved signatures in their predicted amino acid sequences. Presently, several serine proteases have been purified from hyperthermophiles with a variety of approaches and characterized at the enzyme and, in most cases, also the gene level (Table I). Those that have been genetically characterized all belong to the subtilisin-like family of serine proteases, also known as subtilases.[1,2] Moreover, a variety of homologs of subtilases can be detected in the present databases of sequenced genomes of hyperthermophiles (see later). The first of these serine proteases to be characterized at the enzyme and gene level has been pyrolysin from the hyperthermophilic archaeon *Pyrococcus furiosus,* which is the most thermostable protease to date and retains half of its activity after boiling for several hours.[3,4] A characteristic feature of pyrolysin and other proteases is the fact that these enzymes are substrates of their proteolytic activity and degrade themselves in a process termed autoproteolysis. This has for a long time hampered the purification of pyrolysin and prevented its further characterization at the molecular level.

[1] R. J. Siezen, W. M. de Vos, J. A. M. Leunissen, and B. W. Dijkstra, *Protein Eng.* **4,** 719 (1991).
[2] R. J. Siezen and J. A. M. Leunissen, *Protein Sci.* **6,** 501 (1997).
[3] R. Eggen, A. Geerling, J. Watts, and W. M. de Vos, *FEMS Microbiol. Lett.* **71,** 17 (1990).
[4] W. G. B. Voorhorst, R. I. L. Eggen, A. C. M. Geerling, C. Platteeuw, R. J. Siezen, and W. M. de Vos, *J. Biol. Chem.* **271,** 20426 (1996).

TABLE I
SERINE PROTEASES FROM HYPERTHERMOPHILIC MICROORGANISMS[a]

Organisms	Enzyme	Identification	Classification	Ref.
Archaea				
Pyrococcus furiosus	Pyrolysin	Enzyme–gene	Subtilase F	[b]
Pyrobaculum aerophilum	Aerolysin	Gene	Subtilase A	[c]
Desulfurococcus mucosus	Archaelysin	Enzyme	Unknown	[d]
Thermococcus stetteri	Stetterlysin	Enzyme–gene	Subtilase F	[e]
Staphylothermus marinus	Stable	Enzyme–gene	Subtilase F	[f]
Bacteria				
Fervidobacterium pennovorans	Fervidolysin	Enzyme–gene	Subtilase F	[g]

[a] Serine proteases from hyperthermophilic microorganisms, their common names, and their characterization by purification (enzyme) or at the genetic level (gene), which allowed for their classification according to the known families of subtilisin-like serine proteases (subtilases).[1,2]
[b] Eggen et al.[3]; Connaris et al.[5]; Voorhorst et al.[4]
[c] Voelkl et al.[6]
[d] Cowan et al.[7]
[e] Klingeberg et al.[8]; Voorhorst et al.[9]
[f] Mayr et al.[10]
[g] Klingeberg et al.[11]; Kluskens et al.[12]

This article deals with (1) the procedures for the biochemical and genetic characterization of pyrolysin and other related thermostable serine proteases and (2) the prediction of their properties by homology comparisons, database searches, and molecular modeling.

Detection of Pyrolysin

Most crude methods for detecting protease activity rely on the proteolytic degradation of simple proteins into acid-soluble material.[13] Casein is

[5] H. Connaris, D. A. Cowan, and R. J. Sharp, *J. Gen. Microbiol.* **137,** 1193 (1991).
[6] P. Voelkl, P. Markiewicz, K. O. Stetter, and J. Miller, *Protein Sci.* **3,** 1329 (1995).
[7] D. A. Cowan, K. A. Smolenski, R. M. Daniel, and H. Morgan, *Biochem. J.* **247,** 121 (1987).
[8] M. Klingeberg, B. Galunsky, C. Sjoholm, V. Karsche, and G. Antranikian, *Appl. Environ. Microbiol.* **61,** 2098 (1995).
[9] W. G. B. Voorhorst, A. Warner, W. M. de Vos, and R. J. Siezen, *Protein Eng.* **10,** 905 (1997).
[10] J. Mayr, A. Lupas, J. Kellerman, C. Eckerskorn, W. Baumeister, and J. Peters, *Curr. Biol.* **6,** 739 (1996).
[11] M. Klingeberg, A. Friedrich, and G. Antranikian, *DECHEMA Biotechnol. Conf.* **5,** 173 (1992).
[12] L. D. Kluskens, W. G. B. Voorhorst, G. Antranikian, J. van der Oost, and W. M. de Vos, submitted for publication.
[13] M. Kunitz, *J. Gen. Physiol.* **30,** 291 (1947).

an ideal substrate: it is inexpensive, abundant, and relatively small, it does not show any tertiary structure, and it is easily accessible to proteases even at high incubation temperatures. *P. furiosus* DSM 3638 is able to grow on casein and other intact or predigested proteins at its optimal growth temperature of 100°.[14] Using casein as substrate, protease activity is detected in the cell envelope fraction of this strain of *P. furiosus* grown on synthetic seawater containing peptone and yeast extract (50 and 10 g/liter, respectively) in the absence of elemental sulfur but sparged with nitrogen gas. For large-scale experiments the cell envelope fraction is prepared by carefully resuspending 35 g (wet weight) of cells in 25 ml of 50 mM sodium phosphate buffer, pH 6.5 (P buffer). The cell paste is homogenized by sonication for 1 min with several intervals to keep the temperature low and passed through a French pressure cell at 110 MPa followed by centrifugation for 20 min at 17,500g at 4°. The pellet that contains the cell envelope fraction, consisting of membranes and parts of the S layer, is resuspended in 10 ml of P buffer and stored at room temperature. In a typical assay, an aliquot (up to 5 μl) of this cell envelope fraction is incubated with 0.5 ml of a 0.6% casein solution in capped tubes or sealed glass ampoules for various periods of time in water or oil baths at temperatures up to 120°. Subsequently, the tubes or ampoules are cooled down and opened, and 1 ml of a 5% (w/v) trichloroacetic acid solution is added. Subsequently, the mixture is incubated at room temperature for 30 min, and following centrifugation the absorbance of the supernatant at 280 nm is determined. This assay shows the presence of a protease activity with an extremely high thermal stability (up to a half-life of 4 hr at 100° in the absence of substrate) and an optimal caseinolytic activity at 115°. Hence it is termed pyrolysin, derived from the Greek prefix "pyros" for fire and the Latin suffix "lysin" for the one that cleaves. Inhibitor studies show pyrolysin to be a serine protease that is stimulated by the presence of calcium.[3] This is confirmed in subsequent molecular characterizations (see later) that also predict pyrolysin to contain a signal sequence suggesting it to be present at the outside of the membrane in the cell envelope, and show it to differ from the intracellular protease PfpI also identified in *P. furiosus*.[15] Under specific growth conditions, a serine protease strongly resembling pyrolysin is found to be partly present in the culture supernatant of *P. furiosus*, confirming its extracellular nature.[5]

Pyrolysin Enrichment by Proteolytic Digestion

To analyze the molecular size of pyrolysin, the cell envelope fraction is subjected to electrophoresis on 0.2% SDS–10% polyacrylamide gels ac-

---

[14] G. Fiala and K. O. Stetter, *Arch. Microbiol.* **145,** 56 (1986).
[15] S. B. Halio, I. I. Blumentals, S. A. Short, B. M. Berill, and R. M. Kelly, *J. Bacteriol.* **178,** 2605 (1996).

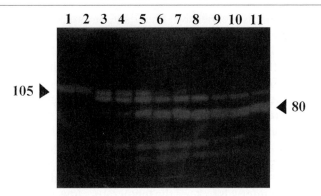

FIG. 1. Predigestion of the *P. furiosus* cell envelope fraction. The pyrolysin fraction (lanes 1 and 2) was incubated in 6 *M* urea at 100° for 0.5 hr (lanes 3 and 4), 1 hr (lane 5), 2 hr (lanes 6 and 7), 3 hr (lanes 8 and 9), and 4 hr (lanes 10 and 11), and proteolytic activity was visualized in a gelatin overlay of a SDS–PAGE gel. The apparent size (in kDa) of the HMW (105) and LMW (80) pyrolysin is indicated.

cording to Laemmli[16] that on staining by Coommassie Brilliant Blue (CBB) R250 reveal a variety of bands, among those two with a very high molecular mass (>200 kDa) that may represent S-layer proteins and a conspicuous band with an apparent size of approximately 120 kDa (data not shown). Because of the low degree of purification, use is made of the vigorous and thermostable proteolytic activity of pyrolysin by self-digesting the cell envelope fraction. Pilot studies show pyrolysin to be resistant to many treatments, including incubation with SDS or high concentrations of urea. To stimulate unfolding of contaminating proteins and hence facilitate proteolytic degradation, the cell envelope fraction is adjusted to 6 *M* urea in P buffer and incubated at 95° for several hours (Fig. 1). During this incubation, the total proteolytic activity remains constant while the protein concentration in the cell envelope fraction decreases approximately hundredfold. Analysis by SDS–polyacrylamide gel electrophoresis (PAGE) confirms that a significant degree of purification is realized by this preincubation, as illustrated by the complete disappearance of the large S-layer-like proteins (not shown).

The resistance of pyrolysin to SDS allows for detecting its activity following SDS–PAGE on gels containing various amounts (0.1–0.01%) of casein or gelatin.[3,5] However, these substrate-containing gels do not give optimal separation, and better and more reproducible results are obtained by detecting the proteolytic activity following SDS–PAGE using overlays

[16] U. K. Laemmli, *Nature* (*London*) **227,** 680 (1970).

of specific substrates. This approach also permits the simultaneous detection of proteins in the SDS gel and their proteolytic activity in the substrate-containing overlay. For this purpose, standard SDS–PAGE is performed and, subsequently, the gel is incubated for 1 hr at room temperature in an excess of 25 mM potassium phosphate buffer (pH 7.5) containing 2.5% Triton X-100 and clamped to a previously prepared gelatin gel consisting of 10% polyacrylamide and 0.1% (w/v) gelatin in a 375 mM Tris–HCl buffer (pH 8.8). The sandwiched gels are wrapped in aluminum foil containing water-saturated tissues to prevent excessive evaporation and incubated for 1 hr at 95° followed by staining of the SDS gel with CBB R250 and the gelatin gel with Amido black [0.1% (v/v) naphthol blue black, 30% (v/v) methanol, and 10% (v/v) acetic acid]. An alternative approach for detecting pyrolysin activity is to cover the SDS gel with a developed Kodak X-OMAT AR X-ray film (Eastman Kodak Co., Rochester, NY) that is gelatin coated. During subsequent incubation at 95° for 10 min, the gelatin partly diffuses into the gel and is degraded by the proteolytic activity. Following removal of the film, the nondegraded gelatin in the SDS gel can be stained by CBB R250, resulting in a blue gel with cleared bands of activity. This is an efficient and convenient method to detect pyrolysin activity, but due to its lower quality than the substrate containing overlay, this approach is specifically suited for the routine analysis of series of gels.

The proteolytic activity of the *P. furiosus* cell envelope fraction is analyzed using the sandwich gel system following its predigestion at 95° (Fig. 1). This reveals that most pyrolysin activity is contained in a band with an apparent size of approximately 120 kDa that on incubation at 95° for 4 hr is found to be present in two major bands: one with a high molecular weight (termed HMW pyrolysin) and one with a low molecular weight (termed LMW pyrolysin) with an apparent size under these conditions of approximately 105 and 80 kDa, respectively.

Purification of Pyrolysin

Initial attempts to purify pyrolysin using a variety of chromatographic procedures were unsuccessful mainly because of the fact that the activity is strongly and irreversibly bound to conventional resins based on agarose. Because SDS–PAGE provides effective separation of pyrolysin and other proteins (see Fig. 1), preparative PAGE is used to achieve further purification. For this purpose, the cell envelope fraction that had been predigested at 95° for 4 hr in P buffer containing 6 M urea is centrifuged (10 min at 17,500g) to remove cell debris and denatured proteins. Subsequently, the supernatant is brought to 100 mM Tris–HCl, 0.9% 2-mercaptoethanol, and 20% glycerol (pH 6.8) and subjected to preparative urea–PAGE using a

Model 491 Prep Cell equipped with a 10 × 2.5-cm gel (Bio-Rad, Hercules, CA). The gel consists of 11% acrylamide containing 6 M urea, and proteins are eluted with native PAGE buffer consisting of 50 mM Tris, 192 mM glycine (pH 8.5) at 45 mA for 48 hr with a flow of approximately 0.3 ml min. Approximately 500 fractions of 1–2 ml are collected, and the aliquot (up to 25 μl) is subjected to SDS–PAGE and analyzed for protease activity by substrate overlay using Kodak X-OMAT films. This preparative urea–PAGE allows for the separation of HMW and LMW pyrolysin, and fractions containing the highest activity of either species are pooled, concentrated by microfiltration on an Amicon (Danvers, MA) Diaflow Cell with a cutoff of 30 kDa (Filtron Technology Corp.), and dialyzed against P buffer. Subsequent purification is realized by anion-exchange chromatography using Fractogel EMD trimethylaminoethyl-650 (M) (Merck, Darmstadt, Germany). This resin is based on acrylamide to which pyrolysin is not bound. Samples of 50 ml are applied to a 1 × 3-cm column containing the resin equilibrated with 50 mM Tris–HCl, pH 8.5, that is subsequently washed with P buffer to remove unbound protein and finally eluted with P buffer containing a gradient of 0–1.5 M NaCl. The proteolytic activity of the HMW pyrolysin elutes at approximately 900 mM and that of the LMW fraction at approximately 750 mM, which are pooled, dialyzed, and concentrated separately by microfiltration as described earlier.

Characterization of Pyrolysin

The purified HMW and LMW pyrolysin fractions appear as single bands on standard SDS–PAGE with apparent molecular sizes of 105 and 80 kDa, respectively. These bands are transferred to Problott paper (Applied Biosystems, Warrington, U.K.) by electroblotting, and their N-terminal amino acid sequences are determined by Edman degradation on a Model 477A gas-phase amino acid sequenator (Applied Biosystems). Remarkably, both pyrolysin fractions yield an identical sequence, suggesting a mutual relation. It is difficult to completely unfold proteins from *P. furiousus* following boiling in the presence of SDS, and pyrolysin is found to retain activity under these conditions (see earlier discussion). Hence, the possibility is addressed that the HMW and LMW bands are differently denatured forms of the same protein by applying alternative denaturation treatments. Initial experiments to denature pyrolysin using SDS at increasing temperatures, e.g., by incubation for 20 min at 120°, yield the same banding patterns on SDS–PAGE. Hence acid denaturation is applied by the addition of formic acid to the samples at a final concentration of 5 M followed by incubation for 45 min at room temperature. Subsequently, the samples are neutralized by the addition of potassium hydroxide to a final concentration of 5 M, followed by analysis by SDS–PAGE. When this procedure is applied to the HMW pyrolysin fraction, a single band with an apparent size of 150 kDa is obtained, whereas the LMW pyrolysin fraction gives rise to two bands with an apparent size of 130 and 105 kDa (not shown). These results confirm that excessive denaturation conditions are required to denature pyrolysin and indicate that the HMW and LMW fractions indeed contain proteins with different sizes. In view of their identical N-terminal sequences and the generation of LMW from the HMW fraction by prolonged incubation (see Fig. 1), it is concluded that both HMW and LMW pyrolysin are different forms of pyrolysin that are generated by C-terminal proteolytic processing.

Identification, Cloning, and Characterization of Genes for Pyrolysin and Other Subtilisin-like Proteases from Hyperthermophiles

A generic approach is followed to isolate the pyrolysin gene that also could be applied for the cloning of other genes for subtilisin-like serine proteases. For this purpose, generic polymerase chain reaction (PCR) primers are designed to amplify genes coding for serine proteases of the subtilisin family.[1,2] Careful alignment of these proteins reveals that the Asp, His, and Ser active site residues are flanked by conserved regions that allow the design of degenerated oligonucleotides to be used in PCR experiments (see Fig. 2 for relative position of these residues). Several oligonucleotides

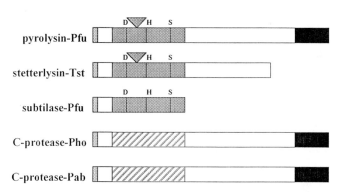

FIG. 2. Predicted molecular architecture of pyrolysin and four other putative proteases from Thermococcales, including stetterlysin from *T. stetteri* (Tst), a subtilase from *P. furiosus* (Pfu-Pf396648; www.genome.utah.edu), and two related thiol proteases (C proteases) from *P. horikoshi* (Pho) and *P. abyssi* (Pab), with accession numbers AP000001 and AJ248288, respectively. The N-terminal prepro sequences (gray and white), the homologous catalytic domain (gray: subtilisin-like serine protease with active site residues; thiol protease: striped) with the insert (triangle), and the conserved C-terminal region (black) are indicated.

with a different degree of degeneration and size are tested for PCR, and the best results are obtained with primers with a combination of both unique and inosine residues at the position of wobble bases. The following primers appear successful in various approaches to clone serine protease-like genes from hyperthermophiles: 5′ GTIGCIGTIMTIGAYACIGG-3′ (BG95) as a forward PCR primer based on the Asp active site region and 5′GTICCIGCIACRTGIGTICCRTG-3′ (BG96) and 5′-GGISYIGCCATI-SWIGTICC-3′ (BG31) as reverse primers based on the His and Ser active site regions, respectively (I, inosine; M, A or C; Y, C or T; S, G or C; W, A or T). These primers are used in various combinations at a concentration of 100 ng in a mixture with a total volume of 100 μl of *Taq* polymerase buffer containing 250 ng chromosomal DNA, 0.2 mM dNTPs, 5 mM MgCl$_2$, and 5 U *Taq* polymerase (Pharmacia, Uppsala, Sweden). After denaturation of the template (5 min, 95°), 35 cycles (1 min, 95°, 2 min, 35°; and 3 min, 72°) are performed followed by a final extension of 7 min at 72° on a DNA thermal cycler (Perkin-Elmer Cetus, Norwalk, CT). The fragments are expected to have different sizes depending on the primer set used and the structure of the subtilisin-like serine protease genes that may have inserts of variable size. With genomic DNA of *P. furiosus,* a fragment with the size of approximately 1.1 kb is obtained with primer pair BG95 and BG31. A variation of this approach is the use of one of the reverse primers in combination with a forward primer, which is based on the determined N-terminal sequence. This approach is also applied for the cloning of the pyrolysin gene where such a forward primer is used in combination with primer BG96, resulting in the selective amplification of a fragment with the size of 0.6 kb.[4] Cloning and sequence analysis of this fragment in combination with a chromosome walking approach result in the character-ization of the complete pyrolysin gene, which has a molecular structure as described in Fig. 2. Similarly, primers BG95 and BG31 amplify a fragment with a size of approximately 1.4 kb from *Thermococcus stetteri.* This frag-ment is used to isolate and characterize the gene for the serine protease from the hyperthermophilic archaeon, called stetterlysin (Table I; Fig. 2).[9] In addition to amplifying fragments of approximately 1.4 kb from *P. furiosus* and *T. stetteri* that are confirmed to be large serine proteases (see Fig. 2), this primer pair BG95 and BG31 also amplify smaller fragments of about 0.6 kb from genomic DNA of several hyperthermophiles. These include *P. furiosus,* which is known to contain a second, small serine protease gene (Fig. 2; see later). Moreover, a 0.6-kb fragment is also obtained from the bacterium *Fervidobacterium pennovorans,* which is used to clone and subse-quently characterize the corresponding gene for fervidolysin, a keratin-degrading serine protease (Table I).[12]

Molecular Architecture of Pyrolysin and Modeling of Its
  Catalytic Domain

The cloning and sequence analysis of the pyrolysin gene allows for the prediction of the molecular architecture of pyrolysin, which is a large serine protease of 1398 amino acid residues (see Fig. 2). Sequence predictions and homology comparisons are performed by using the GCG package version 7.0[17] and the software and databases provided at the NCBI (www.ncbi.nlm.nih.gov). Comparison of the predicted sequence with the N-terminal amino acid sequence determined from the purified pyrolysin allows for the identification of a 149 residue precursor peptide that is predicted to have a pre–pro structure. The presignal is a consensus signal sequence, which is compatible with the extracellular location of pyrolysin. The prosequence of pyrolysin shows little homology to other prosequences that are generally not conserved. However, the prosequences of many subtilisin-like serine proteases are usually removed by autoproteolysis. Experimental support of such an autoproteolytic event is often provided by N-terminal sequence analysis of proteases that have been inactivated by site-specific mutation.[18] Because this requires an efficient expression system that up until now has not been developed for pyrolysin, alternative support for such an autoproteolytic activation can be provided by modeling of the segment containing the predicted cleavage site into the active site of pyrolysin. This active site is located in the catalytic domain of pyrolysin that has a size of approximately 500 residues (Fig. 2). It contains an unusual and unique insert of approximately 150 residues that is also found in stetterlysin and explains the size of the PCR fragment obtained with the primer pair BG95 and BG31, specific for the Asp and His active sites, respectively (approximately 1.6 kb; see earlier discussion).

The homology of the catalytic domain of pyrolysin with other subtilisin-like serine proteases is sufficiently high to allow homology modeling starting from known crystal structures.[1,2] This is realized by using QUANTA 3.2.3 (Molecular Simulations, Cambridge, UK)[19] and CHARMm 22[20] running on a Silicon Graphics 4D25TG workstation. The model is based on sequence homology between pyrolysin and members of the subtilase family and on the known X-ray structures of subtilisin BPN' in complex with the inhibitor

[17] J. Devereux, P. Haeberli, and O. Smithies. *Nucleic Acids Res.* **12**, 387 (1990).

[18] P. Vos, M. van Asseldonk, F. van Jeveren, R. Siezen, G. Simons, and W. M. de Vos, *J. Bacteriol.* **171**, 2795 (1989).

[19] A. Sali and T. L. Blundell, *Mol. Biol.* **234**, 779 (1993).

[20] B. R. Brooks, R. E. Bruccoleri, B. D. Olafson, D. J. States, S. Swaninathan, and M. Karplus, *J. Comp. Chem.* **4**, 187 (1983).

```
pyrolysin-Pfu YSIEEGEYAKYVIITVKFASPVTVTVTYTIYAGPR-VSILTLNFLGYSWYRLYSQKFDELYQKALELGV
C-protease-Pab YYIENG----VIFVVVKQDPTIIAYGSYTKPAVR-ISIPVLNFLGYSWYKLYSQKFEELYNEAIKLGI
C-protease-Pho YYVQNG----IVFVVIKQDPKIVAYGSYAKPAPRRKVSIPTLNFLGYSWYKLYSEKFSKLYEEAVELGV
```
```
                                          *
pyrolysin-Pfu DNETLALALSYHEKAKEYYEKALELSEGNIIQYLGDIRLLPPLRQAYINEMKAVKILEKAIEELEGEE- 1398
C-protease-Pab DNETLGLALKYHEKAAEYYSKVLELTGGNILYHLRDIRLFAPLRGAYVNEMKAVRILEKAIKELQGEES 1204
C-protease-Pho DNETLALALEYHQKAGEYYSKVLELSEGNVIYHLYDIRLLAPLRQAYVNEMKAVRILEKAIKELKGEE- 1155
```

FIG. 3. C-terminal comparisons of pyrolysin from *P. furiosus* (Pfu) and putative thiol proteases (C proteases) from *P. horikoshi* (Pho) and *P. abyssi* (Pab). Identical residues are indicated in bold, and a possible N-glycosylation site is highlighted by an asterisk above the Asn residue. The total number of predicted residues is also indicated.

R45-eglin,[21] thermitase in complex with eglin,[22] and subtilisin in complex with its propeptide.[23] Inserts in pyrolysin of more than 6 residues cannot be modeled. The entire molecule is subjected to energy minimization after constraining the active-site residues (Asp, His, Ser), the Asn of the oxyanion hole, and the two β-sheet strand backbones eI and eIII involved in substrate binding. The atomic coordinates of the two models are available electronically (ftp://ftp.caos.kun.nl/pub/molbio/siezen97/). The backbone conformation of eglin residues 40–47 or subtilisin propeptide residues −6 to −1 (i.e., the segment that interacts with subtilisin or thermitase in the crystal structures) is used to model enzyme–substrate interactions of pyrolysin by substituting the side chains by the appropriate residues of propyrolysin. Because very favorable hydrophobic and electrostatic interactions are predicted, this strongly supports autoproteolytic activation of pyrolysin.

The model of pyrolysin is also used to compare its predicted structure with that of serine proteases with lower thermal stability.[9] Higher thermostability correlates with an increased number of residues involved in pairs and networks of charge–charge and aromatic–aromatic interactions. Pyrolysin is found to have several extra surface loops and inserts with a relatively high frequency of aromatic residues and Asn residues. The latter are often present in putative N-glycosylation sites that have the sequence Asn-X-(Ser/Thr), where X may be any amino acid residue. A total of 32 N-glycosylation sites within the deduced pyrolysin sequence are predicted and the majority of these (29) are located outside the catalytic domain in either the large insert or the carboxy-terminal domain. Experimental evidence for glycosylation at all these residues is difficult to provide, but staining of pyrolysin by periodic acid–Schiff indicates glycosylation, whereas N-terminal sequence analysis shows a gap at the third position

[21] D. W. Heinz, J. P. Priestle, J. Rahuel, K. S. Wilson, and M. G. Grutter, *J. Mol. Biol.* **217**, 353 (1991).

[22] P. Gros, C. Betzel, Z. Dauter, K. S. Wilson, and W. G. J. Hol, *J. Mol. Biol.* **210**, 347 (1989).

[23] T. Gallagher, G. Gilliland, L. Wang, and P. Bryan, *Structure* **3**, 907 (1995).

that is predicted to be an Asn residue conforming to a consensus N-glycosyl-ation site (Asn-Ser-Thr).[4]

Comparison of Pyrolysin to Other Serine Proteases

The catalytic domain of pyrolysin is followed by a very large C-terminal extension, making this archaeal serine protease one of the largest known serine proteases. To discover possible functions it is useful to periodically search the growing number of database entries for any similarities to such a large protein. Comparison of pyrolysin with other proteins in the present NCBI database revealed the highest similarity with (putative) proteases from archaeal hyperthermophiles belonging to the Thermococcales (Fig. 2). This comparison showed that pyrolysin is a mosaic of domains shared by other proteases. The highest similarity is found with stetterlysin that, although not completely sequenced,[9] shares 26% homology in the proregion and the same 150 residue insert in the catalytic domain that shows 39% identity but no homology to any known other proteins. This strongly suggests that pyrolysin and stetterlysin are most closely related in an evolutionary sense. Analysis of the genome sequence of *P. furiosus* indicates that it contains another homologous gene for a subtilisin-like serine protease that only consists of a catalytic domain and lacks the large insert present in pyrolysin (www.genome.utah.edu; Fig. 2). This explains the observed amplification of a second, small (approximately 0.6 kb) DNA fragment using the primer pair BG95 and BG31 (see earlier discussion). Remarkably, genes resembling this putative protease and pyrolysin are not present in the complete genome sequence of *P. horikoshi*.[24] However, this genome and that of *P. abyssi* contain conserved genes for extracellular thiol proteases that are predicted to contain a C-terminal sequence with high similarity to that of pyrolysin (Figs. 2 and 3). Careful inspection of this extension showed the very C-terminal residues to contain a high proportion of mainly charged Lys and Glu residues. Remarkably, a potential N-glycosylation site is conserved. In absence of experimental data it only can be speculated what the function of such a conserved C-terminal domain could be. Because all proteins that contain this extension have a predicted signal sequence and hence are secreted (as is confirmed in the case of pyrolysin), it is very possible that this highly charged and possible glycosylated C-terminal segment is involved in anchoring these proteases to the cell envelope in a yet unknown manner. This illustrates the usefulness of periodic homology searches of database, which in this case suggest the presence of a novel and conserved archaeal topogenic sequence in pyrolysin.

[24] Y. Kawarabayashi, M. Sawada, H. Horikawa, *et al., DNA Res.* **5,** 55 (1998).

[26] Pyrrolidone Carboxylpeptidase from *Thermococcus litoralis*

By MARTIN R. SINGLETON and JENNIFER A. LITTLECHILD

Background

Pyrrolidone carboxylpeptidases (pcps; EC 3.4.19.3) are a group of enzymes that remove pyroglutamate (pGlu) from the amino terminus of peptides and proteins. The enzymatic activity was initially identified in bacteria of the genus *Pseudomonas*[1] and has since been found in a wide variety of prokaryotes and eukaryotes.[2,3] The pcp enzymes form a new class of cysteine peptidase enzymes. During the screening of crude extracts of the hyperthermophilic archaea for esterase/amidase activity of commercial interest, an activity was isolated from the hyperthermophilic archaeon *Thermococcus litoralis* that was later identified as a pcp. The enzyme has been cloned and overexpressed in *Escherichia coli* at high levels.[4] The enzyme has been crystallized from both ammonium phosphate and ammonium sulfate[5] and its structure has been determined to 1.7 Å resolution (PDB code 1A2Z).[6]

The mammalian enzymes have received attention because the soluble human type 1 pcp is known to be involved in the processing of a number of important biological peptides and proteins. There appears to be a strategy of using the pyroglutamyl residue as a protective mechanism against attack by aminopeptidases. In humans, thyrotropin-releasing hormone and luteinizing hormone-releasing hormone have such pGlu N-terminal residues.[7,8] There is also another class of mammalian pcp enzymes that are larger than the type I enzymes and appear to be bound to the external surface of the cell membranes. Studies with inhibitors[9,10] show the type II pcp to be a

[1] R. F. Doolittle and R. W. Armentrout, *Biochemistry* **7,** 516 (1968).
[2] R. F. Doolittle, *Methods Enzymol.* **19,** 555 (1970).
[3] M. Orlowski and A. Meister, *Enzymes* **4,** 123 (1971).
[4] M. R. Singleton, R. Wisdom, A. Bingham, S. Taylor, and J. A. Littlechild, *Extremophiles,* in press (2000).
[5] M. R. Singleton, M. N. Isupov, and J. A. Littlechild, *Acta Cryst. D* **55,** 702 (1999b).
[6] M. R. Singleton, M. N. Isupov, and J. A. Littlechild, *Structure* **7,** 237 (1999c).
[7] N. M. G. Nair, J. F. Barrett, C. Y. Bowers, and A. V. Schally, *Biochemistry* **9,** 1103 (1970).
[8] H. Matsuo, Y. Baba, R. M. G. Nair, A. Arimura, and A. V. Schally, *Biochem. Biophys. Res. Commun.* **43,** 1334 (1971).
[9] S. Wilk and E. Wilk, *Neurochem. Int.* **15,** 81 (1989).
[10] G. Czekay and K. Bauer, *Biochem. J.* **290,** 921 (1993).

metalloenzyme, with a divalent metal ion, probably zinc, essential for catalysis.

The type 1 pcps are all either dimers or tetramers of 22–24 kDa. They show similar substrate specificity, being able to remove the N-terminal pGlu residue from a variety of peptides except when the penultimate residue is a proline. They will also hydrolyze the amide bond in synthetic substrates such as L-pGlu-β-naphthylamide,[11] L-pGlu-p-nitroanilide, and L-pGlu-4-methylcoumarinylamide[12] and may be classed as arylamidases. A putative Cys-His diad was suggested to be involved in the catalytic mechanism.[13]

The archaeal pcp appears to be related to the bacterial pcp enzymes and to the class I mammalian enzymes. The exact function of the enzyme in bacterial and archaeal systems is less well studied. It has been proposed that the pcp activity may serve to reduce toxicity of N-terminally blocked peptides[14] or to play a role in nutrient assimilation.[1]

Thermococcus litoralis Pcp

Pcp Activity Assay

Pcp activity may be monitored by following the hydrolysis of L-pyroglutamyl-β-naphthylamide spectrophotometrically. The enzyme is added to a solution of L-pyroglutamyl-β-naphthylamide (10 mM in potassium phosphate buffer, 50 mM, pH 7.5) and the release of β-naphthylamine followed at 340 nm. An extinction coefficient of $1.78 \times 10^3\ M^{-1}\ cm^{-1}$ is used for the β-naphthylamine. The β-naphthylamine product is extremely carcinogenic and should be disposed of with care.

Cloning of pcp Gene

Recombinant pcp from a variety of bacteria has been successfully cloned and overexpressed in *Escherichia coli.*[15] The *T. litoralis* enzyme was cloned using a direct-screening approach as follows. Genomic DNA is isolated from *T. litoralis* strain NS-C using previously described methods.[16] The DNA is partially digested using *Sau*3A and fragments of approximately

[11] K. E. Patterson, S. H. Hsiao, and A. Keppel, *J. Biol. Chem.* **238,** 3611 (1963).
[12] K. Fufiwara and D. Tsuru, *J. Biochem. (Tokyo)* **83,** 1145 (1978).
[13] T. Gonzales and J. Robert-Baudouy, *J. Bacteriol.* **176,** 2569 (1994).
[14] A. C. Awadé, P. Cleuziat, T. Gonzales, and J. Robert-Baudouy, *Proteins Struct. Funct. Gene.* **20,** 34 (1994).
[15] T. Yoshimoto, T. Shimoda, A. Kitazono, T. Kabashima, K. Ito, and D. Tsuru, *J. Biochem. (Tokyo)* **113,** 67 (1993).
[16] S. Henikoff, *Gene* **28,** 351 (1984).

4 kb isolated by sucrose density gradient centrifugation. The purified fragments are ligated into BamHI-cut, dephosphorylated pTrc-99 vector and transformed into E. coli XL1-Blue. Transformants are plated out on LB agar plates containing ampicillin at 0.1 mg/ml and spread with 50 μl of 0.25 M isopropylthiogalactoside (IPTG). Colonies expressing peptidase activity are screened for by overlaying the plates with 0.5% (w/v) agarose in 100 mM potassium phosphate buffer containing 0.5 mg/ml β-naphthyl acetate and 0.5 mg/ml Fast Blue BB salt. Positives may be identified by rapid formation of a deep red color. These colonies are removed from the plates, grown in overnight minicultures, and plasmids prepared. The plasmid inserts may then be sequenced, for example, as a series of exonuclease III-nested deletions. The coding sequence of the pcp gene can be identified from database homology searches and the gene amplified from genomic DNA using polymerase chain reaction. The gene may then be inserted into a variety of commercially available expression vectors, e.g., pKK223-3.[17]

Overexpression and Purification of pcp

Plasmids containing the pcp gene are transformed into an E. coli expression strain and grown until an A_{600} value of 1.0. Expression is induced by the addition of IPTG to a final concentration of 1 mM. Growth is continued for 4 hr, and the cells are harvested by centrifugation. The cells are sonicated in buffer A [50 mM potassium phosphate, pH 7.5, 10 mM dithiothreitol (DTT)], and the clarified supernatant is heated to 70° for 20 min. After centrifugation to remove precipitated proteins, an ammonium sulfate cut of 40–60% saturation is taken and the precipitated protein resuspended in buffer A. The protein is desalted and purified by application to a Superdex 200 HiLoad 16/60 gel-filtration column (Pharmacia, Piscataway, NJ) followed by passage over a BioScale Q2 anion-exchange column (Bio-Rad, Hercules, CA). Pcp elutes in the wash, with all other impurities retained on the column. Table I shows details of the purification.[4]

Properties of Recombinant Enzyme

The molecular mass of the pcp subunit is estimated from denaturing polyacrylamide gel electrophoresis as ~24,000 Da, consistent with the calculated mass of 24,472. Gel filtration shows the native enzyme to have a mass of 96 ± 12 kDa, suggesting that it is tetrameric as is observed with bacterial pcps that are either dimers or tetramers.[18] The enzyme is inactivated rapidly

[17] T. J. Brosius and A. Holy, Proc. Natl. Acad. Sci. U.S.A. **81**, 6929 (1984).
[18] A. Szewczuk and J. Kwiatkowska, Eur. J. Biochem. **15**, 92 (1970).

TABLE I
PURIFICATION OF RECOMBINANT PYRROLIDONE CARBOXYLPEPTIDASE FROM *T. litoralis*

Step	Total protein (mg)	Total activity (U)	Specific activity (U/mg)	Purification (-fold)	Yield (%)
Crude extract	870	148	0.17	0	100
Heat treatment	152	139	0.9	5.4	94
AS fractionation	110	125	1.1	6.5	85
Gel filtration	56	88	1.6	9.4	59
Anion exchange	24	84	3.5	20.6	57

by thiol-blocking agents such as N-ethylmaleimide, presumable by modification of the active site cysteine residue. In contrast, the presence of thiol-protecting agents such as DTT enhance the activity of the enzyme.[4]

The temperature–activity profile of the enzyme is shown in Fig. 1 and demonstrates a maximum at 70°, slightly lower than the optimum growth temperature of *T. litoralis*. The resistance of the enzyme to thermal denaturation is shown in Fig. 2. Although considerably more thermostable than mesophilic enzymes, the *T. litoralis* pcp loses its activity rapidly at temperatures above 80°. This is likely to be due to destruction of the critical hyper-reactive cysteine in the active site and suggests that the effect may be reduced with suitable concentrations of thiol-protecting agents. However, the enzyme still has a half-life of 1 hr at 70°.

The stereospecificity of the enzyme is demonstrated with the reaction shown in Fig. 3. The formation of the L-pyroglutamate product (**II**) proceeds with a rate constant approximately 450-fold greater than that for the forma-

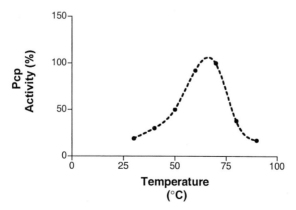

FIG. 1. Temperature–activity relationship for pyrrolidone carboxylpeptidase.

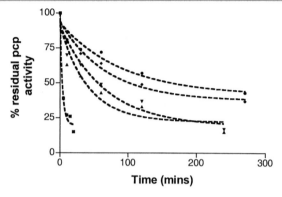

FIG. 2. Temperature–stability curves for pyrrolidone carboxylpeptidase. (■) 90°, (▲) 80°, (▼) 70°, (◆) 60°, (●) 50°.

tion of the D-enantiomer (**I**). In addition, the enzyme is capable of removing L-pyroglutamate from a variety of synthetic esters and amides. The presence of L-pyroglutamate is obligatory; however, there is little discrimination on the other side of the scissile bond.[4]

Structure

The structure of the enzyme has been solved by X-ray crystallography.[5,6] As expected, the native enzyme is a tetramer. Each α/β subunit is folded into a comma-shaped domain with a central β sheet surrounded by five α helices (Fig. 4). The critical conserved cysteine residue is located at the end of helix 4, facing into an exposed solvent channel (Fig. 5). The cysteine is part of a classic catalytic triad with a histidine–glutamate hydrogen-bonded pair, which are totally conserved in primary structure alignments of pcp enzymes (Fig. 6). The cysteine was observed to be oxidized and, therefore, inactive in the crystal structure. Although unusual for an intracellular enzyme, there is a disulfide bridge linking adjacent subunits; this is presumably a thermostabilizing feature and has been seen in some other

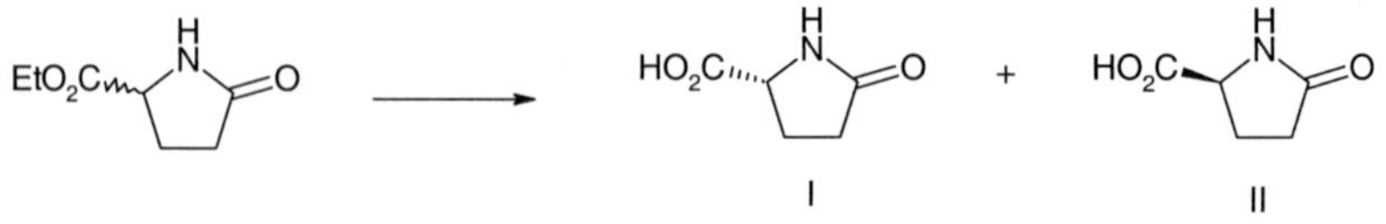

FIG. 3. Reaction catalyzed by pyrrolidone carboxylpeptidase.

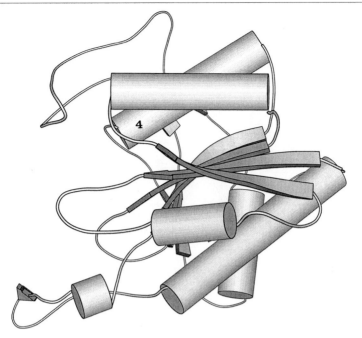

Fig. 4. Ribbon diagram of pcp monomer with the secondary structural elements shown as helices and arrow/sheets. Helix 4, where the active site cysteine residue is located, is labeled. Produced with MOLSCRIPT.[26] P. J. Kraulis, *J. Appl. Cryst.* **24,** 946 (1991).

hyperthermophile-derived proteins.[19–22] This disulfide bridge appears to remain intact, even under strongly reducing conditions. There is also a large central hydrophobic core to the tetramer (Fig. 7), a feature only seen in the *T. litoralis* sequence alignment and absent from the only other pcp structure known, that from *Bacillus amyloliquefaciens*.[23] The *T. litoralis* pcp shows high sequence similarity to the bacterial enzymes apart from the features discussed earlier, as well as the pcp identified in the genome

[19] P. R. E. Mittl and G. E. Schulz, *Protein Sci.* **3,** 1504 (1994).
[20] A. Weichsel, J. R. Gasdaska, G. Powis, and W. R. Montfort, *Structure* **4,** 735 (1996).
[21] M. N. Isupov, G. Crowhurst, A. Dalby, T. Fleming, and J. A. Littlechild, *J. Mol. Biol.* **291,** 651 (1999).
[22] Y. Jiang, S. Nock, M. Nesper, M. Sprinzl, and P. B. Sigler, *Biochemistry* **35,** 10269 (1996).
[23] Y. Odagaki, A. Hayashi, K. Okada, T. Hirotsu, K. Kabashima, K. Ito, T. Yoshimoto, D. Tsuru, M. Sato, and J. Clardy, *Structure* **7,** 399 (1999).

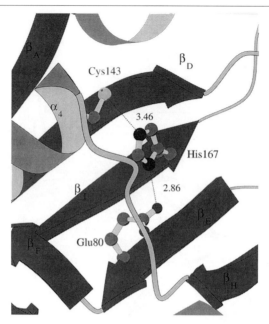

Fig. 5. Diagram showing the active site triad Cys-143, His-167, and Glu-80 of the pcp monomer. The distance (in Å) between Cys-143 Sγ-His-167 Nε and His-167 Nδ-Glu-80 Oε is shown. Secondary structure elements are labeled. Produced with MOLSCRIPT.[26] P. J. Kraulis, *J. Appl. Cryst.* **24,** 946 (1991).

of other archaea *Pyrococcus horikoshii*[24] and *Pyrococcus furiosus.* The highly hydrophobic insertion that forms the core of the tetramer would appear to be a stabilizing feature; however, this is not found in the pcp sequence from the more thermophilic *P. horikoshii* pcp. It appears that hydrophobicity is only an additional stabilizing feature in enzymes that are stable to temperatures of about 70°. At higher temperatures and with other *Pyrococcus* enzymes where structural information is available, additional features, such as salt bridges, appear to play a more important role.[25] The presence of intersubunit disulfide bridges, which is unusual for intracellular proteins, should also enhance the stability of the folded protein. The cysteine residue in the *Thermococcus* pcp enzyme is conserved in the sequence

[24] Y. Kawarabayasi, M. Sawada, H. Horikawa, Y. Haikawa, Y. Hino, S. Yamamoto, M. Sekine, S. Baba, H. Kosugi, A. Hosoyama, Y. Nagai, M. Sakai, K. Ogura, R. Otsuka, H. Nakazawa, M. Takamiya, Y. Ohfuku, T. Funahashi, T. Tanaka, Y. Kudoh, J. Yamazaki, N. Kushida, A. Oguchi, K. Aoki, and H. Kikuchi, *DNA Res.* **5,** 55 (1998).
[25] R. J. M. Russell, J. M. Ferguson, D. W. Hough, M. J. Danson, and G. L. Taylor, *Biochemistry* **36,** 9983 (1994).

```
                        1        10        20        30        40        50        60        70
1:T.litoralis           .MKKVLITGFEPFGGDSKNPTEQIAKYFDRKQIGN.AMVYGRVLPVSVKRATIELKRYLEEIKPEIVINLGLAP
2:P.furiosus            ..MKVLVTGFEPFGGEKINPTERIAKDLDGIKIGD.AQVFGRVLPVVFGKAKEVLEKTLEEIKPDIAIHVGLAP
3:P.horikoshii          ..MKILLTGFEPFGGDDKNPTMDIVEALSER.IP...EVVGEILPVSFKRAREKLLKVLDDVRPDITINLGLAP
4:P.fluorescens         .MRIVLLTGFEPFDQDPVNPSWEAVRQLDGVQLGSDVKIVARRLPCAFATAGECLTRLIDELHPAMVIATGLGP
5:S.pyogenes            ..MKILVTGFDPFGGEAINPALEAIKKLP.ATIHG.AEIKCIEVPTVFQKSADVLQQHIESFQPDAVLCIGQAG
6:L.lactis              ..MKILVTGFDPFGDDKINPAIEAVKRLP.DEIAG.AQIVKLEIPTKFNVSADVVKDAIAKEKPDYVLSIGQAG
7:S.aureus              ..MHILVTGFAPFDNQDINPSWEAVTQLE.NIIGT.HTIDKLKLPTSFKKVDTIINKTLASNHYDVVLAIGQAG
8:B.subtilis            MRKKVLITGFDPFDKETVNPSWEAAKRLNGFETEE.AIITAEQIPTVFRSALDTLRQAIQKHQPDIVICVGQAG
9:B.amyloliquefaciens   MEKKVLLTGFDPFGGETVNPSWEAVKRLNGAAEGP.ASIVSEQVPTVFYKSLAVLREAIKKHQPDIIICVGQAG

                               80        90        100       110       120       130       140
1:T.litoralis           TYSNITVERIAVNIIDARIPDNDGYQPIDEKIEEDAPLAYMATLPVRAITKTLRDNGIPATISYSAGTYLCNYV
2:P.furiosus            GRSAISIERIAVNAIDARIPDNEGKKIEDEPIVPGAPTAYFSTLPIKKIMKKLHERGIPAYISNSAGLYLCNYV
3:P.horikoshii          GRTHISVERVAVNMIDARIPDNDGEQPKDEPIVEGGPAAYFATIPTREIVEEMKKNGIPAVLSYTAGTYLCNFA
4:P.fluorescens         GRSDISVERVAININDARIPDNLGEQPIDTAVVADGPAAFFTTLPIKAMVKAVREAGIAASVSQTAGTFVCNQV
5:S.pyogenes            GRTGLTPERVAINQDDARIPDNEGNQPIDTPIRADGKAAYFSTLPIKAMAKAIHQAGLPASVSNTAGTYVCNHL
6:L.lactis              GRFELTPERVAINLDDGRIQDNAGYQPLNHTIHGDDENAYFTQLPIKAMVAAIREAGVPSAVSNTAGTYVCNHI
7:S.aureus              GRNAITPERVAINIDDARIPDNDDFQPIDQAIHLDGAPRYFSNLPVKAMTQSVINQGLPGALSNSAGTFVCNHV
8:B.subtilis            GRMQITPERVAINLADARIPDNEGHQPIDEEISPDGPAAYWTRLPVKRMTAKMKEHGIPAAVSYTAGTFVCNYL
9:B.amyioliquefaciens   GRMQITPERVAINLNEARIPDNEGNQPVGEDISQGGPAAYWTGLPIKRIVEEIKKEGIPAAVSYTAGTFVCNHL

                          150       160       170       180       190☆      200       210       220
1:T.litoralis           MFKTLHFSKIEGYPLKAGFIHVPYTPDQVVNKFFLLGKNTPSMCLEAEIKAIELAVKVSLDYLE..KDRDDIKI
2:P.furiosus            MYLSLHHSATKGYPKMSGFIHVPYIPEQIIDKIGK.GQVPPSMCYEMELEAVKVAIEVALEELL........
3:P.horikoshii          MYLTLHTSATKGYPKIAGFIHVPYTPDQVLEKKNT....PSMSLDLEIKGVEIAIRVAQSALH..SSQLR...
4:P.fluorescens         FYLLQH..ALAGSGVRSGFIHVPFLPEQVAGSQRP.....SMALDAMVAGLQAAVLTAWHTPV..DVKEAGGQ
5:S.pyogenes            MYQALYLVDKYCPNAKAGFMHIPFMMEQVVDKPN....TAAMNLDDITRGIEAAIFAIVDFKDRSDLKRVGGA
6:L.lactis              FYQVQYMRDKMFPDIKAGFMHIPFLPEQVVTRPE....TPALSLDDDVLGITAAIKAIVSRDGKGDIETIEGK
7:S.aureus              LYHLGYLQDKHYPHLRFGFIHVPYIPEQVVGKSD....TPSMPLEQIVAGLTAAIEAISDHD..DLRIALGT
8:B.subtilis            FYGLMDHISRTSPHIRGGFIHIPYIPQQTIDKT.....APSLSLDTIVRALRIAAVTAAQYDE..DVKSPGGT
9:B.amyloliquefaciens   FYGLMDEISRHHPHIRGGFIHIPYIPEQTLQKS.....APSLSLDHITKALKIAAVTAAVHED..DIETGGGE

1:T.litoralis           PL
2:P.furiosus            ..
3:P.horikoshii          ..
4:P.fluorescens         VS
5:S.pyogenes            TH
6:L.lactis              DH
7:S.aureus              TE
8:B.subtilis            LH
9:B.amyloliquefaciens   LH
```

FIG. 6. Multiple sequence alignment of pcp sequences. Conservative residue changes are shown in bold, and totally conserved residues are blocked out. The cysteine residue involved in the disulfide bridge in the *Thermococcus* pcp is shown with an asterisk. Produced using AMPS and ALSCRIPT.[27] G. J. Barton, *Protein Eng.* **6,** 37 (1993).

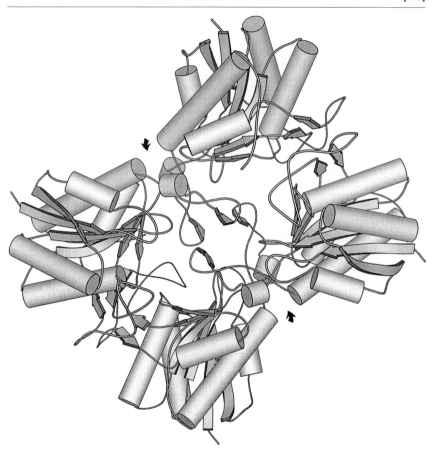

FIG. 7. Pcp tetramer with secondary structural elements shown as helices and arrow/sheets. An arrow denotes the subunit interface where the disulfide bond is found. Produced with MOLSCRIPT.[26] P. J. Kraulis, *J. Appl. Cryst.* **24,** 946 (1991).

of the *P. furiosus* enzyme counterpart but not in the sequence of the *P. horikoshii* pcp enzyme (Fig. 6). A detailed analysis of thermostabilizing features require that the structures of an equivalent enzyme from both thermophilic and mesophilic sources be studied. Because this is not yet possible in the case of pcp, it is difficult to identify unequivocally specific features that contribute to the stability of this structure.

The *pcp* gene from the hyperthermophilic archaeon *T. litoralis* has been cloned and overexpressed in *E. coli*. The recombinant enzyme is a homotetramer of 24-kDa subunits, and as is expected of enzymes derived from such sources, the protein shows enhanced thermostability over mesophilic examples, with an optimum activity at 75° and a half-life of 2 hr at this temperature. The 1.73-Å crystal structure of the enzyme, which is the first for this family (C15) of cysteine proteases, has identified the presence of a catalytic triad and structural comparisons to other enzymes of the open α/β fold. The structural basis of the thermostability of this *Thermococcus* pcp is of interest. Site-directed mutagenesis studies to change the cysteine residues involved in disulfide bridge formation to alanine residues and a deletion mutant lacking the hydrophobic core of the enzyme are currently being prepared. These mutant enzymes will be analyzed to determine their stability and oligomeric properties in order to rationalize the mechanisms involved in thermostability of this *T. litoralis* pcp enzyme.

[26] P. J. Kraulis, *J. Appl. Cryst.* **24,** 946 (1991).
[27] G. J. Barton, *Protein Eng.* **6,** 37 (1993).

[27] Protease I from *Pyrococcus furiosus*

By Lara S. Chang, Paula M. Hicks, and Robert M. Kelly

Introduction

Pyrococcus furiosus is a hyperthermophilic archaeon from the order Thermococcales that is capable of growth on a variety of proteinaceous and carbohydrate-containing substrates.[1] Analysis of gelatin-containing sodium dodecyl sulfate–polyacrylamide gel electrophoresis (SDS–PAGE) gels indicates that at least 11 endoproteinases are active in the cell extracts of this organism[2,3] and the following proteases have been characterized: protease I (PfpI),[2,4,5–7] pyrolysin,[8,9] proteasome,[10] prolyl oli-

[1] G. Fiala and K. O. Stetter, *Arch. Microbiol.* **145,** 56 (1986).
[2] I. I. Blumentals, A. S. Robinson, and R. M. Kelly, *Appl. Environ. Microbiol.* **56,** 1992 (1990).
[3] H. Connaris, D. A. Cowan, and R. J. Sharp, *J. Gen. Microbiol.* **137,** 1193 (1991).
[4] S. B. Halio, I. I. Blumentals, S. A. Short, B. M. Merrill, and R. M. Kelly, *J. Bacteriol.* **178,** 2605 (1996).
[5] S. B. Halio, M. W. Bauer, S. Mukund, M. W. W. Adams, and R. M. Kelly, *Appl. Environ. Microbiol.* **63,** 289 (1997).
[6] P. M. Hicks and R. M. Kelly, *in* "PfpI Protease I (*Pyrococcus furiosus*)" (A. J. Barrett, N. D. Rawlings, and J. F. Woessner, eds.). Academic Press, London, 1998.
[7] P. M. Hicks, Ph.D. Thesis, North Carolina State University, Raleigh, NC, 1998.
[8] R. Eggen, A. Geerling, J. Watts, and W. M. de Vos, *FEMS Mirobiol. Lett.* **71,** 17 (1990).
[9] W. G. B. Voorhorst, R. I. L. Eggen, A. C. M. Geerling, C. Platteeuw, R. J. Siezen, and W. M. de Vos, *J. Biol. Chem.* **271,** 20426 (1996).
[10] M. W. Bauer, S. H. Bauer, and R. M. Kelly, *J. Bacteriol.* **63,** 1160 (1997).

gopeptidase,[11,12] and proline dipeptidase.[13] Blumentals *et al.*[2] found that when *P. furiosus* cell extracts were heated at 98° for 24 hr in the presence of 1% (w/v) SDS, one of the few remaining proteins had proteolytic activity. This proteolytically active sample, which was initially named S66, was approximately 66 kDa in size as determined by SDS–PAGE. However, at least two functional forms of this protease exist and the smallest functional form is a trimer of 19-kDa subunits. Furthermore, it is now recognized that, because of its insensitivity to SDS, the 66-kDa size observed on SDS–PAGE gels does not accurately represent the true molecular mass of this protease.[4,7] It was subsequently renamed PfpI (*P. furiosus* protease I). Figure 1 illustrates the presence of at least two forms of PfpI observed after heat treatment for 24 hr in the presence of SDS. The smallest functional form of the protease (a homotrimer) is the predominant structure after extended incubation under denaturing conditions.

The gene encoding the 19-kDa subunit of PfpI has homologs in nearly every organism and cell examined to date, ranging from *Escherichia coli* to *Homo sapiens*; this ubiquity and evolutionary conservation indicates that it may play a fundamental physiological role.[7] Efforts to study this issue have been exacerbated by difficulties encountered in obtaining significant amounts of a particular assembly of PfpI in either a native or a recombinant form. Native PfpI undergoes autoproteolysis and/or disassembly during direct purification from *P. furiosus* biomass and exists in multiple (single-to multisubunit) forms *in vitro*.[5,6,7,14] The production of a recombinant form of PfpI is also problematic due to its toxicity toward mesophilic hosts.[7] Several methods that have been used to purify PfpI directly from *P. furiosus* cell extracts are described here, together with an assay to detect proteolytic activity, a procedure to determine its molecular mass, and approaches to minimize PfpI-catalyzed proteolysis of other *P. furious* proteins.

Methods

Biochemical Assay

PfpI is specific toward basic and bulky hydrophobic P_1[15] amino acid residues in peptide substrates. In general, peptidase activity (proteolysis at

[11] K. A. Robinson, D. A. Bartley, F. T. Robb, and H. J. Schreier, *Gene* **152**, 103 (1995).

[12] V. J. Harwood, J. D. Denson, K. A. Robinson-Bidle, and H. J. Schreier, *J. Bacteriol.* **179**, 3613 (1997).

[13] M. Ghosh, A. M. Grunden, D. M. Dunn, R. Weiss, and M. W. W. Adams, *J. Bacteriol.* **180**, 4781 (1998).

[14] P. M. Hicks, L. S. Chang, and R. M. Kelly, manuscript in preparation.

[15] B. Keil, "Specificity of Proteolysis." Springer-Verlag, Berlin, 1992.

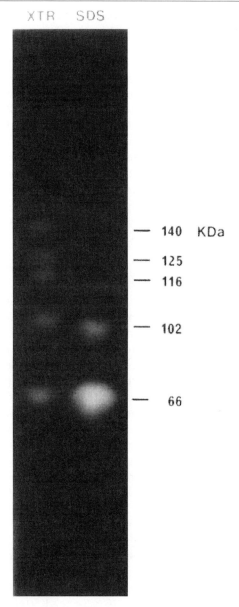

FIG. 1. Gelatin overlay analysis of cell extract (XTR) and of cell extract treated for 24 hr at 98° with 1% SDS (SDS). Numbers on the right correspond to molecular mass in kilodaltons of bands with proteolytic activity.[2]

the C or N terminus of a peptide chain) of PfpI is detected by the release of the fluorescent chemical 7-amido-4-methylcoumarin (MCA) from the carboxyl terminus of amino-terminally blocked substrates (Sigma Chemical Company, St. Louis, MO) using end point assays in microtiter plates. It is most active with the chymotrypsin-like substrate N-succinyl-alanine-alanine-phenylalanine 7-amido-4-methylcoumarin (AAF-MCA).[5] Assay procedures have been described previously by Halio et al.[5] Proteinase activity (proteolysis at the inside of a peptide chain) for PfpI can be followed using gelatin-containing zymograms (Novex, San Diego, CA) or by caseinolytic assay using BODIPY FL casein (Molecular Probes, Inc., Eugene, OR). It should be noted that Tris buffers are often replaced with sodium phosphate buffers at higher temperatures, and that the gelatin-containing gels can only be heated to 70° before autohydrolysis occurs. In addition, the zymograms should be run at 4° to prevent proteolysis during electrophoresis, as PfpI is not denatured or inhibited by SDS.[7]

Purification of PfpI from P. furiosus

All purification steps are carried out at room temperature on a Pharmacia LKB FPLC system (Pharmacia, Uppsala, Sweden). Samples are either clarified with a 0.2-μm filter or centrifuged to remove cellular debris before being subjected to chromatographic separation. All concentration steps are done with stirred-cell concentrators (Amicon, Beverly, MA) using filters of 10-kDa molecular mass cutoff. Elution of PfpI is followed by assaying for hydrolysis of the substrate AAF-MCA, as described in the assay protocol described earlier. Five-minute incubations at 85° are used during all of the assays.[7]

DEAE Chromatography. The cell extract (420 ml) is applied to a 600-ml DEAE Sephadex CL-6B XK50 column (Pharmacia), which is initially equilibrated with 4 liters of 50 mM sodium phosphate buffer, pH 8. After 2 liters of pass-through, an 8-liter linear gradient from 0 to 1 M NaCl in 50 mM sodium phosphate buffer, pH 8, is used to elute the proteins collected in 24-ml fractions at a flow rate of 4 ml/min. The majority of PfpI activity elutes between 168 and 240 mM NaCl.[7]

Hydrophobic Interaction Chromatography (HIC). Active fractions are concentrated to 150 ml, equilibrated with 61 g/liter ammonium sulfate (using repeated dilution and concentration steps with a stirred-cell concentrator), and loaded at 2 ml/min onto a 70 × 5-cm XK50 column (Pharmacia) packed with 750 ml phenyl-Sepharose 650 M (Toso Haas, Montgomeryville, PA). The column is equilibrated with 2 liters of 122 g/liter ammonium sulfate in 50 mM sodium phosphate buffer, pH 7. After a 750-ml pass-through, proteins are eluted with a 250-ml decreasing gradient from 122 to 36 g/liter ammonium sulfate in sodium phosphate buffer, pH 7, followed by a 1-liter decreasing gradient from 36 to 0 g/liter ammonium sulfate in sodium phosphate buffer, pH 7. The flow rate is 5 ml/min and 25-ml fractions are collected. PfpI activity elutes from 17 to 0 g/liter ammonium sulfate in two peaks, which are separately combined into two pools of samples (275- and 200-ml volumes). By SDS–PAGE analysis, the first peak (275 ml) appears as primarily the hexameric form of PfpI, while the second peak (200 ml) corresponds to the trimeric form of PfpI.[7]

Hydroxylapatite Chromatography. The second (trimer) pool from the HIC run is concentrated from 200 to 28 ml and is loaded onto an 80-ml hydroxylapatite (HAP, Calbiochem, La Jolla, CA) XK 16 column (Pharmacia, 1.6×40 cm) equilibrated with 400 ml of 50 mM sodium phosphate buffer, pH 7. Proteins are eluted with a 300-ml linear gradient from 0 to 2 M NaCl in approximately 280 ml. This step is omitted for the hexameric HIC pool.[7]

Strong Anion-Exchange Chromatography: Mono Q. The 2 M NaCl eluant from HAP is desalted and concentrated to 6 ml, and the first pool (hexamer) from the HIC run is concentrated to 7 ml. Both pools are further purified separately by multiple loadings of 0.5 ml onto a 1-ml Mono Q column (HR 5/5, Pharmacia) at a flow rate of 0.1 ml/min. The column is equilibrated with 10–20 ml of 100 mM sodium phosphate buffer, pH 8, prior to loading and washed with 8 ml of the same buffer after loading. Proteins are eluted with sequential linear gradients: 0–120 mM NaCl in 100 mM sodium phosphate buffer, pH 8 (8 ml), 120–130 mM NaCl (9 ml), 130–400 mM NaCl (20 ml), and 0.4–1.5 M NaCl (5 ml). A flow rate of 0.5–1.0 ml/min is used, depending on the back pressure, and 2-ml fractions are collected. Generally, PfpI elutes in the second gradient (120–130 mM NaCl).[7]

Gel Filtration Chromatography. Mono Q eluant from each of the HIC pools is combined separately and concentrated to 1–3 ml before loading (0.5 ml at a time) onto a Superdex 200 HiLoad 16/60 column (Pharmacia) equilibrated with 50 mM sodium phosphate buffer, pH 7, containing 150 mM NaCl. A flow rate of 0.2–0.25 ml/min is used. Peaks corresponding to molecular masses of 107–112 kDa for the hexamer form and 47–56 kDa for the trimer form are typical. Certain proteolyzed fractions may elute much later, as noted in the report by Halio *et al.*[16] Some activity may need to be sacrificed in order to avoid fractions that contain multiple forms of the PfpI enzyme. In past experiments, the void volume of the column was found to be 52.2 ml when measured immediately prior to the run that generated a dodecamer form of PfpI. Blue dextran (2000 kDa) is used to

---

[16] S. B. Halio, Ph.D. Thesis, North Carolina State University, Raleigh, NC, 1995.

determine the void volume, and the following proteins standards are used for the calibration curve: thyroglobulin (669 kDa), apoferritin (443 kDa), α-amylase (200 kDa), and bovine serum albumin (66 kDa).[7]

Affinity Column Purification

Affinity purification of PfpI is used primarily with cell extracts by allowing PfpI to degrade other proteins through storage or heating in small amounts of detergent, or it can be used after an initial ion-exchange step (DEAE or Mono Q). Affinity purification is not efficient for obtaining large amounts of homogeneous PfpI, but it does allow for the larger forms of PfpI (dodecamer or 15-mer) to be isolated. These larger forms are not purified using the standard procedure because the HIC and HAP chromatography steps most likely disrupt hydrophobic interactions between the subunits of PfpI, leading to dissociation of the multimers. Therefore, affinity column purification is the preferred purification method to obtain moderate amounts of the larger forms of PfpI that are not degraded into the smaller forms (trimer or hexamer).[7]

Construction of Affinity Column for PfpI. The affinity column is constructed based on the affinity of PfpI for the peptide AAF-X.[5] Prepacked NHS-activated HiTrap columns (1 ml) are obtained from Pharmacia. AAF-MCA (Sigma Chemical Co.) or AAF (Synpep Corporation, Dublin, CA) is dissolved in 0.2 M NaHCO$_3$, 0.5 M NaCl, pH 8.5, at a concentration of 5 mg/ml. In the case of AAF-MCA, 15% (v/v) dimethyl sulfoxide (DMSO) is used as a cosolvent to increase its solubility. Ligand coupling is performed as directed by vendor's instructions, and excess active groups are deactivated with ethanolamine. The column is stored at 4° in 50 mM sodium phosphate buffer, pH 7, with 0.05% sodium azide. Storage in ethanol has been found to be detrimental to the column. Immediately prior to use, the affinity column is washed, alternating between binding buffer (50 mM sodium phosphate buffer, pH 7, 150 mM NaCl) and elution buffer (50 mM sodium acetate, pH 3, 0.5 M NaCl).[7]

Use of Affinity Chromatography for PfpI. The first affinity column for PfpI was based on AAF-MCA, which has a binding constant of 152 μM.[7] Elution was done at 4° using a Luer-Lok syringe to deliver the elution buffer described in the previous section. It was thought that binding, but not hydrolysis, would occur at this temperature. However, the enzyme would not elute from the column and a steady amount of MCA was detectable in the wash. It was subsequently determined that PfpI does, in fact, have measurable activity at 4°. Therefore, a second column was constructed using only the peptide AAF. One milliliter of cell extract, which had already been enriched by PfpI with long-term storage (the same can be accom-

TABLE I
SPECIFIC ACTIVITY OF AFFINITY CHROMATOGRAPHY POOLS[a]

Pool	Units/ml[b] ($\times 10^{-4}$)	mg/ml	Units/mg ($\times 10^{-3}$)	Purification (-fold)
Incubated cell extracts	227	28	80.4	1.0
Fraction 4	33.9	2.6	131	1.6
Fraction 5	10.3	0.30	342	4.2
Fractions 7 and 8	0.620	0.018	344	4.3
Fractions 10 and 11	0.635	0.010	655	8.1
Fractions 19–22	0.580	0.0080	753	9.4

[a] P. M. Hicks, Ph.D. Thesis, North Carolina State University, Raleigh, NC, 1998.

[b] A unit of activity is defined as the change in fluorescence reading at a sensitivity of 3 after a 5-min incubation at 85°.

plished with heating and small amounts of denaturants), was loaded at 0.1 ml/min to the column (at 25°). Table I shows PfpI activity that eluted at the end of the pass-through (fractions 4 and 5), the beginning of the pH gradient (fractions 7 and 8), at pH 3 (fractions 10 and 11), and after several column volume exposures to low pH (fractions 19–22). The resulting overall purification fold is lower than published previously for PfpI[5,7] as the protein samples were already enriched with PfpI by incubation of the cell-free extract. The specific activity is also lower than published previously, perhaps due to prolonged exposure of the protein to low pH, changes in buffer composition, and the age of the protein sample. As shown in Fig. 2, the

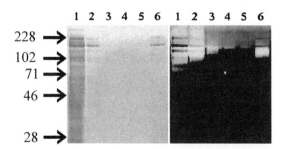

FIG. 2. Zymogram (10% PAGE containing 0.1% gelatin) of PfpI purified with affinity chromatography. The left side of the gel has been destained. The presence of larger forms is thought to be related to purification procedures that avoided hydrophobic steps. Arrows indicate electrophoretic mobility of GIBCO-BRL-prestained high molecular weight markers with the indicated molecular masses: lane 1, fraction 4 from affinity column (pass-through); lane 2, fraction 5 from affinity column (tail of pass-through); lanes 3–5, eluted fractions off of affinity column; and lane 6, PfpI purified by gel filtration and Mono Q chromatography. Adapted from Hicks.[7]

eluted protein is predominantly the larger form of the protein. Furthermore, it is of similar purity to the PfpI that had been purified from the same pool by both ion-exchange and gel filtration chromatography. The zymogram in Fig. 2 shows that although later fractions are very dilute, they still have detectable gelatinase activity.[7]

This affinity purification approach is particularly useful for proteases with affinity for substrates toward which other proteases in the cell extract do not have activity. In the future, it would be desirable to use reversible aldehyde inhibitors to construct the column; if the K_i is sufficiently low, it may be necessary to use chaotropic agents such as guanidinium chloride for elution. In some cases, it may be possible to elute with a substrate for which the protease has a lower K_m. Separation from the substrate can be achieved easily using ultrafiltration or gel filtration chromatography.[7]

Estimation of Molecular Mass by Cross-Linking

Because PfpI occurs in multiple active forms, it is difficult to assess the amounts of these forms even after several purification procedures. One approach that can be used is to cross-link the various PfpI forms with dimethyl suberimidate (Sigma Chemical Co.).[17] Cross-linker (10 mg/ml) is dissolved in 200 mM triethanolamine, pH 8.5, and mixed with various amounts of concentrated protein to yield final total protein concentrations around 0.5 mg/ml. The mixtures are incubated at room temperature for 3 hr, heated at 90° in 1% SDS and 1% 2-mercaptoethanol for 1 hr, and analyzed by 7.5–10% SDS–PAGE. These conditions are such that the predominant amount of cross-linking occurs only between structures that are already associated through noncovalent interactions. Aldolase from rabbit muscle (Boehringer Mannheim, Indianapolis, IN), a homotetramer of 39.2-kDa subunits, is used as a control. This method has demonstrated that PfpI occurs in several active forms (multiples of three up to a 15-mer),[7] in addition to the trimeric and hexameric structures reported previously.[5]

Use of Irreversible Inhibitors to Prevent PfpI from Destroying Intracellular Proteins

PfpI has been shown to be capable of degrading nearly all other intracellular proteins upon extended incubation of the cell-free extracts in either the presence[2] or the absence[7] of SDS and other denaturants. It is a particular problem during attempts to purify other proteins from *P. furiosus*. Specifi-

---

[17] G. E. Davies and G. R. Stark, *Proc. Natl. Acad. Sci. U.S.A.* **66,** 651 (1970).

cally, PfpI interferes directly with the purification of the PEPase[11,12] if the two proteins are not separated immediately after cell disruption.[7] Titration studies with the inhibitor AAF-CMK (chloromethyl ketone) showed that active forms of PfpI (trimer, hexamer, nonamer) contain approximately two reactive sites per monomer.[7] Purification studies indicate that PfpI can account for 0.4–0.5% of the total protein in *P. furiosus* cell extracts under certain conditions.[5,7] Therefore, for every 200 mg of cellular protein, there is approximately 1 mg or 0.053 μmol of PfpI monomer. To inhibit PfpI activity during the purification of *P. furiosus* proteins, it is useful to add a threefold excess (0.16 μmol per 0.053 μmol PfpI) of inhibitor to account for other AAF-active proteases and aminopeptidases in the cell extracts (an amino-terminally blocked inhibitor would be more desirable, if available commercially). When the inhibition procedure was done previously, there was 62% inhibition of PfpI activity after incubating the cell extract with inhibitor AAF-CMK for 15 min at 70°. A complete reaction with inhibitor occurs between 4 and 12 hr at 50–65°. More inhibitor can be used to expedite the reaction. This same protocol can be used with an excess of reversible inhibitor in the case of storing autoproteolytic enzymes.[7]

Cloning and Expressing the pfpI Gene

The gene encoding PfpI (GenBank accession number U57642) can be cloned and expressed as a fusion protein with a histidine tag.[4] The *pfpI* gene is amplified by polymerase chain reaction (PCR) using Vent DNA polymerase, primers complementary to each end of the gene, and *P. furiosus* genomic DNA. The upstream and downstream primers are designed with an *Nde*I restriction site at the N-terminal methionine of the *pfpI* gene and a *Bam*HI restriction site immediately following the *pfpI* stop codon, respectively. The PCR product is gel purified and ligated to a pET-15b plasmid (Novagen, Madison, WI). The resulting vectors are electrotransformed into *E. coli* BL21 (Novagen) and induced with CE6 λ phage, after growth at 30° to an optical density at 600 nm of 0.6. The *E. coli* host cells are harvested by centrifugation and disrupted by sonication. The recombinant protein is separated from other soluble proteins using the HisBind resin and buffer kit (Novagen). After purification, the product is dialyzed and concentrated, and the histidine tag is removed by addition of thrombin protease.[4]

The histidine fusion protein expression product that was obtained by Halio *et al.*[4] was less stable and had different properties than the native PfpI. In addition, only small amounts of the recombinant protein were produced. The temperature optimum and half-life values for the recombinant protease were 95° and 19 min (at 95°), suggesting that the recombinant form was less stable than the native form. As a result, several other expression methods have been attempted with minimal success, including the T7 promoter in pET-22b[+] (inducer CE6 phage), the P_{BAD} promoter in pBAD24 (inducer arabinose), the ϕ31 promoter in pTRK360 (inducer ϕ31 phage), and several other systems.[7] The search to find a better expression system for PfpI is continuing.

Biochemical and Biophysical Properties of PfpI

The PfpI monomer contains 166 amino acid residues, is approximately 19 kDa in size, but is proteolytically inactive.[4] However, *in vitro* these subunits assemble into proteolytically active trimeric, hexameric, and higher trimer-based forms.[5,7] The largest structure seen by Western blots and zymogram gels is 275 kDa, the approximate size of a 15-mer.[7] The trimer form (the smallest active form) is the smallest assembly observed under nondenaturing conditions[7] and is the primary form present after prolonged heating in the presence of SDS.[2,4,5] The trimer (59 ± 3 kDa) and hexamer (124 ± 6 kDa) forms of PfpI are not fully denatured by SDS and, on denaturing gels, have apparent masses of 66 and 86 kDa, respectively.[5] The trimer form self-assembles into the hexamer on incubation at catalytically suboptimal temperatures (4°).[5]

PfpI has both chymotrypsin- and trypsin-like activities and is affected by serine inhibitors.[2] To date, this protease is most active toward the synthetic substrate *N*-succinyl-alanine-alanine-phenylalanine 7-amido-4-methylcoumarin (AAF-MCA).[5] Large proteins, such as azocasein and gelatin, are also degraded by PfpI.[6,7] Using AAF-MCA, the temperature and pH optima of PfpI are 86° and pH 6.3, respectively.[5] The experimentally determined p*I* values of the trimer and hexamer are 3.8 and 6.1, respectively, whereas the calculated p*I* of the 19-kDa subunit is 6.1. If all asparagine and glutamine residues are changed to their corresponding acidic residues, the p*I* of the monomer is calculated to be 4.8. At high temperatures it is possible that the protein is deamidated, especially in its trimeric form, as more thermally labile amino acid residues (i.e., glutamine and asparagine) are left unprotected. This could explain the difference between experimentally determined and calculated p*I* values.[5]

The role of PfpI in *P. furiosus* has yet to be determined, although its ability to hydrolyze the majority of cell protein during the extended incubation of cell extracts under denaturing conditions suggests that it is a predominant protease. It is not known whether the multiple active forms seen *in vitro* also exist *in vivo*. Efforts are underway to explore the role of PfpI in *P. furiosus* as well as in other organisms that contain PfpI homologs.

Acknowledgments

This work was supported in part by grants from the Energy Biosciences Program of the Department of Energy and the National Science Foundation. LSC acknowledges the support of the Department of Education GAANN Fellowship.

[28] Archaeal Proteasomes

By ROBERT D. BARBER and JAMES G. FERRY

Introduction

In recent years, the significance of proteases whose proteolytic activity is spatially confined within the respective enzyme has been revealed throughout biology. Several examples of such enzymes exist, including Lon protease, Clp protease, HslUV, and bleomycin hydrolase. However, the importance of the 20S proteasome in eukaryotic cell physiology has made this enzyme a paradigm for the study of self-compartmentalized proteases.[1,2] Identification and subsequent characterization of proteasomes in both Bacteria and Archaea domains have led to significant contributions in understanding the assembly, maturation, and catalytic activity of the 20S proteasome.[3] The benefit of studying prokaryotic 20S proteasomes lies in the relative simplicity of these enzymes. All 20S proteasomes are composed of four seven-membered rings, two exterior rings consisting of α subunits, and two interior rings composed of β subunits. Immunoelectron microscopy studies reveal that an archaeal 20S proteasome is composed of solitary α and β subunits; a finding in vast contrast with the complexity of 14 distinct α and β subunits found in eukaryotic 20S proteasomes.[4,5] This simplicity in terms of subunit composition has greatly facilitated the cloning and purification of archaeal proteasomes for experimental analysis.

Purification of Archaeal Proteasomes

To date, 20S proteasomes have been purified from three archaeal species: *Thermoplasma acidophilum*, *Pyrococcus furiosus,* and *Methanosarcina*

[1] O. Coux, K. Tanaka, and A. Goldberg, *Annu. Rev. Biochem.* **65,** 801 (1996).

[2] A. L. Goldberg, T. N. Akopian, A. F. Kisselev, D. H. Lee, and M. Rohrwild, *Biol. Chem.* **378,** 131 (1997).

[3] R. DeMot, I. Nagy, J. Walz, and W. Baumeister, *Trends Microbiol.* **7,** 88 (1999).

[4] A. Grziwa, W. Baumeister, B. Dahlmann, and F. Kopp, *FEBS Lett.* **290,** 186 (1991).

[5] B. Dahlmann, F. Kopp, P. Kristensen, and K. B. Hendil, *Arch. Biochem. Biophys.* **363,** 296 (1999).

Archaeal α proteasomal subunits

```
PHO  - MAFVPPQAGYDRAITVFSPDGRLFQVNYAREAVKRGATAVGVKCNEGVVLAVEKRITSRLIEPESYEKIFQIDDHIAAAS
PF   - MAFVPPQAGYDRAITVFSPDGRLFQVNYAREAVKRGATAVGVKCGEGVVLAVEKRITSRLIEPDSYEKIFQIDDHIAAAS
MJ   - MQMVPPS.AYDRAITVFSPEGRLYQVEYAREAVRRGTTAIGIACKDGVVLAVDRRITSKLVKIRSIEKIFQIDDHVAAAT
TA   - MQQGQ..MAYDRAITVFSPDGRLFQVEYAREAVKKGSTALGMKFANGVLLISDKKVRSRLIEQNSIEKIQLIDDYVAAVT
AF   - MHLPQ..MGYDRAITVFSPDGRLFQVEYAREAVKRGATAIGIKCKEGVILIADKRVGSKLLEADTIEKIYKIDEHICAAT
MTH  - MAFVPPQAGYDRAITVFSPDGRLFQVNYAREAVRGATAVGVKWKDGVVLAVEKRITSKLIEPSSYEKIFLIDDHIAAAP
MT   - MAPQ...MGYDRAITVFSPDGRLFQVEYAREAVKRGTTAVGIKAADGVVLLVDKRITSRLVEAESIEKIFQIDDHIGAAT
TKS  - MAFVPPQAGYDRAITVFSPDGRLFQVNYAREAVKRGATAVGVKWKDGVVLAVEKRITSKLIEPSSYEKIFLIDDHIAAAP

PHO  - SGIIADARVLVNRARLEAQIHRLTYGEPAPLAVIVKKICDLKQMHTQYGGVRPFGAALLMAGVNEK.PELYETDPSGAY
PF   - SGIIADARVLVNRARLEAQIYRLTYGEPAPVSVIVKKICDLKQMHTQYGGVRPFGAALLMAGINDR.PELYETDPSGAY
MJ   - SGLVADARVLIDRARLEAQIYRLTYGEEISIEMLAKKICDIKQAYTQHGGVRPFGVSLLIAGIDKNEARLFETDPSGAL
TA   - SGLVADARVLVDFARISAQQEKVTYGSLVNIENLVKRVADQMQQYTQYGGVRPYGVSLIFAGIDQIGPRLFDCDPAGTI
AF   - SGLVADARVLIDRARIEAQINRLTYDEPITVKELAKKICDFKQQYTQYGGVRPFGVSLLIAGVDEV.PKLYETDPSGAL
MTH  - SGIIADARVLVDRARLEAQIYRLTYGEPVPLTVLVKKICDLKQAHTQYGGVRPFGAALLMAGVNEK.PELFETDPSGAY
MT   - SGLVADARALVDRARVEAQVNRVSYDELIGVEVISKKICDHKQTYTQYGGVRPYGTALLIAGVDDKRPRLFETDPSGAL
TKS  - SGIIADARVLVDRARLEAQIYRLTYGEPVPLTVLVKKICDLKQAHTQYGGVRPFGAALLMAGVNEK.PELFETDPSGAY

PHO  - FAWKAVAIGSGRNTAMAIFEEKYKDDMSLEEAIKLAIFALAKTMEKPSAENIEVA.IITVKDKKFRKLSREEIEKYLNE
PF   - FAWKAVAIGSGRNTAMAIFEEKYRDDMNLEDAIKLALMALAKTMENPSADNIEVA.VITVKDKKFRKLTRDEIEKYLSE
MJ   - IEYKATAIGSGRPVVMELLEKEYRDDITLDEGLELAITALTKANEDIKPENVDVC.IITVKDAQFKKIPVEEIKKLIEK
TA   - NEYKATAIGSGKDAVVSFLEREYKENLPEKEAVTLGIKALKSSLEEGEELKAPEIASITVG.NKYRIYDQEEVKKFL
AF   - LEYKATAIGMGRNAVTEFFEKEYRDDLSFDDAMVLGLVAMGLSIESELVPENIEVGYVKVDDRTFKEVSPEELKPYVER
MTH  - FEWKAVAIGSGRNTAMAIFEEHYRDDIGKDDAIKLAILALAKTLEEPTAEGIEVAYI.TMDEKRWKKLPREELEKYLNE
MT   - LEYKATAIGAGRNAVVEVFEADYKEDMNIEAAILLGMDALYKAAEGKFDAGTLEVGVVSLQDKKFRKLEPEEVGNYVQQ
TKS- FEWKAVAIGSGRNTAMAIFEEHYRDDIGKDDAIKLAILALAKTLEEPTAEGIEVAYI.TMDEKRWKKLPREELEKYLNE

PHO  - VMKEVEEEEVKEKEEDYSELDSHY
PF   - VLKEVEEEEVKEKEEDYSELDSNY
MJ   - VKKKLNEENKKEEENREETKEKQEE
AF   - ANERIRELLKK
MTH  - ILQEVKEEEVEEKQEDYSELDQNY
MT   - ILEKHKETENKE
TKS  - ILQEVKEEEVEEKQEDYSELDQNY
```

FIG. 1. Amino acid sequence alignment of characterized and putative archaeal proteasomal subunits. Bold-faced type denotes residues conserved among the entire family of proteins. β subunit amino acids of particular significance to the catalytic mechanism of 20S proteasome are numbered relative to the *Thermoplasma acidophilum* sequence. PHO, *Pyrococcus horikoshii;* PF, *P. furiosus;* MJ, *Methanococcus jannaschii;* TA, *Thermoplasma acidophilum;* MTH, *Methanobacterium thermoautotrophicum;* TKS, *Thermococcus* species KS-1; MT, *Methanosarcina thermophila;* and AF, *Archaeoglobus fulgidus.*

thermophila.[6–8] The approximate abundance of proteasome relative to total protein in cells used for purification is 0.2% for *T. acidophilum,* 0.4% for *P. furiosus,* and 0.02% for *M. thermophila,* although growth *of M. thermophila* cells in media with a high saline content increases proteasome levels greatly.[9] To facilitate biochemical characterization of these enzymes,

[6] G. Puhler, S. Weinkauf, L. Bachmann, S. Muller, A. Engel, R. Hegerl, and W. Baumeister, *EMBO J.* **11,** 1607 (1992).

[7] M. W. Bauer, S. H. Bauer, and R. M. Kelly, *Appl. Environ. Microbiol.* **63,** 1160 (1997).

[8] J. A. Maupin-Furlow and J. G. Ferry, *J. Biol. Chem.* **270,** 28617 (1995).

[9] R. Barber, personal communication.

Archaeal β proteasomal subunits

```
              1           17          33
PHO1  -  MNRKT.....GTTTVGIKVKDGVILAADTQASLDHMVETLNIRKIIPITDRIAITTAGSVGDVQMIARILEAEARYYFA
PHO2  -  MLQLTEKFK.GTTTVGIVCKDGVVLAADRRASLGNIIYARNVTKIHKIDEHLAIAGAGDVGDILNLVRLLRAEAKLYYSQ
PF1   -  .........VLAADTQASLDHMVETLNIKKIIPITDRIAITTAGSVGDVQMLARYLEAEARYYFT
PF2   -  .................VLAADTRASLGNIIYAKNVTKIHKIDEHLAIAGAGDVGDILNLVRLLKAEANLYKST
MJ    -  MDVMK.....GTTTVGLICDDAVILATDKRASLGNLVADKEAKKLYKIDDYIAMTIAGSVGDAQAIVRLLIAEAKLYKMR
TA    -  MNQTLET...GTTTVGITLKDAVIMATERRVTMENFIMHKNGKKLFQIDTYTGMTIAGLVGDAQVLVRYMKAELELYRLQ
AF    -  MSMIEEKIYKGTTTVGLVCKDGVVMATEKRATMGNFIASKAAKKIYQIADRMAMTTAGSVGDAQFLARIIKIEANLYEIR
MTH   -  MNDKNTLK..GTTTVGITCKDGVVFATERRASMGNLIAHKATDKIFKIDEHIAATIAGSVADAQSLMKYLKAEAALYRMR
MT    -  MDNDKYLK..GTTTVGVVCTDGIVLASEQRATMGHFIASKTAKKVYQIDDLVGMTTAGSVGDAQQLVRLVSVESQLYKMR


PHO1  -  WGRPMTTKAMANLLSNILNENKWFPYLVQIIIGGYVDEPT..IANLDPYGGLIFDNYTAT.GSGTPFAIAILEEGYKENL
PHO2  -  SGKRMSVKALATLLANIMNGAKYFPYLAWFLVGGYDEKPK..LYSVDMVGGITEDKYATA.GSGMEFAYSILDSEYKDNL
PF1   -  WGRPMTTKAMANLLSNILNENRWFPYLVQIIIGGYVDEPT...ANLDPFGGLIFDDYTAT.GSGTPFAIAVLEEGYREDL
PF2   -  VGKEMSVKALATLLANILNGSKYFPYLGWFLVGGYDEKPR..LFSVDMVGGITEDNYAAA.GSGMEFAYSILDSEYREEM
MJ    -  TGRNIPPLACATLLSNILHSSRMFPFLTQIIIGGYDLLEGAKLFSLDPLGGMNEEKFFTATGSGSPIAYGVLEAGYDRDM
TA    -  RRVNMPIEAVATLLSNMLNQVKYMPYMVQLLVGGIDTAPH..VFSIDAAGGSVEDIYAS.TGSGSPFAYGVLESQYSEKM
AF    -  RERKPTVRAIATLTSNLLNSYRYFPYLVQLLIGGIDSEGKS.IYSIDPIGGAIEEKDIVATGSGSLTAYGVLEDRFTPEI
MTH   -  NSEKISIEAAAALAANILHSSRFYPFIVQTLLGGVDEN.GAKIYSLDPSGGMIPDKFVS.TGSGSPVAYGVLEDRYSDEL
MT    -  RDESMTIKGITTLMSNFLSRNRYYPMMVQLLIGGVDKNGPG.IYSLDAMGGSIEETRISATGSGSPMAYGVLEDQYRENM


                166
PHO1  -  GIEEAKELAIKAIKAAGSRDVYTGSKKVQVVTITKEGMQEEFIKLMKAKPKKKTTKRSRRKSK
PHO2  -  TLEEGIKLAVKAINTAIKRDVFSGDG.ILVVTITKEGYKELSDSELEATLKQ
PF1   -  TIEEAKELAIRAVRAAGRRDVYTGSKKVQVVTITKDGMKEEFV
PF2   -  SVNDGIKLAVKAINVAIKRDVFTGDG.LLVVTITKDGYKE
MJ    -  SVEEGIKLALNALKSAMERDTFSGNG.ISLAVITKDGVKIFEDEEIEKILDSL
TA    -  TVDEGVDLVIRAISAAKQRDSASGGM.IDVAVITRKDGYVQLPTDQIESRIRKGLGIFRK
AF    -  GVDEAVELAVRAIYSAMKRDSASGDG.IDVVKITEDEFYQYSPEEVEQILAK
MTH   -  YVDEAVDVAIRAIKSAMERDTYSGNG.ILVATVTEEEGFRMLSEEEIQKRIENLN
MT    -  TVKEGLDLAIRAIHNATKRDSASGEN.IDVVVVITKEAFKRLDPEEVKSRRALLN
```

FIG. 1. (*continued*)

genes encoding proteasomal subunits from both *T. acidophilum* and *M. thermophila* have been cloned into expression vectors and mature, active proteasomes can be purified from *Escherichia coli*.[10,11] Genes encoding archaeal proteasome subunits have been identified both by subsequent analysis of purified enzymes and through genomic sequencing. A compilation of open reading frames encoding either characterized or putative archaeal proteasome subunits (Fig. 1) reveals that *Pyrococcus horikoshii* and *Pyrococcus furiosus* have multiple (two) genes encoding β subunits. The significance of multiple β subunits in particular archaeal species has yet to be determined. Purification schemes for archaeal proteasomes from both authentic sources and *E. coli* employ gel-filtration and anion-exchange chromatography, generally following an ammonium sulfate precipitation. Gel filtration is performed routinely utilizing Superose 6B, Superdex 200, or Sephacryl S300, whereas anion-exchange chromatography is performed

[10] P. Zwickl, F. Lottspeich, and W. Baumeister, *FEBS Lett.* **312,** 157 (1992).
[11] J. A. Maupin-Furlow, H. C. Aldrich, and J. G. Ferry, *J. Bacteriol.* **180,** 1480 (1998).

with a variety of supports, including Q-Sepharose, DEAE-Sepharose, hydroxylapatite, and Mono Q. Typically, proteasome activity is monitored throughout purification by measuring the hydrolysis of a "chymotryptic-like" fluorogenic synthetic peptide. This assay is discussed later in a section detailing the catalytic mechanism of the 20S proteasome.

Purification of M. thermophila 20S Proteasome and Independent Proteasome Subunits by Expression in E. coli

High-level expression of *M. thermophila* 20S proteasome can reduce or prevent growth in overnight cultures of *E. coli* cells. As a result, the proteasome is produced routinely in *E. coli* BL21(DE3) using the bacteriophage T7 RNA polymerase–promoter system. Freshly transformed cells are inoculated into Luria–Bertani medium supplemented with ampicillin (100 mg/ liter) and grown at 30° and 200 rpm until cells reach an A_{600} of approximately 0.7. Isopropyl-β-D-thiogalactopyranoside is added to 0.4 mM, and cultures are maintained under growth conditions for 3 hr. Cells are harvested by centrifugation at 5000g for 15 min at 4° and then stored at $-70°$.

Step 1. Preparation of Cell Extracts. Frozen cell pellets are thawed and resuspended in six volumes (w/v) of 20 mM Tris–HCl buffer (pH 7.2) containing 1 mM dithiothreitol (DTT). The suspension is passed through a French pressure cell at 20,000 lb/in^2 twice, followed by centrifugation at 16,000g for 30 min at 4°.

Step 2. Ammonium Sulfate Precipitation. The supernatant is equilibrated at 4° for 30 min with ammonium sulfate added to a final concentration of 60%. The precipitate is removed by centrifugation at 10,000g for 15 min at 4°. The supernatant is equilibrated at 85% ammonium sulfate for 1 hr, and proteins are centrifuged at 10,000g for 30 min at 4°. Resulting protein pellets are resuspended in 5–10 ml of Tris buffer (pH 7.2) and dialyzed twice against 2 liters of Tris buffer at 4° for 18 hr. Note that attempts to concentrate the cell extract prepared in step 1 can cause inadvertent precipitation of the proteasome by the initial ammonium sulfate precipitation.

Step 3. Anion-Exchange Chromatography. The protein suspension is applied to a Q-Sepharose (Pharmacia) column (2.5 × 28.5 cm) equilibrated previously with Tris buffer (pH 7.2) containing 200 mM NaCl. Following sample loading, the column is developed with a linear NaCl gradient from 200 to 350 mM in 80 ml of Tris buffer. Q-Sepharose fractions of interest are pooled, precipitated with 85% ammonium sulfate, resuspended in a minimal volume of Tris buffer, and dialyzed overnight against 2 liters of Tris buffer.

Step 4. Gel Filtration. The protein suspension is applied to a Superose 6 HR 10/30 column (Pharmacia) equilibrated with Tris buffer (pH 7.2) containing 150 mM NaCl. Protein fractions are assayed for proteolytic activity fluorometrically using succinyl-LLVY-7-amido-4-methylcoumarin as a substrate (see later). Protein concentrations are determined by the bicinchoninic acid method with bovine serum albumin as a standard.[12]

Purification of independently expressed α subunits follows the identical protocol as used for the 20S proteasome holoenzyme, however, slight modifications are made when purifying independent β subunits. In step 2, the independent β subunit is precipitated between ammonium sulfate concentrations of 30 to 60%. In step 3, the β subunit-containing protein suspension is applied to a Q-Sepharose column equilibrated with Tris buffer and developed with a linear NaCl gradient (0 to 50 mM NaCl in 20 ml of Tris buffer). Since neither α nor β subunits exhibit proteolytic activity alone, protein purification is monitored by SDS–PAGE electrophoresis.

Assembly and Maturation

20S proteasomes undergo a complex assembly and maturation process; a process considered important for preventing spurious proteolysis within the cell. Studies using both *T. acidophilum* and *M. thermophila* enzymes have elucidated intermediates in the assembly and maturation pathway of the archaeal 20S proteasome.[11,13] Assembly of the archaeal 20S proteasome is dependent on α subunits forming heptameric rings that serve as scaffolding for the subsequent assembly of heptameric rings containing β subunits. Deletion studies show that a highly conserved sequence located among the N-terminal 34 amino acids of archaeal α subunits (Fig. 1) is essential for formation of the α subunit ring.[13] β subunits are synthesized in an inactive form with a prosequence at the N terminus (pro-β subunit) that is removed on assembly. Maturation of the 20S proteasome involves autocatalyzed removal of the β subunit propeptide, which exposes the active site nucleophilic threonine residue that is a defining characteristic of all 20S proteasomes.[14] Although the β subunit propeptide is important for assembly of the eukaryotic 20S proteasome, deletion of the propeptide from archaeal enzymes reveals that these sequences are dispensable for assembly and maturation.[11,13]

The enzymatic requirements for progressing through this assembly and maturation pathway are inherent to the amino acid sequence of the archaeal

[12] P. K. Smith, R. I. Krohn, G. T. Hermanson, A. K. Mallia, F. H. Gartner, M. D. Provenzano, E. K. Fujimoto, N. M. Goeke, B. J. Olson, and D. C. Klenk, *Anal. Biochem.* **150,** 76 (1985).
[13] P. Zwickl, J. Kleinz, and W. Baumeister, *Struct. Biol.* **1,** 765 (1994).
[14] E. Seemuller, A. Lupas, D. Stock, J. Lowe, R. Huber, and W. Baumeister, *Science* **268,** 579 (1995).

proteasome as the enzyme can be assembled *in vitro* from purified α and pro-β subunits. Equimolar ratios of independently purified *M. thermophila* α and pro-β subunits assemble spontaneously into active 20S proteasomes in Tris buffer at pH 7.2.[11] Similar experiments with *T. acidophilum* 20S proteasome require the subunits to be treated at low pH followed by dialysis at neutral pH for assembly and maturation.[15]

Catalytic Mechanism of Archaeal 20S Proteasome

Defined classically as a multicatalytic proteinase complex, the 20S proteasome exhibits multiple substrate specificities as observed with both peptide and protein substrates. The multicatalytic nature of this enzyme coupled with sequence data and inhibitor studies suggested that the proteasome is a novel protease.[16,17] As a result, the catalytic mechanism of the 20S proteasome has been an area of intense research since this enzyme was discovered.

Site-directed mutagenesis and structure determination have implicated the N-terminal threonine exposed on autocleavage of the β subunit as the active site nucleophile involved in proteasome-mediated peptide bond hydrolysis.[11,14,17,18] Furthermore, these studies have suggested a function for β subunit residues Glu(Asp)-17, Lys-33, and Asp-166 in the catalytic network of this enzyme; however, their precise roles are undefined. These experimental insights have been useful in constructing models for how the 20S proteasome degrades proteins; however, several specifics of proteasome-catalyzed proteolysis remain unclear.

For instance, one novel aspect of the proteasome is the processive nature of this enzyme.[19] Most proteases release substrates following peptide bond cleavage, only to bind the substrate again for subsequent cleavage reactions. However, the 20S proteasome does not relinquish a protein until the entire molecule is threaded through the proteasome. The mechanism for this processive protein degradation by the archaeal proteasome and how substrate is presented in the catalytic core of 14 active sites of this enzyme remains undetermined.

Cleavage of Synthetic Peptides

A variety of commercially available short synthetic peptide substrates can be used to detect proteasome activity. Substrate preferences of this

[15] E. Seemuller, A. Lupas, and W. Baumeister, *Nature* **382,** 468 (1996).
[16] B. Dahlmann, L. Kuehn, A. Grziwa, P. Zwickl, and W. Baumeister, *Eur. J. Biochem.* **208,** 789 (1992).
[17] L. Ditzel, D. Stock, and J. Lowe, *Biol. Chem.* **378,** 239 (1997).
[18] J. Lowe, D. Stock, B. Jap, P. Zwickl, W. Baumeister, and R. Huber, *Science* **268,** 533 (1995).
[19] T. N. Akopian, A. F. Kisselev, and A. L. Goldberg, *J. Biol. Chem.* **272,** 1791 (1997).

enzyme are discussed in terms of "chymotrypsin-like," "trypsin-like," and "postglutamyl" depending on the nature of short synthetic peptide substrates used under assay conditions. Routinely, succinyl-LLVY-7-amino-4-methylcoumarin (Suc-LLVY-AMC), *N*-carbobenzyloxy-LLE-β-naphthylamide (Cbz-LLE-βNa), and *N*-benzoyl-FVR-7-amino-4-methylcoumarin (Benz-FVR-AMC) are used as substrates indicative of chymotrypsin-like, postglutamyl, and trypsin-like activities, respectively. The assignment of labels corresponding to substrate preferences has been an unfortunate consequence, as experiments monitoring proteasome-mediated protein degradation contradict the suggested substrate preference that is observed with short synthetic peptides.[20]

Fluorescence Detection of 7-Amino-4-methylcoumarin Formation in Assays for Proteasome Activity[21]

Stock solutions of peptide substrates are made at 50 mM in dimethyl sulfoxide and stored at $-20°$ with desiccant. Substrate stock solutions are diluted into 20 mM Tris–HCl, pH 7.2, to a fixed concentration between 10 and 50 μM. Final dimethyl sulfoxide concentrations in the assay should not exceed 0.4%. Assays (300 μl volume) are preincubated in a 1.5-ml Eppendorf tube at 60° for 15 min. Temperature optima vary for archaeal proteasomes ranging from 70–75° for the *M. thermophila* enzyme to above 90° for the *T. acidophilum* and *P. furiosus* proteasomes; however, temperatures between 37 and 65° are routinely used for convenience. Reactions are initiated by the addition of 1–2 μg of enzyme and terminated by the addition of 100 μl of a stop solution (0.25 g of sodium acetate trihydrate dissolved in 4.375 ml of 1 M acetic acid diluted to 25 ml with water). Reactions are diluted at least 1 : 10 in water, and fluorescence is measured at an excitation wavelength of 350 nm and an emission wavelength of 460 nm. Blanks are prepared without the addition of enzyme, and a standard curve is generated with 7-amino-4-methylcoumarin.

Use of Colorimetric Substrates for Detection of Proteasome Activity

Several commercially available peptide substrates produce β-naph-thylamide when hydrolyzed, which is measured either colorimetrically by coupling with a diazonium salt or fluorometrically.[22,23] Assay preparation

[20] D. Leibovitz, Y. Koch, M. Fridkin, F. Pitzer, P. Zwickl, A. Dantes, W. Baumeister, and A. Amsterdam, *J. Biol. Chem.* **270,** 11029 (1995).

[21] A. J. Rivett, P. J. Savory, and H. Djaballah, *Methods Enzymol.* **244,** 331 (1994).

[22] R. G. Martinek, L. Berger, and D. Broida, *Clin. Chem.* **10,** 1087 (1964).

[23] J. K. McDonald, S. Ellis, and T. J. Reilly, *J. Biol. Chem.* **241,** 1494 (1966).

is essentially identical to that used for substrates conjugated to 7-amino-4-methylcoumarin. For colorimetic determination of β-naphthylamine production, the assay is terminated by the addition of 100 μl of 10% trichloroacetic acid and dilution to 1.0 ml with buffer. Following assay termination, 0.5 ml of 0.2% sodium nitrite is added, mixed well, and incubated at room temperature for 3 min. Subsequently, 1.0 ml of 0.5% ammonium sulfamate is added, mixed well, and incubated at room temperature for 2 min. Finally, 2.0 ml of 0.05% N-(1-naphthyl)ethylenediamine hydrochloride is added, mixed well, and incubated at room temperature for at least 18 min. Blanks are prepared without the addition of enzyme, and a standard curve is generated using β-naphthylamide. The color formed is stable up to 2 hr and measurements are made at 585 nm. The generation of β-naphthylamine can also be measured fluorometrically using an excitation wavelength of 335 nm and an emission wavelength of 410 nm. However, preparation of a standard curve using β-naphthylamine in LAP calibration standard (Sigma) shows that linearity is only achieved over a short range of low concentrations.

20S Proteasome Is a Conformationally Dynamic Enzyme

Kinetic studies using short synthetic peptide substrates have documented the conformationally dynamic nature of the 20S proteasome. Similar to its eukaryotic counterpart, the archaeal 20S proteasome exhibits both substrate inhibition and hysteresis when kinetic properties of its chymotryptic-like activity are assayed with Suc-LLVY-AMC (Fig. 2).[24] Initial rates of velocity begin to decrease at Suc-LLVY-AMC concentrations above 50 μM. In addition, reaction progress curves become biphasic at similar substrate concentrations and are characterized by initial velocities that decay by a first-order process before finally reaching steady-state velocities. These results suggest that the 20S proteasome is a dynamic enzyme adopting new conformations as ligands bind to multiple active sites.

The fluidity of the 20S proteasome is also evident in its response to various agents such as salt and detergent when assayed with short synthetic peptides. In the past, the eukaryotic 20S proteasome was purified in a latent state, but could be activated on exposure to low concentrations of SDS or other treatments.[1] Although active as purified, archaeal proteasomes respond to similar treatments with observable differences in peptide hydrolysis.[11,16] Monovalent cation salts, most divalent cation salts, and low concentrations of detergent stimulate the 20S proteasome-mediated hydrolysis of succinyl-LLVY-7-amino-4-methylcoumarin. However, similar treatments

---

[24] R. L. Stein, F. Melandri, and L. Dick, *Biochemistry* **35,** 3899 (1996).

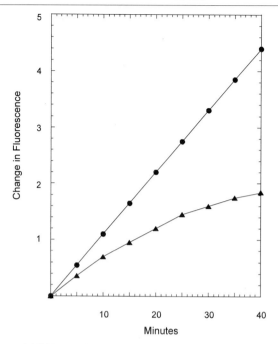

Fig. 2. Substrate inhibition and hysteretic behavior of *M. thermophila* 20S proteasome at elevated Suc-LLVY-AMC concentrations. 20S proteasome was assayed under continuous conditions for chymotrypsin-like activity at 42° in a temperature-controlled jacketed cuvette holder using a Hitachi F-2000 fluorescence spectrophotometer. Changes in fluorescence due to the formation of free 7-amino-4-methylcoumarin are measured in arbitrary units: (●) 50 μM or (▲) 150 μM Suc-LLVY-AMC assayed.

inhibit the "postglutamyl" activity of the archaeal proteasome. One particular divalent cation that has a drastic effect on 20S proteasome activity is zinc. Using the fluorogenic probe, 8-anilino-1-naphthalenesulfonic acid, the archaeal 20S proteasome responds to the presence of zinc by increasing its exposed hydrophobic surface, presumably inactivating the enzyme.[9] Analogous experiments with similar results have been performed using the chaperonin GroESL.[25] These distinct responses to various agents offer further support that the archaeal proteasome is a conformationally flexible protein that adapts to the binding of ligands.

Protein Degradation

20S proteasome proteolysis is measured routinely by the rate of appearance of new amino termini detected using either ninhydrin or fluorescamine.

[25] B. T. Brazil, J. Ybarra, and P. M. Horowitz, *J. Biol. Chem.* **273,** 3257 (1998).

This strategy has proven particularly useful in determining substrate degradation rates for the archaeal proteasome.[26] One consideration in using these detection reagents is that determination of new amino termini requires use of buffers lacking primary amines. Thus, for these assays, 50 mM HEPES, pH 7.2, replaces Tris, which has been described here for use in previous assays.

Preparation of Substrate[27]

Although proteolytic activity can be measured for any protein substrate, casein is used routinely as it has little secondary structure. Stock solutions of either α- or β-casein are made dissolving 25 mg of casein with 700 μl of water and 100 μl of 0.2 N NaOH. The solution is vortexed until clear, and the solution is adjusted to pH 7.8 with 0.1 N HCl. A final working stock of 25 mg casein/ml is made by diluting to 1.0 ml and heating at 90° for 15 min prior to use. Casein stock solutions must be made fresh each time to ensure reliability.

Use of Ninhydrin for Determination of Proteolytic Rates[28,29]

Additional reagents required for use in ninhydrin-based assays include (i) potassium cyanide/sodium acetate buffer solution prepared by diluting 2.0 ml of a 0.01 M KCN solution into 200 ml of 4 M sodium acetate buffer (pH 5.2), (ii) 3% solution of ninhydrin dissolved in methyl cellosolve (ethylene glycol monomethyl ether), (iii) 1 : 1 mixture of 2-propanol and water, and (iv) 10% trichloroacetic acid (TCA). Reaction mixtures containing 50 mM HEPES, pH 7.2, and 0.1 to 1.0 mg of casein substrate are preincubated at 60° for 15 min. The assay is initiated by adding 5–10 μg of enzyme to the reaction mixture (final volume of 150 μl) and is terminated by the addition of 50 μl of 10% TCA. An assay where added enzyme has been excluded is prepared for use as a blank. Precipitated protein is removed by centrifugation, and 100 μl of the supernatant is mixed with 200 μl of the potassium cyanide/sodium acetate buffer solution in a glass test tube. One hundred microliters of 0.3% ninhydrin solution is added, and the tubes are placed in a boiling water bath for 10 min. The samples are cooled and diluted to 1 ml with the 1 : 1 mixture of 2-propanol/water. The absorbance of ninhydrin-reacting material is determined at 570 nm. The determination of degradation product amounts is provided by a standard curve generated under the same conditions using various concentrations of leucine.

[26] A. F. Kisselev, T. N. Akopian, and A. L. Goldberg, *J. Biol. Chem.* **273,** 1982 (1998).
[27] E. H. Reimerdes and H. Klostermeyer, *Methods Enzymol.* **45,** 26 (1976).
[28] H. Rosen, *Arch. Biochem. Biophys.* **67,** 10 (1957).
[29] M. Orlowski and C. Michaud, *Biochemistry* **28,** 9270 (1989).

Use of Fluorescamine for Determination of Proteolytic Rates[30]

Additional reagents required for use of fluorescamine include 100 mM borate buffer (pH 9.2) and fluorescamine dissolved in acetone (0.3 mg/ml). Assays are prepared essentially as described earlier; however, the reactions are stopped by immersing tubes in ice and then adding 1.35 ml of borate buffer. Fluorescamine (200 μl) is added and the solution is vortexed. Fluorescence is measured with an excitation wavelength of 395 nm and an emission wavelength of 475 nm. Similar to the ninhydrin assay, a standard curve is generated using various amounts of leucine.

Isolation of Degradation Products

High-performance liquid chromatography (HPLC) has been used to separate peptide products generated by proteasome using a reversed-phase C_{18} column. This procedure allows an independent means of determining peptide-bond hydrolysis rates by integrating absorbance peaks resulting from HPLC-mediated separation. Identities of peptide products can be determined by either amino acid sequencing or use of matrix-assisted laser desorption ionization time of flight (MALDI-TOF). Studies examining products formed by archaeal proteasome-mediated protein degradation have both supported and contradicted a model suggesting the catalytic mechanism of the 20S proteasome functions as a molecular ruler.[20,26,31]

Cellular Function of Archaeal Proteasomes

In contrast to the central role of the 20S proteasome in eukaryotes, the physiological role of the archaeal 20S proteasome is unclear. The absence of amenable genetic systems has prevented the construction of loss of function alleles in archaeal species. However, use of a proteasome-specific inhibitor, carboxybenzylleucylleucylleucine vinyl sulfone, suggests that the 20S proteasome is essential for *T. acidophilum* under stress conditions.[32] Also, expression patterns of the *M. thermophila* 20S proteasome imply that this enzyme may have a cellular function in cells grown in the presence of high salt.[9] Whether the 20S proteasome is responsible for the turnover of damaged proteins or degrades specific target proteins under these growth conditions remains to be seen.

In addition to the role of 20S proteasome in archaea, the physiological form of this enzyme is also unknown. In eukaryotes, the 20S proteasome

[30] A. L. Goldberg, R. P. Moerschell, C. H. Chung, and M. R. Maurizi, *Methods Enzymol.* **244,** 350 (1994).
[31] T. Wenzel, C. Eckerskorn, F. Lottspeich, and W. Baumeister, *FEBS Lett.* **349,** 205 (1994).
[32] A. Ruepp, C. Eckerskorn, M. Bogyo, and W. Baumeister, *FEBS Lett.* **425,** 87 (1998).

interacts with various cellular factors that modulate the proteolytic activity of this enzyme. Most notably, eukaryotic 20S proteasomes interact with 19S protein complexes containing various ATPases to form the 26S proteasome, an enzyme that is responsible for the turnover of ubiquinated proteins. It has been expected that a similar targeting mechanism would be found in Archaea, particularly since open reading frames with significant sequence similarity to these ATPases have been found in archaeal genomes. However, stable higher order proteasome-containing complexes have yet to be identified either *in vitro* or *in vivo*. Additionally, reports of ubiquitin in *T. acidophilum,* ubiquitin-like sequences in *M. thermophila,* and a role for ubiquitin in archaeal 20S proteasome function remain unsubstantiated *in vivo.*[8,33,34] Thus, how proteins are targeted for archaeal 20S proteasome degradation together with identifying physiological substrates remain important questions that need to be addressed.

Acknowledgments

This work was supported by grants from the National Science Foundation. RDB acknowledges the support of the National Institutes of Health National Research Service Award.

[33] S. Wolf, F. Lottspeich, and W. Baumeister, *FEBS Lett.* **326,** 42 (1993).
[34] T. Wenzel and W. Baumeister, *FEBS Lett.* **326,** 215 (1993).

[29] Thiol Protease from *Thermococcus kodakaraensis* KOD1

By Masaaki Morikawa and Tadayuki Imanaka

Introduction

Thiol proteases (cysteinyl proteinases) are grouped into three families.[1] The largest and most familiar is the papain superfamily. Mammalian lysosomal thiol protease, cathepsins B, L, and H, are homologous with papain.[2] Calcium-dependent thiol proteases, calpains, are also a member of this group.[3] A second superfamily is composed of bacterial proteases, including

[1] E. Shaw, *Adv. Enzymol.* **63,** 271 (1990).
[2] A. Ritonja, T. Popovic, M. Kotnik, W. Machleidt, and V. Turk, *FEBS Lett.* **228,** 341 (1988).
[3] S. Ohno, Y. Emori, S. Imajoh, H. Kawasaki, M. Kisaragi, and K. Suzuki, *Nature* **312,** 566 (1984).

Streptococcus proteinase and *Clostridium* proteinase, clostripain.[4,5] Their primary structures are different from papain. A third group of thiol proteases are represented by viral coded proteases, 2A and 3C.[6] These are relatively small proteases, 17 and 21 kDa, respectively. The hyperthermophilic archaeon, *Thermococcus kodakaraensis* KOD1 (previously classified as *Pyrococcus*), was isolated from Kodakara Island, Kagoshima, Japan.[7] This strain produces at least three kinds of extracellular proteases with different molecular weights. Described here is the isolation and characterization of thiol protease from *T. kodakaraensis*.

Culture Conditions

Thermococcus kodakaraensis KOD1 is cultivated in a half-strength (0.5×) 2216 marine broth.[8] The medium contains (per liter) 18.7 g 2216 marine broth (Difco, Detroit, MI), 3.48 g PIPES (Dojindo, Kumamoto, Japan) buffer, 0.725 g $CaCl_2 \cdot 2H_2O$, 0.4 ml 0.2% resazurin, 475 ml (w/v) artificial seawater (ASW), and 500 ml distilled water, and the pH is adjusted to 7.0 with NaOH. ASW consists of (per liter) 28.16 g NaCl, 0.7 g KCl, 5.5 g $MgCl_2 \cdot 6H_2O$, and 6.9 g $MgSO_4 \cdot 7H_2O$. The broth, in screw-capped culture bottles, is autoclaved and left standing in an anaerobic chamber (filled with anaerobic gas mixture) immediately after autoclaving. $Na_2S \cdot 9H_2O$ (final concentration of 400 μM) and 10 g of sulfur (sterilized separately at 105° for 20 min) are added prior to inoculation. The bottles are sealed and placed in an incubator overnight, usually at 90°. Other buffers, such as MES (pH 5–6.5), HEPES (pH 7–8.5), or CHES (pH 9) (Dojindo), are used instead of PIPES to prepare media at various pH values, where appropriate.

Continuous Culture

For the preparation of enzyme samples, cells are grown in a five-neck round-bottom flask (working volume, 1 liter) maintained at 94°, with a heating mantle (Daiken Electric Co., Tokyo, Japan). The flask is sparged continuously with nitrogen gas.[9] The exhaust gas is connected to a condenser to avoid water loss, and H_2S is removed on exiting the reactor with a

[4] J. Y. Tai, A. A. Kortt, T.-Y. Liu, and S. D. Elliott, *J. Biol. Chem.* **251,** 1955 (1976).
[5] A.-M. Gilles, A. Lecroisey, and B. Keil, *Eur. J. Biochem.* **145,** 469 (1984).
[6] R. T. Libby, D. Cosman, M. K. Cooney, J. E. Merriam, C. J. March, and T. P. Hopp, *Biochemistry* **27,** 6262 (1988).
[7] M. Morikawa, Y. Izawa, N. Rashid, T. Hoaki, and T. Imanaka, *Appl. Environ. Microbiol.* **60,** 4559 (1994).
[8] S. Belkin and H. W. Jannasch, *Arch. Microbiol.* **141,** 181 (1985).
[9] S. H. Brown and R. M. Kelly, *Appl. Environ. Microbiol.* **55,** 2086 (1989).

solution containing 3 N NaOH. Cells are pregrown in a batch mode. When the growth reaches the late-log phase (ca. 10^8 cells/ml), fresh medium is added and then excess culture is removed through a Masterflex Teflon tubing pump (Barnant Co., Barrington, IL) at a dilution rate of 0.15 hr^{-1}. A reservoir tank for culture is maintained in an ice bath. Approximately 100 liters of culture is collected for protease purification.

Assay of Proteolytic Activity

Proteolytic activity is measured by the hydrolysis of azocasein.[10] The reaction mixture consists of 5 mg/ml azocasein in 600 μl buffer (50 mM Tris–HCl buffer, pH 7.5, 5 mM CaCl$_2$) and 120 μl enzyme solution. The reaction is completed at 80° for 20 min, unless otherwise stated. A heating block is used at temperatures higher than 60°. The reaction is stopped by adding 480 μl of 15% trichloroacetic acid and then the sample is placed on ice. The sample is centrifuged at 18,500g at 4° for 10 min to remove the precipitate. After adding 30 μl of 10 N NaOH to the supernatant, A_{440} is measured with a UV-160 spectrophotometer (Shimadzu, Kyoto, Japan). A sample in which trichloroacetic acid was added prior to the enzyme solution is used as a reference. One unit of proteolytic activity is determined as a change of 0.1 absorbance unit at 80° for 20 min. The amount of protein in the sample is estimated by the bicinchoninic acid assay (Pierce, Rockford, IL).

Proteolytic Activities in Cell Extract and Culture Supernatant

Prior to purification efforts, the proteolytic activities of culture superna-tant and cell extracts are measured at various temperatures. The proteolytic activity of the culture supernatant (extracellular fraction) is higher at 100° than that at 80 or 90°, whereas the activity of the cell extract (intracellular fraction) at 100° is lower than that at 80 or 90° (data not shown). This suggests that proteases in the culture supernatant are more thermostable than those in the cell extract. The temperature profile of proteolytic activity in the extracellular fraction shows a maximum value at 105° with a clear shoulder at 80°. The temperature profile implies that there are at least two proteases whose optimum temperatures are lower and higher than 100°, respectively.

[10] I. I. Blunmentals, A. S. Robinson, and R. M. Kelly, *Appl. Environ. Microbiol.* **56,** 1992 (1990).

Activity Staining of Protease

After conventional sodium dodecyl sulfate (SDS)–polyacrylamide gel electrophoresis, the gel is washed in 2.5% Triton X-100, 50 mM sodium phosphate buffer (pH 7.0) for 1 hr to remove (SDS). The gel is put on filter paper that has been soaked in a 50 mM sodium phosphate buffer (pH 7.0). Twelve percent polyacrylamide gel (1 mm in thickness) containing 0.5% gelatin is placed on the original gel. Then, a filter paper and a pack of absorbent papers are placed on the top gel. After being wrapped with a hybridization bag, this gel sandwich is pressed with a 5-mm-thick glass plate. Protease is transferred onto the gelatin gel, and the proteolytic reaction is carried out in a 75° incubator for 16 hr. The gel is stained with 0.1% amido black in 100 ml of 30% methanol, 10% acetic acid, and 60% water. Protease bands are visualized as clear zones resulting from the hydrolysis of gelatin.[10] It was demonstrated that the strain produces at least three kinds of extracellular proteases (Fig. 1b).

Purification of Protease from Strain KOD1

Because the protease activity in the culture supernatant is more thermostable than that of the cell extract, it is chosen for further studies. The culture supernatant is prepared as follows. The broth is centrifuged (10,000g at 4° for 30 min) and concentrated to 1 liter (ca. 100-fold) by ultrafiltration (UF-10PS, Tosoh Co., Tokyo, Japan). Cell extracts are prepared by sonication and subsequent centrifugation (27,000g for 30 min) to remove cell debris. Proteolytic activities are measured as mentioned earlier. Purification starts with 1 liter of concentrated culture supernatant. This crude enzyme sample is brought to 60% ammonium sulfate saturation and kept at 4° overnight (Fig. 1a). The precipitate is collected by centrifugation at 27,000g at 4° for 30 min, dissolved in 25 mM Tris–HCl (pH 7.5) containing 5 mM CaCl$_2$, and dialyzed overnight against the same buffer. The dialyzate is chromatographed using TSKgel Toyopearl HW-55F (Tosoh Co., Tokyo, Japan) (2.0 diameter by 80 cm) at a flow rate of 1 ml/min. The elution pattern is monitored by A_{280}, and proteolytic activity is determined. Fractions containing significant proteolytic activity are pooled and applied to the second gel filtration column FPLC Superose 12 (Pharmacia, Uppsala, Sweden) at a flow rate of 0.5 ml/min. Final purification is performed by passage through a hydroxyapatite column in an isocratic mode (particle size 2 μm; 7.5 mm diameter by 100 mm, Tonen Co., Tokyo, Japan), which has been equilibrated with 5 mM sodium phosphate (pH 6.8). Impurities corresponding to larger molecular weight proteins in the sample are eluted by linear gradient of sodium phosphate from 5 to 200 mM. After hydroxyapatite

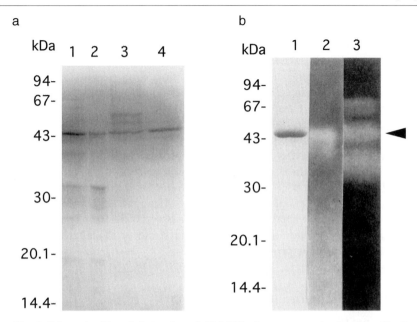

Fig. 1. Polyacrylamide gel electrophoresis (PAGE) of enzyme sample at each purification step. (a) Analysis by SDS–PAGE. The gel was stained by Coomassie Brilliant Blue R-250. Lane 1, after precipitation with ammonium sulfate; lane 2, after gel filtration chromatography on Toyopearl HW-55F; lane 3, after gel-filtration chromatography on Superose 12; and lane 4, after hydroxyapatite chromatography. (b) SDS–PAGE and activity staining. Lane 1, Coomassie Brilliant Blue staining of TT protease after hydroxyapatite chromatography; lane 2, activity staining of the same sample after blotting to a gelatin gel; and lane 3, activity staining of crude culture supernatant sample (only after precipitation with ammonium sulfate).

chromatography, the single 44-kDa protease band is confirmed to be one of the three major proteolytic enzymes of culture supernatant (Fig. 1b). The molecular size of this protease, namely TT protease, is estimated to be 45 kDa by gel-filtration chromatography, indicating that the enzyme is a monomer in its native form. Further biochemical analysis of TT protease is performed using this sample. Although rarely, a faint band corresponding to a protein of 42 kDa may be observed on an SDS–PAGE gel. The 42- and 44-kDa proteins can be separated using the Prosieve gel system (Takara Shuzo Co., Kyoto, Japan) or by reverse-phase HPLC. The two proteins have been proven to have identical N-terminal amino acid sequences, indicating that the 42-kDa protein is derived from the 44-kDa protein, with a C-terminal truncation. TT protease does not seem to be a glycoprotein because it can not be stained by periodate–Schiff reagent (data not shown). The specific activity is determined as 2160 U/mg protein at 80°.

Effect of pH on Proteolytic Activity

Purified enzyme and substrate are dissolved in 50 mM sodium acetate buffer (pH 4.0 to 6.0), sodium phosphate buffer (pH 6.5 to 8.0), or glycine–sodium hydroxide buffer (pH 8.5 to 11.0). Other reaction conditions are the same as described earlier. The enzyme is found to be a typical neutral protease whose optimum pH is 7.0 (Fig. 2a), and it retains high proteolytic activity even at 120° (Fig. 2b). Azocasein is found to be somewhat labile

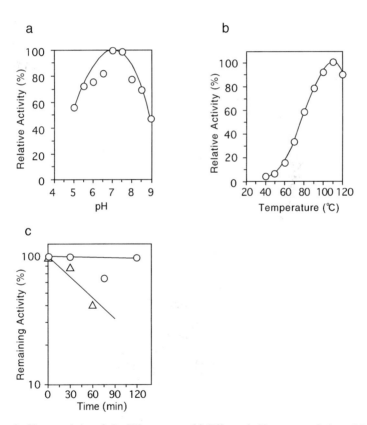

Fig. 2. Characteristics of the TT protease. (a) Effect of pH on proteolytic activity. (b) Effect of temperature on proteolytic activity. (c) Thermostability of TT protease. Sample was incubated at 90° (○) or 100° (△) and the remaining activity was then assayed at 80°.

TABLE I
EFFECT OF INHIBITOR ON PROTEOLYTIC ACTIVITY
OF TT PROTEASE

Inhibitor	Inhibitor class	Concentration (mM)	Inhibition (%)
pCMB	Cysteine	0.1	82
		1	80
E-64	Cysteine	0.1	25
		1	82
PMSF	Serine	5	0
DFP	Serine	10	38
EDTA	Metal ion	10	0

at temperatures over 100° after 20 min. Therefore, the A_{440} value of the reference sample is subtracted from that of each reaction mixture. The optimum temperature for TT protease activity is 110°, as shown in Fig. 2b.

Thermostability of Protease

The purified protease sample in buffer (50 mM Tris–HCl, pH 7.5, 8 mM cysteine, 5 mM CaCl$_2$) is incubated at 90 or 100° for various periods of time and is then put on an ice bath. Residual activity is measured at 80°. The enzyme is found to be very stable at 90° and has a half-life of 60 min at 100° (Fig. 2c).

Effect of Inhibitors on Proteolytic Activity

Enzyme samples are incubated at 37° for 15 min with several inhibitors: diisopropyl fluorophosphate (DFP), phenylmethanesulfonyl fluoride (PMSF), p-chloromercuribenzoic acid (pCMB), E-64 (thiol protease inhibitor produced by *Aspergillus japonicus* TPR-64[11]), or EDTA. Residual activity is then determined. Both pCMB and E-64, which are specific inhibitors for thiol proteases, strongly inhibit the proteolytic activity of this enzyme even at 1 mM (Table I). The enzyme is partially inhibited by 10 mM DFP (an inhibitor of serine protease). EDTA (a metalloprotease inhibitor) and PMSF (a serine protease inhibitor) are ineffective inhibitors for this protease. Proteolytic activity is enhanced three times when 8 mM of cysteine is added to the reaction mixture. These experimental results indicate that the

[11] K. Hanada, M. Tamai, M. Yamagishi, S. Ohmura, J. Sawada, and I. Tanaka, *Agric. Biol. Chem.* **42**, 523 (1978).

enzyme is a thiol protease. Some thiol proteases have been found to be partly inhibited by PMSF or DFP, indicating a similar active site configuration to serine proteases.[12,13] In fact, DFP partially inhibits TT protease.

Evolutional Relationship of Proteases

Most of the S^0-dependent hyperthermophilic heterotrophic archaea require complex proteinaceous substrates as a carbon source. This arises from their requirement for several essential amino acids[14] and suggests that the protease production is important for their growth. Several thermostable proteases have been reported for hyperthermophilic archaea and hyperthermophilic bacteria as shown in Table II. Most of the thermostable proteases, especially those that are extracellular or membrane bound, are serine proteases. In addition to the TT protease, strain KOD1 also produces a subtilisin-like serine protease, and the deduced amino acid sequence from the gene displays highest similarity (40% identity) with subtilisin *Carlsberg* from *Bacillus licheniformis* (unpublished data, 2000). This indicates that a close relationship exists between bacteria and archaea with respect to the evolution of subtilisin-like serine proteases. However, intracellular multicatalytic proteases have been purified from three thermophilic archaea: *P. furiosus,*[15] *Methanosarcina thermophila,*[16] and *Thermoplasma acidophilum.*[17] In *P. furiosus,* a major intracellular proteolytic activity has been identified as PfpI, a homomultimer consisting of 18.8-kDa subunits. An immunological relationship was observed between PfpI and eukaryotic proteasomes,[18] indicat-

[12] K. F. Bazam and R. J. Fletterick, *Proc. Natl. Acad. Sci. U.S.A.* **85,** 7872 (1988).

[13] I. G. Kamphuis, J. Drenth, and E. N. Baker, *J. Mol. Biol.* **182,** 317 (1985).

[14] T. Hoaki, M. Nishijima, M. Kato, K. Adachi, S. Mizobuchi, N. Hanzawa, and T. Maruyama, *Appl. Environ. Microbiol.* **60,** 2898 (1994).

[15] M. W. Bauer, S. H. Bauer, and R. M. Kelly, *Appl. Environ. Microbiol.* **63,** 1160 (1997).

[16] J. A. Maurpin-Furlow and J. G. Ferry, *J. Biol. Chem.* **270,** 28617 (1995).

[17] B. Dahlmann, L. Kuehn, A. Grziwa, P. Zwickl, and W. Baumeister, *Eur. J. Biochem.* **208,** 789 (1992).

[18] L. J. Snowden, I. I. Blumentals, and R. M. Kelly, *Appl. Environ. Microbiol.* **58,** 1134 (1992).

[19] M. Klingeberg, B. Galunsky, C. Sjoholm, V. Kasche, and G. Antranikian, *Appl. Environ. Microbiol.* **61,** 3098 (1995).

[20] R. Eggen, A. Geerling, J. Watts, and W. M. de Vos, *FEMS Lett.* **71,** 17 (1990).

[21] H. Connaris, D. A. Cowan, and R. J. Sharp, *J. Gen. Microbiol.* **137,** 1193 (1991).

[22] S. B. Halio, I. I. Blumentals, S. A. Short, B. M. Merrill, and R. M. Kelly, *J. Bacteriol.* **178,** 2605 (1996).

[23] R. Dib, J.-M. Chobert, M. Dalgalarrondo, G. Barbier, and T. Haertle, *FEBS Lett.* **431,** 279 (1998).

[24] Y. Sako, P. C. Croocker, and Y. Ishida, *FEBS Lett.* **415,** 329 (1997).

[25] D. A. Cowan, K. A. Smolenski, R. M. Daniel, and H. W. Morgan, *Biochem. J.* **247,** 121 (1987).

TABLE II

COMPARISON OF TT PROTEASE AND OTHER PROTEASES FROM HYPERTHERMOPHILIC ARCHAEA AND HYPERTHERMOPHILIC BACTERIA

Protease	Molecular mass (kDa)	Optimum pH	Optimum temperature (°)	Inhibition by				Active site	Thermostability	Origin	Refs.
				EDTA	PMSF	pCMB	Pepstatin				
TT protease	44	7.0	110	−	−	+	NT	Cysteine	100°, 60 min	*Thermococcus kodakaraensis*	7
	68	8.5	85	−	+	NT	NT	Serine	NT	*Thermococcus stetteri*	19
Pyrolysin	140, 130, 115, 100, 65	6.5–10.5	115	−	+	−	NT	Serine	100°, 4 hr	*Pyrococcus furiosus*	20, 21
PfpI (S66)	18.8n	7.0	105	−	+	NT	NT	Serine	98°, 33 hr	*Pyrococcus furiosus*	10, 22
	640 (25, 22)	6.7	95	NT	NT	NT	NT	?	NT	*Pyrococcus furiosus*	15
	150, 105, 60	9.0	95	−	+	NT	NT	Serine	100°, 165 min	*Pyrococcus abyssi*	23
Aeropyrolysin	52	6–8	110	+	+/−	−	NT	Metal ion?	120°, 2.5 hr	*Aeropyrum pernix*	24
Archaelysin	52	7.0	98	−	+	−	NT	Serine	95°, 70–90 min	*Desulfurococcus species*	25
STABLE	150	9.0	85	−	+/−	−	NT	Serine	100°, 10 min	*Staphylothermus marinus*	26
Thermopsin	45	2.0	75	−	−	−	+	Aspartate	NT	*Sulfolobus acidocaldarius*	27
	108 (54)	6.5–8	90	−	+	−	−	Serine	98°, 20 min	*Sulfolobus solfataricus*	28
	320 (80)	6.5	75	+	−	NT	NT	Metal ion?	86°, 15 min	*Sulfolobus solfataricus*	29
	650	8.5	90	−	NT	−	−	Serine?	NT	*Thermoplasma acidophilum*	17
	43	9.0	95	NT	+	NT	NT	Serine	105°, 6 hr	*Aquifex pyrophilus*	30
	>669 (31)	7.1	90	+/−	NT	NT	NT	?	95°, 36 min	*Thermotoga maritima*	31

NT, not tested; +, positive; −, negative; +/−, partially positive.

ing a close relationship between archaeal and eukaryotic proteasomes. Elucidation of the primary structure of the TT protease is expected to add further information on the evolution of thiol proteases in eukarya, bacteria, and viruses.

[26] J. Mayr, A Lupas, J. Kellermann, C. Eckerskorn, W. Baumeister, and J. Peters, *Curr. Biol.* **6,** 739 (1996).

[27] M. Fusek, X. Lin, and J. Tang, *J. Biol. Chem.* **265,** 1496 (1990).

[28] N. Burlini, P. Magnani, A. Villa, F. Macchi, P. Tottora, and A. Guerritore, *Biochim. Biophys. Acta* **1122,** 283 (1992).

[29] M. Hanner, B. Redl, and G. Stoffler, *Biochim. Biophys. Acta* **1033,** 148 (1990).

[30] I.-G. Choi, W.-G. Bang, S.-H. Kim, and Y. G. Yu, *J. Biol. Chem.* **274,** 881 (1999).

[31] P. M. Hicks, K. D. Rinker, J. R. Baker, and R. M. Kelly, *FEBS Lett.* **440,** 393 (1998).

[30] Proline Dipeptidase from *Pyrococcus furiosus*

By AMY M. GRUNDEN, MOUSUMI GHOSH, and MICHAEL W. W. ADAMS

Introduction

Prolyl residues are unique among the 20 common amino acid residues in that they confer a conformational constraint on peptide chains due to the cyclic nature of their pyrrolidine side group.[1,2] This conformational constraint prevents cleavage of bonds adjacent to proline by most proteases; however, several enzymes have been characterized that are capable of hydrolyzing these bonds. These enzymes include (a) proline-specific endopeptidases, which cleave peptides on the carboxyl side of prolyl residues located internally within a polypeptide (-X-Pro-/-X-); (b) prolyl aminopeptidases, which hydrolyze the bond between any N-terminal amino acid and a penultimate prolyl residue in peptides of varying lengths (NH_2-X-/-Pro—X-); (c) proline iminopeptidases, which promote cleavage of unsubstituted N-terminal prolyl residues from dipeptides, tripeptides, and polypeptides (Pro-/-X-); (d) proline-specific C-terminal exopeptidases, which release an amino acid from the C terminus of a polypeptide having a penultimate proline residue (-X—Pro-/-X-COOH); and (e) prolidases (proline dipeptidases), which only hydrolyze dipeptides with proline at the C terminus (NH_2-X-/-Pro-COOH). This cadre of proline-specific peptidases is thought to participate, in conjunction with other endo- and exopeptidases,

[1] D. F. Cunningham, *Biochim. Biophys. Acta* **1343,** 160 (1997).

[2] R. Walter, W. H. Simmons and T. Yoshimoto, *Mol. Cell. Biochem.* **30,** 111 (1980).

in the terminal degradation of cytoplasmic proteins and may also further aid in the recycling of proline.

Prolidases have been isolated from mammalian,[3–5] bacterial (*Lactobacillus*[6,7] and *Xanthomonas*[8]), and archaeal (*Pyrococcus*[9]) sources and appear to be ubiquitous in nature. Although the physiological role of the enzyme is unclear in both bacteria and archaea, its absence in humans results in abnormalities of the skin and other collagenous tissues.[10] Furthermore, prolidases have several potential biotechnological applications. One possible use is in the dairy industry as a cheese-ripening agent,[11] as removal of proline from proline-containing peptides in cheese reduces bitterness. A second application is as an enzyme to degrade organophosphorus acid-containing compounds present in a number of chemical warfare agents and pesticides. The organophosphorus acid-hydrolyzing capabilities of prolidase have been suggested by the discovery that the protein characterized as an organophosphorus acid anhydrolase (OPAA) and prolidase are one in the same.[12,13]

Assay

The prolidase assay described here is based on the quantitation of proline released from the hydrolysis of dipeptides that contain proline at the C terminus. The free proline concentration is determined by modification of the colorimetric ninhydrin method of Yaron and Mlynar.[14] The ninhydrin reagent used in this assay is prepared by adding ninhydrin [3% (w/v)] to a solution of 60% (v/v) glacial acetic acid and 40% (v/v) phosphoric acid. This mixture is heated at 70° for 30 min and should be straw-colored after cooling. The prepared ninhydrin reagent is light sensitive and should be

[3] P. Browne and G. O'Cuinn, *J. Biol. Chem.* **268,** 6147 (1983).
[4] F. Endo, A. Tanoue, H. Nakai, A. Hata, Y. Indo, K. Titani, and I. Matsuda, *J. Biol. Chem.* **264,** 4476 (1989).
[5] H. Sjostrom, O. Noren, and L. Josefsson, *Biochim. Biophys. Acta* **327,** 457 (1973).
[6] M. Booth, V. Jennings, I. N. Fhaolain, and G. O'Cuinn, *J. Dairy Res.* **57,** 245 (1990).
[7] M. D. Fernández-Esplá, M. C. Martín-Hernández, and P. F. Fox, *Appl. Environ. Microbiol.* **63,** 314 (1997).
[8] K. Suga, T. Kabashima, K. Ito, D. Tsuru, H. Okamura, J. Kataoka, and T. Y. Oshimoto, *Biosci. Biotechnol. Biochem.* **59,** 2087 (1995).
[9] M. Ghosh, A. M. Grunden, D. M. Dunne, R. Weiss, and M. W. W. Adams, *J. Bacteriol.* **180,** 4781 (1998).
[10] R. C. Scriver, R. J. Smith, and J. M. Phang, *Metab. Basis Inher. Dis.* 360 (1983).
[11] W. Bockelmann, *Int. Dairy J.* **5,** 977 (1995).
[12] T. C. Cheng, S. V. Harvey, and A. N. Stroup, *Appl. Environ. Microbiol.* **59,** 3138 (1993).
[13] T. C. Cheng, S. V. Harvey, and G. L. Chen, *Appl. Environ. Microbiol.* **62,** 1636 (1996).
[14] A. Yaron and D. Mlynar, *Biochem. Biophys. Res. Commun.* **32,** 658 (1968).

stored in a dark-colored glass bottle. The prolidase assay mixture has a total volume of 500 μl and contains 50 mM MOPS [3-(N-morpholino)propanesulfonic acid] buffer (pH 7.0), 4 mM Met-Pro (substrate), and 1.2 mM CoCl$_2$. This mixture is incubated at 100° for 5 min prior to initiating the reaction with the addition of pure prolidase or extract. The reaction is incubated at 100° for another 10 min and is stopped with the addition of 500 μl glacial acetic acid and 500 μl ninhydrin reagent. This solution is heated in a boiling water bath for 15 min to develop the red color. The assay mixtures are cooled to 23° before measuring the absorbance at 515 nm. The molar absorption coefficient for the ninhydrin–proline complex is 4570 M^{-1} cm^{-1}. One unit of prolidase activity is defined as the amount of enzyme that releases 1 μmol of proline per minute under the assay conditions.

Purification of Native and Recombinant Prolidase

Native Prolidase Purification

For purification of the native prolidase enzyme, 500 g wet weight of frozen *Pyrococcus furiosus* cells is thawed in 1800 ml of 50 mM Tris–HCl buffer (pH 8.0) containing lysozyme (1 mg/ml) and DNase (10 μg/ml). The cells are lysed by incubation at 37° for 2 hr followed by sonication (Branson 8200 sonicator) for 1 hr. The cell extract is prepared by ultracentrifugation of the broken cell suspension at 50,000g for 2 hr at 4°. The resultant supernatant (1800 ml) is loaded onto a column (10 × 14 cm) of DEAE Fast Flow (Pharmacia, Piscataway, NJ), equilibrated with 50 mM Tris (pH 8.0) containing 10% (v/v) glycerol. The column is eluted at a flow rate of 10 ml/min with a 10-liter linear gradient of 0 to 1.0 M NaCl in the same Tris–glycerol buffer. Prolidase activity is detected in fractions eluted with 0.25 to 0.40 M NaCl. The fractions containing activity are pooled (1500 ml) and solid ammonium sulfate is added to a final concentration of 1.5 M. This solution is applied to a column (3.5 × 10 cm) of phenyl-Sepharose (Pharmacia) equilibrated with Tris–glycerol buffer containing 1.5 M ammonium sulfate. This column is eluted with a 1-liter gradient of 1.5 to 0 M ammonium sulfate in the Tris–glycerol buffer at a flow rate of 7 ml/min. Prolidase elutes as 0.45 to 0.78 M ammonium sulfate is applied to the column. The prolidase-containing fractions are pooled (250 ml) and concentrated to a volume of 7 ml by ultrafiltration (PM-30 membrane filter, Amicon, Beverly, MA) before applying to a column (3.5 × 60 cm) of Superdex 200 (Pharmacia) equilibrated with 50 mM Tris (pH 8.0) containing 0.5 M NaCl at a flow rate of 0.5 ml/min. Fractions with prolidase activity are then pooled and applied to a column (1.6 × 2.5 cm) of HiTrap Q (Pharmacia), equili-

TABLE I

PURIFICATION OF NATIVE AND RECOMBINANT *P. furiosus* PROLIDASE

Step	Activity[a] (U)	Protein (mg)	Specific activity (U/mg)	Purification (-fold)	Recovery (%)
Native					
Cell-free extract	48,400	21,060	2.3	1	100
DEAE Sepharose	25,600	4,500	5.7	2.5	53
Phenyl-Sepharose	10,400	855	12	5.3	21
Superdex 200	2,180	14	153	66.4	4
HiTrap-Q	1,890	3	630	274	4
Recombinant					
Cell-free extract	45,400	498	91	1	100
Phenyl-Sepharose	31,800	105	302	3.3	70
Heat treatment	27,200	24	1,130	12	60
HiTrap-Q	24,400	18	1,360	15	54

[a] Measured using Met-Pro (4 mM) as the substrate.

brated with 50 mM Tris (pH 8.0). The enzyme is eluted with a 100-ml gradient from 0 to 0.5 M NaCl in the same buffer at a flow rate of 4 ml/min. After being judged >95% pure by sodium dodecyl sulfate–polyacrylamide gel electrophoresis (SDS–PAGE) analysis, the prolidase-containing fractions (10 ml) are frozen in small aliquots using liquid nitrogen and stored at −80° until required. Results of this purification are presented in Table I.

Expression of Recombinant Prolidase

The gene encoding *P. furiosus* prolidase has been successfully cloned and expressed in *Escherichia coli*.[15] The gene has been polymerase chain reaction (PCR) amplified with engineered *Bam*HI and *Not*I recognition sites and cloned into the T7 polymerase-driven expression vector pET-21b (Novagen, Madison, WI). For PCR amplification of the *P. furiosus* prolidase gene, a forward primer with the sequence 5′-ATAGGATCCGGTGAG-GAGGTTGTATGAAAGACTTGAA-3′, which contains an engineered *Bam*HI site and spans from −21 to 6 on the coding strand, is used, along with the reverse primer 5′-ATAGGATCCGGTGAGGAGGTTGTAT-GAAAGAAAG-3′ that has an engineered *Not*I site and is complementary to sequences ranging from 1541 to 1511 on the coding strand. The amplification is performed using native *P. furiosus* DNA polymerase, which has

[15] M. Ghosh, A. M. Grunden, D. M. Dunne, R. Weiss, and M. W. W. Adams, *J. Bacteriol.* **180,** 4781 (1998).

proofreading activity, and a Robocycler 40 (Stratagene, La Jolla, CA) thermocycler programmed for 39 cycles, with each cycle consisting of dena- turation for 1 min at 95°, annealing at 52° for 1.5 min, and extension at 72° for 3 min. PCR amplification of the *P. furiosus* prolidase gene results in production of an approximately 1.1-kb DNA fragment, which is subse- quently gel purified and digested with *Bam*HI and *Not*I restriction enzymes and finally cloned into compatible sites in the expression vector pET-21b.

Expression of recombinant *P. furiosus* prolidase is performed in *E. coli* strain BL21(DE3) (Novagen), which contains the T7 RNA polymerase gene under control of an isopropylthiogalactoside (IPTG)-inducible *lacUV* promoter. For consistently high yields of expression of recombinant proli- dase, only freshly transformed *E. coli* cells are used, as a significant decrease in expression levels has been observed for cells stored in frozen glycerol stocks. The expression of recombinant prolidase has been accomplished by inoculating 100 liters in a 100-liter capacity fermentor of Luria–Bertani (LB) medium with freshly transformed *E. coli* cells (0.1% inoculum). The LB medium is supplemented with ampicillin (100 μg/ml) to ensure mainte- nance of the expression plasmid. The culture is incubated at 37°, and its growth is monitored by measuring absorbance at 600 nm. Once an op- tical density of 1.0 is reached by the culture, crystalline IPTG is added (1 mM). The culture is then incubated an additional 4 hr before harvesting the cells.

Recombinant Prolidase Purification

Recombinant prolidase can be purified in three steps by taking advan- tage of the intrinsic heat stability of *P. furiosus* prolidase enzyme compared to most contaminating *E. coli* proteins. For recombinant purification, cell paste (10 g wet weight) is suspended in 10 ml of 50 mM Tris–HCl, pH 8.0, containing the protease inhibitor benzamidine hydrochloride (0.5 mg/ml). The cell suspension is broken by passage of the cell suspension through a French pressure cell (20,000 lb/in^2) twice. Alternatively, the cell suspension may be lysed using a standard lysozyme–DNase treatment. The broken cell suspension is centrifuged at 39,000g for 1 hr at 4° to remove unbroken cells and other cellular debris. The supernatant is then diluted to 300 ml with 50 mM Tris–HCl, pH 8.0. Solid ammonium sulfate is then slowly added with stirring to a final concentration of 1.5 M, and the solution is applied to a column (3.5 × 10 cm) of phenyl-Sepharose (Pharmacia) equilibrated with the same buffer at a flow rate of 7 ml/min. Bound protein is eluted with a 1-liter gradient of 1.5 to 0 M ammonium sulfate in 50 mM Tris–HCl, pH 8.0. The recombinant prolidase elutes when 0.67 to 1.0 M ammonium sulfate is applied. The prolidase-containing fractions are pooled and then incubated at 100° for 2.5 hr. Denatured *E. coli* proteins are removed by centrifugation at 27,000g for 30 min at 4°. The supernatant is then diluted threefold with 50 mM Tris–HCl, pH 8.0, before loading onto a column (1.6 × 2.5 cm) of HiTrap Q (Pharmacia) equilibrated with 50 mM Tris–HCl, pH 8.0. A 100-ml gradient from 0 to 0.5 M NaCl in the same buffer is applied to the column. Recombinant prolidase elutes between 0.25 and 0.37 M NaCl and is judged greater than 95% pure by SDS–PAGE. Purified protein is snap-frozen with liquid nitrogen and stored at −80°. Results of the purification of the recombinant prolidase are shown in Table I.

Properties of Purified Prolidase

A comparison of representative catalytic and biochemical properties of native and recombinant *P. furiosus* is presented in Table II.

Molecular Properties and Stability

The *P. furiosus* prolidase gene consists of 1047 bp and encodes a protein of 349 amino acid residues with a calculated molecular mass of 39.4 kDa. However, both the native and recombinant forms of prolidase, as purified, migrate as a 42-kDa protein on SDS–12% polyacrylamide gels after protein

TABLE II
PROPERTIES OF NATIVE AND RECOMBINANT PROLIDASE

Property	Native	Recombinant
Apparent molecular mass (kDa)[a]	2 × 42	2 × 42
Metal content (g-atoms/subunit)	1.0 ± 0.3 Co	1.0 ± 0.2 Co
	2.0 ± 0.5 Zn	2.0 ± 0.5 Zn
Thermal stability[b] ($t_{1/2}$ at 100°)	4 hr	1 hr
Optimum temperature	100°	100°
Optimum pH	7.0	7.0
Apparent metal association		
$\quad$ Co^{2+} (mM)	0.24	0.50
$\quad$ Mn^{2+} (mM)	0.62	0.66
K_m (mM)	2.8	3.3
V_{max} (μmol/min/mg)	645	1,250
k_{cat} (sec^{-1})	271	525
k_{cat}/K_m (mM^{-1} sec^{-1})	97	45

[a] As determined from gel-filtration analyses and the migration of the protein after SDS–PAGE. Note that the calculated molecular mass of the subunit is 39.4 kDa.

[b] For heat stability determination, a protein concentration of 0.003 mg/ml is used.

samples are boiled in denaturing buffer at 100° for 30 min. The native enzyme has been shown to migrate at even greater masses (>51 kDa) on SDS–polyacrylamide gels when lower temperatures (80°) and shorter incubation times (10 min) are used for denaturation. Thus, it appears that neither native or recombinant prolidases are entirely denatured when subjected to standard denaturing conditions for SDS–PAGE. Furthermore, when prolidase is eluted from a gel filtration column, a molecular mass of 100 ± 10 kDa is obtained, which is higher than that expected (78.8 kDa) for a homodimeric protein. The difficulty in completely denaturing *P. furiosus* prolidase and its nonideal migration through a gel-filtration column are most likely due to the extreme stability of the enzyme. To avoid complications associated with the anomalous size of prolidase as determined by standard techniques, the mass of 39.4 kDa is used for the prolidase subunit size for all calculations.

An analysis of the metal content of native and recombinant prolidase reveals that only Co and Zn are present in significant amounts (>0.1 g-atom/subunit). Both enzymes contain 1.0 ± 0.3 g-atoms of Co/subunit and 2.0 ± 0.5 g-atoms of Zn/subunit. When purification of the recombinant prolidase is conducted with the inclusion of 1 mM CoCl$_2$ in all buffers, except for those used in the last size-exclusion gel-filtration step (Superdex 200) to remove unbound Co, 1.0 ± 0.2 g-atoms of Co/subunit is detected. However, because both native and recombinant prolidases, as purified, require the addition of Co^{2+} to the assay mixture for activity, it appears that there is at least one additional occupancy site for Co per subunit and that these sites have very different affinities. However, it is shown that the removal of Zn from recombinant prolidase by chelation using ethylenediaminetetraacetic acid (EDTA) (resulting in 0.3 ± 0.1 g-atoms of Zn/subunit) does not affect enzyme activity adversely.

As expected, native prolidase is very thermostable, with negligible loss of activity when a sample (0.3 mg/ml in 100 mM MOPS, pH 7.0) is incubated in a sealed vial at 100° for 12 hr. However, this heat stability is dependent on the protein concentration used, as the same enzyme preparation loses 50% of its activity after 4 hr at 100° when sampled at a concentration of 0.003 mg/ml. The recombinant prolidase also demonstrates concentration-dependent thermostability with a $t_{1/2}$ of 3 and 1 hr at 100° for protein concentrations of 0.3 and 0.003 mg/ml, respectively. The addition of Co^{2+} ions (1 mM) does not significantly affect the thermostability of either the native or the recombinant prolidase.

Catalytic Properties

The catalytic activities of the native and recombinant prolidase are virtually indistinguishable. Both enzymes show a pH optimum of 7.0 and temperature optimum at $\geq 100°$. Negligible activity is seen for assay temperatures below $40°$. All assays used to determine catalytic properties of prolidase included 1.2 mM Co^{2+} and these ions could not be replaced with other divalent (Mg^{2+}, Ca^{2+}, Fe^{2+}, Zn^{2+}, Cu^{2+}, or Ni^{2+}) or monovalent (Na^+ or K^+) cations except with Mn^{2+}. Optimum activity for both native and recombinant forms of the enzyme is obtained when 1.2 mM $CoCl_2$ is included in the assay mix. Supplementation of the assay mix with 1.6 mM $MnCl_2$ results in 75% activity compared to that seen with $CoCl_2$ addition. When either cation is added to the assay mix above its optimum concentration, inhibition of activity can be observed. Association constants for Co^{2+} of 0.24 and 0.5 and for Mn^{2+} of 0.62 and 0.66 can be derived for native and recombinant prolidase, respectively.

Pyrococcus furiosus prolidase was first identified by its ability to hydrolyze the dipeptide Met-Pro, and this dipeptide has subsequently been used as the substrate in all routine assays. With Met-Pro as the substrate, a K_m of 2.8 and 3.3 mM, a V_{max} of 645 and 1250 μmol/min/mg, a k_{cat} of 271 and 525 sec^{-1}, and k_{cat}/K_m of 97 and 45 mM^{-1} sec^{-1} can be derived using linear double-reciprocal plots for the native and recombinant versions of prolidase, respectively. Substrate specificity studies of *P. furiosus* prolidase confirm that this enzyme is only capable of hydrolyzing dipeptides with Pro at the C terminus. In addition, the nature of the amino acid residue in the N-terminal position is critical, with substantial activity occurring only in the cases where substrates contain nonpolar amino acids (Table III). As indicated, only Met-Pro, Leu-Pro, Val-Pro, Phe-Pro, and Ala-Pro support significant activity by either enzyme and that the relative percentage activities for native and recombinant enzyme using these substrates are quite similar. Inhibition studies using the thiol or serine protease inhibitors (each at 1 mM) iodoacetate, E-64 (L-*trans*-epoxysuccinylleucylamido(4-guanido)-butane), N-ethylmaleimide, phenylmethylsulfonyl fluoride (PMSF), or diisopropylfluorophosphate (DFP) show little affect when either the native or the recombinant enzyme is treated with the agents at $25°$ for 30 min prior to assaying.

Structural Properties

Enzyme activity and metal affinity studies suggest that *P. furiosus* prolidase requires the presence of at least two Co atoms per subunit for activity. This finding is consistent with results gathered for a broad class of binuclear metallohydrolases, represented by N-terminal exopeptidases, whose active sites contain two metal ions that typically differ in their exchange kinetics.[16]

---

[16] D. E. Wilcox, *Chem. Rev.* **96**, 2435 (1996).

TABLE III
SUBSTRATE SPECIFICITY OF NATIVE AND RECOMBINANT PROLIDASE

Substrate[a]	Relative activity (%)[b]	
	Native prolidase	Recombinant prolidase
Met-Pro	100	100
Leu-Pro	75	79
Val-Pro	46	10
Phe-Pro	25	24
Ala-Pro	23	17
Lys-Pro	4	10
Gly-Pro	0	1
Pro-Ala	0	0
Pro-Hydroxypro-Pro	0	0
Lys-Trp-Ala-Pro	1	0
Gly-Arg-Gly-Asp-Thr-Pro	0	0
Pro-Pro-Gly-Phe-Ser-Pro	1	0
Pro-Pro-Gly-Phe-Ser-Pro-Phe-Arg	0	0
N-Acetyl-Pro	0	0

[a] All substrates are used at a final concentration of 4 mM.

[b] The rate of hydrolysis is expressed as a percentage of the activity compared to that obtained when using Met-Pro as the substrate. An activity of 100% corresponds to 600 U/mg for native and 1350 U/mg for recombinant prolidase.

Specific examples of members of this family are bovine leucine aminopeptidase[17] and *Aeromonas proteolytica* aminopeptidase,[18] each of which have two Zn^{2+} per catalytic unit. Furthermore, it has been shown that similar to the case of *P. furiosus* prolidase, the *A. proteolytica* enzyme can be purified such that only one of the two Zn^{2+} is present per catalytic subunit.[19] Of the enzymes comprising binuclear metallohydrolases, the only other class of enzymes that naturally contain Co are the methionine aminopeptidases,[20] and these enzymes also bind two Co^{2+} ions per active site. *E. coli* methionine aminopeptidase has been crystallized and its structure has been solved.[21] From its structure, the following five amino acid residues responsible for liganding the two Co^{2+} ions can be identified as Asp-97, Asp-

[17] S. K. Burley, P. R. David, R. M. Sweet, A. Taylor, and W. N. Lipscomb, *J. Mol. Biol.* **224,** 113 (1992).

[18] B. Chevrier, C. Schalk, H. Dorchymont, J. M. Rondeau, C. Tarnus, and D. Moras, *Structure* **2,** 283 (1994).

[19] B. Bennett and R. C. Holz, *J. Am. Chem. Soc.* **119,** 1923 (1997).

[20] S. M. Arfin, R. L. Kendall, L. Hall, L. H. Weaver, A. L. Stewart, B. W. Matthews, and R. A. Bradshaw, *Proc. Natl. Acad. Sci. U.S.A.* **92,** 7714 (1995).

[21] S. L. Roderick and B. W. Mattews, *Biochemistry* **32,** 3907 (1993).

P. furiosus prolidase
L. delbrukei prolidase
E. coli prolidase
Alteromonas OPAA
E. coli MAP

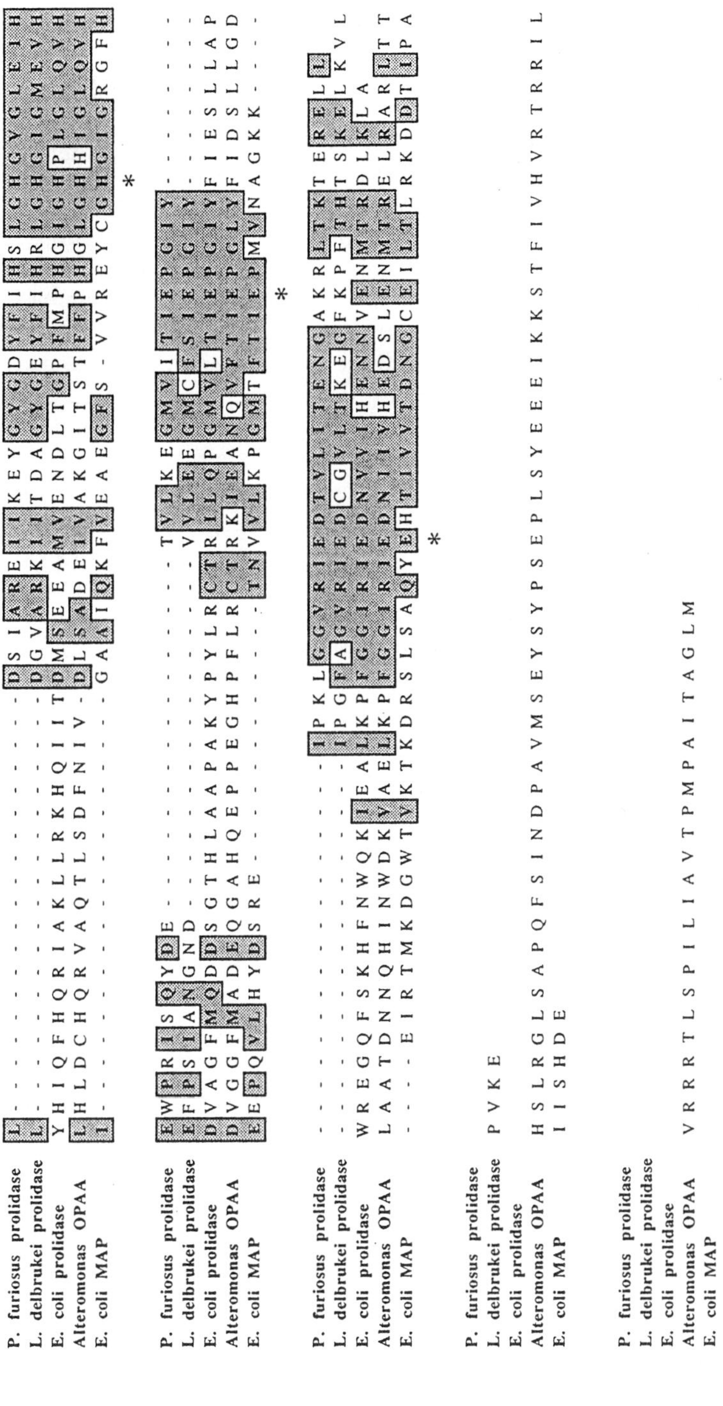

FIG. 1. Alignment of prolidases and related enzymes. OPAA, organophosphorus acid anhydrolase; MAP, methionine aminopeptidase. Similar residues are designated by dark shading and boldface type. The five conserved residues that are thought to coordinate the binuclear metal site are indicated by an asterisk.

108, His-171, Glu-204, and Glu-235. The His-171 and Glu-204 residues participate solely in binding the first Co^{2+} ion, Asp-97 functions solely to ligand the second Co^{2+} ion, and Asp-108 and Glu-235 serve as bidentate ligands for both sites. Notably, these Co-binding amino acid residues are conserved in *P. furiosus* prolidase (Asp-209, Asp-220, His-280, Glu-313, and Glu-327), as indicated by the asterisks in the alignment in Fig. 1, and it is predicted that these amino acid residues will ligand its two Co^{2+} ions in a manner similar to that found for *E. coli* methionine aminopeptidase.

A comparison of *P. furiosus* prolidase with other known prolidases isolated from *Methanococcus jannschii, Lactobacillus delbrukei, Haemophilus influenzae, E. coli,* and *Homo sapiens* reveals similarities of 69, 61, 58, 56, and 53%, respectively. In all cases, these enzymes share the five conserved metal-liganding residues. However, *P. furiosus* prolidase is the only one characterized to date that has been shown to contain cobalt. Prolidases from *L. casei,*[22] *Xanthomonas maltophilia,*[23] and human beings[24] are all activated by Mn^{2+} ions, although their metal contents have not been reported. Whereas the prolidases isolated from *Aureobacterium estararomaticum,*[25] *L. lactis,* guinea pig brain,[26] and bovine intestine[27] have not been shown to contain metal centers or require metal for activation, an analysis of their sequences indicates that these enzymes are quite similar to *P. furiosus* prolidase. All share the five conserved metal-binding amino acid residues, and so presumably also have the binuclear metal centers.

The sequence of *P. furiosus* prolidase is 51% similar to that of OPAA from *Alteromonas,* and whereas OPAA is a monomeric enzyme and *P. furiosus* prolidase is dimeric, both enzymes contain the five putative metal-binding residues[28] (Fig. 1). Furthermore, OPAA was first identified for its ability to hydrolyze P-F, P-C, and P-O bounds but it also has prolidase-type activity and cleaves Leu-Pro and Ala-Pro in a Mn^{2+}-dependent manner. OPAA, like *P. furiosus* prolidase, is unable to hydrolyze tri- or tetrapeptides or dipeptides with Pro at the N terminus. Thus, it appears that OPAA and *P. furiosus* prolidase are very similar, if not identical, types of enzyme and

[22] M. D. Fernández-Esplá, M. C. Martín-Hernández, and P. F. Fox, *Appl. Environ. Microbiol.* **63,** 314 (1997).

[23] K. Suga, T. Kabashima, K. Ito, D. Tsuru, H. Okamura, J. Kataoka, and T. Y. Oshimoto, *Biosci. Biotechnol. Biochem.* **59,** 2087 (1995).

[24] F. Endo, A. Tanoue, H. Nakai, A. Hata, Y. Indo, K. Titani, and I. Matsuda, *J. Biol. Chem.* **264,** 4476 (1989).

[25] M. Fujii, Y. Nagaoka, S. Imamura, and T. Shimizu, *Biosci. Biotechnol. Biochem.* **60,** 1118 (1996).

[26] P. Browne and G. O'Cuinn, *J. Biol. Chem.* **268,** 6147 (1983).

[27] T. Yoshimoto, F. Matsubara, E. Kawano, and D. Tsuru, *J. Biochem.* **94,** 1889 (1983).

[28] T. C. Cheng, S. V. Harvey, and G. L. Chen, *Appl. Environ. Microbiol.* **62,** 1636 (1996).

that *P. furiosus* prolidase, along with its established function of cleaving prolyl dipeptides, may also have utility in the degradation of organophosphorus compounds.

Acknowledgment

This research was supported by grants from the Department of Energy (FG05-95ER20175 and SW994-19/RXE-7-17039).

[31] Prolyl Oligopeptidase from *Pyrococcus furiosus*

By VALERIE J. HARWOOD and HAROLD J. SCHREIER

Introduction

Prolyl oligopeptidase (EC 3.4.21.26; also called prolyl endopeptidase, postproline cleaving enzyme) (POPase) is a serine protease that cleaves peptides at the carboxyl side of proline residues. The enzyme is widely distributed in nature and is found in organisms ranging from mammals to bacteria. However, it is not ubiquitous and POPase homologs have not been found in many of the genomes whose DNA sequence have been determined to date. Similarly, in a screen of over 500 bacterial isolates, only one POPase-positive organism was found.[1] Methods for analysis of the *Sus scrofa* (porcine) POPase have been presented in an earlier volume of this series.[2] This article focuses on the POPase from the hyperthermophilic archaeon, *Pyrococcus furiosus.* Although the *P. furiosus* POPase has many characteristics that are similar to other POPases, the enzyme possesses properties unique in this class of serine proteases, including its thermal stability and its autoproteolytic activity.[3]

Assay

POPase activity is measured using the dipeptide substrate carboxybenzoxylglycylprolyl-*p*-nitroanilide, Z-Gly-Pro-*p*NA (Bachem Biosciences, King of Prussia, PA), which possesses a blocked amino-terminal and the carboxy-terminal *p*-nitroanilide group that is detectable on cleavage. Hy-

[1] T. Yoshimoto and D. Tsuru, *Agric. Biol. Chem.* **42,** 2417 (1978).
[2] L. Polgar, *Methods Enzymol.* **244,** 188 (1994).
[3] V. J. Harwood, J. D. Denson, K. A. Robinson-Bidle, and H. J. Schreier, *J. Bacteriol.* **179,** 3613 (1997).

drolysis of Z-Gly-Pro-pNA is monitored at 410 nm by either continuous or fixed-time assays. Z-Gly-Pro-pNA is suspended in 10 mM glycylglycine buffer (pH 7.5) to a final concentration of 200 μM, and the first-order rate constant (V_{max} conditions) is measured at 85°. The buffering capacity of glycylglycine changes relatively little over broad temperature ranges; when glycylglycine is adjusted to pH 8.0 at room temperature the pH at 85° is 7.5. A unit of activity is defined as the amount of enzyme required to produce 1 μmol of product per minute. Although POPase is maximally active at a temperature above 85° (see later), assays are carried out at this temperature since there is little difference in activity between 85 and 95° and evaporation is minimal compared to the higher temperatures.

Continuous Assay

Examination of POPase activity as a function of time requires a spectrophotometer outfitted with a heated sample chamber, which is achieved by water circulation or via a Peltier-regulated temperature control system (e.g., the system available for use with a Beckman Instruments DU-60 spectrophotometer). If a water-heated chamber is used, the loss of heat between the water bath and the sample chamber must be taken into account. The assay mixture (buffer and substrate) is preheated to 85° in covered tubes in a heating block. One milliliter of assay mixture is transferred to a cuvette, and 5 to 20 μl of the enzyme preparation (approximately 1.0 μg purified enzyme) is added to initiate the reaction. Absorbance is monitored at 410 nm and is linear up to 5 min.

Fixed Time Assay

One milliliter of assay mixture is preheated for 5 min in a capped microcentrifuge tube that is placed into a heating block set at 85°. Enzyme is added (1 μg purified enzyme in 5 to 20 μl) and incubation is allowed to proceed at 85°. After 5 min the reaction is terminated by the addition of 100 μl ice-cold 1 M trichloroacetic acid. Samples are then cooled on ice, centrifuged at 10,000g for 5 min at room temperature, whereupon the absorbance at 410 nm is measured immediately.

Enzyme Purification

Purification from Escherichia coli Expression System

Cloning. The structural gene for the *P. furiosus* POPase, *prpA*, was first identified from a cDNA library[4] and subsequently sequenced (GenBank

---

[4] K. A. Robinson, D. A. Bartley, F. T. Robb, and H. J. Schreier, *Gene* **152,** 103 (1995).

accession number U08343). Expression of the *prpA* product is obtained by cloning the entire open reading frame contained in a 1.9-kb *Nco*I–*Bam*HI DNA fragment placed into expression vector pET11d (Novagen, Inc.) (Amp^r). The fragment is constructed by polymerase chain reaction (PCR) amplification using oligonucleotide primers 5′-TGATCTCCATGGAA-GATCCCTAC-3′ (upstream primer) and 5′-CGTGGATCCAAAGAA-TAATAG-3′ (downstream primer), which introduces *Nco*I and *Bam*HI restriction sites at the 5′ and 3′ ends of the *prpA* coding region, respectively. *Nco*I provides a cleavage site within the presumed *prpA* initiator codon, 11 bases downstream from the *P. furiosus* transcription start point.[4] Selection of the AUG codon at this site is based on a comparison of the deduced *P. furiosus prpA* sequence with several POPase sequences in the database.[3]

Expression. The *prpA*-containing pET11d construct is introduced into *E. coli* strain BL21(DE3)pLysS (Novagen, Inc.), which is grown in Luria–Bertani broth[5] supplemented with chloramphenicol (34 μg/ml) and ampicillin (50 μg/ml) with shaking at 37°. To achieve maximum POPase production, isopropyl-β-D-thiogalactoside (IPTG) is added to a final concentration of 0.4 mM when cultures reach an OD$_{600}$ of 0.7 to 0.8. After incubation at 37° for an additional 2 to 3 hr, cultures are harvested by centrifucation at 5000g for 5 min at 4°. The cell pellets are then placed at −80°.

Purification. Lysis of frozen cells is accomplished by suspending cell pellets from 100 ml of culture in 5 ml of 50 mM Tris–HCl (pH 7.5) and passing the suspension twice through a French press cell at 12,000 lb/in². After removal of cell debris by centrifugation at 10,000g for 10 min at 4°, nucleic acids are removed by treating the supernatant fraction with streptomycin sulfate at a final concentration of 1% (w/v) for 30 min with slow stirring at 4°. The sample is then centrifuged for 10 min at 10,000g at 4° and the supernatant fraction is heated to 90° for 10 min followed by centrifugation to remove insoluble material. The 90° treatment is repeated and the clarified sample is then added to a Sephacryl S-200 size-exclusion column eluting with 50 mM Tris–HCl (pH 8.0) in the presence of 10% (v/v) glycerol. POPase collected at this point is greater than 95% homogeneous and can be stored at −20° for months with little loss in activity. A 100-ml culture of *E. coli* BL21(DE3)pLysS harboring the recombinant pET11d construct yields approximately 1 mg of purified POPase. The specific activity of purified *P. furiosus* POPase using Z-Gly-Pro-*p*NA is 4U/mg protein.

[5] J. Sambrook, F. Fritsch, and T. Maniatis, "Molecular Cloning: A Laboratory Manual." Cold Spring Harbor Laboratory Press, Cold Spring Harbor, NY, 1989.

Purification from P. furiosus

Pyrococcus furiosus strain DSM3638 is grown to late exponential at 95° in medium described by Robb *et al.*[6] supplemented with 10 m*M* maltose (filter sterilized and added to autoclaved medium). Cells are harvested by centrifugation (5000*g* for 10 min at 4°) and washed once in 10 m*M* Tris–HCl (pH 8.0)–100 m*M* KCl at which point cells may be stored at −80° or used directly for purification. Cells are suspended in 50 m*M* Tris–HCl (pH 8.0) and lysed by twice passage through a French press cell at 12,000 lb/in². After centrifugation (10,000*g*, 4°, 10 min), nucleic acids are removed by the addition of streptomycin sulfate to a final concentration of 1% (w/v) as described earlier. The supernatant fraction is diluted 10-fold in 50 m*M* Tris–HCl buffer (pH 8.0) and applied to a DEAE-Sephacel anion-exchange column previously equilibrated with the same buffer. POPase is eluted using a linear gradient of 0 to 1.0 *M* NaCl in 50 m*M* Tris–HCl (pH 8.0); fractions containing POPase activity appear at a NaCl concentration of approximately 0.3 *M*. POPase-containing fractions are concentrated at 4° by ultrafiltration with a Cenriprep 3 microconcentrator (nominal molecular mass cutoff = 3 kDa; Amicon, Danvers, MA) or by dialysis against polyethylene glycol 6000. The preparation is then added to a Sephacryl S-200 column as described earlier, previously equilibrated with 50 m*M* Tris (pH 8.0), and elution is carried out with the same buffer. Fractions containing POPase are then pooled and added to a hydroxylapatite column, equilibrated with 10 m*M* Na₂HPO₄ (pH 6.8); POPase activity elutes from the column in the presence of 200 m*M* Na₂HPO₄ (pH 6.8). The pooled fractions are dialyzed against 1000 volumes of 50 m*M* Tris–HCl (pH 8.0) and 10% glycerol. The enzyme at this point is no greater than approximately 30 to 40% purified. Procedures for further purification have not been developed as the partially purified enzyme displays the same characteristics found for homogeneous enzyme prepared from the overproducing *E. coli* strain, including thermal stability, specificity, and molecular weight under denaturing and nondenaturing conditions.

Physical Characteristics

Based on the *prpA* sequence, POPase from *P. furiosus* has a molecular weight of 70,867, with a theoretical extinction coefficient of 105,400 and a p*I* of 5.76. The enzyme prepared either from the *E. coli* expression system or from *P. furiosus* extracts displays a molecular weight of approximately 72,000 when subjected to denaturing polyacrylamide gel electrophoresis

---

[6] F. T. Robb, J.-B. Park, and M. W. W. Adams, *Biochim. Biophys. Acta* **1120,** 267 (1992).

(Fig. 1A), which is consistent with that predicted from the gene sequence. As is the case for POPase from all sources examined,[7–9] gel permeation studies have shown that the *P. furiosus* enzyme functions as a monomer. However, a molecular weight of approximately 59,000 was consistently obtained from these studies when chromatography was done at 4° or room temperature (Fig. 1B).[3] This lower-than-expected size was observed for POPase prepared from *P. furiosus* extracts as well as from the *E. coli* overexpressor. Furthermore, chromatography in the presence of 150 m*M* KCl or 25 m*M* proline did not influence the native molecular weight of the protein. It is unlikely that the difference in size between native and denatured protein is due to proteolysis, as a molecular weight of 72,000 was observed for the denatured protein before and after gel-permeation chromatography and POPase activity at 4° is minimal.[3] Thus, while the 12,000 molecular weight difference may be a chromatography artifact, it is conceivable that the disparity is due to a physical property of the enzyme that allows it to alternate between a compact structure and one that is more open, perhaps for catalytic activation. Such unfolding has been observed for enzymes from hyperthermophiles.[10,11]

Specificity

In general, POPase activity is unique among proline-specific peptidases, cleaving at the carboxyl end of proline in peptides having the sequence X-Pro-/-Y, where "-/-" denotes the scissile bond. Although the X-Pro bond in peptides can be found in either the *cis* or *trans* conformation, this bond must be in the *trans* conformation in order for it to be recognized by POPase.[12] Several proline-specific enzymes are known to target the imino side of proline, cleaving an amino-terminal residue from the peptide, e.g., prolyl aminopeptidase (EC 3.4.11.9) and proline dipeptidase (EC 3.4.13.9). Aside from POPase, enzymes that cleave the carboxyl side of proline include proline iminopeptidase (EC 3.4.11.5), which separates an amino-terminal proline from long or short peptides (NH$_2$-Pro-/-Y), and dipeptidyl

[7] S. Chevallier, P. Goeltz, P. Thibault, D. Banville, and J. Gagnon, *J. Biol. Chem.* **267,** 8192 (1992).

[8] A. Kanatani, T. Yoshimoto, A. Kitazono, T. Kokubo, and D. Tsuru, *J. Biochem.* **113,** 790 (1993).

[9] D. Rennex, B. A. Hemmings, J. Hofsteenge, and S. R. Stone, *Biochemistry* **30,** 2195 (1991).

[10] J. DiRuggiero, F. T. Robb, R. Jagus, H. H. Klump, K. M. Borges, M. Kessel, X. Mai, and M. W. W. Adams, *J. Biol. Chem.* **268,** 17767 (1993).

[11] H. Klump, J. DiRuggiero, M. Kessel, J.-B. Park, M. W. W. Adams, and F. T. Robb, *J. Biol. Chem.* **267,** 22861 (1992).

[12] L. Lin and J. Brandt, *Biochemistry* **22,** 4480 (1983).

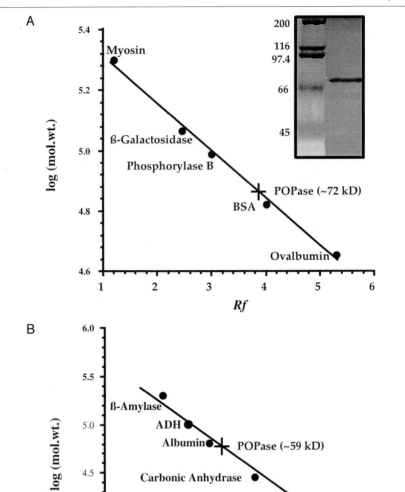

peptidase IV (EC 3.4.14.5), which hydrolyzes two amino-terminal residues from a peptide when the penultimate residue is proline (NH_2-X-Pro-/-Y). Prolyl carboxypeptidase (EC 3.4.16.2) also cleaves on the carboxyl side of proline when proline is the penultimate amino acid (-X-Pro-/-Y-COOH). The unique specificity displayed by POPase and its lack of homology with other serine proteases such as chymotrypsin and subtilisin have placed POPases into their own family, designated S9.[9,13] Of the proline-specific peptidases, only dipeptidylpeptidase IV belongs to the S9 family.[14]

POPases cleave most efficiently at proline residues, and proline-containing dipeptides with a blocked amino-terminal make excellent substrates. The *P. furiosus* enzyme displays high affinity for Z-Gly-Pro-*p*NA (apparent K_m of 53 μM).[3] Dipeptides possessing a free amino-terminal (e.g., H-Gly-Pro-*p*NA) are cleaved at 1/10 to 1/20 the rate of the corresponding carboxy-benzoyl-blocked substrates,[15] suggesting that the S_3 subsite of the enzyme (according to the Schecter and Berger[16] nomenclature) must be occupied for maximum catalytic activity.

The requirement for proline at the P_1 position is not absolute for the *P. furiosus* POPase. The enzyme cleaves Z-Ala-Ala-*p*NA at approximately 1/10 the rate of Z-Ala-Pro-*p*NA.[15] Cleavage at alanine has been noted for bacterial and eukaryal POPases; however, the rate for these enzymes is 1/100 to 1/1000 that at proline.[17,18] Z-Gly-Gly-Leu-*p*NA, a substrate for proteinase yscE, subtilisins, and neutral endopeptidase 24.5, cannot be cleaved by the *P. furiosus* POPase, indicating that leucine cannot substitute for proline at the scissile bond.[3]

Most POPases cleave oligopeptides and not proteins; the functional

---

[13] N. D. Rawlings, L. Polgar, and A. J. Barret, *Biochem. J.* **279**, 907 (1991).

[14] N. Rawlings and A. Barrett, *Methods Enzymol.* **244**, 19 (1994).

[15] V. J. Harwood, unpublished results (1998).

[16] I. Schecter and A. Berger, *Biochem. Biophys. Res. Commun.* **27**, 157 (1967).

[17] R. Walter and T. Yoshimoto, *Biochemistry* **17**, 4139 (1978).

[18] T. Yoshimoto, R. Walter, and D. Tsuru, *J. Biol. Chem.* **255**, 4786 (1980).

FIG. 1. Analysis of *P. furiosus* POPase by SDS–polyacrylamide gel electrophoresis and gel-filtration chromatography. (A) Analysis by 0.1% SDS–10% polyacrylamide gel electrophoresis of the purified protein using the molecular weight standards (Bio-Rad, Hercules, CA) shown on the gel (inset). R_f is relative mobility. The molecular weight of the POPase subunit was determined to be 72,000 ± 3500. (B) Size determination of POPase under nondenaturing conditions. A Sephacryl S-200 column, equilibrated with 50 mM Tris–HCl (pH 8.0) at a flow rate of 0.5 ml/min, was calibrated with molecular weight standards: β-amylase (200,000), alcohol dehydrogenase (ADH, 100,000), albumin (66,000), carbonic anhydrase (29,000), and cytochrome *c* (12,400), as shown. V_e/V_o is the ratio of eluted volume to void volume. The apparent molecular weight of purified POPase was 59,000 ± 2700.

TABLE I
EFFECT OF PROTEIN CONCENTRATION ON
POPase ACTIVITY[a]

[POPase] (μg/ml)	Activity (%)
1	100
5.4	80
27	42
67	17
136	4.5
669	1

[a] Samples were incubated at 95° in the presence of 25% glycerol at the protein concentrations shown. After 20 hr, aliquots were removed and POPase activity was determined by the standard assay. Percentage activity is the ratio of activity before and after incubation multiplied by 100. The standard deviation for each is ±5%.

limit for porcine POPase was placed at 3000 Da.[19] The *P. furiosus* POPase, however, differs in that it has the capacity to hydrolyze large peptides such as azocasein.[3] Furthermore, it is the only POPase for which concentration-dependent autoproteolysis has been demonstrated (Table I).[3] Such proteolysis appears to occur randomly at any of the 30 Pro residues scattered throughout the 616 amino acid sequence.[20] Under standard conditions, autoproteolysis is not a significant problem as the enzyme has a greater affinity for the short, synthetic Z-Gly-Pro-pNA substrate than it does for itself, and assays are done using protein concentrations that do not stimulate proteolysis (nM levels).

Inhibitors

POPase from *P. furiosus* does not contain Cys residues and is not inhibited by tosyl-L-Lys chloromethyl ketone (TLCK) and *N*-[*N*-(L-3-*trans*-carboxirane-2-carbonyl)-L-Leu]agmatine (E-64). The dipeptidylpeptidase inhibitor diprotin A (H-Ile-Pro-Ile-OH) is effective in competitively inhibiting Z-Gly-Pro-pNA hydrolysis (apparent K_I of 343 μM).[3] The serine protease inhibitor phenylmethylsulfonyl fluoride (PMSF) is effective in

[19] A. Moriyama, M. Nakanishi, O. Takenaka, and M. Sasaki, *Biochim. Biophys. Acta* **956,** 151 (1988).
[20] H. J. Schreier, unpublished results (1997).

inactivating the *P. furiosus* enzyme; activity is almost completely abolished when POPase is pretreated in the presence of 5 mM PMSF.[3] PMSF sensitivity has been shown for only one other POPase, the enzyme from *Sphingomonas capsulata*,[21] and is not a feature of other POPases, which are resistant to PMSF but sensitive to serine protease inhibitor diisopropyl fluorophosphate.[2,7]

Like other POPases,[18,22] metal chelators do not inhibit the *P. furiosus* enzyme. Furthermore, activity is unaffected by thiol reagents and the enzyme is not sensitive to NaCl when present in assays at concentrations as high as 1 M.[15]

As is the case for many *P. furiosus* proteases,[23–25] POPase is moderately resistant to denaturation by SDS; at a concentration of 150 μg/ml, the enzyme retained 40% of its activity when heated to 100° for 10 min and 1% SDS.[3]

Temperature Characteristics

Temperature Optimum

Pyrococcus furiosus POPase has a temperature optimum between 85 and 90°; at 20 and 105° activity is 5 and 30% of the maximum, respectively.[3]

Thermostability

POPase from *P. furiosus* retains a significant amount of activity when incubated at elevated temperatures for extended periods of time. Thermostability can be enhanced by the inclusion of stabilizing agents, e.g., glycerol, in storage buffer.[3] For instance, approximately 40 and 15% of the initial POPase activity remained after incubation for 24 hr at 65 and 95°, respectively, at an enzyme concentration of 86 μg/ml. Inclusion of glycerol to the incubation buffer [final concentration of 25% (v/v)] resulted in no significant loss of activity at 65°. At 95°, however, glycerol appeared to stabilize POPase activity during the first hour of incubation, but a rapid decrease in activity was observed (15% of initial activity) that could be attributed to autoproteolysis rather than thermal denaturation.[3] At low enzyme concentrations,

[21] T. Kabashima, M. Fujii, Y. Meng, K. Ito, and T. Yoshimoto, *Arch. Biochem. Biophys.* **358,** 141 (1998).

[22] T. Yoshimoto, T. Nishimura, T. Kita, and D. Tsuru, *J. Biochem.* (*Tokyo*) **94,** 1179 (1983).

[23] I. Blumentals, A. S. Robinson, and R. M. Kelly, *Appl. Environ. Microbiol.* **56,** 1992 (1990).

[24] H. Connaris, D. A. Cowan, and R. J. Sharp, *J. Gen. Microbiol.* **137,** 1193 (1991).

[25] S. B. Halio, I. I. Blumentals, S. A. Short, B. M. Merrill, and R. M. Kelly, *J. Bacteriol.* **178,** 2605 (1996).

however, no decrease in activity was observed after 20 hr of incubation at 95° in the presence of 25% glycerol (see Table I).

Physiological Function

Despite its wide distribution, little is known about the physiological role(s) of POPase in any organism. The presence of the enzyme in mammalian brain tissue[22,26] and its ability to cleave proline-containing neuroactive peptides[27,28] have led to the suggestion that it has a role in regulation of the half-life of these peptides. In mammals, evidence is emerging that suggests that POPase plays a role in memory as POPase inhibitors have antiamnesiac properties.[29] The function of POPase in prokaryotes is even less clear, although it has been suggested that peptidases capable of cleaving at proline catalyze the hydrolysis of larger peptides to allow for their use as substrates for fermentation.[30] For *P. furiosus,* POPase activity and mRNA levels are elevated approximately 10-fold[3,4] when growth is carried out in peptide-containing media that is supplemented with maltose compared to the same medium lacking maltose. Thus, it is reasonable to consider that POPase in *P. furiosus* has a greater role than that required for peptide fermentation, perhaps as a regulator of the level of specific intracellular peptides, as is the case for the mammalian enzyme. The nature of the role of POPase awaits the development of genetic tools for this microorganism.

Acknowledgments

Research from the authors' laboratories was supported, in part, by grants from the Department of Commerce Advanced Technology Program (H.J.S.) and the National Institutes of Health, NIH 1R15GM55902-01 (V.J.H.).

[26] T. Yoshimoto, W. H. Simmons, T. Kita, and D. Tsuru, *J. Biochem. (Tokyo)* **90,** 325 (1981).
[27] S. Blumberg, V. I. Teichberg, J. L. Charli, L. B. Hersh, and J. F. McKelvy, *Brain Res.* **192,** 477 (1980).
[28] L. B. Hersh and J. F. McKelvy, *Brain Res.* **168,** 553 (1979).
[29] T. Yoshimoto, K. Kado, F. Matsubara, N. Koriyama, H. Kaneto, and D. Tsuru, *J. Pharmacobio-Dyn.* **10,** 730 (1987).
[30] M. Ghosh, A. Grunden, D. Dunn, R. Weiss, and M. Adams, *J. Bacteriol.* **180,** 4781 (1998).

[32] Homomultimeric Protease and Putative Bacteriocin Homolog from *Thermotoga maritima*

By Paula M. Hicks, Lara S. Chang, and Robert M. Kelly

Introduction

Thermotoga maritima is an anaerobic heterotroph belonging to the bacterial order Thermotogales that grows optimally at 80° by fermentation of carbohydrates and proteins, including starch, glucose, galactose, glycogen, and yeast extract.[1-4] The bacterium is also able to reduce thiosulfate to sulfide, with an improved growth rate.[5] Although the microorganism is a facultative sulfur reducer, the reduction of sulfur does not provide an energetic boost as seen by the lack of effect on growth yields and fermentation balances.[1] *T. maritima* appears to be motile, migrating at a speed proportional to the temperature.[6]

In 1998,[7] a multisubunit protease from *T. maritima* was discovered and, later, was tentatively named maritimacin. The origins of this name come from its relationship to a putative bactericidal homolog from the mesophilic bacterium *Brevibacterium linens*. This antilisterial bacteriocin, linocin M18,[8] has a 32% amino acid sequence identity and a 53% amino acid similarity to maritimacin. The details of this consensus, along with a comparison to an antigen from *Mycobacterium tuberculosis* and a putative protein from *Aquifex aeolicus,* are shown in Fig. 1. In addition, transmission electron microscopy (TEM) analyses of the globular form of maritimacin, illustrated in Fig. 2, show similarities to linocin M18 in size and shape. TEM analyses from separate purification steps also suggest that maritimacin has structures similar to those of phage tail-like bacteriocins.[7]

This article describes the purification protocols used to isolate maritimacin, as well as the biochemical assays used to measure its activity.

[1] R. Huber, T. A. Langworthy, H. Konig, M. Thomm, C. R. Woese, U. B. Sleytr, and K. O. Stetter, *Arch. Microbiol.* **144**, 324 (1986).

[2] K. D. Rinker, Ph.D. Thesis, NCSU, Raleigh, NC, 1998.

[3] R. Rachel, A. M. Engel, R. Huber, K. O. Stetter, and W. Baumeister, *FEBS Lett.* **262**, 64 (1990).

[4] K. O. Stetter, *FEBS Lett.* **452**, 22 (1999).

[5] G. Ravot, B. Ollivier, M.-L. Fardeau, B. K. C. Patel, K. T. Andrews, M. Magot, and J.-L. Garcia, *Appl. Environ. Microbiol.* **62**, 2657 (1996).

[6] M. F. Gluch, D. Typke, and W. Baumeister, *J. Bacteriol.* **177**, 5473 (1995).

[7] P. M. Hicks, K. D. Rinker, J. R. Baker, and R. M. Kelly, *FEBS Lett.* **440**, 393 (1998).

[8] N. Valdés-Stauber and S. Scherer, *Appl. Environ. Microbiol.* **60**, 3809 (1994).

```
                        1                                                          50
T. maritima      mvnMEFLkRs fAPLTEkqWq EIDNrAREiF KtQlyGRKFV DVEGPYGWEy
M. tuberculosis  ~~~MnnLyRd lAPvTEaaWa EIeleAartF KrhiaGRrvV DVsdPgGpvt
B. linens        ~~~MnnLyRe lAPipgpaWa EIeeeARrtF KrniaGRriV DVaGPtGfEt
A. aeolicus      ~~~MEFLqRd qAPLTaeeWe qIDktAyEvF KstvvcRKFm pVvGPfGagh
Consensus        ---MEFL-R- -APLTE--W- EIDN-ARE-F K-Q--GRKFV DVEGPYGWE-
                  •  •  •         ••     •    •   •   •   •      •    • •

                        51                                              88
T. maritima      AAhpl..... ...GEVEVlS DEn....EvV KWGLRKSLPL IELRAtFTLd
M. tuberculosis  AAVst..... ...Grlidvk apt....ngV iahLRaSkPL vrLRvpFTLs
B. linens        sAVtt..... ...Ghirdvq sEt....sgl qvkqRivqey IELRtpFTvt
A. aeolicus      qvVsydvlyg vepGvcEVkp gqeykvcEpV rtGeRKhvPv ptLykdFvis
Consensus        AAV------- ---GEVEV-S DE-----E-V KWGLRKSLPL IELRA-FTL-
                                    •                       •      •  •

                        89                                              138
T. maritima      lWELDnlERG kPnVDlSsle EtvRKvAEfE DeVIFRGCEk sGvkGlLSfe
M. tuberculosis  rnEiDdvERG skdsDwepVk EAAkKlAfvE DrtIFeGysa asIeGirSas
B. linens        rqaiDdvaRG sgdsDwqpVk dAAttiAmaE DraIlhGlda aGIgGivpgs
A. aeolicus      wrdLehwrqf nlpVDttgVa aAAsslAvaE DklIlfGnqe mGIeGfLtak
Consensus        -WELD--ERG -P-VD-S-V- EAARK-AE-E D-VIFRGCE- -GI-G-LS--
                                  •            •  • •  •          •

                        139                                             185
T. maritima      er.kIEcgsT PK..DLLeAI VrALSIfsKD GieGPYtLvi NTDRwinFlk
M. tuberculosis  snpaltlped Pr..eipdvI sqALSeLrla GvdGPYsvll saDvYtk.vs
B. linens        snaavaipda ve..DfadAv aqALSvLrtv GvdGPYsLll ssaeYtk.vs
A. aeolicus      gtlreElsdw eKvgnafqdv VkgiSrLvek GfytnYyLiv NpkRYfllnr
Consensus        ----IE---T PK--DLL-AI V-ALSIL-KD G---GPY-L-- NTDRY--F--
                                                   •     • •

                        186                                             232
T. maritima      EEAgH.YPle krvEeCLRGG kIItTPrI.. EdALVVSeRG GDFkLILGQD
M. tuberculosis  EtsdHGYPir Ehlnr.LvdG dIIwaPaI.. dgAfVlttRG GDFdLqLGtD
B. linens        EstdHGYPir Ehlsrq LgaG eIIwaPal.. EgALlVStRG GDyeLhLGQD
A. aeolicus      ihdntGllel Eqikkvvk.. evyqTPiIpe divLlVSasp anFdLaialD
Consensus        EEA-HGYP-- E--E-CLRGG -II-TP-I-- E-ALVVS-RG GDF-LILGQD
                                                             •    • •

                        233                                 268
T. maritima      LSIGYEdreK DaVRLfitET FTFqvvnPEA lilLkf
M. tuberculosis  vaIGYashdt DteRLylqET lTFlcyTaEA SVaLSh
B. linens        LSIGYyshds etVeLylqET FgFlalTdEs SVpLSl
A. aeolicus      vnvafvetsn mnhtfrvmEm vvprikrPEA ilifSs
Consensus        LSIGYE---K D-VRL---ET FTF---TPEA SV-LS-
                                                 •
```

FIG. 1. Consensus regions (capital letters) for maritimacin and putative homologs from bacteria. The areas of consensus indicated with a dot below each residue represent conserved amino acid residues in all sequences compared. Numbering is according to the maritimacin amino acid sequence (bold).[8a] The accession numbers for the sequences are AE001747 (*T. maritima* maritimacin), Y12820 (*Mycobacterium tuberculosis* antigen), X93588 (*Brevibacterium linens* linocin M18), and AE000754 (*Aquifex aeolicus* protein). Computations were performed by the BLAST 2.0 Program (basic blastp) at the National Center for Biotechnology Information Web Site (www.ncbi.nlm.nih.gov).

[8a] Modified from P. M. Hicks, K. D. Rinker, J. R. Baker, and R. M. Kelly, *FEBS Lett.* **440,** 393 (1998); K. E. Nelson, R. A. Clayton, S. R. Gill, M. L. Gwinn, R. J. Dodson, D. H. Haft, E. K. Hickey, J. D. Peterson, W. C. Nelson, K. A. Ketchum, L. McDonald, T. R. Utterback, J. A. Malek, K. D. Linher, M. M. Garrett, A. M. Stewart, M. D. Cotton, M. S. Pratt, C. A. Phillips, D. Richardson, J. Heidelberg, G. G. Sutton, R. D. Fleischmann, J. A. Eisen, O. White, S. L. Salzberg, H. O. Smith, J. C. Venter, and C. M. Fraser, *Nature* **399,** 323 (1999).

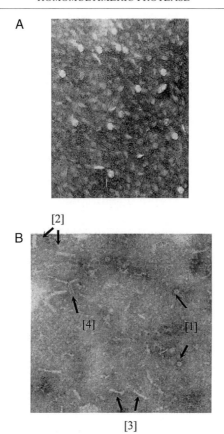

Fig. 2. TEM analysis of *Thermotoga maritima* homomultimeric protease, maritimacin. (A) S200 pool, taken at 54,800× and enlarged 2.8×; the average diameter of the multisubunit protease is estimated to be 20 nm. (B) Affinity pool, taken at 31,000× and enlarged 2.8×. [1] spherical form of maritimacin as in A, [2] tail with contracted sheath, average length of 70 nm, [3] putative fibers of maritimacin, and [4] (faint) hollow core. From P. M. Hicks, K. D. Rinker, J. R. Baker, and R. M. Kelly, *FEBS Lett.* **440,** 393 (1998). With permission from Elsevier Science.

Methods

Biochemical Assay

The proteolytic activity of maritimacin is detected by the release of 7-amido-4-methylcoumarin (MCA) from the carboxyl terminus of amino-terminally blocked peptides (synthetic peptides, Sigma Chemical, St. Louis, MO). Maritimacin shows the highest activity with the following synthetic substrates: AAF-, AFK-, and VKM-7-amido-4-methylcoumarin (MCA).[7] The assay is carried out by heating (in a thermal cycler) both the enzyme or buffer (for blank samples) and the substrate in 50 mM sodium phosphate buffer (SPB), pH 7, at 75–80° for 20–30 min. Substrate concentrations of 0.5 mM and enzyme concentrations of 0.5–5.0 μg per 100 μl assay volume are used. Enzyme concentrations should be adjusted so that the fluorescence measured at a sensitivity setting of 4 is within linear range of the fluorometric plate reader and at least 50% above background levels.[7] A fluorimeter is used to follow the increase in fluorescence at 360 nm excitation and 460 nm emission. Fluorescence units are converted to picomoles of MCA released through the use of a standard curve prepared with known dilutions of MCA in 5% dimethyl sulfoxide (DMSO) and 50 mM SPB, pH 7.[9]

Processing of Thermotoga maritima Cells

Thermotoga maritima (DSM 3109) is grown in continuous culture (80–85°) on RDM-based media containing 25 g/liter NaCl.[10] Cell pellets are resuspended in 50 mM SPB, pH 8 (2 ml per gram wet weight), and treated with 1 mg/ml lysozyme overnight at 4°. The cells are disrupted in a French-pressure cell at 18,000 psi and spun at 14,000g for 30 minutes at 4°.[7] The supernatant is used in the following purification steps.

Purification of Maritimacin

The purification is carried out at room temperature. Before each chromatographic step, samples are clarified with a 0.2 μm filter or by centrifugation to remove cellular debris. Samples are concentrated at room temperature with stirred-cell concentrators, using filters of 10-kDa molecular mass cutoff.[7]

DEAE Chromatography. The supernatant from the cell-free extract is loaded (25 ml of a ~40-mg/ml sample per run) at a flow rate of 0.1 ml/min on a 120-ml DEAE column (Pharmacia, Piscataway, NJ; DEAE Sepharose CL-6B 36/40) equilibrated with 50 mM SPB, pH 8. Note that a relatively slow flow rate is used during loading because of the high viscosity of the samples. The column is washed with three column volumes of 50 mM SPB, pH 8, and elution buffer is applied (1 M NaCl in 50 mM SPB, pH 8) at a flow rate of 1.5 ml/min. During the original set of purification experiments, a gradient from 0 to 1 M of elution buffer was applied to the column. However, maritimacin does not bind to the column and will, in-

[9] S. B. Halio, M. W. Bauer, S. Mukund, M. W. W. Adams, and R. M. Kelly, *Appl. Environ. Microbiol.* **63**, 289 (1997).
[10] K. D. Rinker and R. M. Kelly, *Appl. Environ. Microbiol.* **62**, 4478 (1996).

stead, elute in the pass-through fraction. Therefore, the elution buffer gradient can be ignored for the case where maritimacin is the only protein of interest.[7]

Hydroxylapatite (HAP) Chromatography. The active samples from the series of DEAE column runs (pass-through) are pooled (volume of ~2 liter) and concentrated to obtain approximately 600 ml of a 0.55-mg/ml sample. This is applied at 0.75 ml/min onto an 80-ml hydroxylapatite column (Pharmacia XK 16) equilibrated with 3.75 column volumes of 25 mM potassium phosphate buffer (PPB), pH 7. The column is washed with two column volumes of 25 mM PPB, pH 7, and a linear gradient (3.75 column volumes) from 25 to 250 mM PPB, pH 7, is applied.[7]

Affinity Chromatography. The active HAP fractions (those eluting between 70 and 200 mM PPB) are pooled (volume of 175 ml) and concentrated to 25 ml of a 5-mg/ml sample. The protein sample is then equilibrated to loading buffer (50 mM Tris, pH 8, 150 mM NaCl) using an Amicon (Danvers, MA) stirred-cell concentrator. Approximately 5 ml per run of sample is loaded at 0.2 ml/min onto a column (Pharmacia XK 16/20) containing 25 ml of benzamidine Sepharose 6B and previously equilibrated with loading buffer. After washing the column with 2.5 column volumes of loading buffer, the flow rate is increased to 1.6 ml/min and elution buffer (10 mM HCl, pH 2, 0.5 M NaCl) is applied. During the original purification procedure, a gradient of elution buffer was applied. However, maritimacin does not bind to this column and, instead, elutes with the pass-through. If maritimacin is the only protein of interest, the gradient can be ignored. If other proteins eluting with the elution buffer are to be studied, those fractions should be immediately neutralized with 1/10th volume of 100 mM Tris, pH 8.5, to bring the solution pH to values near neutral.[7]

Gel-Filtration Chromatography. The active affinity fractions (pass-through) are pooled and concentrated to 10 ml of a 5-mg/ml sample. Approximately 1 ml per run of sample is loaded at 0.25 ml/min onto a Superdex 200 HiLoad (Pharmacia 16/60) equilibrated to 50 mM SPB, pH 7, containing 150 mM NaCl. Protease activity elutes with apparent molecular masses of ≥2000 kDa and approximately 780 kDa. Blue dextran is used to determine void volume, and protein standards that can be used to calibrate the column include thyroglobulin (669 kDa), apoferritin (443 kDa), α-amylase (200 kDa), alcohol dehydrogenase (150 kDa), and bovine serum albumin (66 kDa).[7]

Biochemical and Biophysical Properties

Maritimacin is a ≥780-kDa homomultimeric structure of 31-kDa subunits. It exhibits chymotrypsin- and trypsin-like activities optimally between

90 and 93° and at pH 7.1 (active range pH 6–9). It has a half-life of 36 min on VKM-MCA at a temperature of 95° in the same conditions outlined in the biochemical assay listed previously.[7]

The relationship between maritimacin and other mesophilic bacteriocins implies that certain antibacterial activity could involve proteolysis. Furthermore, the possibility that maritimacin may be a bacteriocin or may carry some bactericidal activity raises the prospect for antimicrobial ecological strategies in hyperthermophilic niches. Efforts were made to screen maritimacin for antibacterial activity against *Thermotoga* species and members of the Thermococcales but no significant growth inhibition was noted.[11] This may have been related to the titer (amount of maritimacin vs amount of cells) used. Furthermore, preliminary data of maritimacin against *Listeria* species, *Bacillus* species, *Escherichia coli,* and other hyperthermophiles has not shown any antibacterial action by maritimacin.[12] In order to study the possible antibacterial affects of maritimacin further, efforts to clone and express the gene encoding the protease in *E. coli* are underway.[13] Although bacteriocins have been found in thermophiles (growth $T_{opt} = 45$–$80°$) such as *Thermoactinomyces* species[14] and *Streptococcus thermophilus,*[15] there has been no previous mention of a bacteriocin or other antimicrobial·agent from a hyperthermophilic microorganism (growth $T_{opt} \geq 80°$). It remains to be seen if maritimacin is in fact a bacteriocin and if proteolysis represents its mode of action.

Acknowledgments

This work was supported in part by the Department of Energy and the National Science Foundation. LSC acknowledges support from the Department of Education GAANN Fellowship.

[11] K. D. Rinker, P. M. Hicks, and R. M. Kelly, 1998, unpublished data.
[12] P. M. Hicks and R. M. Kelly, 1998, unpublished data.
[13] L. S. Chang, A. M. Grunden, M. W. W. Adams, and R. M. Kelly, 1999, unpublished data.
[14] A. A. Makawi, *Zentralbl. Bakteriol.* [*Naturwiss*] **135,** 12 (1980).
[15] D. J. Ward and G. A. Somkuti, *Appl. Microbiol. Biotechnol.* **43,** 330 (1995).

[33] Carboxylesterase from *Sulfolobus solfataricus* P1

By A. C. SEHGAL, WALTER CALLEN, ERIC J. MATHUR, J. M. SHORT, and ROBERT M. KELLY

Introduction

Esterases (EC 3.1.1.1) catalyze the hydrolysis of ester bonds and have been found in all domains of life.[1,2] Most of the esterases that have been studied to date have been from mesophilic microorganisms, such as *Pseudomonas fluorescens*,[3,4] *Arthrobacter globiformis*,[5] and *Aspergillus niger*.[6,7] Although these enzymes have long been studied, their exact physiological role in many cases has yet to be defined.[8] In some instances, esterases have been implicated in the degradation of substituted xylans for nutritional purposes, whereas in other cases they have been associated with the blood–brain barrier, regulation of free choline levels, and lipid metabolism.[9–11] The catalytic mechanism used by this group of enzymes is characterized by a reaction sequence composed of an acylation and a deacylation step,[2] typically involving serine, glutamic or aspartic acid, and histidine residues, located in the esterase active site.[12] These residues comprise a catalytic triad, a common feature of the serine hydrolase superfamilies, and are responsible for the hydrolysis of many endogenous and exogenous compounds.[2]

Although a significant amount of information regarding esterase biocatalysis has been reported since the 1970s, a comprehensive classification scheme has been elusive.[2] This is mostly due to the fact that esterases

[1] E. Ozaki, A. Sakimae, and R. Numazawa, *Biosci. Biotechnol. Biochem.* **59**, 1204 (1995).
[2] T. Satoh and M. Hosokawa, *Annu. Rev. Pharmacol. Toxicol.* **38**, 257 (1998).
[3] M. Suguira, T. Oikawa, K. Hirano, and T. Inukai, *Biochim. Biophys. Acta* **488**, 353 (1997).
[4] A. Nakagawa, T. Tsujita, and H. Okuda, *J. Biochem.* **95**, 1047 (1984).
[5] M. Nishizawa, H. Gomi, and F. Kishimoto, *Biosci. Biotechnol. Biochem.* **57**, 594 (1993).
[6] S. Okumura, M. Iwai, and Y. Tsujisaka, *Agric. Biol. Chem.* **47**, 1869 (1983).
[7] J. Fukumoto, M. Iwai, and Y. Tsujisaka, *J. Gen. Appl. Microbiol.* **9**, 353 (1963).
[8] G. Manco, S. D. Gennaro, M. D. Rosa, and M. Rossi, *Eur. J. Biochem.* **221**, 965 (1994).
[9] W. Shao and J. Wiegel, *Appl. Environ. Microbiol.* **61**, 729 (1995).
[10] T. Yamada, M. Hosokawa, T. Satoh, I. Moroo, M. Tkahashi, H. Akatsu, and T. Yamamota, *Brain Res.* **658**, 163 (1994).
[11] E. Schmidt and F. W. Schmidt, *in* "Esterases, Lipases, and Phospholipases from Structure to Clinical Significance" (M. I. Mackness and M. Clerc, eds.), p. 13. Plenum Press, New York, 1994.
[12] M. Haruki, Y. Oohashi, S. Mizuguchi, Y. Matsuo, M. Morikawa, and S. Kanaya, *FEBS Lett.* **454**, 262 (1999).

METHODS IN ENZYMOLOGY, VOL. 330

typically exhibit a broad substrate specificity range, thus rendering any attempts to classify these enzymes on this basis ambiguous and contradictory.[13,14] One commonly used method of esterase classification is based on the interaction of these enzymes with organophosphates.[15] "A-esterases" hydrolyze organophosphate compounds, "B-esterases" are inhibited by them, and "C-esterases" neither hydrolyze nor are they inhibited by these substrates.[16] Although this method of classification is somewhat limited, it avoids the problem of substrate specificity overlap. This classification scheme, however, takes little account of the biochemical properties and structural features. Efforts to address this limitation have been described that utilize the increasing availability of amino acid sequence information.[2,17,18] However, it remains to be seen whether amino acid sequence information for esterases will provide much insight into catalytic function and substrate preference.

Although esterases are responsible for hydrolysis reactions, they also possess the ability to catalyze several other types of biotransformations (e.g., esterifications and transesterifications) in environments with minimal water content.[19-21] This distinctive feature has raised the biotechnological prospects of this enzyme class for a number of industrially significant biotransformations, either as a hydrolytic or as a synthetic catalyst. Potential applications include blocking or unblocking of catalytic groups in peptide chemistry, modification of sugars, synthesis of flavor esters for the food industry, and the resolution of racemic mixtures to produce optically active compounds.[1,22,23] Despite these opportunities, esterase biocatalysis has yet to support large-scale industrial technology.[24] A significant limitation in this regard has been the instability of biological catalysts in harsh, industrially relevant reaction environments (e.g., high temperatures, organic solvents, strong alkalinity or acidity). Hyperthermophilic enzymes, however, have been shown to be inherently more resistant to a variety of enzyme denatur-

[13] C. H. Walker and M. I. Mackness, *Biochem. Pharmacol.* **32**, 3265 (1983).

[14] R. Mentlein, M. Suttorp, and E. Heymann, *Arch. Biochem. Biophys.* **228**, 230 (1984).

[15] W. N. Aldridge, *Biochem. J.* **53**, 110 (1953).

[16] F. Bergmann, R. Segal, and S. Rimon, *Biochem. J.* **67**, 481 (1957).

[17] H. Hemila, T. T. Koivula, and I. Palva, *Biochim. Biophys. Acta* **1210**, 249 (1994).

[18] E. Krejci, N. Duval, A. Chatonnet, P. Vincens, and J. Massoulie, *Proc. Natl. Acad. Sci. U.S.A.* **88**, 6647 (1991).

[19] R. Lortie, *Biotechnol. Adv.* **15**, 1 (1997).

[20] N. N. Gandhi, *JAOCS* **74**, 621 (1997).

[21] Boland W., C. Frößl, and M. Lorenz, *Synthesis* **12**, 1049 (1991).

[22] A. P. Wood, R. Lafuente, and D. Cowan, *Enzyme Microb. Technol.* **17**, 816 (1995).

[23] A. Kademi, N. A. Abdelkader, L. Fakhreddine, and J. C. Baratti, *Enzyme Microb. Technol.* **24**, 332 (1999).

[24] K. Faber and M. C. R. Franssenn, *TIBTECH* **11**, 461 (1993).

ants and, thus, represent promising alternatives for the development of industrial biocatalytic processes.[25] Along these lines, hyperthermophilic esterases represent a group of enzymes with the potential for catalyzing an array of industrially significant biotransformations that otherwise would need to be carried out less effectively through organic synthesis.

To date, relatively few investigations regarding the purification and characterization of thermostable ($T_{opt} > 60°$) esterases have been conducted. Thus far, esterases from *Bacillus acidocaldarius*,[8,26] *Pyrococcus furiosus*,[27] *Bacillus stearothermophilus*,[22,28,29] *Sulfolobus shibatae*,[30] *Thermoanaerobacterium* sp.,[9] *Pyrococcus abyssi*,[31] and *Sulfolobus acidocaldarius*[32,33] have been studied to varying extents (see Table I). Those thermostable esterases that have been evaluated vary significantly with respect to molecular mass, although their biochemical properties, such as pH optimum and substrate specificity, are quite similar. Table I also shows that only limited kinetic data have been reported for thermostable esterases. The most detailed kinetic information has been from efforts by Manco *et al.*[26] and Wood *et al.*,[22] who have reported catalytic efficiencies of 600,000 and 104,000 sec^{-1} mM^{-1} for esterases from *B. acidocaldarius* and *B. stearothermophilus*, respectively.

This article describes the cloning, expression, purification, and characterization of a recombinant carboxylesterase (Sso P1 carboxylesterase, U.S. Patent 5,942,430 and Patents Pending, Diversa Corporation) from the extreme thermoacidophile *Sulfolobus solfataricus* P1. *S. solfataricus*, which is a member of the Crenarchaeota, can be found in sulfurous caldrons and volcanic muds.[34,35] This organism is an obligate aerobe and grows between 50 and 87° and at pH values of 3.5–5.0. Its genome is approximately 3.1

[25] D. A. Cowan, *Comp. Biochem. Physiol.* **118A,** 429 (1997).

[26] G. Manco, E. Adinolfi, F. Pisani, G. Ottolina, G. Carrea, and M. Rossi, *Biochem. J.* **332,** 203 (1998).

[27] M. Ikeda and D. S. Clark, *Biotechn. Bioeng.* **57,** 624 (1998).

[28] D. Simoes, D. McNeill, B. Kristiansen, and M. Mattey, *FEMS Microbiol. Lett.* **147,** 151 (1997).

[29] A. P. Wood, R. Fernandez-Lafuente, and D. A. Cowan, *Biotechnol. Appl. Biochem.* **21,** 313 (1995).

[30] S. Huddleston, C. A. Yallop, and B. M. Charalambous, *Biochem. Biophys. Res. Commun.* **216,** 495 (1995).

[31] L. Cornec, J. Robineau, J. L. Rolland, J. Dietrich, and G. Barbier, *J. Mar. Biotechnol.* **6,** 104 (1998).

[32] H. Sobek and H. Görisch, *Biochem. J.* **250,** 453 (1988).

[33] H. Sobek and H. Görisch, *Biochem. J.* **261,** 993 (1989).

[34] W. Zillig, *Nucleic Acids Res.* **21,** 5273 (1993).

[35] S. Cady, M. W. Bauer, W. Callen, M. A. Snead, E. J. Mathur, J. M. Short, and R. M. Kelly, *Methods Enzymol.* **330** [22] (2001) (this volume).

TABLE I

BIOCHEMICAL PROPERTIES OF THERMOSTABLE ESTERASES

Source	Crude or purified	Molecular mass (kDa)	pH_{opt}	T_{opt}	Preferred p-NP substrate	K_m (mM)	k_{cat} (sec)$^{-1}$	k_{cat}/K_m (sec mM)$^{-1}$	Refs.
Bacillus acidocaldarius	Purified	34	7.1	70	*p*-NP caproate	0.011	6610	600,000	[a]
Pyrococcus abyssi	Crude	nr[l]	5.8–7.8	65–74	*p*-NP valerate	nr	nr	nr	[b]
B. stearothermophilus	Purified	45–50	7.0	60[d]	*p*-NP caprylate	0.012	1250	104,000	[c]
Sulfolobus shibatae	Crude	nr	6.0	90	nr	nr	nr	nr	[e]
Thermoanaerobacterium sp.	Purified[g]	195	7.0	80	nr	0.45[k]	nr	nr	[f]
		106	7.5	84		0.52[k]			
S. acidocaldarius	Purified	117–128	7.5–8.5	nr	*p*-NP valerate	0.157	nr	nr	[h]
P. furiosus	Crude	nr	7.6	100	nr	nr	nr	nr	[i]
S. solfataricus	Purified	33	7.7	95–100	*p*-NP caproate	0.45[k]	1000[k]	2200[k]	[j]

[a] G. Manco, E. Adinolfi, F. Pisani, G. Ottolina, G. Carrea, and M. Rossi, *Biochem. J.* **332**, 203 (1998).
[b] L. Cornec, J. Robineau, J. L. Rolland, J. Dietrich, and G. Barbier, *J. Mar. Biotechnol.* **6**, 104 (1998).
[c] A. P. Wood, R. Lafuente, and D. Cowan, *Enzyme Microb. Technol.* **17**, 816 (1995).
[d] Temperatures above 60° were not evaluated because of elevated rates of autohydrolysis.
[e] S. Huddleston, C. A. Yallop, and B. M. Charalambous, *Biochem. Biophys. Res. Commun.* **216**, 495 (1995).
[f] W. Shao and J. Wiegel, *Appl. Environ. Microbiol.* **61**, 729 (1995).
[g] Two esterases were evaluated in this work.
[h] H. Sobek and H. Görisch, *Biochem. J.* **250**, 453 (1988).
[i] M. Ikeda and D. S. Clark, *Biotechnol. Bioeng.* **57**, 624 (1998).
[j] A. C. Sehgal and R. M. Kelly, manuscript in preparation.
[k] Kinetic parameters were calculated with 4-methylumbelliferyl acetate. Other kinetic values reported for other enzymes were calculated with *p*-NP substrates.
[l] Not reported.

Mb in size, one of the largest yet reported in the Archaea. The genome of the closely related strain, *S. solfataricus* P2, is currently being sequenced (http://niji.imb.nrc.ca/sulfhome/).

Recombinant Carboxylesterase from *Sulfolobus solfataricus* P1

Microbial Strain and Gene Cloning

Sulfolobus solfataricus strain P1 (DSM 5354) has been provided by Karl O. Stetter, University of Regensburg, Germany. Its growth characteristics are essentially the same as *S. solfataricus* P2 (DSM 1617). Both of these organisms can be obtained from Deutsche Sammlung von Mikroorganismen, Braunschweig, Germany. Cultivation conditions for *S. solfataricus* are described by Zillig *et al.*[36] Genomic DNA is prepared from *S. solfataricus* P1 (Sso P1) using a method similar to that described by Cady *et al.*[35] The gene encoding Sso P1 carboxylesterase is expressed using the vector pQE30 (Qiagen Inc., Valencia, CA) in the *Escherichia coli* host strain, M15 pREP4 (Qiagen Inc.).

Production of Recombinant Sso P1 Carboxylesterase

Colonies of the recombinant *E. coli* are plated on LB media (Bacto-tryptone 10 g/liter, Bacto-yeast extract 5 g/liter, and NaCl 10 g/liter) with 1.5% agarose (w/v) and the appropriate antibiotics [ampicillin (100 μg/ml) and kanomycin (80 μg/ml)]. The pH of the LB medium is adjusted to 7.0 with 1 M NaOH. The plated colonies are grown overnight at 37°. Single colonies are transferred to 5-ml test tubes containing LB media, ampicillin (100 μg/ml), and kanomycin (80 μg/ml) and allowed to grow overnight in a shaking incubator at 37° and 250 rpm. Five milliliters of culture is then transferred to 200 ml of a LB media/antibiotic solution and allowed to grow in a shaking incubator (37°, 250 rpm) until the absorbance at 600 nm is between 0.6 and 1.0. The culture is transferred to 1800 ml of a LB media/ antibiotic solution and allowed to grow for 12–16 hr at 37° and 250 rpm. Isopropyl-β-D-thiogalactopyranoside (IPTG, 1 mM final concentration) is added to the 2-liter culture for gene expression. Cells are then grown for an additional 4 hr at 37° and 250 rpm, after which they are removed from the incubator. Cells are then centrifuged at 4° for 15 min at 8300g. The supernatant is discarded and the cell paste is resuspended in 50 ml of 50 mM sodium phosphate buffer, pH 8.0. Lysozyme (1.5 ml of a 0.01-g/ml

---

[36] W. Zillig, K. O. Stetter, S. Wunderl, W. Schulz, H. Priess, and I. Scholz, *Arch. Microbiol.* **125**, 259 (1980).

stock) is added to the suspended cells and incubated in a shaking incubator for 1 hr at 37° and 150 rpm. The cells are removed and twice passed through a French press (SLM Instruments Inc., Urbana, IL) at 9000 psig and 4°. This is followed by centrifugation at 30,000g at 4° for 30 min. The supernatant is incubated at 80° for 25 min in order to denature a majority of the proteins from *E. coli*. The suspension is then centrifuged for 30 min at 30,000g at 4°, and the resulting supernatant (crude enzyme) is stored at 4°.

Purification of Recombinant Sso P1 Carboxylesterase

In order to purify the Sso P1 carboxylesterase, 5-ml samples of crude enzyme (4 mg/ml) are loaded onto a DEAE Sepharose Fast Flow (Amersham Pharmacia Biotech, Piscataway, NJ) chromatography column (2.6 × 40 cm). Column flow rates are maintained at 5 ml/min, with a linear NaCl gradient (0–2 *M*) being introduced after 1 hr. Column fractions (10 ml) are collected and assayed for esterase activity. Fractions containing esterase activity are concentrated with a Microcon-10 centrifugal concentrator (Millipore Corp., Bedford, MA) and checked for purity by SDS–PAGE. Sso P1 carboxylesterase has virtually no affinity for the anion-exchange column used. This simplifies the purification procedure as most of the Sso P1 carboxylesterase can be recovered as a homogeneous product with the column pass-through (Fig. 1).

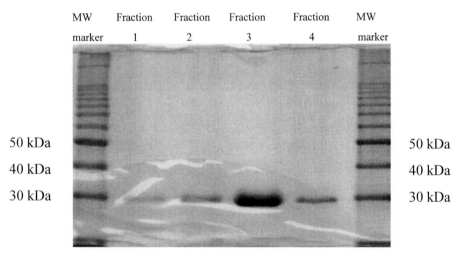

FIG. 1. SDS–PAGE gel of initial column eluant. The Sso P1 carboxylesterase had little affinity for the weak anion-exchange column resin, thus pure protein was captured in the initial column eluant (fractions 1–4).

Enzyme Assay Methods for High Temperature Esterase Activity

Two standard assays are used in the evaluation of the Sso P1 carboxylesterase. The *p*-nitrophenyl caproate assay can be used at temperatures up to 70°. However, an increase in substrate instability with increasing temperature mandates the need for a more temperature-stable compound at higher assay temperatures. 4-Methylumbelliferyl derivatives are less vulnerable to autohydrolysis at higher temperatures and should be used for experiments performed at temperatures greater than 70°. Note that at temperatures below 60°, most 4-methylumbelliferyl substrates have limited solubility in water. It is interesting to note that depending on the type of chromogenic substrate used (i.e., *p*-NP or 4-MU derivatives), significantly different catalytic efficiencies can be obtained (see Table I).

For many hyperthermophilic enzymes, substrate lability is a significant problem in high temperature biochemical assays. This can be especially true for synthetic substrates that are based on the chemical coupling of a chromogenic or fluorometric moiety, which is released on hydrolysis and monitored readily by standard laboratory analytical instruments. Here, two different assay methods are used for determining the biochemical properties of the Sso P1 carboxylesterase, which are effective over different temperature ranges.

p-Nitrophenyl Caproate Assay. Esterase activity at temperatures up to 70° is determined by monitoring the release of *p*-nitrophenol from *p*-nitrophenyl caproate at 405 nm. Spectrophotometric measurements are made with a Lambda 3 spectrophotometer (Perkin-Elmer Corp., Norwalk, CT), equipped with a thermostatted six-cell transport system. The temperature of the cell holder is controlled by a liquid circulating temperature bath (Model 1130: VWR Scientific, Philadelphia, PA) containing 100% ethylene glycol. Routine assays represent modifications of existing published techniques[37] and are performed as follows: 900 μl of 1 mM *p*-NP caproate is added to two 2-ml quartz cuvettes followed by 100 μl of 50 mM sodium phosphate buffer, pH 8.0. The cuvettes are incubated for 4 min in the thermostatted cell holder to allow the substrate and buffer to reach the assay temperature. After 4 min, 30 μl of enzyme (0.015 ng/μl) is introduced into the sample cell to initiate the reaction. (The enzyme solution is preincubated for 30 sec at the working temperature of the assay prior to addition to minimize fluctuations in temperature on enzyme addition.) This procedure is then repeated with a new reference solution being made each time to properly account for background hydrolysis. One unit of enzyme activity is defined as the amount of enzyme required to release 1 μmol of *p*-

[37] E. Heymann and R. Mentlein, *Methods Enzymol.* **77**, 333 (1981).

nitrophenol/min under the standard assay conditions. For example, for Sso P1 carboxyesterase, standard assay conditions are 70° and pH 8.0 for p-NP substrates.

4-Methylumbelliferyl Acetate Assay. Esterase activity at temperatures greater than 70° is determined by monitoring the release of 4-methylumbelliferone from 4-methylumbelliferyl acetate at 354 nm. Spectrophotometric measurements are made with a Lambda Bio 20 spectrophotometer (Perkin-Elmer Corp.) equipped with two PTP-1 digital temperature controllers (Perkin-Elmer Corp.). Routine assays represent modifications of existing published techniques[9,27] and are performed as follows: 780 μl of 50 mM sodium phosphate buffer (pH 7.0) is added to two 2-ml quartz cuvettes followed by 10 μl of 50 mM 4-methylumbelliferyl acetate in dimethyl sulfoxide (DMSO). The cuvettes are incubated for 4 min in the thermostatted cell holder to allow the substrate and buffer to reach the assay temperature. After 4 min, 30 μl of enzyme (0.053 ng/μl) is introduced into the sample cell to initiate the reaction. Again, the enzyme solution is preincubated for 30 sec at the working temperature of the assay, prior to addition to minimize fluctuations in temperature on enzyme addition. This procedure is then repeated with a new reference solution being made each time to properly account for background hydrolysis. One unit of enzyme activity is defined as the amount of enzyme required to release 1 μmol of 4-methylumbelliferone/min under the standard assay conditions.

pH Optimum. The effect of pH on enzyme activity is evaluated using p-NP caproate as the substrate. However, because the absorbance of p-nitrophenol at 405 nm is pH dependent, the release of product is monitored at 348 nm (isosbestic wavelength of p-nitrophenol and p-nitrophenoxide). The 4-methylumbelliferyl assay can also be used to evaluate the effect of pH on enzyme activity; however, because the extinction coefficient of 4-methylumbelliferone varies with pH, a separate calibration curve is constructed for each pH value evaluated. In addition, the 4-methylumbelliferyl esters are unstable at pH values much greater than eight, thus care should be taken when evaluating enzyme activity at alkaline pH values.

Differential Scanning Calorimetry

The melting temperature of the Sso P1 carboxylesterase can be determined using a CSC Nano Differential Scanning Calorimeter (Calorimetry Sciences Corporation, Salt Lake City, UT). Purified enzyme (2 ml of 1.67 mg/ml) is first filter-sterilized using a 0.2-μm filter and dialyzed overnight against 50 mM sodium phosphate buffer, pH 8.1. The sample is then put into a 15-ml plastic conical vial for storage at 4° until the completion of the buffer scans. Six to eight separate buffer scans are performed (25–125°)

overnight with the dialyzed buffer to ensure a stable baseline. The enzyme sample is then degassed and introduced into the DSC for evaluation over the temperature range of 25–125° using a scan rate of 1°/min.

Biochemical Properties

The molecular mass of the Sso P1 carboxylesterase is calculated to be 33 kDa from the amino acid sequence (see Fig. 2). This correlates well with SDS–PAGE results, which show the purified enzyme band to be in the 30-kDa range (Fig. 1). Note the location of putative catalytic residues of importance for esterase biocatalysis. The pH versus activity profile for the Sso P1 carboxylesterase shows an optimum of 7.7 as determined by

```
  1 -M  P  L  D  P  R  I  K  K  L  L  E  S  A  L  T  I  P  I  G -20

 21 -K  A  P  V  E  E  V  R  K  I  F  R  Q  L  A  S  A  A  P  K -40

 41 -V  E  V  G  K  V  E  D  I  K  I  P  G  S  E  T  V  I  N  A -60

 61 -R  V  Y  F  P  K  S  S  G  P  Y  G  V  L  V  Y  L  H  G  G -80

 81 -G  F  V  I  G  D  V  E  S  Y  D  P  L  C  R  A  I  T  N  A -100

101 -C  N  C  V  V  V  S  V  D  Y  R  L  A  P  E  Y  K  F  P  S -120

121 -A  V  I  D  S  F  D  A  T  N  W  V  Y  N  N  L  D  K  F  D -140

141 -G  K  M  G  V  A  I  A  G̲  D̲  S̲  A̲  G̲  G  N  L  A  A  V  V -160

161 -A  L  L  S  K  G  K  I  N  L  K  Y  Q  I  L  V  Y  P  A  V -180

181 -S  L  D  N  V  S  R  S  M  I  E  Y  S  D  G  F  F  L  T  R -200

201 -E  H  I  E  W  F  G  S  Q  Y  L  R  S  P  A  D  L  L  D  F -220

221 -R  F  S  P  I  L  A  Q  D  F  N  G  L  P  P  A  L  I  I  T -240

241 -A  E  Y  D  P  L  R  D  Q  G  E  A  Y  A  N  K  L  L  Q  A -260

261 -G  V  S  V  T  S  V  R  F  N  N  V  I  H  G  F  L  S  F  F -280

281 -P  L  M  E  Q  G  R  D  A  I  G  L  I  G  S  V  L  R  R  V -300

301 -F  Y  D  K  I
```

FIG. 2. Amino acid sequence of the Sso P1 carboxylesterase. The G-X-S-X-G motif (underlined) likely contains the conserved esterase active site serine.

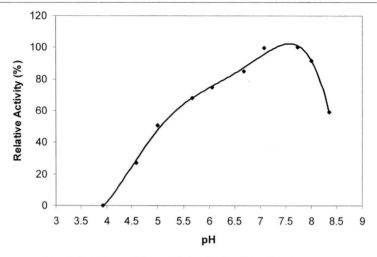

FIG. 3. Relative activity vs pH for the Sso P1 carboxylesterase.

the 4-methylumbelliferyl butyrate assay at 70° (Fig. 3). Recall the need to account for changes in extinction coefficient with pH for 4-methylumbelliferone so that the pH activity profile is accurate. The activity of the esterase increases with increasing acyl chain length (up to a carbon chain length of 6) at 70° and pH 8.0 for a variety of p-NP esters. Maximum activity is seen with p-NP caproate (carbon chain length 6) with close to a ninefold drop in activity being observed with p-NP caprate (carbon chain length 10). Michaelis–Menten kinetic parameters, along with the specific activity (SA) of the Sso P1 carboxylesterase for 4-methylumbelliferyl acetate at 80° and pH of 7.0, are as follows: V_{max} = 32.9 mM/sec mg, K_m = 0.45 mM, k_{cat} = 1000 sec^{-1}, k_{cat}/K_m = 2200 sec^{-1} mM^{-1}, and SA = 1600 U/mg. The melting temperature of the Sso P1 carboxylesterase is 97°, with no reversible denaturation being observed on rescanning under the given operating conditions. Finally, relative initial reaction rates of the Sso P1 carboxylesterase increase with decreasing solvent log P values in cosolvent experiments. In addition, the Sso P1 carboxylesterase shows good activity in various DMSO cosolvent systems [5, 10, and 20% (v/v)], which is encouraging as DMSO is an excellent dipolar aprotic solvent.[38]

Concluding Remarks

The Sso P1 carboxylesterase represents the first thermostable esterase to be cloned, expressed, and characterized from a strain of *S. solfataricus.*[39]

---

[38] M. G. Loudon, "Organic Chemistry." Benjamin/Cummings, Menlo Park, CA, 1988.
[39] A. C. Sehgal and R. M. Kelly, manuscript in preparation.

The recombinant enzyme is purified readily from *E. coli* cell extracts as it is mainly present in anion-exchange column pass-through. This may be a consequence of hydrophobic surface properties of the esterase. It remains to be seen whether other hyperthermophilic esterases will be recovered as readily from mesophilic hosts. Care must be taken to evaluate biochemical properties of hyperthermophilic esterases because of the thermolability of certain standard synthetic substrates. Here, even though the T_{opt} for the enzyme is probably between 95 and 100° (A. C. Sehgal and R. M. Kelly, unpublished data, 1999), the p-NP-based substrates show poor stability for assays much above 70°. Function at elevated temperatures may prove to be a key advantage for thermostable esterases in catalyzing specific biotransformations in aqueous and nonaqueous systems. Future work will be conducted to evaluate the ability of this enzyme to function in a variety of chemical environments.

Acknowledgments

RMK acknowledges the National Science Foundation for financial support of this work. We also thank Mr. Kevin Epting for his assistance in performing experiments with the differential scanning calorimeter.

Author Index

Numbers in parentheses are footnote reference numbers and indicate that an author's work is referred to although the name is not cited in the text.

A

Aagaard, C., 336
Abdelkader, N. A., 462
Abe, S., 268
Abroskina, O. N., 246
Achenbach, L. A., 331
Achenbach-Richter, L., 169, 221
Adachi, K., 431
Adam, M. J., 379
Adams, M. D., 151, 167, 170, 171, 171(9), 172, 172(9, 10), 199, 454
Adams, M. W. W., 25, 29(6–26), 30, 31, 32, 35(10), 36(4, 10), 50, 135, 165, 166, 264, 268, 348, 362, 403, 404, 404(5), 406(5), 409(5), 410(5), 412(5), 433, 436, 448, 449, 458, 460
Adinolfi, E., 463, 464
Adolphson, R., 348
Ador, L., 149
Aduse-Opoku, J., 248, 258(17), 259(17)
Aebersold, R., 211, 225
Aguilar, C. F., 212, 214(23), 367, 372(17)
Akagawa-Matsushita, M., 49
Akatsu, H., 461
Akeba, T., 275
Akiba, T., 35(44), 40, 275
Akkermans, A. D., 338, 367
Akopian, T. N., 413, 418, 422, 423(26)
Albertini, A. M., 172
Albright, C., 324
Aldredge, T., 48, 150, 172, 199
Aldrich, H. C., 415, 417(11), 418(11), 420(11)
Aldridge, W. N., 462
Alkin, B. L., 162
Allison, R. W., 304, 306(14), 312(14)
Alloni, G., 172
Altenbuchner, J., 334

Alting-Mees, M. A., 160
Altman, S., 166
Altschul, S. F., 44, 47, 50, 162, 171, 173(15), 184, 233, 244
Amann, E., 334
Amellal, N., 310
Amils, R., 51
Ammerman, J. W., 124, 135
Amsterdam, A., 419, 423(20)
Anderson, I. J., 40
Ando, S., 133
Andrade, C., 275, 277(15)
Andrade, M., 43
Andrews, K. T., 455
Anfinsen, C. B., 275, 279(18), 355, 358(12), 361(12), 363(8)
Ankai, A., 199
Anraku, Y., 128
Antranikian, G., 128, 202, 260, 262, 262(1), 266(10), 268, 268(10), 269, 269(10), 270, 275, 276, 277, 277(15), 278, 279, 279(17, 30, 35), 281(30, 35), 285, 329, 346, 353(2), 355, 384(11), 390, 431, 432(19)
Aoki, K., 167, 172, 176(19), 177(19), 199, 400
Aono, S., 29(24, 26), 30
Apeweiler, R., 162
Apweiler, R., 45, 46(34)
Arai, M., 244
Araki, T., 226
Aravind, L., 51, 71, 87, 89, 94, 95, 101, 103, 120, 121, 150, 201
Arcand, N., 235
Arfin, S. M., 441
Argos, P., 215
Arimura, A., 394
Aristan-Atac, I., 260
Armentrout, R. W., 394, 395(1)
Arnold, F. H., 372

473

O

P

Subject Index

A

Aminoacyl-tRNA synthetase
classes in *Pyrococcus*, 149–150
missing sequences in hyperthermophilic
  Archaea, 150–152
overview of function, 148–149
Pyrococcus furiosus
  AlaX, 152–154
  CysRS, 152
  dinucleotide bias and implications
    for recombinant expression, 145,
    147
  GluRS, 152
  LysRS, 152
  MetX, 152–154
  PheRS, 152
Amylopullulanase, *Pyrococcus furiosus*
assays, 357
cell culture, 355
denaturing agent studies, 361
function, 354
gene cloning, 355–356, 358
industrial application, 354–355, 363
metal chelation, 357–358
metal ion effects, 359–360
purification of native and recombinant
  proteins, 356
sequence analysis, 361–363
substrate specificity, 358–359
Anaerobic purification, redox enzymes and
  proteins from *Pyrococcus furiosus*
anion-exchange chromatography, 28–
  29
extract preparation, 28
α-Amylases, hyperthermophiles
assay, 284
classification, 272
expression in recombinant *Escherichia
  coli*, 278–279, 281
gel electrophoresis, 285
hydrolysis product characterization,
  285–288

industrial applications, 289
properties from various species, 274–276
Pyrococcus furiosus enzyme
assays, 357
cell culture, 355
denaturing agent studies, 361
function, 354
gene cloning, 355–356, 358
industrial application, 354–355, 363
intracellular enzyme, 363
metal ion effects, 359–360
purification, 357
sequence analysis, 361–363
substrate specificity, 358–359
β-Amylases, hyperthermophiles
assay, 284
classification, 272
gel electrophoresis, 285
hydrolysis product characterization,
  285–288
industrial applications, 289
Annotation, *see* Genomic annotation
Aquifex, phylogeny, 5, 13
Aquifex aeolicus
genome sequencing
  fosmid sequencing, 160–161
  overview, 158–159, 168–169
  phases, 159–160
  polymerase chain reaction, 161
  shotgun sequencing, 159–160
genomic annotation
  arrangements of biosynthetic genes,
    163–165
  his genes, 163–164
  missing genes
    catalase, 165–166
    chemotaxis genes, 167–168
    ribonuclease P, 166–167
  open reading frame assignment,
    162–163
  trp genes, 164
phenotype, 158
Aquificales, hyperthermophilic genera, 8

M